Chemical Thermodynamics

Reversible and Irreversible
Thermodynamics

Second Edition

Chemical Thermodynamics

Reversible and Irreversible Thermodynamics

Second Edition

Byung Chan Eu

McGill University, Canada

Mazen Al-Ghoul

American University of Beirut, Lebanon

NEW JERSEY · LONDON · SINGAPORE · BEIJING · SHANGHAI · HONG KONG · TAIPEI · CHENNAI · TOKYO

Published by

World Scientific Publishing Co. Pte. Ltd.

5 Toh Tuck Link, Singapore 596224

USA office: 27 Warren Street, Suite 401-402, Hackensack, NJ 07601

UK office: 57 Shelton Street, Covent Garden, London WC2H 9HE

Library of Congress Cataloging-in-Publication Data

Names: Eu, B. C. (Byung Chan), 1935– author. | Al-Ghoul, M. (Mazen), author.

Title: Chemical thermodynamics: reversible and irreversible thermodynamics / Byung Chan Eu,
    McGill University, Canada, Mazen Al-Ghoul, American University of Beirut, Lebanon.

Description: Second edition. | [Hackensack] New Jersey : World Scientific, [2018] |
    Includes bibliographical references and index.

Identifiers: LCCN 2017048946 | ISBN 9789813226050 (hardback : alk. paper)

Subjects: LCSH: Thermochemistry. | Thermodynamics.

Classification: LCC QD511 .E8 2018 | DDC 541/.36--dc23

LC record available at https://lccn.loc.gov/2017048946

British Library Cataloguing-in-Publication Data

A catalogue record for this book is available from the British Library.

For any available supplementary material, please visit
http://www.worldscientific.com/worldscibooks/10.1142/10599#t=suppl

Desk Editor: Amanda Yun

Typeset by Stallion Press
Email: enquiries@stallionpress.com

Printed in Singapore

*Dedicated to students aspiring to attain a liberated
perspective to thermodynamics.*

Contents

Preface to the Second Edition

The second edition of Chemical Thermodynamics now consists of two parts: with Part I addressing equilibrium thermodynamics and Part II exploring irreversible (or nonequilibrium) thermodynamics generally far removed from equilibrium. The two parts are designed to blend smoothly without disruption. In Part I minor revisions were made to various chapters from the equilibrium part of the first edition, except for Chapter 17 on galvanic cells. In the new Chapter 17, we have added discussions on electric cells of modern technological applications, such as photovoltaic cells and betavoltaic cells as well as lithium batteries. Old figures have been remade afresh throughout the book, and some new ones have been added. In Part II, the nonequilibrium part, three additional chapters have been added and the discussion on chemical oscillations has been expanded. It now consists of a general nonequilibrium theory of irreversible thermodynamics, and its applications to linear irreversible processes, nonlinear irreversible processes, such as non-Newtonian flows as a typical nonlinear flow process in nonequilibrium liquids, and discussions on Liesegang precipitation phenomena. Selected topics of these subjects constitute chapters 19 to 22. In the last chapter, Chapter 22, irreversible phenomena giving rise to oscillatory structures in time and space, which are often observed in laboratory experiments on chemical systems, are discussed. Such oscillatory structures in space and time are quite interesting and challenging scientifically because they are related to natural phenomena we often observe in everyday experience in systems far removed from equilibrium, but they are not only generally poorly understood theoretically, especially, from the standpoint of the existing irreversible thermodynamics, but also seem to demand revisions of the latter. In Part II, we have not paid as much attention to the phenomenological theory of thermodynamics as to the hydrodynamic aspects of irreversible thermodynamics since it is possible to carry on phenomenological studies almost parallel to Part I, based on the fundamental relations for calortropy or nonequilibrium internal energy, enthalpy, and free energies. For this purpose one has to develop phenomenological constitutive relations. We hope such phenomenological constitutive relations for nonequilibrium fluxes (variables) will be further developed and refined in the future beyond what we know about them at present.

January, 2017   B.C.E. & M.A.-G.

Preface to the First Edition

After a quick glance at it, one is liable to pass the judgment that thermodynamics appears to be an old subject in science, yet it is hardly true since, when closely examined, it still is a developing subject. It is regarded by many in science as an old subject because its basic principles were enunciated almost 160 years ago and the theory of reversible processes was basically completed in the hands of J. W. Gibbs within the space of 25 years from the time when the thermodynamic laws were stated by Clausius and Kelvin. It thus appears that there is nothing new to add to the subject. Nevertheless, the general theory of thermodynamic processes that include irreversible processes has been in an arrested state of evolution and has remained incomplete until L. Onsager, J. Meixner, I. Prigogine, formulated a theory of linear irreversible processes. This theory is still incomplete because irreversible processes about which the second law of thermodynamics is basically concerned have not received a fully satisfactory theoretical treatment. Fuller treatments of the subject have been given serious attention in recent years. Therefore thermodynamics in the generalized sense is still worth serious study, if possible, from a more general viewpoint than that taken in the traditional approach to the subject and, in particular, equilibrium thermodynamics, where only reversible processes are studied.

Since reversible processes are the idealized limits of irreversible processes observed in nature, the thermodynamics of reversible processes — namely, equilibrium thermodynamics — must be the limiting form of a more general theory of thermodynamic processes, and as such it is worth having a fresh examination in what manner it is a limiting theory. We examine some basic aspects of equilibrium thermodynamics under such a motivation in this textbook which also contains traditional treatments of various topics taught in courses in thermodynamics at an advanced undergraduate level and at the graduate level. The treatments given of the second law of thermodynamics and related topics, such as equilibrium conditions and stability of equilibrium, in this work are different from those in the conventional textbooks on equilibrium thermodynamics available at present, and in fact they are extended versions of them. A couple of examples of application of the extended treatment of the second law is discussed in the last chapter to show how one might apply the concepts to study irreversible processes far from equilibrium.

The materials on the conventional topics in thermodynamics in this book, excluding those related to the second law, have been taught by one of the authors over many years in the courses in thermodynamics at McGill University. The treatments given to the second law of thermodynamics and related topics in this book, being new and more recent fruits of labor on the part of one of the authors, have not been exposed to the classes in the past. However, we believe that they should be an integral part of a course on equilibrium thermodynamics. The reason is that not only the new results of research add to the science of thermodynamics the mathematical representation of the second law of thermodynamics that adequately covers irreversible processes, and provide clarifications of various related topics about which the conventional treatment of the subjects leaves us uncomfortable, but also the equilibrium thermodynamics of reversible processes emerges as the limiting case of a more general theory of thermodynamic processes as it should be. Furthermore, the generalized mathematical representation of the second law of thermodynamics removes some nagging conceptual features that arise from the fact that the entropy was originally defined for reversible processes only, yet one still thinks of the entropy as if it is a nonequilibrium quantity, when it comes to consideration of systems in the vicinity of an equilibrium state as in the case of equilibrium conditions and stability of equilibrium states. Therefore, when equilibrium thermodynamics is considered from the generalized theory — namely, generalized thermodynamics, we can form more harmonized viewpoints towards the subject and contemplate on a formulation of a more comprehensive theory of thermodynamic processes, reversible or irreversible. In this potential lies our desire to examine equilibrium thermodynamics from an angle different from the traditional viewpoint. The examples discussed in the last chapter, albeit brief, illustrate the utility of the concept of calortropy in study of irreversible processes. The reader interested in more involved discussions on irreversible phenomena and hydrodynamics in the nonlinear regime is referred to monographs dealing with the subjects. Such monographs are available at present. It is the hope of the authors that this work is a help and a stepping stone toward a more complete theory of irreversible phenomena.

September, 2009   B.C.E. & M.A.-G.

Chapter 1

Introduction

Thermodynamics in the generalized sense is a branch of natural science in which we study heat, work, energy, their interrelationships, and the modes by which systems exchange heat, matter, and energy with each other and with the surroundings, and convert heat into work, and vice versa. Since all human activities and natural phenomena involve matter and energy of one form or another, the importance of such a science is obvious, and for this reason, it is in the foundations of physical, biological, and engineering sciences. As a matter of fact, thermodynamics owes its genesis to the urgent need at the early stage of the Industrial Revolution in the first half of the 19th century to understand how steam engines work and improve their efficiencies since the efficiencies of such engines had significant economic implications. Such questions are still relevant even to this day in our everyday economic and industrial activities. On the one hand, such a need motivated scientists and engineers to study the properties of steam in particular and gases in general and construct, for example, the steam table. On the other hand, it culminated in the idealization of engines with a reversible cycle by S. Carnot, who made a lasting contribution through his penetrating analysis of how heat engines operate, and his study resulted in his famous principle now known as Carnot's theorem, although his analysis was based on the caloric theory of heat which was proven to be an incorrect notion of heat. Later pioneers such as R. Clausius and W. Thomson (Lord Kelvin) adopted the correct notion of heat — in which heat is regarded as a form of energy — and developed a correct theory by retaining truthful features in Carnot's exposition and adding something new. Through the efforts by them and their followers, the science of thermodynamics was born in the second half of the 19th century. The subject was refined, especially in an important way, by J. W. Gibbs through his well-known work on heterogeneous equilibria. The modern form of the science of thermodynamics, laid on the foundations shaped by the efforts by S. Carnot, Count Rumford (Benjamin Thompson), J. R. Mayer, R. Clausius, W. Thomson, H. Helmholtz, and J. W. Gibbs, among others, has been shaped by numerous other researchers, but its applications to chemical and chemical engineering problems owe a great deal to the works by M. Planck, W. Nernst, F. Haber, and G. N. Lewis and his school to name a few. The works of Max Born and C. Caratheodory have given equilibrium thermodynamics another mathematical aspect through Caratheodory's theorem, which opens up a geometrical viewpoint to thermodynamics. However, we will not discuss this line of thought in this work.

It is now generally believed that all natural macroscopic phenomena occur in full conformation to the laws governing thermodynamics. Although the subject of thermodynamics has been around in science over 160 years by now, since the pioneers in thermodynamics limited the development to reversible processes and thus to systems in equilibrium, it is not closed, but is still developing. It is therefore worth a serious study, especially since irreversible phenomena are not sufficiently well understood as yet from the standpoint of the laws of thermodynamics, especially, if irreversible processes occur far removed from equilibrium.

From the viewpoint of thermodynamics of irreversible processes, equilibrium thermodynamics — which, more precisely, should be called thermostatics and we are going to study here — is merely dealing with systems at a singular state of thermodynamic equilibrium. Since there are no macroscopically discernable processes occurring in equilibrium systems, equilibrium thermodynamics deals with idealized reversible processes and thus is not capable of describing what is happening in the real system over a finite time span and over space; rather, it is only able to tell us of the possibilities of that particular event as far as the laws of thermodynamics are concerned. The description of the process over a finite span of time and space is in the realm of irreversible thermodynamics, which is still in the developing stage at present. The basic reason that the subject of equilibrium thermodynamics is useful and powerful despite the idealized reversible processes studied in it is that some of macroscopic thermodynamic properties of a system, which is going through irreversible processes, can be related to the complementary quantities computed from the reversible processes as will be shown later as we develop the subject in greater detail because there is a state function of thermodynamic state variables for the system even if the processes are irreversible. This state function extends the notion of the equilibrium entropy into the domain of irreversible processes. The existence of such a state function — called calortropy — lifts thermodynamics from the level of studying only idealized reversible processes, as in equilibrium thermodynamics, to a level of using a more insightful and powerful mathematical tool for studying various aspects of irreversible behavior of the system. We will elaborate on this point in the chapters dealing with the second law of thermodynamics and in nonequilibrium part consisting of Chapters 19–22 of this book.

Since we are going to use various terms in our study, we fix their meanings by introducing the following system of terminology. A *system* is that part of the physical world which is under consideration, and the rest of the physical world is called the *surroundings*.

When a system exchanges mass, heat, work, and any other forms of energy with the surroundings, it is called an *open system* (c). When a system does not exchange matter but energies with the surroundings, it is said to be a *closed system* (b). If a system has no interaction whatsoever with the surroundings, it is called an *isolated system* (a). It is possible to regard the system and the surroundings together as an isolated system. We will find it convenient to do so for some cases (Fig. 1.1).

Thermodynamics is concerned with gross observables of a macroscopic system and their interrelationships. Since according to the atomic theory

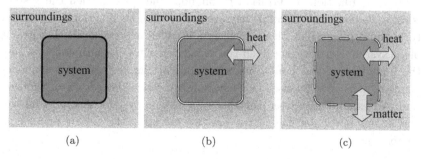

Fig. 1.1. The system and surroundings — universe. Panel (a) is an isolated system; (b) is a closed system; (c) is an open system.

of matter, a macroscopic system consists of an enormous number of atoms
and molecules not counting elementary subnuclear particles, a microscopic
description of the macroscopic system would entail a knowledge of an
enormous number of microscopic variables. However, a large number of
molecules in an assembly exhibit as a rule a collective behavior, which may
be described by a small number of variables called *macroscopic variables* or
macroscopic coordinates. A macroscopic variable (coordinate) is an observable whose determination requires only measurements, over time spans long
compared with periods of thermal motion, that take averages of microscopic
variables over regions containing a large number of molecules and involving
energies large compared with individual energies of atoms and molecules.
Pressure, volume, temperature, internal energy, and so on, are examples of
such macroscopic variables.

We often speak of *thermodynamic properties*. These are termed as the
properties of the system which describe its macroscopic coordinates. Thermodynamic properties are classified into two categories. If a thermodynamic property is independent of the mass of the system, then it is called
an *intensive property*. Examples are pressure, temperature, concentrations,
and molar properties. If a thermodynamic property depends on the mass
of the system, it is called an *extensive property*. Examples are the volume, energy, entropy, and so on, of a system which increase in proportion
to the mass of the system. Intensive and extensive properties (variables)
often appear as conjugate pairs of variables in thermodynamics. We may
take the examples of pressure and volume, and temperature and entropy
for such conjugate pairs of thermodynamic variables. Intensive properties,
however, may vary with position in the space as do the extensive variable
if the system is not homogeneous in space. If the intensive properties are
continuous functions of position throughout the system, then the system
is called *homogeneous*, and if they are not continuous functions of position
throughout the system, then the system is called *heterogeneous*. For example, a system is heterogeneous if the density changes discontinuously across
the boundary of two homogeneous regions of the system. Such homogeneous regions of a heterogeneous system are called the *phases* of the system. An example for heterogeneous systems is a system of water and ice,
and in this particular case, there are two phases in the system of a single
component.

We will often speak of a *thermodynamic state*. The thermodynamic
state of a system is defined by its intensive properties. They may or may

not change over space and time. A system is said to be in a state of *thermodynamic equilibrium* if (a) the thermodynamic state of the system does not change in the time duration of the observation performed, and (b) no material or energy flux exists in its interior or at the boundaries with the surroundings. Otherwise, the system is in a state of *nonequilibrium*. In equilibrium thermodynamics, we examine systems which are in a state of thermodynamic equilibrium.

A system may be composed of a single component or more than one component. In the former case, the system is called a *pure system* and in the latter case, it is called a *mixed system* or simply a *mixture*. In a mixture, there may arise the question as to the number of *independent components* of the system: it is defined by the minimum number of chemical species from which the system can be prepared in each phase of the system by a specified set of physico-chemical procedures. A practical way of determining the number of independent components is the total number of components minus the number of distinct chemical and other restrictive conditions such as chemical reactions and charge neutrality conditions. For example, consider a system formed when NaCl, KBr, and H_2O are mixed. If KCl, NaBr, NaBr·H_2O, KBr·H_2O, and NaCl·H_2O are isolated on chemical analysis, the distinct chemical reactions are

$$NaCl + KBr \rightarrow NaBr + KCl,$$

$$NaCl + H_2O \rightarrow NaCl \cdot H_2O,$$

$$KBr + H_2O \rightarrow KBr \cdot H_2O,$$

$$NaBr + H_2O \rightarrow NaBr \cdot H_2O.$$

If the concentration of species i is denoted by c_i then since there hold the relations for concentrations

$$c_{NaCl}^{initial} = c_{NaBr} + c_{NaCl} \cdot H_2O + c_{NaBr} \cdot H_2O,$$

and

$$c_{KCl} = c_{NaCl}^{initial} - c_{NaCl} \cdot H_2O,$$

where $c_{NaCl}^{initial}$ is the initial concentration of NaCl, we find

$$c_{KCl} = c_{NaBr} + c_{NaBr} \cdot H_2O.$$

This means that the number of independent components is $(8 - 1) - 4 = 3$ in this case.

In Chapter 2 of this book, the notions of temperature, work, and heat are discussed. In Chapter 3, the first law of thermodynamics is discussed together with thermochemistry, which deals with measurements of heat released or absorbed by the system. In Chapter 4, the second law of thermodynamics is discussed. This important principle, which was literally enunciated by Lord Kelvin and R. Clausius, was given a mathematical representation in the form of inequality now known as the Clausius inequality. In this work, the Clausius inequality will be replaced by an equation as a general mathematical representation of the second law of thermodynamics, which remains valid even if there are irreversible processes present in the system. The said mathematical representation of the second law of thermodynamics permits us to develop the thermodynamics of irreversible processes in a general context. For this purpose, we introduce the notion of *calortropy*. This part of the treatment of the second law of thermodynamics sets the present work apart from the conventional methods used in other works available in the literature on thermodynamics. The distinctive point of the new quantity is that it is the extension to nonequilibrium of the notion of equilibrium entropy that was originally introduced by Clausius for *reversible processes only*. By the accomplished extension, we are now provided with the starting point of a theory of irreversible processes in a general form, and even the equilibrium thermodynamics of Clausius is provided with a window through which we can glimpse into the world of irreversible phenomena even if one studies just the reversible process associated with the irreversible process in question. The nonequilibrium extension of the Clausius entropy is given the new term *calortropy*, which means *heat evolution*. This extension frees us from the shackles of entropy defined for equilibrium only, and equilibrium thermodynamics consequently becomes easier to comprehend than otherwise.

In the rest of the book, we treat the conventional subjects of equilibrium thermodynamics, which are commonly discussed in courses on thermodynamics. The subjects covered are thermodynamics of gases, liquids, and solutions, heterogeneous equilibria, chemical equilibria, strong electrolytes, galvanic cells, and the Debye–Hückel theory of strong electrolytes, which is the only concession we make to discuss a statistical treatment of macroscopic phenomena. We will also discuss the thermodynamics of systems subject to electromagnetic fields and the thermodynamics of interfacial phenomena.

In nonequilibrium part of this book, we present a summary of a general theory of thermodynamics of irreversible processes in systems removed from equilibrium at arbitrary degree in Chapter 19. In Chapters 20–22, we provide examples for the applications of the general theory to linear irreversible processes occurring in the vicinity of equilibrium; to nonlinear irreversible phenomena such as non-Newtonian flow, non-Fourier heat conduction, nonlinear electrical conduction; and to nonlinear phenomena involving spatial and temporal oscillatory phenomena in chemically reacting medium. These are some examples that we encounter in recent chemical and physical experiments and in engineering and biological sciences, for which thermodynamics of irreversible phenomena is not only relevant but also of importance. These chapters are meant to introduce the readers to the thermodynamics of irreversible phenomena in macroscopic material systems.

We believe that thermodynamics is a subject that should be studied without an intrusion by a molecular theory approach (i.e., statistical mechanics) because it is a subject that allows us to make deductions with regard to macroscopic properties of matter without reference to the molecular picture of matter and eventually serves as the ultimate aim of molecular theory of macroscopic matter, which is developed by means of statistical mechanics on the basis of molecular theories of matter. Mixing thermodynamics and statistical mechanics tends to confuse important issues regarding thermodynamics, which is distinctive from the molecular theoretic treatment of macroscopic properties of matter used in statistical mechanics and the various issues involved therein, and possible confusions on the issues of separate distinctive nature would hinder a logical development of the continuum theory of irreversible processes. Nevertheless, in this book, a concession is made for the pedagogical importance of the Debye–Hückel theory indispensable in the physical chemistry of electrolytes and plasmas, including the theory of electrolytic conductance.

The major portion of this book is based on the materials taught by the first author in the courses on thermodynamics, on and off, over a period of over two decades at McGill University. The same materials have been taught by the second author in American University of Beirut for a few academic terms.

Part I

Thermodynamics of Reversible Processes — Equilibrium Thermodynamics

In Part I, we discuss the thermodynamics of reversible processes, which is conventionally known as equilibrium thermodynamics. It consists of Chapters 2–18. However, in the discussion of the first and second laws of thermodynamics irreversible processes are taken into consideration and a generalized notion of equilibrium entropy is introduced. It appears in the form of calortropy.

Chapter 2

Temperature, Work, and Heat

Temperature, work, and heat are three basic concepts underlying the science of thermodynamics, the scientific quantification of which traces back to the very beginning of thermodynamics. Feeling hot or cold is a physiological sensation we have when we touch an object, but such sensation was not given an objective measure until the times of Galileo. Furthermore, the relation of temperature to heat and the relation of work and heat have evolved through the history of science, their evolution embodying our struggle to understand their nature. In this chapter, we discuss their quantification, so that they can be made use of in the subsequent study of thermodynamics in a scientific and logical manner.

2.1. Temperature

Temperature is not only one of the most important quantities in thermodynamics but also one of the basic state variables, in terms of which thermal properties of matter can be characterized and reckoned with. Its quantification had taken a long winding process of evolution in thoughts before the concept took the form currently in universal use. It is introduced as a quantifiable quantity by the zeroth law of thermodynamics stated below.

If two bodies of different degrees of hotness are put into contact, the difference in hotness eventually disappears between the two bodies, and we say that they are in thermal equilibrium. What in fact happens when such two bodies are put into thermal contact is that heat flows from the hotter to the colder body until there is no difference in hotness or coldness. Such a phenomenon is experienced in our everyday life. In order to develop the science of thermodynamics, it is necessary to quantify the sense of hotness and coldness. The desired quantification is essentially achieved by the following law.

The zeroth law of thermodynamics.   *If two systems A and B are in thermal equilibrium with system C, then the systems A and B are also in thermal equilibrium.*

In other words, if there exists thermal equilibrium between A and C, there is a property (parameter of state) called temperature θ such that

$$\theta_A = \theta_C,$$

and similarly for systems B and C,

$$\theta_B = \theta_C.$$

Therefore, there follows the equality

$$\theta_A = \theta_B. \tag{2.1}$$

These relations are illustrated in Fig. 2.1. They supply a scientific means to quantify temperature.

Let us examine how we might achieve the desired quantification of temperature. Imagine a balloon containing air is heated. The air inside the

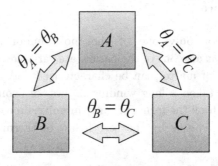

Fig. 2.1.   Thermal equilibrium and the definition of temperature. Pairs of systems A, B, and C in thermal equilibrium no longer exchange heat or energy in another form, and the property characterizing the thermal equilibrium is the temperature of the systems.

balloon then gets hotter than it was before heating and as the volume of the air in the balloon increases, the volume of the balloon increases. We thus see that the measure of hotness, i.e., the temperature of the air inside the balloon, may be quantified in terms of the volume of the balloon. Put differently, the physical quality, temperature, of the air is a function of volume and vice versa. Variables such as volume or pressure, which can be used for measuring temperature, are called the *thermometric properties* of the system, and the instruments to quantify temperature are commonly called thermometers. There are other useful thermometric properties, such as electrical resistance, magnetic susceptibility, wavelengths of radiation emitted by the body, sound wave velocity transmitted by matter, dielectric constants, and so on that can be used to quantify the notion of temperature.[1]

Let us call P a thermometric property. Then, by the consideration made earlier, we may assume θ is a function of P:

$$\theta = \theta(P). \tag{2.2}$$

In order for the quantity P to qualify as a useful and practical thermometric property, the function $\theta(P)$, however, *is preferable to be monotonic and linear with respect to* P *in the temperature range where the property is utilized for measuring temperature*[2]:

$$\theta = a + bP, \tag{2.3}$$

where a and b are constants, which generally depend on the thermometric material used. The linearity and monotonicity of the relation are

---

[1]See, for example, J. F. Schooley, *ed.*, *Temperature* (American Institute of Physics, New York, 1982) for methods of measurement of temperature.

[2]The concept of temperature has taken a long time to acquire its present form. Galileo Galilei is credited to have devised for the first time in history (ca. 1592–1603) a thermometer, which was then called a thermoscope. The instrument was merely a glass tube with a bulb attached at one end, which contained air, and the other end of the tube immersed in water. As the temperature of the air in the bulb changes, the water level moved up or down as the air inside either contracted or expanded. In fact, it was a "barothermoscope" since the water level also depended on the pressure, although the pressure effect was not recognized until much later. His thermoscope had no fixed point. Thermometers with one fixed point were proposed in 1665 by Robert Boyle, Robert Hooke, and Christiaan Huygens. Thermometers with two fixed points were made in 1669 by Honoré Fabri, who adopted the melting point of snow for the lower fixed point, but rather vague "greatest summer heat" for the upper fixed point. In 1694, Carlo Renaldini proposed to take the freezing and boiling temperatures of water as the two fixed point temperatures. The present day centigrade scale is credited to Anders Celsius (1742), although it is believed to have been first suggested in 1710 by a Swede named Elvius.

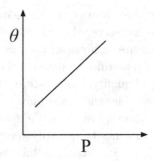

Fig. 2.2. Temperature is linear with respect to thermometric property P.

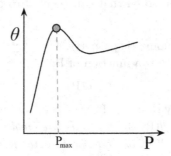

Fig. 2.3. Temperature is nonlinear with respect to thermometric property P. For $P \geq P_{max}$, there is no one-to-one correspondence between θ and P.

preferable for one-to-one correspondence between the temperature and the thermometric property as indicated in Fig. 2.2 because the relation is the simplest possible. Obviously, nonlinear relations as in Fig. 2.3 cannot be useful, since there is no guaranty of one-to-one correspondence between θ and the thermometric property P in the range of $P > P_{max}$ (the point of maximum θ) where the temperature cannot be uniquely determined. Since the linear relationship (2.3) between the thermometric property and temperature contains two parameters a and b, which must be obtained empirically, they are determined in reference to two typical points of θ. Since choice of such points can be arbitrary, there are various temperature scales possible.

2.1.1. *Centigrade (Celsius) Scale*

In the centigrade (Celsius) scale, two reference points of temperature are respectively at the ice point and the boiling point of pure water at 1 atm pressure. Let us denote the values of thermometric property P at the ice and boiling points by P_0 and P_{100}, respectively. The corresponding values of θ at the ice and boiling points are taken to be equal to 0 and 100. Thus, there are 100 degrees between the two reference points. We then find from Eq. (2.3)

$$0 = a + bP_0,$$

$$100 = a + bP_{100}.$$

These two equations are solved for a and b to obtain the relations

$$a = -\frac{100P_0}{P_{100} - P_0},$$

$$b = \frac{100}{P_{100} - P_0}.$$

Substituting back in Eq. (2.3), we obtain

$$\theta = \frac{1}{\alpha} \cdot \frac{(P - P_0)}{P_0}, \tag{2.4}$$

where α is defined by

$$\alpha = \frac{P_{100} - P_0}{100P_0}. \tag{2.5}$$

Therefore, α is the mean relative change in P per degree between θ_0 and θ_{100}. For example, if the volume of a gas is taken for P, then α represents a mean expansion coefficient of the gas (thermometric substance) between the two reference points of temperature. We therefore see that temperature can be measured by observing the variation of P relative to its value at one of the reference points. Well-known examples are alcohol and mercury thermometers.

The temperatures measured by two different thermometric materials generally do not necessarily agree at the temperatures between two reference points θ_0 and θ_{100}. That is, thermometers are not universal and the temperature measured depends upon the thermometric materials and properties used for the purpose of measuring temperature.[3] Since it is difficult to carry on the science of thermodynamics without a universal scale of temperature, it is necessary to devise one. This is achieved with the ideal

---

[3]L. A. Guildner, *Phys. Today*, **December, 1982**, p. 24.

gas thermometer, and as will be seen, it provides a universal temperature scale.

2.1.2. Fahrenheit Scale

In some parts of the world, the Fahrenheit temperature scale is used in daily life by custom. It is a temperature scale in which the ice point of water is set at 32 degree (F), the boiling point of water at 212 degree (F), and the difference is divided by 180 for the measure of a degree. In science, the Fahrenheit scale of temperature is not used. We simply note to help conversion from one scale to another that the relation between the t degree (C) in the centigrade scale of temperature and t' degree (F) in the Fahrenheit scale of temperature is given by the equation

$$t°C = \tfrac{5}{9} \left[t'°F - 32 \right].\tag{2.6}$$

2.1.3. Absolute Temperature Scale and Ideal Gas Thermometer

To devise a universal temperature scale, the ideal gas thermometer employs the ideal gas as the thermometric material. The absolute temperature scale is obtained therewith.

If volume V is taken for the thermometric property P, then α represents a mean expansion coefficient of the thermometric substance between the two reference points of temperature. The study of behaviors of gases under the influence of heat and pressure traces back to the early stages in the development of modern science, and the names like R. Boyle, J. Gay-Lussac, and H. V. Regnault were associated with it. It was found through extensive experiments that gases show certain universal behaviors towards heat and pressure as they become sufficiently diluted so that the pressure is low. The state of such gases is then called ideal. According to the modern molecular theory interpretation of the ideal gas behavior, there are no interactions between the molecules in ideal gases. Such a universal behavior independent of materials is obviously a desired property to exploit in devising a temperature scale.

If the volume is taken as the thermometric property for the ideal gas thermometer, then the volume V at temperature θ is known to be given by

$$V = V_0(1 + \alpha\theta), \tag{2.7}$$

where if the Celsius scale of temperature is used, α is given by

$$\alpha = \frac{V_{100} - V_0}{100V_0}. \tag{2.8}$$

Equation (2.7) is the result of rearranging Eq. (2.4) after setting the thermometric property $P := V$, and α is the mean expansion coefficient of the gas. It is empirically found that as the gas pressure p decreases to a sufficiently low value, α approaches a limit, i.e.,

$$\lim_{p \to 0} \alpha = \alpha^*, \tag{2.9}$$

where α^* is a universal constant independent of gases and has the value[4]

$$\alpha^* = \frac{1}{273.15}.$$

We will put

$$\alpha^* = \frac{1}{T_0}, \tag{2.10}$$

i.e., $T_0 = (273.15 \pm 0.02)°C$ for all ideal gases.

Equation (2.9) together with Eq. (2.4) implies that

$$\lim_{p \to 0} \theta = \theta^* \tag{2.11}$$

and thus in the low pressure limit,

$$V = V_0(1 + \alpha^*\theta^*). \tag{2.12}$$

Equation (2.10) therefore indicates that there exists a temperature scale which does not depend on the material employed and consequently is *universal*. Such a thermometer is called the ideal gas thermometer, and the scale of temperature based thereon the ideal gas temperature scale.

By using Eq. (2.10), we may rewrite Eq. (2.12) in the form

$$T = T_0 \frac{V}{V_0}, \tag{2.13}$$

---

[4]Gay-Lussac found $\alpha = 1/267$ approximately, but Regnault in 1847 obtained $\alpha = 1/273$ by using an improved experimental procedure.

where

$$T = T_0 + \theta^*. \tag{2.14}$$

Since the value of T_0 is subject to experimental errors, to fix T_0 universally and unequivocally, it is agreed by the *Tenth Conference of the International Committee on Weights and Measures* that at the triple point of water, the temperature is

$$T = 273.16 \text{ K exactly.}$$

The temperature scale so determined puts the temperature of ice point at 1 atm pressure at

$$T_0 = 273.15. \tag{2.15}$$

This temperature scale is called the absolute temperature scale and the temperature in such scale is expressed with unit K, meaning Kelvin.

Let us now consider an immediate implication of the thermodynamic temperature scale devised with an ideal gas. It is empirically established through investigations by Boyle, Gay-Lussac, and others that for ideal gases, there holds the relation

$$\lim_{p \to 0} pV = \beta(\theta) \tag{2.16}$$

and

$$\frac{\beta(\theta_1)}{\beta(\theta_2)} = \text{a universal constant independent of gases,} \tag{2.17}$$

where p is the pressure and β is a function of θ, which is independent of gases. Since as $p \to 0$

$$V = \frac{\beta(\theta)}{p},$$

it follows from Eq. (2.16) that

$$\frac{T}{T_0} = \frac{\beta(\theta)}{\beta(\theta_0)} = c, \tag{2.18}$$

where c is a universal constant according to Eq. (2.17). Since Eq. (2.18) means that

$$T \propto \beta(\theta),$$

we may write

$$\beta(\theta) = nRT, \tag{2.19}$$

where n denotes the number of moles of the gas contained in V and R is a universal constant. It is called the gas constant and its value is $8.3143 \, \mathrm{J \, K^{-1} \, mol^{-1}}$. Therefore, Eq. (2.13) now may be written as[5]

$$\lim_{p \to 0} pV = nRT \qquad (2.20)$$

or, more simply,

$$pV = nRT. \qquad (2.21)$$

For a mole of an ideal gas, it may be written as

$$pv = RT,$$

where $v = V/n$ denotes the molar volume of the gas (in fact, ideal gas). This is the ideal gas equation of state. When the equation of state is written in the form of Eq. (2.21), it must be understood that it holds only when p is sufficiently low so that the gas behaves like an ideal gas. The formulation of the ideal gas equation of state as in Eq. (2.21) indicates that the ideal gas equation of state is intimately tied up with the ideal gas temperature scale and, as will be seen later, all equations of state and thermodynamic properties for real substances are expressed in terms of the temperature scale devised on the basis of the ideal gas and reckoned in the absolute temperature scale.

Having elucidated the existence of a quantity called temperature and having devised the absolute temperature scale based on the ideal gas, we now come to the question of whether it is possible to speak of temperature if a system is not in equilibrium internally.

The zeroth law requires that two bodies in thermal contact be in thermal equilibrium for the concept of temperature to be meaningful. This does not mean that the two bodies have to be internally in equilibrium. The

---

[5]This limiting formula may be written in another form

$$T = \lim_{n \to 0} \frac{PV}{nR} = \lim_{n \to 0} \frac{pv}{R}.$$

Since the pressure–volume work per mole, pv, on the right-hand side may be calculated by using its molecular theory representation — i.e., statistical mechanical formula — this limiting form may be taken as the statistical mechanical formula for the temperature of fluids. Since the pressure–volume work in question is also extendable to nonequilibrium fluids, the formula makes it possible to extend the notion of temperature to the case of nonequilibrium fluids. If we denote the measure of nonequilibrium by ϵ, from the mathematical standpoint, the extension of T mentioned may be regarded as an analytic continuation of T to $\epsilon > 0$.

only condition required for parameter θ to exist is that there is no heat exchange between the two bodies over a characteristic span of observation time. If one of the two bodies is a thermometer calibrated against the ideal gas thermometer, then the characteristic span of time has to do with the resolution power of the thermometer, which is the measure of how fast the thermometer responds to a heat transfer between the body and the thermometer. Even in such a case, one still speaks of the temperature of the body. For example, if the body is animate like the human body, which is obviously undergoing complex irreversible processes within itself, we routinely speak of the temperature of the body, but it is recorded by the temperature at the point of thermal equilibrium between the body and the thermometer in contact with the body. Similarly, one may insert a thermometer in a stream of a liquid which is obviously not in equilibrium, yet speaks of the temperature of the liquid. Depending on the processes that are going on in the liquid, the temperature may be at a constant value over time or may be varying in time. The temperature scale devised based on the ideal gas thermometer introduced earlier can be still used to record the thermal state of systems where some irreversible processes are in progress. However, it should be kept in mind that the temperature so measured of a system in a nonequilibrium state may be only the temperature of the local volume sampled, but not the temperature of the global system. The latter would be the case if the system is globally maintained at a fixed temperature even if it is in a nonequilibrium state, as is the case for human or animate bodies. See footnote 5.

2.2. Pressure

Pressure is a mechanical quantity relevant to and, in fact, indispensable in, developing thermodynamics of matter. It is defined as the force exerted on a unit area of a surface:

$$\text{pressure} = \frac{\text{force}}{\text{area}} = \frac{F}{A}.$$

Therefore, the dimension of pressure is N m^{-2} = kg m/s^2, where N is the unit of force, Newton, in the SI system of units. This unit of pressure is called pascal and denoted by Pa. In practice, since this SI unit is too small, other more practical units are used. For example,

$$1 \text{ bar} = 10^5 \text{ N m}^{-2} = 10^5 \text{ Pa},$$

and alternatively,

$$1\,\text{atm} = 1.01325\,\text{bar}$$

with 1 atm of pressure defined as the pressure exerted by the 0.760 m column of mercury at 0°C at the sea level. The standard value of the gravitational acceleration (g) is used for this calculation:

$$g = 9.80665\,\text{m}\,\text{s}^{-2}.$$

Pressure is occasionally expressed in the units of Torr, meaning Torricelli. The conversion factor of the units of Torr to the units of atm is

$$1\,\text{atm} = 760\,\text{Torr}.$$

The definition of pressure given earlier serves to provide a means to measure it. The simplest way to measure pressure is the use of a manometer. The pressure of a body (e.g., a fluid) is recorded as mechanical equilibrium is established between the body and the manometer. It is important to recognize that as in the case of temperature, which is measured when the system is in thermal equilibrium with the thermometer, the measurement of pressure requires mechanical equilibrium between the system and the manometer — the measuring device. Therefore, equilibrium of a sort is the prerequisite for measuring temperature and pressure of a system.

2.3. Work

The meaning of the thermodynamic concept of work is similar to that in mechanics. When an object is displaced from one point in space to another under the influence of a force $\mathbf{F}$, we say that a work is done on the system, and for an infinitesimal displacement $d\mathbf{s}$, the differential work is

$$dW = \mathbf{F} \cdot d\mathbf{s}. \tag{2.22}$$

The work done to move the object (system) from point 1 to point 2 along the curve $\mathbf{s}$ is then obtained by integrating Eq. (2.22) over the path along which the work is performed:

$$W_{12} = \int_1^2 \mathbf{F} \cdot d\mathbf{s}. \tag{2.23}$$

It should be noted that the integration is a line integral. This integral may be put in the form indicating the path dependence of the value of the

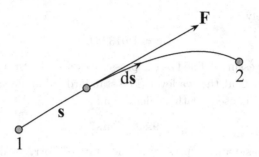

Fig. 2.4.   Work by a force along path s.

integral

$$W_{12} = \int_{1(\text{path})}^{2} dW \tag{2.24}$$

with the subscript "path", meaning that the integration is path-dependent, i.e., it is a line integral.[6] We use a slash on the differential symbol đ to denote that, for example, đW is not an exact differential of W. The notation đW simply means that it is an infinitesimal quantity, namely, an infinitesimal work in this case, which arises from an infinitesimal displacement of a body by ds under the action of a force. This nonexactness property will be further elaborated shortly. It is important to note that work depends on the path and therefore, the works done from point 1 to point 2 along two different paths are not generally equal, as we empirically well recognize the truthfulness of the statement from our daily life experience. In mechanics, we know that they are equal only if the force field has a potential energy, namely, is conservative. In thermodynamics, forces usually do not have a potential (Fig. 2.4).

Since it is necessary to distinguish the work done on the system from the work done by the system and they are opposite in sign, we introduce a sign convention.

Sign convention.   *The work done on the system by the surroundings is taken as positive whereas the work done on the surroundings by the system is taken as negative.*

---

[6]The term "line integral" is a mathematical terminology, which means that the value of the integral depends on the path of integration performed in the space of integration variable.

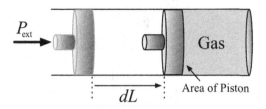

Fig. 2.5. Work by a gas in the piston under an external pressure. The gas in the piston is compressed.

For example, if an external pressure P_{ext} is applied on area A of the container of a gas, which is capable of displacement and as a consequence, the surface A is displaced by an infinitesimal distance dL in the negative direction (see Fig. 2.5), then the work done on the system is

$$đW = -P_{ext}dL \times A = -P_{ext}dV, \tag{2.25}$$

where $dV = AdL$. The total work done on the system arising from the volume change from V_1 to V_2 is then

$$W = -\int_{V_1}^{V_2} P_{ext}dV. \tag{2.26}$$

The integral is clearly path-dependent. The knowledge of P_{ext} as a function of V will therefore be necessary to perform the integration for W. If the pressure remains constant, then the integration is immediate, and the work is given by

$$W = P_{ext}(V_1 - V_2). \tag{2.27}$$

Since $V_1 - V_2 > 0$, the right-hand side, namely, the work, is positive in agreement with the convention we have adopted (Figs. 2.6 and 2.7).

As another example, suppose an object of mass M falls a distance dh under the influence of the gravitational field. Then the work done by the system on the surroundings (Earth) is given by

$$đW = -gMdh. \tag{2.28}$$

For yet another example, consider a force exerting on a rubber band. If a rubber band of length L is stretched dL by a tensile force f, then the work done on the system is

$$đW = fdL. \tag{2.29}$$

Other kinds of work relevant in thermodynamics are listed in Table 2.1. The symbols in Table 2.1 are as follows: ϕ is the electric potential, q is

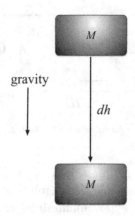

Fig. 2.6. Work by gravitational force.

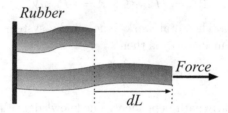

Fig. 2.7. Work by an extensional force.

Table 2.1. Other types of work.

type	dW
Ohmic	$\phi dq = \phi I dt$
surface	γdA
electric	$\mathbf{E}d\mathbf{D}$
magnetic	$\mathbf{H}d\mathbf{M}$

the charge, I is the current (i.e., $I = dq/dt$), γ is the surface tension, A is the area, $\mathbf{E}$ is the electric field, $\mathbf{D}$ is the electric displacement, $\mathbf{H}$ is the magnetic field, and $\mathbf{M}$ is the magnetization.

If we denote by X_i the force and by x_i the conjugate variable, namely, displacement, the work done on the system by a set of forces is given by

$$dW = \sum_i X_i dx_i, \tag{2.30}$$

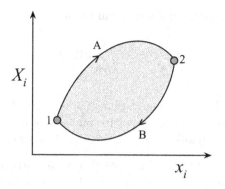

Fig. 2.8. Work performed under a force X_i in a cyclic process. The shaded region denotes the net work performed in the process.

where the sum is over all types of work performed. The total work corresponding to finite displacements from x_i^1 to x_i^2 is then given by

$$W = \sum_i \int_{x_i^1}^{x_i^2} X_i dx_i, \qquad (2.31)$$

where the integrals are line integrals along the paths of work performed since $X_i dx_i$ is not an exact differential (see Fig. 2.8).

In connection with the line integrals like the work defined in Eq. (2.31), it is useful to add a mathematical note on nonexact differentials for which đW is an example. If a differential dz is exact, then the value of the integral depends only on the values of z at the endpoints as in the expression

$$\int_1^2 dz = z(2) - z(1).$$

However, if it is not exact, then

$$\int_1^2 đz \neq z(2) - z(1),$$

and the value of the integral depends on the line along which the integration is performed. The differential work đW has the latter property, and it is important to understand the basic difference between an exact differential, say, dz and a nonexact differential such as đW. For example, when integrated along the two different paths as in Fig. 2.8, its integral may be broken up into two parts and written in the form

$$\int_{1(A)}^{2} \mathrm{d}W - \int_{1(B)}^{2} \mathrm{d}W \neq 0. \tag{2.32}$$

Since a line integral is the area under line (curve) defined by the path of integration between two endpoints, the left-hand side of Eq. (2.32) corresponds to the net work (the shaded area in Fig. 2.8) performed in the cyclic process.

This net work would have been equal to zero if $\mathrm{d}W$ were an exact differential and hence there existed a potential. In mechanics of a conservative field, the work over a cycle is equal to zero, since the force has a potential energy.[7] As an illustration of this point, let us consider Eq. (2.30) and assume that there exists a function $\varphi(x_1, x_2, \ldots, x_n)$ such that the force is given by the gradient of φ

$$X_i = -\frac{\partial \varphi}{\partial x_i} \quad (i = 1, 2, \ldots, n). \tag{2.33}$$

Then Eq. (2.30) can be written as an exact differential of φ:

$$\mathrm{d}W = -\sum_i \frac{\partial \varphi}{\partial x_i} dx_i = -d\varphi, \tag{2.34}$$

and

$$W = \int_{1}^{2} \mathrm{d}W = -\int_{1}^{2} d\varphi = \varphi(1) - \varphi(2). \tag{2.35}$$

The differential work $\mathrm{d}W$ in this case becomes an exact differential, and the function $\varphi(x_1, x_2, \ldots, x_n)$ is a potential energy mentioned earlier. In this case, in which the force is conservative, the differential form $\mathrm{d}W$ is *monogenic* in the mathematical terminology. In thermodynamics, the differential form for work generally is not monogenic and the work over a cycle is not equal to zero for the reason to be elaborated later when the notion of energy is discussed.

---

[7]In mathematical terminology, it is said that the force is monogenic if it is derivable from a scalar function, whereas it is said to be polygenic if it is given by a differential form that is not expressible as a derivative of a scalar function. A typical example is the differential form (2.30), which is not a derivative of a scalar function.

2.4. Heat

The concept of heat has evolved throughout the scientific history. Before Galilei (1623) and Francis Bacon (1620), it was believed that heat was a kind of matter, and a transfer of such a matter between bodies was the cause for a body being heated up by another. Thus, it was believed that a body gets warmer when it receives a certain amount of such matter and gets colder when it gives it up. This concept was put on a scientific basis by Joseph Black (1760) in the late 18th century, who developed the caloric theory of heat. Galilei appears to be the first to realize that heat is not a matter, but a manifestation of the motion of particles and therefore a form of energy.[8] As was evident by the Black's caloric theory of heat contrary to Galilei's idea, the revolutionary concept of Galilei and Bacon, however, lay dormant for over 200 years[9] until the observations made by Count Rumford, Humphrey Davy, Robert Mayer, and James P. Joule, which indicated that heat cannot be a matter, but is a form of energy. Joule finally established through his famous experiments (e.g., paddle wheel experiment; see Fig. 3.1 in Chapter 3) that heat is a form of energy, namely, the mechanical equivalence of heat.

We have seen that there is a parameter that can be used as a gauge for measuring the degree of hotness of a body (system). We have called that measure the temperature of the body. Now, we can make use of this parameter to devise a way of quantifying heat.

When two systems A and B at different temperatures are put into thermal contact, there is a transfer of energy between them in the form of heat. The amount of heat transfer for a given temperature difference between A and B depends on the substances making up the systems, and we define an extensive property called the heat capacity of a substance.

---

[8] According to Joseph Black in his *Lectures in the Elements of Chemistry* (reproduced in *The Early Development of the Concepts of Temperature and Heat* by D. Roller) (Harvard University Press, 1950), Francis Bacon (1620) proposed, on the basis of the consideration of means by which heat is produced, or made to appear, in bodies, such as the percussion of iron, the friction of solid bodies, the collision of flint and steel, that heat is motion. Considering this, it appears that Francis Bacon's notion of heat had more to do with the cause and generation of heat than the notion of heat itself unlike the case with Galilei's.
[9] The idea that heat is motion was used by Daniel Bernoulli in ca. 1780 when he derived an ideal gas equation of state in his book *Hydrodynamica*, but this also did not draw attention.

Heat was used to be measured in units of calorie. It was initially defined by two different ways:

Mean gram calorie. The mean gram calorie q_m is defined as the mean amount of heat required to raise by 1°C the temperature of 1 gram of water at 1 atm pressure between 0 and 100°C, namely,

$$q_m = \frac{Q(0°\text{C} \to 100°\text{C}) \text{ of water at 1 atm pressure}}{100\, w \text{ grams}}.$$

15° gram calorie. The 15°gram calorie $q_{15°}$ is defined as the amount of heat required to raise by 1°C the temperature of 1 gram of water at 14.5°C and at 1 atm pressure, namely,

$$q_{15°} = \frac{Q(14.5°\text{C} \to 15.5°\text{C}) \text{ of water at 1 atm pressure}}{w \text{ grams}}.$$

The values of q_m and $q_{15°}$ do not exactly agree with each other, but form a certain ratio:

$$q_{15°} = 1.00024\, q_m. \tag{2.36}$$

For this reason, the *Ninth International Conference on Weights and Measures* recommended that the Joule be used as the unit of heat: thermo-mechanical calorie or simply calorie is defined as follows:

$$1 \text{ thermomechanical calorie} = 1\, \text{cal}$$

$$= 4.184\, \text{Joules}$$

$$= 4.184\, \text{Nm}.$$

The change in the heat content of a system after a process of change in state depends on the manner in which the change has occurred; in other words, it depends on the path (mode) along which the change has been effected on the system. Suppose, for example, a mass of a liquid is heated from temperature T_A to temperature T_B under constant pressure in one case, and under constant volume in another. It is experimentally known that the amounts of heat required for the same change in temperature are different for the two cases. These two modes of change in temperature are examples of different paths of change referred to earlier. If an infinitesimal heat change is denoted by đQ, then the total heat change arising in a change in the state of the system from A to B along the path Γ (in as-yet-unspecified state space) is

$$Q = \int_{A(\Gamma)}^{B} đQ. \tag{2.37}$$

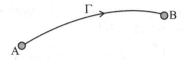

Fig. 2.9. Path of heat change from state A to state B along path Γ.

Note that dQ is not an exact differential and therefore, the value of Q is path-dependent (Fig. 2.9).

Before proceeding further on the discussion of heat, it is useful to introduce the sign convention for heat, since it is necessary to distinguish the heat absorbed from the surroundings by the system from the heat given up to the surroundings by the system.

Sign convention on heat. *Heat given up by the system to the surroundings is counted negative whereas heat absorbed by the system from the surrounding is counted positive.*

Clearly, because of the equivalence of heat to work and energy, this sign convention is consistent with the sign convention adopted earlier for work, as it should.

The heat capacity of a system is the amount of heat required to raise its temperature by 1°C. The mean heat capacity may be defined by

$$\bar{C} = \frac{Q}{T_2 - T_1}, \tag{2.38}$$

where Q is the amount of heat absorbed by the system to raise its temperature from T_1 to T_2. The more useful is the differential heat capacity, which is defined by

$$C = \frac{dQ}{dT} = \lim_{\Delta T \to 0} \frac{\Delta Q}{\Delta T}. \tag{2.39}$$

The heat capacities measured at constant pressure and at constant volume are different. In order to distinguish them, we will affix a subscript p or v to C:

$$C_p = \left(\frac{dQ}{dT}\right)_p, \quad C_v = \left(\frac{dQ}{dT}\right)_v. \tag{2.40}$$

The integral in Eq. (2.37) may be rewritten in terms of a heat capacity. If the path is that of constant pressure, then

$$Q_p = \int_{T_1}^{T_2} \left(\frac{dQ}{dT}\right)_p dT = \int_{T_1}^{T_2} C_p(T)\, dT, \qquad (2.41)$$

and if the path is that of constant volume, then

$$Q_v = \int_{T_1}^{T_2} \left(\frac{dQ}{dT}\right)_v dT = \int_{T_1}^{T_2} C_v(T)\, dT. \qquad (2.42)$$

We have attached a subscript p or v to Q to distinguish the paths — that is, the modes of heat transfer. Since $C_p \neq C_v$, the heats Q_p and Q_v are not the same, indicating the path-dependence of the integral of the inexact differential dQ.

2.5. Reversible Processes and Reversible Work

Processes in real systems occur at a finite rate over a finite duration of time. Therefore, work and heat transfer are generally performed irreversibly. The study of such irreversible processes is not a subject of equilibrium thermodynamics discussed in Part I of this book. In equilibrium thermo-dynamics, a rather special class of processes is studied, namely, reversible processes. A process is called *reversible* only if there is continuous equilib-rium between the system and its surroundings throughout the process. We speak of a *reversible change* when the system passes through a sequence of internal equilibrium states without necessarily being in equilibrium, but in quasi-equilibrium, with its surroundings. In Part II, some topics of thermo-dynamics of irreversible processes are discussed.

Therefore, it follows that a reversible process is an idealized limit pro-cess, which can be achieved by making the actual process infinitesimally slow and thus taking an infinite duration of time. This way, the causes making a process irreversible have died away in the time span of interest, and equilibrium is established continuously between the system and its sur-roundings over the course of the entire process. Such processes obviously are not realizable in practice, since the processes are usually finished in a finite duration of time in real systems. Nevertheless, this idealization provides us with a very useful conceptual device to carry out calculations in thermody-namics, unencumbered by the knowledge required of the energy dissipation accompanying an irreversible process. In equilibrium thermodynamics of reversible processes, thermodynamic potentials can be obtained from differential forms which admit integrals, namely, potentials. This feature can be useful for studying thermodynamics of irreversible processes, since a reversible process can be imagined to complete a cycle with an irreversible process, and the change in a thermodynamic potential over an irreversible process can be computed in terms of the thermodynamic potential over the reversible process complementary to the irreversible process making up the cycle. In this sense, the thermodynamics of reversible processes becomes a powerful theoretical tool for understanding some aspects of irreversible processes. It then is no longer a subject studying idealized situations of physical systems, but a theoretical tool to study some aspects of real processes in physical systems undergoing irreversible processes.

As is for reversible processes, we call a work reversible if it is performed through a reversible process which maintains the system in continuous equilibrium with the surroundings over the entire process of work. *The reversible work is a maximum work.* This statement will be proved later when we are mathematically better equipped in Chapter 4. What the statement implies is that there is an amount of *unavailable work* for the given task of work if the process is performed irreversibly, and this unavailable work is regarded as energy dissipation for the given task, which the work in question aims to achieve. Any irreversible process is accompanied by some sort of energy dissipation, and reversible processes may be defined as those in which energy dissipation is absent for the task in question. Therefore, a reversible work is seen to be a maximum work for the given task.

The meaning of reversible processes can be made more precise in mathematical terms if the concept of thermodynamic forces is taken advantage of. If a system is not in thermodynamic equilibrium, there are intensive variables which are inhomogeneous in space or time, or both. That is, in a nonequilibrium system, the intensive variables may vary over a distance or in time. The measure of spatial variations in intensive variables are given by their gradients, and *thermodynamic forces are defined as spatial gradients of intensive variables*. These forces are considered to be driving macroscopic irreversible processes in the system and to cause the latter to dissipate energy. In fact, the sum of squares of the thermodynamic forces is related to the energy dissipation associated with the irreversible processes. As these driving forces are spent in the course of time, the system reaches thermodynamic equilibrium under the given constraint and the energy dissipation arising from the driving forces vanishes. The reversible processes may then be defined as the limiting processes that occur under vanishingly small thermodynamic forces. They are, therefore, processes in which energy dissipation is absent. As a matter of fact, this was basically the definition used by Clausius when he introduced the notion of reversible processes in his formulation of equilibrium thermodynamics. However, this notion of Clausius has not been used in most textbooks on equilibrium thermodynamics to the detriment of a clearer understanding of the true meaning of reversible processes. We prefer this definition to the definition of reversible processes made with quasi-equilibrium processes maintained throughout the processes performed over an infinite duration of time. This aspect will be discussed in more detail in Chapter 4.

Chapter 3

The First Law of Thermodynamics

3.1. Equivalence of Heat and Energy

A scientific concept often trails a long tortuous path of evolution to attain the shape it currently assumes. Such a path of evolution necessarily reflects the evolution of our own thinking toward the phenomena underlying the concept, and an evolution means a modification of what is currently prevalent to a new and better suited form. Such a modification is usually prompted by our inherent desire to come up with a more encompassing, comprehensive viewpoint and theory, when faced with new empirical evidence which renders invalid or inappropriate the concept that has so far served us well and thus has been universally accepted as truthful.

In physical science, we empirically observe the states of mechanical objects and their changes, and discern that their states change in rather intricate but seemingly haphazard manners. We then look for rules and laws governing the manners in which the states change, so as to find an order in the state of affair that appears complicated and complex.

Constancy is a quality that stands out when contrasted to qualities that change. The concept of energy was born out of our desire to find a constant quality of mechanical systems that is preserved over the course of time irrespective of some changes of state that the systems have gone through.

In 1669, Christiaan Huygens discovered that the mass times the velocity squared, mv^2, was conserved during elastic collisions of mechanical bodies. Leibnitz called it *vis viva*. Much later, W. Thomson (Lord Kelvin) called $mv^2/2$ the kinetic energy. It was found that the kinetic energy was not conserved in inelastic collision even if friction was absent, but it was found that if the potential energy was added to the kinetic energy the sum was conserved. It is well known in mechanics that the energy of a system is conserved if the force is conservative. Thus, the kinetic energy of two elastically colliding billiard balls is conserved. But it is no longer conserved if an inelastic collision or friction is allowed between the two balls, since then the kinetic energy is transformed into another form of energy associated with the inelasticity of collision. In the case of inelastic collision, the total energy is conserved if that part of the energy associated with the inelastic excitation of internal states is added to the energy of the relative motion. However, if friction takes effect, heat is generated and the mechanical energy is proportionately diminished. Therefore, the energy conservation law that holds for the elastic and inelastic collisions appears to be no longer valid.

Count Rumford (Benjamin Thompson[1]) observed that the amount of heat generated during the boring of a cannon is proportional to the mechanical energy expended. This observation was further developed conceptually and put on a firm scientific basis by J. P. Joule, who performed the first determination of the mechanical equivalent of heat. It was observed that heat generated by mechanical work is independent of the materials involved and strictly proportional to the work done, when the system is subjected to a cyclic process whose net result is merely the conversion of work into heat with the system returning to its initial state. This experimental discovery, coupled with the realization through Joule's experiments that *heat is another form of energy*, led to the first law of thermodynamics as a broad, generalized enunciation of the mechanical energy conservation law.

Joule found in his famous paddle-wheel experiment (Fig. 3.1) that mechanical energy is converted into heat at a universal ratio independent of substance and the processes of conversion. He was thereby able to determine

---

[1] An American, who was in service to Elector of Bavaria, Ludwig. He was a colorful figure who was also a scientist, statesman, and businessman in addition to being briefly the husband to Madame Lavoisier, the widow of Antoine Laurent Lavoisier.

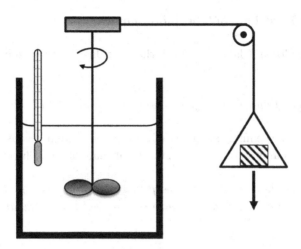

Fig. 3.1. Joule's paddle-wheel experiment. As the weight on the right falls a height h, the paddle-wheel turns performing a work, which is converted into heat and thus raises the temperature of the water. The temperature change is then recorded by the thermometer on the left.

the mechanical equivalent of heat:

$$W = JQ, \tag{3.1}$$

where W is the work, Q is the heat, and J is a constant called the mechanical equivalent of heat. Its unit is called Joule, and 1 cal = 4.1840 Joules. Its value may be determined, for example, by measuring the amount of heat generated owing to the work done when a body (e.g., a ball) falls distance h under the influence of the gravitational field:

$$J = \frac{Mgh}{Q}.$$

Joule's experiment implies that heat must be regarded as a form of energy. Therefore, the energy conservation law must be considered with mechanical energy and heat together. Computed in the units of Joule, the total internal energy change dE due to a process then must be the sum of the change in work dW and the change in heat dQ:

$$dE = dQ + dW. \tag{3.2}$$

We call E the internal energy of the system. This way, the energy conservation law is generalized so that thermal processes as well as mechanical motions are brought under its aegis in a unified form.

3.2. The First Law of Thermodynamics

In 1850, Rudolph Clausius stated the first law of thermodynamics as follows:

"The energy of an isolated system (universe) is constant."

In other words, the internal energy of an isolated system is conserved. This may be put in another equivalent statement made by M. Planck:

"It is impossible to construct a perpetual machine of the first kind — a machine that, working in a cycle, expends no heat to produce an equivalent work."

Let us examine a consequence of this statement. Suppose there is a cycle of process $A \xrightarrow{1} B \xrightarrow{2} A$, which takes the system from state A to state B through path 1 and then returns it back to state A through path 2 as schematically depicted in Fig. 3.2.

For the process $A \xrightarrow{1} B$, the internal energy change is given by the expression

$$\int_{A(1)}^{B} dE = \int_{A(1)}^{B} đQ + \int_{A(1)}^{B} đW.$$

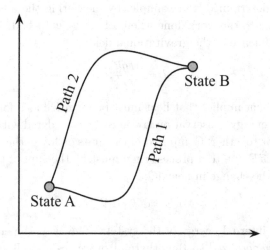

Fig. 3.2.   A cyclic process from state A to state B consisting of two segments.

Similarly, for the process $B \overset{2}{\to} A$, it is given by

$$\int_{B(2)}^{A} dE = \int_{B(2)}^{A} dQ + \int_{B(2)}^{A} dW.$$

Addition of these two equations yields a relationship between circular integrals as follows:

$$\oint dE = \oint dQ + \oint dW, \tag{3.3}$$

where the circle on the integral sign means an integration over the cycle performed in the space of state variables. According to the first law of thermodynamics, energy cannot be created over the cycle since, upon the system's returning to its original state through a cyclic process, the internal energy of the system must assume the same value. Therefore, expressed mathematically, the cyclic integral of the internal energy must vanish:

$$\oint dE = 0. \tag{3.4}$$

This implies that

$$Q + W = 0. \tag{3.5}$$

Here,

$$Q = \oint dQ, \quad W = \oint dW. \tag{3.6}$$

Equation (3.4) embodies in a mathematical form the experimental fact that *heat is an equivalent form of work or energy* as was enunciated by Joule. It follows from Eq. (3.4) that the energy change ΔE_{AB} along path 1 must equal the energy change ΔE_{AB} along path 2:

$$\Delta E_{AB(1)} = \Delta E_{AB(2)}, \tag{3.7}$$

where

$$\Delta E_{AB(i)} = \int_{A(i)}^{B} dE = E(B) - E(A) \quad (i = 1, 2).$$

That is, the change in E is independent of the path along which the change of state is effected; it depends only on the initial and final states.

Such a function is called a *state function*. The differential of a state function is said to be exact in the macroscopic state space of the system: dE is an *exact differential in such a space*. We see that the exactness of dE is demanded by the first law of thermodynamics as a general energy conservation law. Equation (3.2) is a mathematical statement of the first law of thermodynamics in a differential form for an infinitesimal process. A rigorous formulation of the first law in local form is given in Appendix C at the end of the book.

In obtaining relation (3.7), there was no mention made of whether the process is reversible or not. If a process is not reversible, there can be dissipative effects on work or heat transfer that we will not be able to deduce if the irreversible process is not explicitly examined. Suppose process 2 is performed reversibly whereas process 1 is irreversible. In this case, relation (3.7) makes it possible to deduce the effects of the irreversible process on the internal energy change simply in terms of $\Delta E_{AB(2)}$, which is the internal energy change accompanying the reversible segment of the cycle. And this $\Delta E_{AB(2)}$ is rather simple to compute by means of equilibrium thermodynamics as will be shown. This is the point alluded to earlier in Chapter 2 with regard to the theoretical utility of equilibrium thermodynamics in connection with irreversible processes. This aspect will receive further examination in the course of development of the theory.

For a closed system with no external forces except for a uniform normal pressure $p = p(V, T)$ on the system, the differential work is given by the pressure–volume work

$$đW = -pdV, \tag{3.8}$$

and Eq. (3.2) takes the form

$$dE = đQ - pdV. \tag{3.9}$$

This is the differential form of the first law of thermodynamics in the case of pressure–volume work alone, when there is no dissipative internal work, which we do not assume to be present in equilibrium thermodynamics, the subject of interest in this work. If the pressure–volume work and the heat transfer are performed reversibly, expression (3.9) is for a reversible process. In this case, the calculation of changes in the internal energy is fairly simple to perform as will be seen.

3.3.   Enthalpy

The first law of thermodynamics has been shown to give rise to the conclusion that the internal energy E of the system is a state function. There is another function related to the internal energy which is also a state function depending only on the initial and final states of a process. It sometimes is more convenient to use this new function than the internal energy in thermodynamic considerations. We introduce the new function below.

Suppose the volume of a system is changed at constant pressure from V_1 to V_2. We shall designate the initial and final states 1 and 2, respectively. The heat change accompanying the process is

$$Q = \int_1^2 đQ. \tag{3.10}$$

By using Eq. (3.9) in the integral, if the pressure–volume work is the only kind of work, we find

$$
\begin{aligned}
Q &= \int_1^2 dE + \int_{V_1}^{V_2} pdV \\
&= E_2 + pV_2 - (E_1 + pV_1),
\end{aligned} \tag{3.11}
$$

where we have made use of the fact that p is constant throughout the change. It is convenient to define a new state function

$$H = E + pV, \tag{3.12}$$

which is called the enthalpy. It represents the heat content of the system. With this definition, the heat change Q at constant p is given by

$$Q = \int_1^2 dH = H_2 - H_1 = \Delta H, \tag{3.13}$$

which confirms the notion that enthalpy is a measure of the heat content of the system. Equation (3.13) appears to suggest that Q is a state function, but it must be noted that Q is not necessarily equal to ΔH if p is not kept constant. This point becomes evident if we examine the differential form for enthalpy.

3.3.1.   *Differential Forms for Enthalpy*

Since the first law of thermodynamics is expressible in the differential form (3.2), an equivalent differential form useful for the enthalpy defined can be obtained. It is easy to deduce the desired form

$$dH = \text{d}Q + \text{d}W + d(pV). \tag{3.14}$$

If the work term is split into the pressure–volume work and the rest as in

$$\text{d}W = -pdV + \text{d}W', \tag{3.15}$$

where $\text{d}W'$ represents the totality of the other kinds of work than the pressure–volume work, then dH is given by

$$dH = \text{d}Q + V dp + \text{d}W'. \tag{3.16}$$

If $\text{d}W' = 0$, then

$$dH = \text{d}Q + V dp. \tag{3.17}$$

These differential forms for enthalpy may be regarded as expressions equivalent to the first law of thermodynamics, which was expressed in the differential form of internal energy

$$dE = \text{d}Q - pdV. \tag{3.18}$$

Having introduced state functions, internal energy (E) and enthalpy (H), it is useful to reexamine heat capacities in terms of either the internal energy or the enthalpy just defined. If we assume that there is a pressure–volume work only, then from Eq. (3.18), the constant volume heat capacity defined in Eq. (2.40) may be written as

$$C_v = \left(\frac{\partial E}{\partial T}\right)_V, \tag{3.19}$$

and similarly, the constant pressure heat capacity defined in Eq. (2.40) as

$$C_p = \left(\frac{\partial E}{\partial T}\right)_p + p\left(\frac{\partial V}{\partial T}\right)_p.$$

By the definition of enthalpy, we find

$$\left(\frac{\partial E}{\partial T}\right)_p = \left(\frac{\partial H}{\partial T}\right)_p - p\left(\frac{\partial V}{\partial T}\right)_p. \tag{3.20}$$

Substitution of this into the equation above yields

$$C_p = \left(\frac{\partial H}{\partial T}\right)_p. \tag{3.21}$$

This also follows more directly from Eq. (3.17). These two heat capacities are clearly not equal to each other. The difference of the two heat capacities may be computed as follows:

$$C_p - C_v = \left(\frac{\partial H}{\partial T}\right)_p - \left(\frac{\partial E}{\partial T}\right)_V$$

$$= p\left(\frac{\partial V}{\partial T}\right)_p + \left(\frac{\partial E}{\partial T}\right)_p - \left(\frac{\partial E}{\partial T}\right)_V. \tag{3.22}$$

Since the internal energy may be regarded as a function of T and V — in other words, if the state space is assumed to be spanned by T and V — we obtain the relation

$$dE = \left(\frac{\partial E}{\partial T}\right)_V dT + \left(\frac{\partial E}{\partial V}\right)_T dV. \tag{3.23}$$

Differentiation of Eq. (3.23) with respect to T at constant p yields

$$\left(\frac{\partial E}{\partial T}\right)_p = \left(\frac{\partial E}{\partial T}\right)_V + \left(\frac{\partial E}{\partial V}\right)_T \left(\frac{\partial V}{\partial T}\right)_p. \tag{3.24}$$

When this is substituted into the right-hand side of Eq. (3.22), there follows the equation

$$C_p - C_v = \left[p + \left(\frac{\partial E}{\partial V}\right)_T\right]\left(\frac{\partial V}{\partial T}\right)_p. \tag{3.25}$$

Since the right-hand side of Eq. (3.25) is not equal to zero, the two specific heats are clearly not identical. Mathematically speaking, this non-vanishing difference is a consequence of the nonexactness of dQ since heat transfer depends on the path along which it is performed, and a constant pressure path is certainly different from a constant volume path. We will see later that the right-hand side of Eq. (3.25) can be expressed in another form more convenient for measurement. Since, as will be shown later,

$$\left(\frac{\partial E}{\partial V}\right)_T = -p + T\left(\frac{\partial p}{\partial T}\right)_V, \tag{3.26}$$

we find

$$C_p - C_v = T \left(\frac{\partial p}{\partial T} \right)_V \left(\frac{\partial V}{\partial T} \right)_p, \tag{3.27}$$

which can be calculated if the equation of state is given. We will return to this quantity in a later chapter.

3.3.2. The Difference in Isobaric and Isochoric Heat Capacities for Ideal Gases

As established in Chapter 2, the ideal gas is an idealized substance, which obeys the equation of state

$$pV = nRT. \tag{3.28}$$

This was also seen to be intimately connected to the ideal gas temperature scale. From the viewpoint of energy, ideal gases are defined as those that do not cost energy as the gases are isothermally expanded. This is mathematically expressed in the form

$$\left(\frac{\partial E}{\partial V} \right)_T = 0. \tag{3.29}$$

Alternatively put, this second condition means that *the internal energy of an ideal gas is independent of its volume*. This is a consequence of the absence of interactions between the molecules in the ideal gas, which is the behavior attained by the gas as the pressure is diminished to a vanishingly small value.

When Eqs. (3.28) and (3.29) are used in Eq. (3.25), there follows the relation

$$C_p - C_v = nR, \tag{3.30}$$

and the relation of C_p and C_v has become particularly simple and independent of substances. This is another aspect of the universality of the behavior of ideal gases. Recall that the equation of state (3.28) is also independent of substances since there is no material parameter present in it.

3.4. Work and Heat of Isothermal Reversible Expansion

Suppose a vessel containing an ideal gas is maintained at thermal equilibrium with a thermostat at temperature T. Then the gas is reversibly expanded. Since Eq. (3.29) holds for the ideal gas by definition and since $dT = 0$ by the experimental condition, the internal energy of the ideal gas is conserved as the gas is reversibly but isothermally expanded. This conclusion follows from Eq. (3.23) and Eq. (3.29) since then

$$dE = 0. \tag{3.31}$$

It therefore follows from Eqs. (3.9) that the heat change then must be balanced by the work:

$$đQ = pdV.$$

This means that if an ideal gas is isothermally expanded, heat Q, equal in amount to the work W done by the system, must be absorbed from the heat reservoir (surroundings). For a reversible isothermal expansion from V_1 to V_2, the heat absorbed therefore is expressible as

$$Q = -W = \int_{V_1}^{V_2} pdV. \tag{3.32}$$

By using the ideal gas equation of state (3.28), we calculate the integral in Eq. (3.32) to find that the isothermal reversible heat change is

$$Q = -W = nRT \ln\left(\frac{V_2}{V_1}\right) = -nRT \ln\left(\frac{p_2}{p_1}\right). \tag{3.33}$$

In the case of reversible isothermal compression, the sign in Eq. (3.33) is reversed since the work is done on the system.

3.5. Work of Adiabatic Expansion

Suppose a vessel containing an ideal gas is thermally isolated so that the gas neither receives heat from, nor gives it up to, its surroundings. (A thermal insulator may be used to isolate the vessel thermally.) Then the gas is adiabatically and reversibly expanded from V_1 to V_2. This change is accompanied by a temperature decrease from T_1 to T_2. Since $đQ = 0$ owing to the process being adiabatic, we have

$$dE = \left(\frac{\partial E}{\partial T}\right)_V dT = -pdV. \tag{3.34}$$

Therefore, the reversible adiabatic work is given by the expression

$$W = -\int_{V_1}^{V_2} pdV = \int_{T_1}^{T_2} C_v(T)dT. \tag{3.35}$$

The pressure and volume of a gas obey a certain relation during an adiabatic expansion or compression. Let us find the relation. Substitution of Eq. (3.28) into Eq. (3.34) and use of the definition of C_v puts the equation in the form

$$T^{-1}C_v(T)dT = -RV^{-1}dV.$$

Integration of this equation under the assumption of constant C_v yields

$$T_1 V_1^{\gamma-1} = T_2 V_2^{\gamma-1}, \tag{3.36}$$

where γ is the polytropic ratio of the gas:

$$\gamma = \frac{C_p}{C_v}. \tag{3.37}$$

We have made use of Eq. (3.30) for Eq. (3.36). It must be emphasized that Eq. (3.36) holds if the equation of state is given by Eq. (3.28) and C_v is independent of temperature. When the equation of state (3.28) is made use of in Eq. (3.36), it may be written as

$$p_1 V_1^{\gamma} = p_2 V_2^{\gamma}$$

or

$$pV^{\gamma} = \text{constant.} \tag{3.38}$$

The case of real gases will be considered for a relation equivalent to Eq. (3.36) when the thermodynamics of real gases is studied later.

3.6.   Heat Capacity and Heat Change

The heat change accompanying a reversible process in a substance at constant pressure can be computed from the heat capacity data of the substance. Heat capacities are measured in the laboratory by means of calorimetry, that is, by using a calorimeter. Calorimetry is a well-developed experimental discipline widely practiced in thermodynamics and in physical chemistry as a means of studying thermal and molecular properties of substances. Heat capacity data so measured are usually expressed as a power series in T at different pressures. It is found generally useful to summarize heat capacity data in the form of series in T:

$$C_p(T) = a_{-2}T^{-2} + a_0 + a_1 T + \cdots, \tag{3.39}$$

where a_{-2}, a_0, and a_1 are empirical coefficients determined by calorimetry. They are generally dependent on pressure. Such coefficients are tabulated and available in the literature.[2] Since the pressure dependence of heat capacities can be computed if the equation of state is known for the substance, it is useful to tabulate them in a condition at which the heat capacities are independent of pressure. Since it is known that the heat capacities of ideal gases are independent of pressure, the coefficients must tend to limits

$$\lim_{p \to 0} a_{-2} = a_{-2}^0,$$

$$\lim_{p \to 0} a_0 = a_0^0,$$

$$\lim_{p \to 0} a_1 = a_1^0,$$

and therefore,

$$C_p^0 \equiv \lim_{p \to 0} C_p = a_{-2}^0 T^{-2} + a_0^0 + a_1^0 T + \cdots, \qquad (3.40)$$

where a_{-2}^0, a_0^0, and a_1^0 are material constants independent of p. For this reason, the heat capacities of gases are usually tabulated for ideal gases in the literature. When the temperature expansion (3.39) is used for ideal gases, Eq. (3.40) must be understood. In the case of solids and liquids, they are tabulated at 1 atm. Some typical values for the coefficients in Eq. (3.39) are listed in Table 3.1. The heat capacities at other pressure values can then be computed, based on the tabulated values, by means of the thermodynamic procedure that will be developed later in this work.

The heat capacity data can be used to compute other thermodynamic quantities. Here, we consider an example. By using Eq. (3.39), we can compute ΔH for an isobaric process where the temperature of the system at pressure p changes from T_1 to T_2:

$$\Delta H = \int_1^2 dH = \int_{T_1}^{T_2} dT \, C_p(T)$$

$$= a_{-2}(T_1^{-1} - T_2^{-1}) + a_0(T_2 - T_1) + \frac{1}{2} a_1(T_2^2 - T_1^2) + \cdots. \qquad (3.41)$$

The Arabic numerals for the endpoints in the first integral represent the initial and final states (T_1, p) and (T_2, p), respectively.

---

[2]See the NIST Standard Reference Database: http://webbook.nist.gov/chemistry/fluid.

Table 3.1. Heat capacities for some substances.

Substance	$kJ \cdot deg^2 \, mole^{-1}$ a_{-2}	$kJ \, mole^{-1}$ a_0	$kJ \, mole \cdot deg^{-1}$ a_1
Gases (298−2000 K)			
S	0.36×10^5	5.26	-0.10×10^{-3}
H_2	0.12×10^5	6.52	0.78×10^{-3}
O_2	-0.40×10^5	7.16	1.00×10^{-3}
N_2	-0.12×10^5	6.83	0.90×10^{-3}
CO	-0.11×10^5	6.79	0.98×10^{-3}
CO_2	-2.06×10^5	10.57	2.10×10^{-3}
H_2O	—	7.30	2.46×10^{-3}
NH_3	-0.37×10^5	7.11	6.00×10^{-3}
Liquids			
H_2O	—	18.04	—
Solids (298 to T_m or 2000 K)			
C (graphite)	-2.04×10^5	4.03	1.14×10^{-3}
Al	—	4.94	2.96×10^{-3}
Cu	—	5.14	1.50×10^{-3}

Joseph Black of Edinburgh in the 18th century discovered that there is a characteristic amount of heat absorbed or released, for example, when a given amount of a liquid is vaporized or condensed from its vapor. He termed it the *latent heat*. The latent heat is a characteristic property of a substance and differs from substance to substance. Generally, when there is a phase transformation in a substance, there is a latent heat associated with the transformation. The latent heats associated with vaporization, melting, sublimation, and so on are called, respectively, the heat of vaporization, of melting, of sublimation, and so on.

Suppose the temperature of a substance is increased from T_1 to T_p, at which it undergoes a phase transition to another phase, and then the temperature of the substance in the new phase is further increased from T_p to T_2. The enthalpy change for this reversible change consists of three parts:

$$\Delta H = \int_{T_1}^{T_p} dT \, C_p^{(1)}(T) + \Delta H_p + \int_{T_p}^{T_2} dT \, C_p^{(2)}(T), \qquad (3.42)$$

where $C_p^{(1)}$ and $C_p^{(2)}$ are the heat capacities of the low and high temperature phases 1 and 2, and ΔH_p is the latent heat for the phase transition. Note that the reversible change is at constant pressure. We will discuss in a later chapter the pressure dependence of ΔH for a process in which pressure is changed under constant temperature condition. If the temperature dependence is known for the heat capacities in Eq. (3.42), the integrals may be evaluated more explicitly as in Eq. (3.41).

3.7. Thermochemistry

Heat changes involved in chemical systems and, in particular, chemical reactions are important quantities. In the case of chemical reactions, they can be used to deduce some thermodynamic properties of the reacting systems. Here, we consider some of such quantities, which will be useful for computing other thermodynamic quantities, such as equilibrium constants and free energy changes, discussed in subsequent chapters.

3.7.1. *Heat of Reaction*

Changes in heat content accompany chemical reactions occurring at constant pressure and temperature. Such a heat content change is referred to as the heat of reaction. Since a change in heat content at constant pressure is found to be equal to a change in enthalpy, it is sufficient to find the latter. More precisely, let us consider the chemical reaction written in a general form

$$\sum_{i \in R} \nu_i^{(R)} X_i = \sum_{i \in P} \nu_i^{(P)} Y_i,$$

where X_i and Y_i are the reactants and products, the superscripts R and P stand for the reactants and products, respectively, and $\nu_i^{(R)}$ and $\nu_i^{(P)}$ are stoichiometric coefficients. The heat of reaction for the chemical reaction is defined as the difference between the total enthalpies of the products and reactants of the chemical reaction at constant pressure and temperature:

$$\Delta H = \sum_{i \in P} h_{Pi} - \sum_{i \in R} h_{Ri}. \tag{3.43}$$

Here, h_{Pi} and h_{Ri} are the molar enthalpies of products and reactants, respectively. They are available in the literature, generally tabulated at 298.15 K and 1 atm pressure. When ΔH is evaluated according to Eq. (3.43), an assumption is implicit that the reactants are completely converted into the products.

Since heats of reaction can differ if the states of aggregation of the constituents are different even if the reaction involves the same constituents, it is customary to indicate the states of aggregation of the constituents. For example, we write

$$C(s) + O_2(g) = CO_2(g) \tag{3.44}$$

and

$$CH_4(g) + 2O_2(g) = CO_2(g) + 2H_2O(l), \tag{3.45}$$

where s, l, and g, respectively, refer to the solid, liquid, and gaseous state of the constituents. Thus, if we take the first of the aforementioned reactions as an example for illustration of Eq. (3.43), its heat of reaction is

$$\Delta H = h\,(CO_2, g) - h\,(C, s) - h\,(O_2, g)$$
$$= -393.41\ \text{kJ mol}^{-1}\ (\text{at } 298.15\ \text{K and 1 atm}). \tag{3.46}$$

In the case of the second reaction (3.45),

$$\Delta H = h\,(CO_2, g) + 2h\,(H_2O, l) - h\,(CH_4, g) - h\,(O_2, g)$$
$$= -890.36\ \text{kJ mol}^{-1}\ (\text{at } 298.15\ \text{K and 1 atm}).$$

The IUPAC convention used in expressing ΔH for chemical reactions is that ΔH refers to the enthalpy change for a mole of the chemical reaction as expressed.

3.7.2. *Standard States and Heat of Formation*

It is of particular interest to find the heats of reaction for chemical reactions in which compounds are formed from their constituent elements, since such information can be used to deduce heats of reaction for other related chemical reactions as will be shown shortly. Since heats of reaction are the differences in enthalpies, it is possible to fix heats of reaction in reference to the standard state where the enthalpies for elements are set equal to zero. It is customary to choose the state of aggregation as that form of an element which is most stable at the temperature under consideration.

Particularly, *for gaseous elements the standard state is chosen as the state of the hypothetical ideal gas at 1 atm pressure and 298.15 K. For solid and liquid elements, the standard state is chosen at 1 atm pressure and 298.15 K.* With the standard states so chosen and the enthalpies of stable elements set equal to zero, the heat of reaction for a mole of a compound formed from its elements in the standard state is simply the standard heat

Table 3.2.   Standard heats of formation.

Compound	State	ΔH^0_{298} kJ mol^{-1}
H_2O	g	-239.75
H_2O	l	-283.38
H_2O_2	g	-132.03
HF	g	-258.01
HCl	g	-91.52
CO	g	-109.57
HBr	g	-35.92
HI	g	-25.72
HIO_3	c	-236.56
H_2S	g	-19.97
H_2SO_4	l	-804.34
SO_2	g	-294.34
SO_3	g	-391.78
CO_2	g	-390.13
$SOCl_3$	l	-204.08
S_2Cl_2	g	-23.64

Table 3.3.   Standard heats of formation.

	Substance	Formula	ΔH^0_{298} kJ mol^{-1}
	methane	CH_4	-74.10
	ethane	C_2H_6	-83.75
	propane	C_3H_5	-102.66
	n-butane	C_4H_{10}	-123.26
Paraffins	isobutane	C_4H_{10}	-130.04
	n-pentane	C_5H_{12}	-140.10
	2-methylbutane	C_5H_{12}	-152.11
	tetramethylmethane	C_5H_{12}	-163.46
	ethylene	C_2H_4	52.08
	propylene	C_3H_6	20.56
	1-butene	C_4H_5	1.59
Monolefins	cis-2-butene	C_4H_3	-5.76
	trans-2-butene	C_4H_5	-9.70
	2-methylpropene	C_4H_5	-13.29
	1-pentene	C_5H_{10}	-19.26

of formation, which will be indicated by ΔH^0_f. To illustrate this, we consider reaction (3.44). Since $h(C, s) = h(O_2, g) = 0$ at the standard states, we find

$$\Delta H^0_f = h(CO_2, g) \ (\text{at } 298.15 \, \text{K and 1 atm}),$$

which is now the standard heat of formation of CO_2. A few values are listed in Tables 3.2–3.4 for standard heats of formation.

Table 3.4.   Standard heats of formation.

	Substance	Formula	ΔH_{298}^0 kJ mol^{-1}
		C_3H_4	191.00
Allenes	1,3-butadiene	C_4H_6	111.44
	1,3-pentadiene	C_5H_5	78.33
	1,4-pentadiene	C_5H_5	106.04
Acetylenes	Acetylene	C_2H_2	224.94
	methylacetylene	C_3H_4	183.79
	dimethylacetylene	C_4H_6	146.10

3.7.3.   *Hess's Law*

Heats of reaction can be determined by direct measurement of changes in heat content. However, it is not necessary to measure the heat of reaction for every chemical reaction we encounter, since it is often possible to deduce it from the enthalpy changes of other reactions. Such deductions are possible because enthalpy is a state function which is determined by its values at the initial and final states of the system and that the mass as well as the energy is conserved in a chemical reaction. Suppose ΔH is the enthalpy change for a reversible process from state a to state z. It is then possible to write it in terms of ΔH's for other reversible processes:

$$\Delta H = H_z - H_a$$
$$= (H_z - H_y) + (H_y - H_x) + (H_x - H_b) + \cdots + H_b - H_a$$
$$= \sum_i \Delta H_i, \tag{3.47}$$

where ΔH_i stands for $H_z - H_y$, $H_y - H_x$, and so on. This is a mathematical consequence of the exactness of the differential dH since it is possible to write

$$\Delta H = \int_a^z dH = \int_y^z dH + \int_x^y dH + \int_b^x dH + \cdots + \int_a^b dH$$
$$= \sum_i \Delta H_i.$$

In the case of chemical reactions, this decomposability of ΔH is known as Hess's law. We now translate this mathematical expression into the case of chemical reactions. Suppose there are m chemical reactions, which we compactly express by the equation

$$\sum_{i=1}^{r} \nu_i^{(\sigma)} X_i = 0 \quad (\sigma = 1, 2, \ldots, m), \tag{3.48}$$

where X_i denotes species (compound) i and $\nu_i^{(\sigma)}$ is the stoichiometric coefficient for species i of reaction σ in a set of chemical reactions. It is counted positive for the products and negative for the reactants. For example, if the reaction

$$\frac{1}{2} H_2 + \frac{1}{2} Cl_2 = HCl$$

is designated the σth reaction of the set, then $\nu_{HCl}^{(\sigma)} = 1$, $\nu_{Cl_2}^{(\sigma)} = -\frac{1}{2}$, and $\nu_{H_2}^{(\sigma)} = -\frac{1}{2}$. The enthalpy change for reaction σ may be expressed in the form

$$\Delta H^{(\sigma)} = \sum_{i=1}^{r} \nu_i^{(\sigma)} h_i, \tag{3.49}$$

where h_i is the molar enthalpy of substance i. We have earlier discussed a method of calculating ΔH when pressure is kept constant. We will discuss how to compute them in general from the basic thermodynamic data in a later chapter.

Suppose there is a chemical reaction that can be composed by taking a linear combination of the m reactions in the set of chemical reactions in Eq. (3.48). We will denote the reaction by

$$\sum_{i=1}^{r} \nu_i X_i = 0. \tag{3.50}$$

The enthalpy change for this reaction may be given in the form

$$\Delta H = \sum_{i=1}^{r} \nu_i h_i. \tag{3.51}$$

Since reaction (3.50) is a linear combination of reactions (3.48), there exists a set of constants ω_σ such that

$$\sum_{\sigma=1}^{m} \sum_{i=1}^{r} \omega_\sigma \nu_i^{(\sigma)} X_i = \sum_{i=1}^{r} \nu_i X_i.$$

By comparing both sides of this equation, we find the relation of the stoichiometric coefficients ν_i in reaction (3.50) to those in the set of reactions (3.48):

$$\nu_i = \sum_{\sigma=1}^{m} \omega_\sigma \nu_i^{(\sigma)}. \tag{3.52}$$

Inserting it into Eq. (3.51) and using Eq. (3.49), we obtain, for reaction (3.50), the enthalpy change ΔH in terms of $\Delta H^{(\sigma)}$ for the set of reactions in Eq. (3.48):

$$\Delta H = \sum_{\sigma=1}^{m} \omega_\sigma \Delta H^{(\sigma)}. \tag{3.53}$$

This result embodies Hess's law. Note that ΔH here has exactly the same coefficients ω_σ of the linear combination as in the linear combination of the stoichiometric coefficients in Eq. (3.52). Equation (3.53) shows how the enthalpy change of a reaction might be obtained from the sources of information on $\Delta H^{(\sigma)}$ for the chemical reactions making up the set in Eq. (3.48). Hess's law holds at any temperature and pressure as long as the reactions involved are for the substances in the same conditions, namely, if the temperature, pressure, and the states of aggregation are the same for the compounds in reaction (3.50) and those in the set of reactions (3.48). If the conditions are altered at the end of a process, a proper account should be made of the changes in the heat content arising from the altered conditions. A partial account is already given for such a correction in the previous section. See Eq. (3.42).

We consider an example for the application of Hess's law below. Consider the reaction

$$C_2H_4(g) + H_2(g) = C_2H_6(g). \tag{3.54}$$

The heat of reactions for the following reactions are known in the units of kJ mol^{-1}:

(1) $C_2H_4(g) + 3O_2 = 2CO_2(g) + 2H_2O(l)$, $\Delta H^{(1)} = -1411.26$,
(2) $2H_2(g) + O_2(g) = 2H_2O(l)$, $\Delta H^{(2)} = -571.70$,
(3) $2C_2H_6(g) + 7O_2(g) = 4CO_2 + 6H_2O(l)$, $\Delta H^{(3)} = -3119.59$.

Take $\omega_1 = 1$, $\omega_2 = 1/2$, and $\omega_3 = -1/2$. Then reaction (3.54) is obtained from Reactions (1)–(3) above and the heat of reaction for reaction (3.54) is accordingly obtained as follows:

$$\Delta H = \Delta H^{(1)} + \frac{1}{2}\Delta H^{(2)} - \frac{1}{2}\Delta H^{(3)}$$
$$= -137.31\,\text{kJ}\,\text{mol}^{-1}.$$

If $\Delta H^{(\sigma)}$ are taken for the standard state heats of reactions (1)–(3) above, then ΔH will be the standard heat of reaction for the reaction of interest.

3.7.4. *Kirchhoff's Equation*

The heat of reaction varies with temperature and pressure, and there often arise cases where the heat of reaction at states other than the standard state is of interest. Particularly, the variation of ΔH with temperature can be of interest. It can be derived in the following manner. Differentiating Eq. (3.43) with respect to T at constant p, we obtain

$$\left(\frac{\partial}{\partial T}\Delta H\right)_p = \sum_{i \in P}\left(\frac{\partial}{\partial T}H_{Pi}\right)_p - \sum_{i \in R}\left(\frac{\partial}{\partial T}H_{Ri}\right)_p, \qquad (3.55)$$

but since

$$C_{p\,Pi} = \left(\frac{\partial}{\partial T}H_{Pi}\right)_p, \quad C_{p\,Ri} = \left(\frac{\partial}{\partial T}H_{Ri}\right)_p,$$

the specific heat change for the reaction is

$$\Delta C_p = \sum_{i \in P}C_{p\,Pi} - \sum_{i \in R}C_{p\,Ri}. \qquad (3.56)$$

If we make use of Eq. (3.53) instead of Eq. (3.43), then the specific heat change is

$$\Delta C_p = \sum_{\sigma=1}^{m}\omega_\sigma c_p^{(\sigma)}, \qquad (3.57)$$

where $c_p^{(\sigma)}$ is a molar heat capacity. This formula may be considered Hess's law applied to specific heats. From Eqs. (3.55) and (3.57) follows the equation

$$\left(\frac{\partial \Delta H}{\partial T}\right)_p = \Delta C_p. \qquad (3.58)$$

This is called the Kirchhoff equation. If ΔH is for a single chemical reaction, then ΔC_p is for the reaction, but if ΔH is for the set of chemical reactions defined by Eq. (3.43), then ΔC_p is given by Eq. (3.57). In either case, the Kirchhoff equation formally remains the same. By integrating the Kirchhoff equation and using calorimetric data on C_p, it is possible to calculate ΔH at temperature T in terms of the enthalpy change at T_0:

$$\Delta H(T,p) = \Delta H(T_0,p) + \int_{T_0}^{T} dT' \, \Delta C_p(T',p). \qquad (3.59)$$

If the calorimetric data are represented as in Eq. (3.39), the integration can be easily performed, and the result is similar to that in Eq. (3.42). The integration is left to the reader as an exercise.

3.8.   Mathematical Notes

Here, we discuss some mathematical notions and relations which are handy in studying and performing calculations in thermodynamics. They are quite well known in calculus and the readers will versed in calculus may skip this section.

3.8.1.   *Exact Differentials*

We elaborate a little more on exact differentials in the case where the number of independent variables is 2. If it is larger than 2, the conditions for the differentials to be exact differentials, namely, the integrability conditions, are rather complicated. The reader is referred to the theory of differential forms and differential manifolds[3] for the topic.

Consider a differential dz of a function $z(x,y)$:

$$dz(x,y) = M(x,y)dx + N(x,y)dy. \qquad (3.60)$$

---

[3]See, for example, I. N. Sneddon, *Elements of Partial Differential Equations* (McGraw-Hill, New York, 1957); H. Flanders, *Differential Forms* (Academic Press, New York, 1963).

The necessary and sufficient condition for dz to be an exact differential is

$$\frac{\partial M}{\partial y} = \frac{\partial N}{\partial x}. \tag{3.61}$$

Since

$$M(x,y) = \frac{\partial z}{\partial x}, \quad N(x,y) = \frac{\partial z}{\partial y}, \tag{3.62}$$

condition (3.61) implies that

$$\frac{\partial^2 z}{\partial y \partial x} = \frac{\partial^2 z}{\partial x \partial y}. \tag{3.63}$$

Thus, if a function of two variables $z(x,y)$ satisfies this condition, its differential is exact.

With this mathematical preparation, let us examine if đQ is indeed inexact. Since the internal energy E may be regarded as a function of T and V, we may write

$$dE = \left(\frac{\partial E}{\partial T}\right)_V dT + \left(\frac{\partial E}{\partial V}\right)_T dV. \tag{3.64}$$

By substituting it into Eq. (3.9) and rearranging terms, we find

$$đQ = \left(\frac{\partial E}{\partial T}\right)_V dT + \left[p + \left(\frac{\partial E}{\partial V}\right)_T\right] dV. \tag{3.65}$$

Since

$$\frac{\partial^2 E}{\partial x \partial y} = \frac{\partial^2 E}{\partial y \partial x}$$

by the exactness of dE, it is necessary for đQ to be exact that

$$\left(\frac{\partial p}{\partial T}\right)_V = 0,$$

but this is not true in general. Therefore, it is concluded that đQ is not an exact differential.

3.8.2. Chain Relations

A functional relation between two variables x and y may be put in the implicit form

$$f(x, y) = C, \tag{3.66}$$

where C is a constant and $f(x, y)$ is a differentiable function of x and y. For example, consider the relation for a circle of radius R

$$x^2 + y^2 = R^2.$$

We wish to calculate the derivative dy/dx with the implicit function. Taking the differential of f, we obtain

$$df = \left(\frac{\partial f}{\partial x}\right)_y dx + \left(\frac{\partial f}{\partial y}\right)_x dy = 0.$$

If f is kept constant and the equation is differentiated with dx, we obtain

$$\left(\frac{\partial f}{\partial x}\right)_y + \left(\frac{\partial f}{\partial y}\right)_x \left(\frac{\partial y}{\partial x}\right)_f = 0, \tag{3.67}$$

which may be put in the form[4]

$$\left(\frac{\partial y}{\partial x}\right)_f = -\frac{\left(\frac{\partial f}{\partial x}\right)_y}{\left(\frac{\partial f}{\partial y}\right)_x}. \tag{3.68}$$

This relation can be generalized to a situation of three variables x, y, and z. Let there be a function f such that

$$f(x, y, z) = C, \tag{3.69}$$

---

[4]It is useful to devise a mnemonic scheme for this useful relation. To construct the right-hand side of Eq. (3.68), we can proceed as follows: for the numerator, we construct a derivative from the derivative on the left-hand side, starting from the subscript and moving clockwise as the derivative is formed with the third variable in the chain $f \to x \to y$ becoming the subscript in the new derivative formed. That is, from the chain $f \to x \to y$, form the derivative $\left(\frac{\partial f}{\partial x}\right)_y$. For the denominator, we start from f in the counterclockwise chain $f \to y \to x$ and form the derivative $\left(\frac{\partial f}{\partial y}\right)_x$ with the third member of the chain becoming the subscript in the derivative formed, and finally put a negative sign on the ratio of the derivatives. This procedure mechanically produces the right-hand side of Eq. (3.68) from the derivative on the left-hand side without having to go through the steps used for obtaining it from Eq. (3.67).

where C is a constant. One variable among the three variables x, y, and z may then be considered a function of the other two. Take the differential of f:

$$df = \left(\frac{\partial f}{\partial x}\right)_{y,z} dx + \left(\frac{\partial f}{\partial y}\right)_{x,z} dy + \left(\frac{\partial f}{\partial z}\right)_{x,y} dz = 0. \qquad (3.70)$$

If one variable, say, z is kept constant in addition to f, then this differential may be written as

$$\left(\frac{\partial y}{\partial x}\right)_{f,z} = -\frac{\left(\frac{\partial f}{\partial x}\right)_{y,z}}{\left(\frac{\partial f}{\partial y}\right)_{x,z}}, \qquad (3.71)$$

which is in the same form as (3.68). If y is kept constant instead of z, then there follows the relation

$$\left(\frac{\partial z}{\partial x}\right)_{y,f} = -\frac{\left(\frac{\partial f}{\partial x}\right)_{z,y}}{\left(\frac{\partial f}{\partial z}\right)_{x,y}}. \qquad (3.72)$$

The aforementioned mnemonic scheme also applies to these chain relations of derivatives if y is kept invariant. A similar relation can be obtained if variable x is kept constant. If we make the functional relation explicit and write

$$z = g(x, y), \qquad (3.73)$$

and take the differential of z, we obtain

$$dz = \left(\frac{\partial g}{\partial x}\right)_y dx + \left(\frac{\partial g}{\partial y}\right)_x dy. \qquad (3.74)$$

If z is kept constant, then this differential form may be written as

$$\left(\frac{\partial y}{\partial x}\right)_z = -\frac{\left(\frac{\partial y}{\partial x}\right)_y}{\left(\frac{\partial g}{\partial y}\right)_x}, = -\frac{\left(\frac{\partial z}{\partial x}\right)_y}{\left(\frac{\partial z}{\partial y}\right)_x}, \qquad (3.75)$$

which is the same as the relation in Eq. (3.68). This relation will be useful for some calculations in thermodynamics.

3.8.3. *Jacobians*

Suppose there is a transformation of a two-dimensional space to another: $(x, y) \rightarrow (u, v)$. This may be expressed by a pair of functional relations:

$$x = u(x, y), \quad y = v(x, y). \tag{3.76}$$

Under this transformation, the elements of area in the two spaces are related to each other in the following manner:

$$dxdy = |J| \, dudv,$$

where J is called the Jacobian[5] of transformation and defined by the determinant of partial derivatives

$$J(x, y, u, v) = \frac{\partial(x, y)}{\partial(u, v)} = \left| \begin{array}{cc} \left(\dfrac{\partial x}{\partial u}\right)_v & \left(\dfrac{\partial x}{\partial v}\right)_u \\ \left(\dfrac{\partial y}{\partial u}\right)_v & \left(\dfrac{\partial y}{\partial v}\right)_v \end{array} \right|. \tag{3.77}$$

The absolute value must be taken for the determinant. Since the value of the determinant is invariant to the interchange of rows and columns, the Jacobian may be written as

$$J = \left| \begin{array}{cc} \left(\dfrac{\partial x}{\partial u}\right)_v & \left(\dfrac{\partial y}{\partial u}\right)_v \\ \left(\dfrac{\partial x}{\partial v}\right)_u & \left(\dfrac{\partial y}{\partial v}\right)_u \end{array} \right|. \tag{3.78}$$

The Jacobian has the following properties which are derived from those of determinants:

$$\frac{\partial(x, y)}{\partial(u, v)} = - \frac{\partial(x, y)}{\partial(v, u)},$$

$$\frac{\partial(x, x)}{\partial(u, v)} = 0,$$

$$\frac{\partial(x, y)}{\partial(u, u)} = 0, \tag{3.79}$$

$$\frac{\partial(k, y)}{\partial(u, u)} = 0,$$

---

[5]For Jacobians, see H. Margenau and G. M. Murphy, *The Mathematics of Physics and Chemistry* (van Nostrand, Princeton, NJ, 1956), pp. 17–24, Chapter 1.

where k is a constant. The aforementioned identities can be obtained from the fact that a determinant changes its sign when two rows or columns are interchanged and a determinant vanishes if two rows or columns are equal. From the rule for the product of two determinants follows the identity

$$\frac{\partial(x,y)}{\partial(u,v)} \cdot \frac{\partial(u,v)}{\partial(t,s)} = \frac{\partial(x,y)}{\partial(t,s)}. \tag{3.80}$$

The definition of the Jacobian yields the identity

$$\frac{\partial(x,y)}{\partial(x,u)} = \left(\frac{\partial y}{\partial u}\right)_x. \tag{3.81}$$

We apply the aforementioned properties to the following example. It was previously shown that from the implicit relation (3.69) follows the chain relation

$$\left(\frac{\partial x}{\partial y}\right)_z = -\frac{\left(\frac{\partial z}{\partial y}\right)_x}{\left(\frac{\partial z}{\partial x}\right)_y}. \tag{3.82}$$

This relation also follows if the properties of Jacobians presented earlier are made use of

$$\left(\frac{\partial x}{\partial y}\right)_z = \frac{\partial(x,z)}{\partial(y,z)} = \frac{\partial(x,z)}{\partial(x,y)}\frac{\partial(x,y)}{\partial(y,z)}$$

$$= -\left(\frac{\partial z}{\partial y}\right)_x \left(\frac{\partial x}{\partial z}\right)_y$$

$$= -\frac{\left(\frac{\partial z}{\partial y}\right)_x}{\left(\frac{\partial z}{\partial x}\right)_y}. \tag{3.83}$$

The properties in Eq. (3.79) can be generalized if the dimension of the space is larger than 2, and the relations similar to Eq. (3.83) can be easily obtained by taking advantage of the properties of Jacobians.

Problems

(1) Show that for the following reversible cycle performed by an ideal gas

$$(p_1, V_1, T_1) \to (p_1, V_2, T_1) \to (p_1, V_2, T_2) \to (p_2, V_1, T_2) \to (p_1, V_1, T_1),$$

the internal energy is conserved, namely,

$$\oint dE = 0,$$

whereas

$$\oint dW \neq 0, \quad \oint dQ \neq 0.$$

Show also that

$$\oint dH = 0.$$

(2) Prove that

$$C_p = C_v + \left[V - \left(\frac{\partial H}{\partial p}\right)_T\right]\left(\frac{\partial p}{\partial T}\right)_V.$$

(3) Show that for an ideal gas

$$\left(\frac{\partial C_v}{\partial V}\right)_T = 0, \quad \left(\frac{\partial C_p}{\partial V}\right)_T = 0.$$

(4) Show that

$$\left(\frac{\partial C_v}{\partial V}\right)_T = T\left(\frac{\partial^2 p}{\partial T^2}\right)_V.$$

(5) Calculate the change in ΔH for reaction

$$H_2(g) + \frac{1}{2}O_2(g) = H_2O(g)$$

for the temperature change from T to $2T$ at constant pressure. The following calorimetric data are available:

$$C_p(H_2O, g) = \left(30.20 + 9.933 \times 10^{-3}\, T + 1.117 \times 10^{-6}\, T^2\right)\, J\,mol^{-1},$$
$$C_p(H_2, g) = \left(29.07 - 0.837 \times 10^{-3}\, T + 2.012 \times 10^{-6}\, T^2\right)\, J\,mol^{-1},$$
$$C_p(O_2, g) = \left(25.50 + 1.361 \times 10^{-2}\, T - 4.255 \times 10^{-6}\, T^2\right)\, J\,mol^{-1}.$$

(6) Show that for an ideal gas

$$\left(\frac{\partial H}{\partial V}\right)_T = 0.$$

(7) The equation of state for gases at relatively low pressures may be expressed in the form

$$pV = RT + Bp,$$

where B is independent of p or V. By using this equation of state, derive an expression for the quantity

$$\left(\frac{\partial H}{\partial V}\right)_T - \left(\frac{\partial E}{\partial V}\right)_T.$$

(8) Let the equation of state be given by $pV = RT + Bp$ as in Problem 5. Obtain expressions for Q and W of isothermal reversible expansion by using the equation of state and compare them with those of an ideal gas.

(9) Use Eqs. (3.39) and (3.59) to compute ΔH at T and p.

Chapter 4

The Second Law of Thermodynamics

The first law of thermodynamics elucidates the nature of heat and what we mean by the energy of thermal systems as well as the mutual relationship between work and heat. It also requires that all natural processes conserve energy at the end of a cyclic change. This conservation law of energy then demands that the internal energy be a state function, and as a consequence, its differential is an exact differential in thermodynamic state space. However, the first law does not tell us whether it is possible for a natural process to occur spontaneously or not. It merely imposes a restriction with regard to the energy of the system at the end of a process.

In nature, we see that some processes occur spontaneously whereas some others do not. It is an indisputable fact that the river flows downstream, not upstream, unless an external agency intervenes with it to make the reverse process possible by some sort of compensation in energy. Similarly, a vacuum is filled by air if it is connected with a container filled with air, but the vacuum is not created spontaneously in a container filled with gas without a compensation. There are numerous examples of this kind in nature that while some processes can happen spontaneously, their reverse processes are not spontaneously possible, even if they are energetically possible, not being in violation of the first law. Such a directional preference by natural phenomena is made an axiom in the form of the second law of thermodynamics. The second law of thermodynamics was preceded by Carnot's theorem enunciated by Sadi Carnot who, on examination of an idealized cyclic operation characteristic of steam engines, discovered a universal principle for thermal macroscopic processes in 1824. It is perhaps useful to observe that the study of cyclic processes is on the basis of the Carnot theorem and also the second law of thermodynamics. Therefore, it is indispensable to consider them in order to gain a proper understanding of the essence and evolution of the idea of the second law of thermodynamics.

Carnot originally formulated his theorem on the basis of the caloric theory of heat which was later discarded as an incorrect concept, but the essence of his theorem[1] was found to be correct if the notion of heat (distinguished as *chaleur* from *calorique* by him) was identified with a form of a mechanical energy, as enunciated by Joule, Davy, Count Rumford, and others mentioned in Chapter 3. After examination of Carnot's work, R. Clausius (1850) and W. Thomson (1851) independently were able to modify the Carnot theorem and enunciate the second principle of thermodynamics when systems undergo cyclic processes. Their universal principle is now known as the second law of thermodynamics, and it was stated by them as follows.

Clausius Principle. *It is impossible to transfer heat from a colder to a hotter body without converting at the same time a certain amount of work into heat at the end of a cycle of changes.*

W. Thomson (Lord Kelvin) equivalently stated it as follows.

Kelvin Principle. *In a cycle of processes it is impossible to transfer heat from a heat reservoir and convert it all into work, without transferring at the same time a certain amount of heat from a hotter to a colder body.*

These statements can be also put in another still equivalent form by M. Planck.

Planck Principle. *It is impossible to construct a perpetual machine of the second kind which transfers heat from a colder to a hotter body without a compensation.*

---

[1]In Carnot's theory, there appears the notion of *calorique*, which is distinctive from *chaleur*. He did not clarify its true nature, but it is safe to assume it is a quantity associated with heat transfer and equivalent (for reversible processes) to entropy or, more precisely, calortropy in the present work.

These principles can be shown to be equivalent, for an equivalence proof is available in the literature.[2]

Schematically, the second law states that reversing the following process without compensation is not possible unless some work is done: Let the temperatures of two systems be θ_1 and θ_2, respectively, and $\theta_2 < \theta_1$. Let the amount of heat Q_1 given up by the system (heat reservoir) at θ_1 to the engine (E), which performs work W therewith and then gives up heat Q_2 to the lower temperature system (heat reservoir) at θ_2. It then is impossible to convert all of Q_1 to work making Q_2 equal to zero, i.e., $Q_2 = 0$. In other words, it is impossible to return all of Q_1 to the system at θ_1 without an equivalent amount of work W performed in reverse, the end result being no work performed by the engine. Similar examples can be constructed with the gravitational force and other kinds of force. Another schematic way of looking at the essence of the second law is to examine the essential feature of a heat engine depicted in Fig. 4.1. The engine (E) takes heat Q_1 from the high temperature heat reservoir at θ_1 and performs work W returning some heat Q_2 to the lower temperature heat reservoir at θ_2. It is impossible

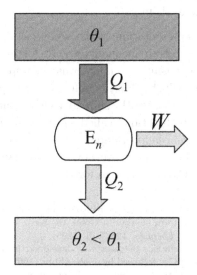

Fig. 4.1. A schematic rendering of the essence of the second law of thermodynamics, which states that Q_2 cannot be equal to zero for a realizable process (engine).

---

[2]See, for example, J. G. Kirkwood and I. Oppenheim, *Chemical Thermodynamics* (McGraw-Hill, New York, 1961).

to make an engine which converts all of Q_1 into work W without some heat $Q_2 \neq 0$ emitted to the lower temperature reservoir. That is, it is not possible to have $Q_2 = 0$.

Since the efficiency of the engine in question may be defined by

$$\epsilon = -\frac{W}{Q_1} \qquad (4.1)$$

for work done by the system and by the first law of thermodynamics

$$W = -(Q_1 - Q_2) \qquad (4.2)$$

for the cyclic process, the efficiency will be unity if $Q_2 = 0$. Therefore, the second law of thermodynamics implies that the efficiency of a cycle cannot be unity, because Q_2 is not equal to zero.

However universal and truthful a principle may be, it is not possible to formulate a physical theory with only a literal statement of it, and it is no exception for the second law of thermodynamics as stated earlier. One of the statements must be translated into an equivalent mathematical form. Since the two statements are equivalent, it is sufficient to have a mathematical representation for one of them.

A mathematical representation of the second law of thermodynamics was made in the form of an inequality by R. Clausius. The inequality, nowadays called the Clausius inequality, was then specialized to the case of reversible processes which Clausius defined as the processes where the uncompensated heat vanishes. He showed that in such processes, there exists a quantity (state function) called *entropy*.[3] *It, however, is a thermodynamic state function defined only for the case of reversible processes.* In order to develop this concept, we now follow Clausius, who used the Carnot cycle and the Carnot theorem for the purpose.[4] The restriction of

---

[3]It was coined by Clausius from a Greek word 'root tropy' to which prefix en was added so that it looks akin to the term energy. In Greek the term 'entropy' means evolution. See R. Clausius, *Ann. Phys. (Leipzig)* **125**, 313 (1865).

[4]Later, when statistical mechanics of gases was developed initially by J. C. Maxwell (1867) and L. Boltzmann (1872) followed by J. W. Gibbs (1902), the second law of thermodynamics was given statistical mechanical interpretations along with the statistical mechanical formula for the entropy and its inequality. In some textbooks on thermodynamics, the statistical mechanical entropy inequality is used to the exclusion of adequate accounts of its evolution as a law that imposes a limitation upon macroscopic physical processes in nature. Such a treatment of this important physical law is not only unjustifiable but also detrimental to the development of thermodynamic theory of irreversible processes.

reversible processes can be removed, and the notion of entropy can be generalized to irreversible processes. This will be discussed later. It is interesting to note that Clausius was able to give the second law of thermodynamics a mathematical representation by using the Carnot theorem. Therefore, the important position that Carnot's theorem occupies in thermodynamics cannot be overstated.

4.1. Carnot Cycle

In the early 19th century, the industrial revolution, powered by steam engines, was progressing in earnest in England, and it made England dominant, economically and politically, in the world at the time. In the book[5] entitled *Réflexions sur la Puissance motrice du Feu et sur les Machines* published in 1824, Sadi Carnot proposed to study an idealized cyclic process in order to understand steam engines which he called the English machines, since the English were using them to power the industrial revolution and became powerful. This cyclic process, now called the Carnot cycle, does not really represent the working of real engines, but is contrived to capture the essence of their working. As mentioned earlier, Carnot used the notion of heat according to the caloric theory of heat, which was overturned about a quarter century later by Joule, Rumford, and others. Nevertheless, he captured the essential truth for the workings of engines and macroscopic processes, physical and biological, in nature. We discuss the essence of his work in the following.

Imagine a working substance, for example, steam or air, is contained in a cylinder fitted at one end with a movable, adiabatic piston. The other end of the cylinder can be made perfectly diathermal or adiabatic at will, but the cylinder wall is adiabatic. The following sequence of operations is performed on the substance in the cylinder (see Fig. 4.3).

First operation. Put the heat conducting wall in contact with the hot reservoir A at temperature[6] θ_1 and let the piston rise to volume V_2 from

---

[5]English translation: E. Mendoza, *Reflections on the Motor Power of Fire and on the Machines* (Peter Smith, Gloucester, MA, 1977).

[6]In the present discussion, it is not necessary to specify a particular temperature scale; it is sufficient to have two thermal reservoirs at given temperatures. However, the subsequent calculations using the ideal gas equation of state will make it abundantly clear that the absolute temperature scale is meant by θ. We use the symbol θ to distinguish it from

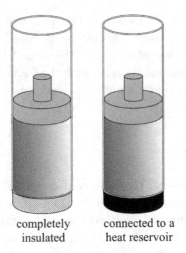

completely   connected to a
insulated   heat reservoir

Fig. 4.2. The cylinder on the left is fitted with an insulator (adiabatic wall) at the bottom, and the bottom wall of the cylinder on the right is diathermal. Both pistons and the cylinder wall are adiabatic.

V_1 by reducing the external pressure from p_1 to p_2 (see Fig. 4.2). Since the process is an isothermal expansion, heat Q_1 is absorbed from A. The change in the state of the system in this process is from (p_1, V_1, θ_1) to (p_2, V_2, θ_1).

Second operation. Remove the cylinder from A and thermally insulate the base of the cylinder so that the cylinder is thermally isolated. Then the pressure is reversibly reduced and as a consequence, the volume is increased. Since the process is adiabatic, the temperature is decreased. This process is continued until the temperature drops to θ_2. Assume that the pressure and the volume at that point are p_3 and V_3, respectively. The change in the state of the system in this process is from (p_2, V_2, θ_1) to (p_3, V_3, θ_2).

Third operation. The insulating base of the cylinder is removed from the cylinder, and the cylinder is put in thermal contact with the cold reservoir B at temperature θ_2. Then the pressure is reversibly increased to p_4 and the volume is decreased to V_4 as a consequence. This pressure is taken such that

---

the thermodynamic temperature T which will be introduced later on the basis of the Carnot theorem and the efficiency of a Carnot cycle. This thermodynamic temperature is eventually made coincident with the absolute temperature.

the system returns to state (V_1, p_1, θ_1) at the end of the fourth operation. Since the process is isothermal and the volume is decreased, the substance gets hotter under compression and hence heat Q_2 is transferred to the cold reservoir B. The change in the state of the system in this process is from (p_3, V_3, θ_2) to (p_4, V_4, θ_2).

Fourth operation. The cylinder is removed from B and the base of the cylinder is thermally insulated, and then the substance is compressed reversibly and adiabatically until the temperature rises to θ_1. Since p_4 was so chosen as to return the system to its original state, the system will return to (V_1, p_1) at θ_1. The change in the state of the system in this process is from (p_4, V_4, θ_2) to (p_1, V_1, θ_1).

Thus, the system is returned to the initial state at the end of a cyclic operation. This reversible cyclic process is depicted in four steps in Fig. 4.3. The diagram as in Fig. 4.3 was introduced for the first time by E. Clapeyron, who gave a mathematical analysis of Carnot's *Réflexion* in which mathematical analysis was avoided because the book was intended for popular consumption.

Let us calculate the net amount of work done by the cycle by using a mole of an ideal gas so as to make calculation simple. We will assume

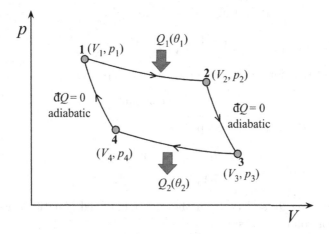

Fig. 4.3. A Carnot cycle in the p–V plane. The four operations are described in the text. The horizontal segments represent isothermal processes and the vertical segments adiabatic processes.

that the specific heat is independent of temperature. The work done by the system in the four steps are as follows:

Step 1:

$$W_{12} = -R\theta_1 \ln\left(\frac{V_2}{V_1}\right),$$

Step 2:

$$W_{23} = \int_{\theta_1}^{\theta_2} d\theta\, c_v(\theta) = c_v(\theta_2 - \theta_1),$$

Step 3:

$$W_{34} = -R\theta_2 \ln\left(\frac{V_4}{V_3}\right),$$

Step 4:

$$W_{41} = \int_{\theta_2}^{\theta_1} d\theta\, c_v(\theta) = c_v(\theta_1 - \theta_2).$$

Here c_v is the molar heat capacity of the working substance, namely, the ideal gas. Note that c_v for the ideal gas is independent of temperature θ. The total work is given by

$$W = W_{12} + W_{23} + W_{34} + W_{41}$$
$$= -R\theta_1 \ln\left(\frac{V_2}{V_1}\right) - R\theta_2 \ln\left(\frac{V_4}{V_3}\right). \tag{4.3}$$

Now recall that for ideal gases in an adiabatic process,

$$\theta V^{\gamma-1} = \text{constant.}$$

More explicitly written for Steps 2 and 4,

$$\theta_1 V_2^{\gamma-1} = \theta_2 V_3^{\gamma-1},$$
$$\theta_1 V_1^{\gamma-1} = \theta_2 V_4^{\gamma-1}.$$

By eliminating θ_1 and θ_2 from the aforementioned equations, we find a relation of volume ratios:

$$\left(\frac{V_1}{V_2}\right)^{\gamma-1} = \left(\frac{V_4}{V_3}\right)^{\gamma-1}. \tag{4.4}$$

Use of this relation in Eq. (4.3) gives rise to the net amount of work in a rather simple form:

$$W = -R(\theta_1 - \theta_2) \ln \left(\frac{V_2}{V_1} \right). \tag{4.5}$$

The heat transfers Q_1 and Q_2 during the isothermal processes (i.e., Steps 1 and 3) are easily calculated in the case of an ideal gas. According to the result obtained in Section 3.4, they are as follows:

$$Q_1 = -W_{12} = R\theta_1 \ln \left(\frac{V_2}{V_1} \right),$$

$$Q_2 = -W_{34} = R\theta_2 \ln \left(\frac{V_4}{V_3} \right) \tag{4.6}$$

$$= R\theta_2 \ln \left(\frac{V_1}{V_2} \right).$$

For the second equality of Q_2, Eq. (4.4) is used.

The efficiency of the Carnot cycle (engine) is defined by the ratio of the work W to the heat supplied Q_1:

$$\eta = \frac{-W}{Q_1} = \frac{|W|}{Q_1}. \tag{4.7}$$

Since for a cycle the internal energy is conserved, i.e.,

$$\oint dE = 0,$$

the work done by the reversible Carnot cycle is

$$-W = Q_1 - Q_2, \tag{4.8}$$

and hence we may express the efficiency in terms of heat alone:

$$\eta = \frac{Q_1 - Q_2}{Q_1} = 1 - \frac{Q_2}{Q_1}. \tag{4.9}$$

Upon use of Q_1 and Q_2 in Eq. (4.6), the efficiency for the reversible cycle may be expressed only in terms of temperatures of the reservoirs and no other parameters:

$$\eta = \frac{R(\theta_1 - \theta_2) \ln (V_2/V_1)}{R\theta_1 \ln (V_2/V_1)} = \frac{\theta_1 - \theta_2}{\theta_1} = 1 - \frac{\theta_2}{\theta_1}. \tag{4.10}$$

The temperatures in this expression is in the absolute temperature scale because of the ideal gas equation of state used for calculations involved. We emphasize that this efficiency formula is shown to be valid in the case of an ideal gas undergoing reversible processes. We will see in Section 4.4 that it holds in general for real substances obeying a fairly general class of equation of states, reinforcing the universality of the result as demanded by the Carnot theorem.

4.2. Carnot's Theorem

As a result of the analysis of the working of a Carnot cycle as an idealization of real engines, Carnot obtained a theorem concerning the efficiency of reversible cycles and its relationship to the efficiency of irreversible cycles. This theorem is now known as Carnot's theorem.

Carnot's theorem. *The efficiency of reversible Carnot cycles is inde- pendent of materials and modes of operation and is maximum. It depends only on the temperatures of the heat reservoirs. In other words, if we denote the efficiencies of reversible cycles by η_{rev}, η'_{rev}, and so on, and that of an irreversible cycle by η_{irr}, then it may be stated that*

$$\eta_{\text{rev}} = \eta'_{\text{rev}} \quad \text{and} \quad \eta_{\text{rev}} \geq \eta_{\text{irr}}.$$

Moreover, η_{rev} depends only on the temperatures θ_1 and θ_2 of the two heat reservoirs, hot and cold.

Proof. In order to prove the first part of the theorem, we first consider two reversible engines operating between two heat reservoirs at temperatures θ_1 and θ_2 ($\theta_1 \geq \theta_2$). Engine En' takes heat Q'_1 from the hotter reservoir and converts part of it into work W and rejects heat Q'_2 to the colder reservoir. The other engine En takes heat Q_1 from the hotter reservoir and converts part of it into work W and rejects heat Q_2 to the colder reservoir. The two engines are now coupled such that engine En' is a prime mover and the other engine En is a refrigerator; see Fig. 4.4. Engine En thus operates reversely to the mode by which the engine En' operates, pumping heat Q_2 from the colder reservoir and returning heat Q_1 to the hotter reservoir at the expense of work W. Now, assume that

$$\eta'_{\text{rev}} > \eta_{\text{rev}}.$$

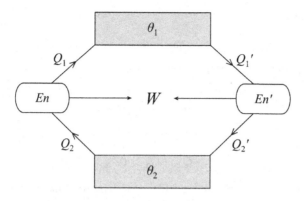

Fig. 4.4. Coupling of a prime mover and a refrigerator as a device to prove Carnot's theorem.

By the definition of efficiency, this may be written equivalently as

$$\frac{-W}{Q_1'} > \frac{-W}{Q_1},$$

which means the inequality

$$Q_1 - Q_1' > 0.$$

This implies that heat $\Delta Q = Q_1 - Q_1'$ is transferred at the end of the cycle from the colder to the hotter reservoir without a compensation. This is against the second law of thermodynamics. Therefore, the assumption is false.

Let us then assume that the reverse is true, i.e.,

$$\eta_{\text{rev}}' < \eta_{\text{rev}}.$$

By reversing the cycle, i.e., by considering En as a prime mover and En' as a refrigerator, we reach the same conclusion as before. Therefore, we must conclude that

$$\eta_{\text{rev}}' = \eta_{\text{rev}}. \tag{4.11}$$

Since the same argument can be carried out with another engine En'' instead of En', the first part of the theorem is proved.

In order to prove the second part of the theorem, we regard En' as an irreversible prime mover and En as a reversible refrigerator. Now, suppose that

$$\eta_{\text{irr}} > \eta_{\text{rev}}.$$

Then, by the definition of efficiency,

$$\frac{-W}{Q'_1} > \frac{-W}{Q_1},$$

which means

$$Q_1 - Q'_1 > 0.$$

This again contradicts the second law of thermodynamics, since it implies that heat is transferred from the colder body to a hotter one at the end of a cycle without a compensation. Therefore, we must conclude that the assumption is false, and

$$\eta_{\text{rev}} \geq \eta_{\text{irr}}. \tag{4.12}$$

In the case of pair of a reversible and an irreversible engine, the reverse process is not possible, since it involves an irreversible engine. Therefore, the first part of the Carnot theorem is proved.

We now prove[7] the last part of the theorem: that the reversible efficiency depends only on the temperatures of the two heat reservoirs, hot and cold.

The reversible Carnot cycle is entirely determined when the adiabatics and isotherms of the cycle are known. If the working substance is specified so that the constitutive relations of the substance are given, the isotherms are determined by the temperatures θ_1 and θ_2, and the adiabatics by some independent variables, say, v_1 and v_2 corresponding to the temperatures. The efficiency of the reversible Carnot cycle is then thought to be a function of θ_1, θ_2, v_1, and v_2 and of the working substance C. Let us therefore assume

$$\frac{-W}{Q_1} = f(\theta_1, \theta_2, v_1, v_2, C). \tag{4.13}$$

This is a continuous function of the variables. The Carnot theorem states that f depends only on θ_1 and θ_2. Consider two bodies C and C' which transform between the same heat reservoirs and describe cycles En and En', respectively, the former being a prime mover and the latter working in the reverse sense — a refrigerator. For this to be possible, the temperatures

---

[7]See H. Poincaré, *Thermodynamique* (Georges Carré, Paris, 1892).

must satisfy certain inequalities. Let θ_1 and θ_2 be the temperatures of two reservoirs, hot and cold, θ'_1 and θ'_2 the temperatures of the isotherms of the cycle En, and θ''_1 and θ''_2 the temperatures of the isotherms of the cycle En'. Then for the coupled cycles to be possible, the following inequalities of temperatures must hold:

$$\theta''_1 > \theta_1 > \theta'_1 > \theta'_2 > \theta_2 > \theta''_2.$$

Denote by W the work produced by the cycle En, Q_1 the heat that is taken from the hot reservoir, and Q_2 the heat that it cedes to the cold reservoir, and by W', Q'_1, and Q'_2 similar quantities corresponding to cycle En'. Then

$$\frac{-W}{Q_1} \leq \frac{-W'}{Q'_1}. \tag{4.14}$$

This is proved as follows. If m and m' are the masses of bodies C and C', which are transformed in the cycles, we have for the heat taken from the hot heat reservoir by the combined cycles

$$mQ_1 - m'Q'_1.$$

Since Q_1 and Q'_1 are positive, we can take for m and m' values such that

$$mQ_1 - m'Q'_1 = 0. \tag{4.15}$$

But then the work

$$-mW + m'W'$$

produced by the two cycles cannot be positive because then we will have a production of work with only one source of heat contrary to the second law of thermodynamics. Therefore,

$$-mW + m'W' \leq 0. \tag{4.16}$$

Replacing m and m' with Q_1^{-1}, and Q'^{-1}_1, we obtain the inequality (4.14).

Consider now two Carnot cycles En and En' which are, respectively, defined by the variables $\theta'_1, \theta'_2, v'_1$, and v'_2 for the former and by $\theta''_1, \theta''_2, v''_1$, and v''_2 for the latter. The cycle En runs in the direct sense, and the cycle En' in the reverse sense between two same heat reservoirs of temperatures θ_1 and θ_2, respectively. For this to be possible, we must have the inequalities

$$\theta'_1 < \theta_1, \quad \theta'_2 > \theta_2 \quad \theta''_1 > \theta_1, \quad \theta''_2 < \theta_2.$$

Then it follows from inequality (4.14) that

$$f\left(\theta'_1, \theta'_2, v'_1, v'_2, C\right) \leq f\left(\theta''_1, \theta''_2, v''_1, v''_2, C'\right). \tag{4.17}$$

If we suppose

$$\theta'_1 > \theta_1, \quad \theta'_2 < \theta_2 \quad \theta''_1 < \theta_1, \quad \theta''_2 > \theta_2,$$

we can describe the cycle En in the reverse direction and the cycle En' in the direct sense. Then we have the inequality

$$f\left(\theta'_1, \theta'_2, v'_1, v'_2, C\right) \geq f\left(\theta''_1, \theta''_2, v''_1, v''_2, C'\right). \tag{4.18}$$

Since f is continuous, we can make the transitions $\theta'_1, \theta''_1 \to \theta_1$ and $\theta'_2, \theta''_2 \to \theta_2$ without changing the sign of inequalities (4.17) and (4.18). We then have, for the aforementioned two cases, the inequalities

$$f\left(\theta_1, \theta_2, v'_1, v'_2, C\right) \leq f\left(\theta_1, \theta_2, v''_1, v''_2, C'\right), \tag{4.19}$$

$$f\left(\theta_1, \theta_2, v'_1, v'_2, C\right) \geq f\left(\theta_1, \theta_2, v''_1, v''_2, C'\right). \tag{4.20}$$

These inequalities cannot be satisfied simultaneously. Therefore, we have

$$f\left(\theta_1, \theta_2, v'_1, v'_2, C\right) = f\left(\theta_1, \theta_2, v''_1, v''_2, C'\right), \tag{4.21}$$

which means that f is independent of variables v_1, v_2, and C. This proves the last part of the Carnot theorem. Therefore, the Carnot theorem is completely proved.

The proof presented is in fact also seen as the proof of equivalence of the second law of thermodynamics and the Carnot theorem. The Carnot theorem provides not only the more transparent mathematical representation of the second law but also an important concept that is not transparent in the Clausius and Kelvin principles, since it shows that *the efficiencies of irreversible cycles cannot exceed the efficiency of the reversible cycles*. Furthermore, we now see that the second law of thermodynamics, which is based on everyday experience and therefore less transparent mathematically than the Carnot theorem, is given a clearer mathematical representation by means of Carnot's theorem. We reiterate that both the second law of thermodynamics and the Carnot theorem are phrased about cyclic processes, *since cyclic processes are the only kind of processes in which the system is assured of returning to the original state without fail and thus the existence of conserved quantities characteristic of the system is implied without recourse to any other measure to ascertain their conserved nature.* Conserved quantities are certainly convenient to use in formulating a theory of macroscopic processes.

From the Carnot theorem, it is possible to deduce that *a reversible work is a maximum work possible.* To show it, let us assume that there are two engines operating between two heat reservoirs and both of them absorb heat Q_1 from the hotter reservoir and emit heat Q_2 to the colder reservoir. The work performed by the reversible engine is W and that by the irreversible engine is W'. Then according to the Carnot theorem, the following inequality holds:

$$\eta_{\text{rev}} = \frac{-W}{Q_1} \geq \frac{-W'}{Q_1} = \eta_{\text{irr}}, \tag{4.22}$$

which means that

$$-W \geq -W'. \tag{4.23}$$

This proves the statement made above. It is seen as a corollary to the Carnot theorem. By the sign convention on work, $-W$ is positive since W is the work performed by the system and similarly for $-W'$. Therefore, inequality (4.23) implies that if a process is irreversible then there is an amount of work unavailable to the given task of the process. This is related to the energy dissipation accompanying an irreversible process. It is convenient to introduce the unavailable work W_{ua} by the relation

$$W = W' + W_{\text{ua}}. \tag{4.24}$$

The precise nature of unavailable work depends on the irreversible process in question and the system as well as the device with which the work is performed. This unavailable work must be elucidated if the theory of irreversible processes is the aim — a task taken up in Part II of this book.

4.3. Thermodynamic Temperature

The Carnot theorem makes it possible to give the second law of thermodynamics a mathematical representation, which would be much desired if the second law were to be described by a precise mathematical machinery based on equations, but not inequalities. To achieve the desired aim, it is necessary to lay foundations for it.

First of all, let us consider the implications of the Carnot theorem. One of the most important is the notion of *thermodynamic temperature* which was first deduced from the Carnot theorem by Lord Kelvin. We consider the gist of it here. Since the efficiency of reversible Carnot cycles is independent of materials and the modes of operation, but depends only on the temperatures θ_1 and θ_2 of the heat reservoirs according to Carnot's theorem, we see that the ratio of Q_2 to Q_1 is a function of θ_1 and θ_2 only:

$$\frac{Q_2}{Q_1} = f(\theta_2, \theta_1). \tag{4.25}$$

Since we may write the left-hand side in the form

$$\frac{Q_2}{Q_1} = \frac{Q_2}{Q_0} \cdot \frac{Q_0}{Q_1} = f(\theta_2, \theta_0) f(\theta_0, \theta_1),$$

we conclude that $f(\theta_2, \theta_1)$ is a function with the following property:

$$f(\theta_2, \theta_1) = f(\theta_2, \theta_0) f(\theta_0, \theta_1). \tag{4.26}$$

Let $\theta_2 = \theta_1$. Then

$$f(\theta_1, \theta_0) = \frac{1}{f(\theta_0, \theta_1)}. \tag{4.27}$$

The properties of f described in Eqs. (4.26) and (4.27) imply that there exists a function of θ, $\phi(\theta)$, such that

$$f(\theta_2, \theta_1) = \frac{\phi(\theta_2)}{\phi(\theta_1)}, \tag{4.28}$$

and, consequently, it is possible to write

$$\frac{Q_2}{Q_1} = \frac{\phi(\theta_2)}{\phi(\theta_1)}. \tag{4.29}$$

The function $\phi(\theta)$ is arbitrary. It will be denoted by T:

$$T = \phi(\theta). \tag{4.30}$$

This is called the thermodynamic temperature. The temperature based on the Carnot theorem must be universal since the Carnot theorem is universal owing to the fact that *the reversible efficiencies are independent of the modes of operation and the working substances.* In the manner shown, the

Carnot theorem can serve as the basis of introducing a universal thermodynamic temperature.[8] Since the efficiency calculated for a reversible cycle in Eq. (4.10) is expressed in absolute temperature, it is appropriate to regard θ here as a temperature in the absolute scale. In this event, the efficiency formula for a reversible cycle in Eq. (4.10) suggests that $\phi(\theta)$ must be linearly proportional to θ, and the proportionality constant may be taken such that θ coincides with the thermodynamic temperature T:

$$T = \theta. \tag{4.31}$$

In this manner, the thermodynamic temperature is made to coincide with the absolute temperature based on the ideal gas temperature, and they now can be interchangeably used. It then is possible to express the efficiency of a reversible cycle in terms of the absolute temperature (or thermodynamic temperature):

$$\eta = \frac{Q_1 - Q_2}{Q_1} = \frac{T_1 - T_2}{T_1}. \tag{4.32}$$

It is important to recall that T_1 and T_2 are originally the temperatures of the heat reservoirs and, only in the case of a reversible cycle, can they be regarded as the temperatures of the system which is in thermal equilibrium with the heat reservoirs in various steps of operation in the cycle.

4.4. Entropy and Calortropy

4.4.1. *Clausius Inequality*

With the Carnot efficiency for reversible cycles expressed in terms of the thermodynamic temperature, the Carnot theorem now acquires a firmer mathematical grip on the second law of thermodynamics. To pursue this line further, which eventually gives rise to the Clausius inequality, it is convenient at this point to adopt a sign convention for heat transfer in a way consistent with the sign convention for work introduced earlier.

Sign convention. *If heat is transferred to the system, Q is taken positive whereas, if it is transferred out of the system, Q is taken negative.*

---

[8]The existence of universal thermodynamic temperature presupposes a reversible cycle or process. Without it, the temperature defined is no longer universal and hence generally dependent on the substance or the system.

With this convention, we may write the efficiency of a reversible cycle in the form

$$\frac{Q_1 - (-Q_2)}{Q_1} = \frac{T_1 - T_2}{T_1}.$$

Rearranging the terms in this equation, we find for the reversible cycle

$$\frac{Q_1}{T_1} + \frac{Q_2}{T_2} = 0. \tag{4.33}$$

If the amounts of heat injected into and rejected by or withdrawn from the system (cycle) are infinitesimal, we may write Eq. (4.33) in a differential form in terms of differential elements dQ_i:

$$\frac{dQ_1}{T_1} + \frac{dQ_2}{T_2} = 0. \tag{4.34}$$

This equation may be further generalized by the following mathematical device. A reversible cycle may be regarded as consisting of a large number of infinitesimal reversible cycles as shown in Fig. 4.5. The horizontal lines represent the isotherms and the vertical lines intersecting the former the adiabatics. The cycle i takes an infinitesimal amount of heat dQ_i from the $(i-1)$th cycle and rejects an infinitesimal amount of heat dQ_{i+1} to the $(i+1)$th cycle. The temperatures of the isotherms are denoted by T_i. For the whole collection (sequence) of the infinitesimal cycles, we may then write

$$\sum_i \frac{dQ_i}{T_i} = 0, \tag{4.35}$$

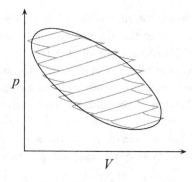

Fig. 4.5. Decomposition of a Carnot cycle into infinitesimal Carnot cycles $\{i\}$. The ith cycle takes heat dQ_i from the adjacent $(i-1)$th cycle and rejects heat dQ_{i+1} to its immediate neighbor, the $(i+1)$th cycle in the sequence.

which is a generalization of Eq. (4.34). If the number of infinitesimal cycles is taken to infinity, the summation may be replaced with an integral over the whole cycle:

$$\oint \frac{dQ_{\text{rev}}}{T} = 0. \tag{4.36}$$

It must be emphasized that this equation holds for reversible cycles only and dQ_{rev} denotes a differential heat transfer in a reversible process. They are cycles in which processes are performed such that the system is in continuous equilibrium with the surroundings or there is no energy dissipation, the meaning of which will be elucidated later more precisely.

Let us now consider a more general case involving an irreversible cycle. In this case, the Carnot theorem reads

$$\eta_{\text{irr}} = \frac{Q_1 - Q_2}{Q_1} \leq \frac{T_1 - T_2}{T_1} = \eta_{\text{rev}}. \tag{4.37}$$

By applying the same method of reasoning as for that leading to Eq. (4.36), we obtain the inequality

$$-\oint \frac{dQ}{T} \geq 0. \tag{4.38}$$

This inequality, first obtained by Clausius, is called the Clausius inequality. It is important to remember that T is the temperature associated with the reversible cycle and thus the infinitesimal cycle in equilibrium, and that even if the process in question is irreversible, the temperature factor T in the inequality is still that of the reversible process corresponding to the irreversible process of interest; see Eq. (4.37).

Clausius called dQ *the compensated heat* since it is a heat transfer between the system and the surroundings (heat reservoirs) that occurs irreversibly. He also identified another basic quantity that he called *the uncompensated heat* and denoted it by N. This quantity, which is always positive and vanishes if the process is reversible, is equal in value to the closed contour integral (simply called the circular integral) in the inequality (4.38):

$$N = -\oint \frac{dQ}{T} \geq 0. \tag{4.39}$$

Before we examine this form of the Clausius inequality in detail, it is useful, in historical deference to the development made by Clausius, to consider a special case of his inequality. *This special case gives rise to equilibrium thermodynamics.*

4.4.2. *Entropy*

Let us return to Eq. (4.36), which holds for reversible processes. Although
Clausius discovered the existence of uncompensated heat N, its precise
nature was not known sufficiently well to him at the time.[9] To make some
progress in the absence of the detailed knowledge of N, Clausius considered
the case of reversible processes[10] for which N = 0 identically. Since the
line integral of the compensated heat thus obtained is nothing other than
Eq. (4.36) and meets the mathematical criterion for exact differentials, it
led him to conclude that there exists an exact differential dS such that

$$dS = \frac{dQ_{\text{rev}}}{T} \qquad (4.40)$$

in the space of thermodynamic variables characteristic of the system for the
reversible process considered. Moreover, it can be shown that there exists
such a function S for any reversible process since the Carnot theorem holds
for all reversible processes. Clausius called S *the entropy*[11] of the system,
and it is a state function in the space of thermodynamic variables for the
system, which is spanned by the internal energy and volume and, in the
case of a mixture, also by the concentrations of species.

The second law of thermodynamics, expressed by Eq. (4.36) for *a
reversible process*, is now represented by the vanishing circular integral

$$\oint dS = 0. \qquad (4.41)$$

It should be emphasized that this vanishing integral is true only if the
process is reversible. It must also be noted that dQ_{rev} denotes a nonexact
differential for heat change although the symbold is not used in place of the
differential symbol d. When dQ_{rev} is divided by the absolute temperature,
the result becomes an exact differential. Such factors which make nonexact
differentials exact are called integrating factors, and $1/T$ is an integrating

---

[9]In connection with this line of work and much before he put forward the notion of
entropy, Clausius had made use of the notion of *disgregation*, which appeared to be akin
to a quantity he later called *entropy*. However, he never developed and quantified the
notion beyond the entropy.

[10]Reversible processes may be defined as the processes in which the uncompensated heat
N vanishes everywhere. This definition is preferable to the definition based on the process
of continuous quasi-equilibrium introduced in Section 2.5.

[11]For the reference, purpose see footnote 3.

factor in the case of dQ_{rev}. Therefore, the entropy change for a reversible process $A \to B$ is obtained on integration of Eq. (4.40):

$$\Delta S = S_B - S_A = \int_A^B dS = \int_A^B \frac{dQ_{rev}}{T}, \qquad (4.42)$$

where the integration is over the reversible path. For an isolated system, there is no heat put in or taken out of the system and hence,

$$dQ_{rev} = 0,$$

which then implies

$$\Delta S = 0.$$

In other words,

$$S_A = S_B$$

for isolated systems at equilibrium with the surroundings. It should be clearly understood that the entropy S is defined for reversible processes.

The differential for S given in Eq. (4.40) can be combined with the first law of thermodynamics to yield the fundamental relation for thermodynamics of reversible processes or equilibrium

$$dE = TdS - pdV \qquad (4.43)$$

in the case of pressure–volume work alone. If there are other kinds of work besides the pressure–volume work, then the differential form for energy may be written as

$$dE = TdS - pdV + \sum_i X_i dx_i, \qquad (4.44)$$

where X_i are forces and x_i are displacements whose examples were given in Chapter 2. We will consider thermodynamics in the cases of such forms of work in later chapters. The differential form (4.43) or (4.44), which are often referred to as the equilibrium Gibbs relation or, more simply, the Gibbs relation, are inappropriately taken as the mathematical expression for the second law of thermodynamics, but it must be emphasized that it is only for reversible processes or processes in equilibrium systems.

4.4.3.   Calortropy

Before we discuss the application of the notion of entropy in thermodynamics, it is useful to examine more closely the form of the Clausius
inequality, (4.39), to learn about the nature of the uncompensated heat N.
It must be associated with the *unavailable (ua) work* appearing in Eq. (4.24)
since the numerator in the right member of Eq. (4.13) is equal to $-W'$
which is less than the work $-W$ performed by the reversible cycle corresponding to the right member of the inequality (4.37). By adding a
term $-W_{ua}/Q_1$ to the left member of the inequality (4.37), which is made
an equation thereby, we can cast the inequality (4.37) in the form of an
equation

$$-\frac{W_{ua}}{T_2} = -\oint \frac{dQ}{T}. \tag{4.45}$$

If the left-hand side is identified with the uncompensated heat

$$N := -\frac{W_{ua}}{T_2} \geq 0, \tag{4.46}$$

then the Clausius inequality (4.39) is recovered and a little more definite
meaning of the uncompensated heat is attained. Clearly, the uncompensated heat so identified is always positive in the sign convention adopted
for work and vanishes only if the process is reversible. This consideration
made here indicates that *the compensated and uncompensated heats are
two independent physical quantities, which exactly balance each other for
the cycle.* In other words, we may write Eq. (4.45) in the form

$$\oint \frac{dQ}{T} + N = 0. \tag{4.47}$$

The compensated and uncompensated heats are characteristic of the system and the surroundings in a given heat transfer process involved between
them. The symbol N clearly is not another way of writing the circular
integral on the right-hand side of Eq. (4.45), which appears to be an identity at a quick glance. Neither is the circular integral on the right-hand
side of Eq. (4.45) the definition of N, which should be elucidated on
its own right if an irreversible process were an object of study. It is
important to recognize that the Clausius inequality is derived from the
Carnot theorem. Therefore, the temperature T appearing in the Clausius inequality or Eq. (4.47) is the temperature expressing the efficiency

of the reversible cycle with which the efficiency of an irreversible cycle is compared.[12]

The uncompensated heat may be expressed as an integral over the path[13] of the cycle

$$N = \oint dN \geq 0, \tag{4.48}$$

where dN is such that $(dN/dt) \geq 0$ everywhere in the cycle, since otherwise it is possible to construct a cycle that contravenes the positivity of the integral and thus the second law of thermodynamics. Then the Clausius inequality can be written as

$$\oint \left(\frac{dQ}{T} + dN \right) = 0. \tag{4.49}$$

This is a mathematical representation for the second law of thermodynamics, which was literally stated in the forms of the Clausius and Kelvin principles. It now is written in a form of equation instead of an inequality initially obtained by Clausius, and such an equation is realized by treating the compensated and uncompensated heats as two independent physical quantities on the equal footing in formulating thermodynamics of irreversible processes. Although it may appear as a simple and trivial rewriting of the Clausius inequality, it not only provides *a fresh way of looking at the thermodynamics of irreversible processes but also generalizes the classical equilibrium thermodynamics to nonequilibrium without additional assumptions*. We now examine how the classical thermodynamics of reversible processes or equilibrium can be recovered and made free from the troublesome features, especially the one related to the concept of entropy, which is defined only when the system is in equilibrium, but, as will be shown below, has been nevertheless used in the context in which the system is away from equilibrium under the hypothesis[14] of local equilibrium.

As in the case of Eq. (4.36) holding for reversible processes, Eq. (4.49) implies that even if the process is irreversible, there still exists a state

---

[12]Later on in Part II, such notion of T is slightly extended (or, rather, mathematically continued) to the cases of irreversible processes.

[13]The path is defined in the manifold of macroscopic thermodynamic variables, which will be more precisely and explicitly defined when thermodynamics of irreversible processes is discussed in Chapter 19 in Part II.

[14]See monographs on linear irreversible processes, for example, S. R. de Groot and P. Mazur, *Non-equilibrium Thermodynamics* (North-Holland, Amsterdam, 1962).

function Ψ in the space of thermodynamic variables whose differential form is given by

$$d\Psi = \frac{dQ}{T} + dN \tag{4.50}$$

and vanishes on integration over the cycle:

$$\oint d\Psi = 0. \tag{4.51}$$

This is reminiscent of the vanishing circular integral of dS in Eq. (4.41). The vanishing circular integral (4.51) implies that $d\Psi$ is an exact differential in the thermodynamic space which may be spanned by the internal energy, volume, concentrations of species, and macroscopic variables characteristic of the nonequilibrium system in which irreversible processes occur. The precise nature of the thermodynamic space should be elucidated upon detailed examination of the irreversible processes in hand, but it is clear that the space should be spanned by macroscopic variables describing the irreversible process in hand in addition to the usual variables characterizing the equilibrium system of temperature T, such as the internal energy, volume, and concentrations of species involved. This state function Ψ is called the calortropy.[15] Since the process is irreversible, Ψ is a generalized form of the Clausius entropy S, which holds for reversible processes only. The vanishing circular integral (4.51) is a mathematical representation of the second law of thermodynamics in the thermodynamic manifold for the irreversible cyclic process under consideration. For the same cyclic process, the first law of thermodynamics is also expressible as a vanishing circular integral

$$\oint dE = 0. \tag{4.52}$$

This pair of circular line integrals is over the same irreversible path in the thermodynamic space for the system. We recall that although dW and dQ are not exact differentials, their sum dE is an exact differential. It is then interesting to see that, just as in the case of the first law of thermodynamics,

---

[15]It is a composite word meaning heat (*calor*) evolution (*tropy*). It is akin to Carnot's *calorique*, which he used in his *Réflexion* to mean a quantity that is similar to heat, but not quite the same as heat. Since *calorie* is already in use, the term *calorique* is inappropriate to use. The term "coined" thus seems to be suitable. Some research workers in irreversible thermodynamics use the term "nonequilibrium entropy" to mean what is essentially the calortropy here, but the latter is simpler and more concise than the former.

although dQ/T and dN are not exact differentials if the process is irreversible, their sum $d\Psi$ is an exact differential, and the first and second laws of thermodynamics are mathematically expressible as a pair of vanishing circular integrals if the process is cyclic in the thermodynamic manifold of variables.[16] We summarize these results as a theorem.

Theorem. *In the case of a cyclic process, regardless of whether it is reversible or irreversible, the first and second laws of thermodynamics are, respectively, expressible as a pair of vanishing cyclic integrals in the thermodynamic manifold.*[17]

$$\oint dE = 0, \tag{4.53}$$

$$\oint d\Psi = 0. \tag{4.54}$$

4.4.4. Inequalities of Entropy and Calortropy

Since $(dN/dt) \geq 0$ where dt is an infinitesimal time interval, for an infinitesimal interval of the cycle, the following inequality holds:

$$d\Psi \geq \frac{dQ}{T}. \tag{4.55}$$

Since $(dN/dt) = 0$ if the process is reversible, not only does the equality hold in Eq. (4.55), but $d\Psi$ also becomes identical with dS, namely,

$$d\Psi_{\text{rev}} = \frac{dQ_{\text{rev}}}{T}$$

$$= dS, \tag{4.56}$$

where Ψ_{rev} and Q_{rev} are the calortropy and heat transfer accompanying the reversible process.

---

[16]The precise nature of the manifold is not as yet clarified. It will be the task of the thermodynamics of irreversible processes. For this aspect of investigation, see, for example, B. C. Eu, *Generalized Thermodynamics* (Kluwer, Dordrecht, 2002).

[17]Thermodynamic manifold is a union of extensive variables such as density or volume, internal energy, concentrations, nonconserved variables (in the case of irreversible processes) and the tangent manifold consisting of intensive variables such as temperature, pressure, chemical potentials, and generalized potentials (in the case of irreversible processes). Note that extensive and intensive variables mentioned are conjugate variables to each other.

If inequality (4.55) is integrated over a finite irreversible process from state 1 to state 2, then there follows the inequality

$$\Delta\Psi \geq \int_{1(\text{irr})}^{2} \frac{dQ}{T}, \tag{4.57}$$

where

$$\Delta\Psi = \int_{1(\text{irr})}^{2} d\Psi. \tag{4.58}$$

Inequality (4.57) also follows directly from Eq. (4.50) if the latter is integrated over the same interval:

$$\Delta\Psi = \int_{1(\text{irr})}^{2} \frac{dQ}{T} + \int_{1(\text{irr})}^{2} dN$$

$$\geq \int_{1(\text{irr})}^{2} \frac{dQ}{T}. \tag{4.59}$$

The equality holds only if the process is reversible in which case we may write

$$\Delta\Psi_{\text{rev}} = \Delta S = \int_{1(\text{rev})}^{2} \frac{dQ_{\text{rev}}}{T}. \tag{4.60}$$

If the system is isolated, then there is no heat exchange between the system and the surroundings. In this case, dQ vanishes identically over the process, and inequality (4.57) gives rise to the inequality for the calortropy

$$\Delta\Psi \geq 0, \tag{4.61}$$

where the equality holds only if the process is reversible. This is a special case of Eq. (4.60) for which

$$\Delta\Psi_{\text{rev}} = \Delta S = 0. \tag{4.62}$$

For irreversible processes in an isolated system,

$$\Delta\Psi > 0. \tag{4.63}$$

Therefore, the calortropy tends toward a maximum and attains a maximum value identical with the entropy at equilibrium. This conclusion should be compared with the famous statement for the second law of thermodynamics made by Clausius[18] in 1865: *The entropy of the world tends towards a maximum.* Since by definition the entropy is an attribute of the system at

---

[18]See the reference cited in footnote 3 of this chapter.

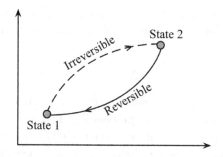

Fig. 4.6. A cyclic path consisting of an irreversible and a reversible segment.

equilibrium, in our opinion, it makes sense that calortropy should replace the term 'entropy' in the statement. We elaborate on this assertion with the following examination.

Suppose a system has irreversibly and spontaneously reached a final state from an initial state as in Fig. 4.6. If the system is not closed, it is always possible to make the system return to the original state through a reversible path, provided that the system is properly compensated to do so. For such a system, it is always possible to construct a cyclic process consisting of two segments, one irreversible and the other reversible, as depicted in Fig. 4.6, and to find from Eq. (4.51), the relation

$$\Delta \Psi = \int_{1(\text{irr})}^{2} d\Psi = \int_{1(\text{rev})}^{2} dS = \Delta S. \tag{4.64}$$

This equation together with Inequality (4.57) implies the inequality

$$\Delta S \geq \int_{1(\text{irr})}^{2} \frac{dQ}{T}. \tag{4.65}$$

This is the Clausius–Duhem inequality known in thermodynamics as a mathematical form of the second law of thermodynamics. This means that Inequality (4.57) implies the Clausius–Duhem inequality. We emphasize that Eq. (4.64) holds only if there exists a reversible segment in the cycle which in this particular case requires that the system should not be closed since some sort of compensation is necessary for the system to return to the original state from which it irreversibly departed. An isolated system that has irreversibly and spontaneously reached a final state cannot be spontaneously returned to the initial state since the necessary compensation can only come from the surroundings for the system to do so.

The Clausius–Duhem inequality (4.65) is mathematically and conceptually impeccable, but it does not mean that for an isolated system, the following inequality holds:

$$\Delta S \geq 0, \tag{4.66}$$

since there does not exist a complementary reversible process Γ_{rev} that makes up a cycle with irreversible process Γ_{irr} if the system is isolated. The existence of such a path Γ_{rev} implies that there is a mode of compensation from the surroundings. If so, the system is no longer isolated and inequality $\Delta S > 0$ is misleading.

Inequality (4.66), in fact, causes a considerable conceptual difficulty when closely examined, since ΔS is not defined if the process is irreversible, and $\Delta S = 0$ for a reversible process in an isolated system since

$$\Delta S = \int_{1(\text{rev})}^{2} \frac{dQ_{\text{rev}}}{T} = 0 \tag{4.67}$$

identically because $dQ_{\text{rev}} = 0$ for an isolated system. Therefore, it cannot be concluded that $\Delta S > 0$ if there is an irreversible process in an isolated system. The confusing part of Inequality (4.66) is the conclusion that $\Delta S > 0$ although the system is isolated so that $dQ_{\text{rev}} = 0$ and thus, by definition, $\Delta S = 0$. This difficulty disappears if ΔS is replaced by $\Delta \Psi$ so that Inequality (4.61) holds for the case of an isolated system undergoing an irreversible process; Inequality (4.61) has no such problem as for Inequality (4.66) for the reason that it is not equal to zero even if the system is isolated and thus $dQ_{\text{rev}} = 0$ because of a nonvanishing uncompensated heat for an irreversible process.

The laxity of the Clausius statement has been a source of debate[19] and confusion in the fine points of thermodynamics, but we now seem to have a way to free us from such confusion. The discussion presented here on the basis of Eq. (4.64) and inequality (4.65) suggests that the conventional Clausius–Duhem inequality for the Clausius entropy should be phrased in terms of the calortropy over the irreversible segment as in inequality (4.57) and, in the case of an isolated system, as in the inequality (4.61). Inequality (4.61) is easier to comprehend than the Clausius–Duhem inequality for an isolated system because it does not present a difficulty that Eq. (4.67)

---

[19]See, for example, J. Kestin, *The Second Law of Thermodynamics* (Dowden, Hutchinson, & Ross, Stroudburg, PA, 1976).

holds for a reversible process in an isolated system. We further elaborate on this point.

Inequality (4.66) may be understood in the following sense. Suppose the irreversible process Γ_{irr} in Fig. 4.6 occurs when the system is isolated whereas the reversible process Γ_{rev} occurs when the system is not isolated. In other words, after the irreversible process Γ_{irr} is completed, the system is allowed to interact with its surroundings and made to return reversibly from state B to state A. In this case,

$$\Delta S = \Delta\Psi(\Gamma_{\text{irr}}) = \int_{A(\text{irr})}^{B} dN \geq 0$$

where the equality holds if Γ_{irr} is reversible. This equation, as Eq. (4.64), allows one to calculate $\Delta\Psi(\Gamma_{\text{irr}})$ in terms of ΔS over a complementary reversible process, but only implies that, if an irreversible process occurs in an isolated system, then the ΔS of a reversible process is positive, which is complementary to the irreversible process in question and occurs in interaction with the surroundings. This, we believe, is probably what Clausius should have meant by inequality (4.66).

If the nature of uncompensated heat is elucidated in detail, it should be possible to calculate $\Delta\Psi$. We briefly comment on this aspect. Let us imagine that there is a finite segment of an irreversible process $1 \to 2$. Then, upon integrating Eq. (4.50), there follows

$$\Delta\Psi = \int_{1(\text{irr})}^{2} \frac{dQ}{T} + \Xi_g, \tag{4.68}$$

where

$$\Xi_g = \int_{1(\text{irr})}^{2} dN \geq 0. \tag{4.69}$$

This inequality is a form of the second law of thermodynamics. If it is possible to take the system to the initial state 1 through a reversible process, then the entropy change for the reversible process ΔS is given by the equation

$$\Delta S = \int_{1(\text{irr})}^{2} \frac{dQ}{T} + \Xi_g \tag{4.70}$$

as was originally suggested by Clausius. Provided we know how to compute Ξ_g for the process, it is possible to compute the integral in Eq. (4.70) by computing ΔS for a reversible process complementary to the irreversible process in question, or Ξ_g from the knowledge of the integral in Eq. (4.70) and ΔS.

Let us examine the significance of Eq. (4.70) with the following example. A metal bar at temperature T_2 is separated by an insulator C from another of the same metal at temperature T_1, which is lower than T_2. The insulator is then removed and the metal bars are allowed to interact thermally. They will reach equilibrium at which point the temperature will be $\frac{1}{2}(T_1 + T_2)$. The systems are subsequently brought to the original states by a reversible process by making them interact with the surroundings. This reversible process is accompanied by an entropy change ΔS in the system. Since the irreversible process occurs in an isolated condition, the integral in Eq. (4.70) is equal to zero and ΔS is given by

$$\Delta S = \Xi_g \geq 0. \tag{4.71}$$

The entropy change ΔS for the reversible process can be easily calculated:

$$\Delta S = C_p \ln \left(\frac{T_1 + T_2}{2\sqrt{T_1 T_2}} \right), \tag{4.72}$$

for which we have assumed, for the simplicity of the result, that the specific heat C_p is independent of temperature. Since there holds the inequality

$$\tfrac{1}{4}(T_1 + T_2)^2 \geq T_1 T_2,$$

we conclude that indeed

$$\Delta S \geq 0$$

and inequality (4.71) is satisfied, indicating that the irreversible process in question is spontaneous with the value of the global uncompensated heat estimated for the irreversible process as in Eq. (4.72). Therefore, we see that Eq. (4.70) can be used for estimating the uncompensated heat Ξ_g from the knowledge of ΔS for a reversible process complementary to the irreversible process in question. We emphasize that ΔS in Eq. (4.72) is for a reversible process in a system interacting with the surroundings (Fig. 4.7).

4.5. Carnot's Theorem and Real Gases

We have previously calculated the efficiency of a reversible Carnot cycle when the working substance is ideal. However, the working substance is not ideal in practice but real, since the molecules do interact with each other and the ideal gas equation of state holds only in the sufficiently low

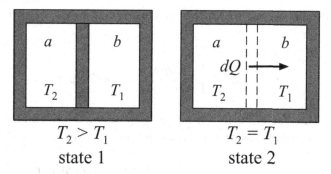

Fig. 4.7. Two metal bars in contact at two different temperatures which go through an irreversible process of heat transfer.

pressure limit. If a real gas is taken for the working substance, some steps in the calculation we have performed for the Carnot cycle are not valid. There then arises the question as to the generality of the result obtained for the efficiency. The result, however, holds in general for real working substances, and we would like to prove this statement for the Carnot cycle depicted in Fig. 4.4 in the following.

The equation of state for 1 mole of a real gas may be given in the form

$$p = \frac{RT}{V} + f(V, T), \tag{4.73}$$

where $f(V, T)$ is a function of V and T which we do not have to specify except that it is a well-behaved function and piecewise integrable. Note that in the case of real gases,

$$\left(\frac{\partial E}{\partial V} \right)_T \neq 0.$$

This requires a new mode of calculation for the heat change accompanying an isothermal expansion, which is different from that used for an ideal gas.

The work of isothermal expansion is

$$W_{ij} = - \int_{V_i}^{V_j} p\, dV \ (i = 1, 3; \ j = 2, 4), \tag{4.74}$$

which may be explicitly calculated if the integration is performed with Eq. (4.73) substituted into it. Such calculation, however, is not necessary for our purpose here.

By regarding the internal energy as a function of V and T, we may write

$$dQ = dE + pdV$$

$$= C_v dT + \left[T \left(\frac{\partial p}{\partial T} \right)_V - p \right] dV + pdV. \qquad (4.75)$$

Therefore, since $dQ = 0$ for an adiabatic process, the work of adiabatic expansion (or compression) may be written in the form

$$W_{23} = -\int_{V_2}^{V_3} pdV = -\int_{T_1}^{T_2} C_v dT + \int_{V_2}^{V_3} \left[p - T \left(\frac{\partial p}{\partial T} \right)_V \right]_{T_2} dV, \qquad (4.76)$$

$$W_{41} = -\int_{V_4}^{V_1} pdV = \int_{T_1}^{T_2} C_v dT + \int_{V_4}^{V_1} \left[p - T \left(\frac{\partial p}{\partial T} \right)_V \right]_{T_1} dV. \qquad (4.77)$$

The total work is then the sum of the four components calculated above:

$$W = W_{12} + W_{23} + W_{34} + W_{41}$$

$$= -\int_{V_1}^{V_2} p(T_1)dV - \int_{V_3}^{V_4} p(T_2)dV + \int_{V_2}^{V_3} \left[p - T \left(\frac{\partial p}{\partial T} \right)_V \right]_{T_2} dV$$

$$+ \int_{V_4}^{V_1} \left[p - T \left(\frac{\partial p}{\partial T} \right)_V \right]_{T_1} dV. \qquad (4.78)$$

By the first law of thermodynamics,

$$\oint dE = 0$$

for the cycle, and we obtain

$$-W = Q_1 - Q_2.$$

Therefore, the efficiency is

$$\eta = \frac{-W}{Q_1} = 1 - \frac{Q_2}{Q_1},$$

which takes the same form as for the ideal working substance. We would like to show that, as is for the ideal gas, the efficiency for the real gas is still given by

$$\eta = 1 - \frac{T_2}{T_1}.$$

For this purpose, we first observe that for an isothermal process,

$$dQ = T \left(\frac{\partial p}{\partial T} \right)_V dV,$$

which follows from Eq. (4.75) if $dT = 0$. Therefore, the heat changes accompanying the isothermal expansion and compression in the cycle are

$$Q_1 = \int_1^2 dQ = \int_{V_1}^{V_2} dV \, T_1 \left(\frac{\partial p}{\partial T} \right)_V (T_1),$$

$$Q_2 = \int_4^3 dQ = \int_{V_4}^{V_3} dV \, T_2 \left(\frac{\partial p}{\partial T} \right)_V (T_2),$$

(4.79)

where we have used the sign convention on heat emission and absorption. Along the adiabatics

$$\frac{C_v}{T} dT + \left(\frac{\partial p}{\partial T} \right)_V dV = 0, \qquad (4.80)$$

which follows from Eq. (4.75) if $dQ = 0$. Moreover, since

$$dC_v(T, V) = \left(\frac{\partial C_v}{\partial V} \right)_T dV \text{ at } T = \text{constant},$$

we obtain

$$C_v(T, V) = C_v^0 + \int^V dV \left(\frac{\partial C_v}{\partial V} \right)_T$$

$$= C_v^0 + \int^V dV \, T \left(\frac{\partial^2 p}{\partial T^2} \right)_V$$

$$= C_v^0 + \int^V dV \, T \left(\frac{\partial^2 f}{\partial T^2} \right)_V (T = \text{constant}), \qquad (4.81)$$

where C_v^0 is generally a function of T, but we will consider the case of constant C_v^0. Substitution of Eq. (4.81) into Eq. (4.80) and use of the equation of state (4.73) yield

$$d\left[R\ln(T^\alpha V) + \int^V dV\left(\frac{\partial f}{\partial T}\right)_V\right] = 0,\tag{4.82}$$

where $\alpha = C_v/R$. Integrating it, we obtain

$$R\ln(T^\alpha V) + \int^V dV\left(\frac{\partial f}{\partial T}\right)_V = \text{constant.}\tag{4.83}$$

Since from the equation of state

$$\left(\frac{\partial p}{\partial T}\right)_V = \frac{R}{V} + \left(\frac{\partial f}{\partial T}\right)_V,$$

using Eqs. (4.79) and (4.83), we find

$$Q_1 = RT_1\ln\left(\frac{T_2}{T_1}\right)^\alpha,\quad Q_2 = RT_2\ln\left(\frac{T_2}{T_1}\right)^\alpha\tag{4.84}$$

and, consequently,

$$\frac{Q_2}{Q_1} = \frac{T_2}{T_1}.\tag{4.85}$$

Finally, we thus find

$$\eta = 1 - \frac{T_2}{T_1}\tag{4.86}$$

even if the working substance is real. This result also holds even if the assumption on constant C_v is removed. The following condition holds true:

$$\int^T \frac{C_v^0}{T}dT + \int^V dV\left(\frac{\partial p}{\partial T}\right)_V(T) = \text{constant.}\tag{4.87}$$

By the calculation performed above, we have proved that *the efficiency of a reversible Carnot cycle of a real fluid for the working substance obeying the equation of state (4.73) still depends only on the temperatures of the heat reservoirs between which the cycle operates regardless of the working material.*

4.6.   Examples of Other Cycles

Carnot's cycle was the precursor to other cycles invented later. Some examples will be shown in this section. They underlie some of the practical engines used in machines including automobiles. The Diesel cycle,

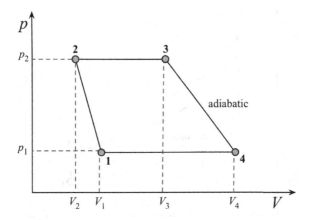

Fig. 4.8. Rankine cycle: step 2-3, isobaric expansion; step 3-4, adiabatic expansion; step 4-1, isobaric compression; step 1-2, adiabatic compression.

for example, is one typical example of the application of the underlying thermodynamic principles described.

4.6.1. Rankine Cycle

The Scottish engineer W. J. M. Rankine, a contemporary of Kelvin and an important pioneer in thermodynamics, invented the Rankine cycle, which basically describes the working of steam engines. It may be schematically represented by the diagram in the p–V plane in Fig. 4.8.

4.6.2. Otto Cycle

The German engineer N. Otto built an engine in 1876 invented by the Frenchman A. Beaude Rochas in 1862. Otto's engine became the prototype of automobile engines made in the United States. Gasoline engines may be represented by the Otto cycle which is described by the diagram in the p–V plane in Fig. 4.9.

4.6.3. Diesel Cycle

In about 1900, the German engineer R. Diesel built engines now known as Diesel engines. The Diesel engines may be described by the p–V diagram

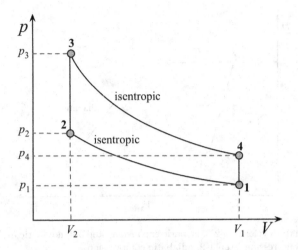

Fig. 4.9. Otto cycle: step 1-2, adiabatic compression (isentropic); step 2-3, constant volume heating; step 3-4, adiabatic expansion (isentropic); step 4-1, constant volume cooling.

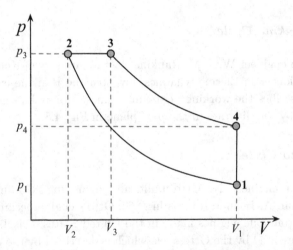

Fig. 4.10. Diesel cycle: step 1-2, adiabatic (isentropic) compression; step 2-3, isobaric expansion (fuel injection and burning at constant pressure); step 3-4, adiabatic (isentropic) expansion; step 4-1, constant volume heat rejection (exhaust and intake of air).

depicted in Fig. 4.10. It is called the Diesel cycle and is currently in use for various automobiles and power generators. The efficiency of the Diesel cycle is greater than the efficiency of an Otto cycle at the same peak pressure.

4.7. Calculation of Entropy Change

In this section, we evaluate the entropy changes accompanying some simple reversible processes. More general methods of evaluation will be discussed in later chapters.

4.7.1. *Phase Transition and Entropy Change*

It is well known that the so-called latent heats are associated with phase transitions, such as the melting of solids or vaporization of liquids and solids. Such heats are more specifically called as the heat of fusion (melting), the heat of vaporization, and so on. There are entropy changes associated with such phase transitions. They are called the entropy of fusion, the entropy of vaporization, and so on. They are easily evaluated from the latent heats. Let us recall that since

$$dQ = TdS,$$

and

$$H = E + pV,$$

if the process involved is reversible, combining them with the first law of thermodynamics, Eq. (3.9), yields

$$dH = TdS + Vdp \qquad (4.88)$$

for the enthalpy change. Using this equation at $p = $ constant, we can calculate the entropy change ΔS_{lat} corresponding to the latent heat ΔH_{lat} of a phase transition at constant p:

$$\Delta S_{\text{lat}} = \frac{\Delta H_{\text{lat}}}{T}, \qquad (4.89)$$

where T is the phase transition temperature. Note that ΔS_{lat} and ΔH_{lat} are discontinuous changes between two phases at T.

4.7.2. *Entropy Changes of an Ideal Gas*

For reversible processes, the entropy change may be written

$$dS = \frac{1}{T}(dE + pdV), \qquad (4.90)$$

which results from fundamental relation (3.9). Since in the case of an ideal gas, E is a function of T only, Eq. (4.90) may be cast in the form

$$dS = \frac{1}{T}c_v dT + RV^{-2}dV. \tag{4.91}$$

Here, we are considering 1 mole of an ideal gas and c_v denotes the molar specific heat at constant volume.

Suppose a reversible process takes the system from state (T_1, V_1) to state (T_2, V_2). The entropy change for this process may be evaluated as follows: The temperature of the system is increased from T_1 to T_2 at $V = V_1$ and then the volume is increased from V_1 to V_2 at $T = T_2$. This process is depicted in Fig. 4.11. The entropy change for the isochoric process $(T_1, V_1) \to (T_2, V_1)$ is

$$(\Delta S)_{\text{isochoric}} = \int_{T_1}^{T_2} dT \, \frac{c_v(T)}{T}$$

$$= c_v \ln \left(\frac{T_2}{T_1} \right), \tag{4.92}$$

where the second line holds if C_v is independent of T. The entropy change for the isothermal process $(T_2, V_1) \to (T_2, V_2)$ is

$$(\Delta S)_{\text{isothermal}} = R \int_{V_1}^{V_2} dV \, \frac{1}{V} = R \ln \left(\frac{V_2}{V_1} \right). \tag{4.93}$$

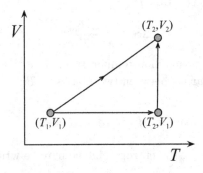

Fig. 4.11. A reversible path in T–V plane and its distortion into an isochoric and an isothermal path.

Combining Eqs. (4.92) and (4.93) yields the entropy change for the process $(T_1, V_1) \to (T_2, V_2)$:

$$\Delta S = (\Delta S)_{\text{isochoric}} + (\Delta S)_{\text{isothermal}}$$

$$= \int_{T_1}^{T_2} dT \, \frac{c_v(T)}{T} + R \ln \left(\frac{V_2}{V_1} \right)$$

$$= c_v \ln \left(\frac{T_2}{T_1} \right) + R \ln \left(\frac{V_2}{V_1} \right), \tag{4.94}$$

where the third line holds if c_v is independent of T. Since $c_p = c_v + R$ for a mole of an ideal gas, we may write Eq. (4.94) to obtain ΔS in the form

$$\Delta S = \int_{T_1}^{T_2} dT \frac{c_p(T)}{T} + R \ln \left(\frac{p_1}{p_2} \right)$$

$$= c_p \ln \left(\frac{T_2}{T_1} \right) + R \ln \left(\frac{p_1}{p_2} \right), \tag{4.95}$$

where the second line holds for the case of c_p independent of T.

Let us now remove the assumption of a constant specific heat, since the specific heats for ideal diatomic and polyatomic gases generally depend on temperature. The temperature dependence of c_v may be represented by the expansion

$$c_v(T) = \sum_{n=-\infty}^{\infty} b_n T^n, \tag{4.96}$$

where b_n are coefficients independent of T. Note that if $b_n = 0$ for $n < -2$, we recover an expansion similar to Eq. (3.39). When Eq. (4.96) is substituted into the first line of Eq. (4.94) and the integration is performed, there follows the isochoric component of ΔS :

$$(\Delta S)_{\text{isochoric}} = \sum_{\substack{n=-\infty \\ (n \neq 0)}}^{\infty} \left[\frac{b_n}{n} T^n + b_0 \ln T \right]_{T_1}^{T_2}.$$

By combining this with the isothermal component of ΔS, we obtain the overall entropy change for ideal diatomic or polyatomic gases.

4.8. Free Energies

4.8.1. *Helmholtz Free Energy*

The fundamental relations introduced earlier such as for dE and dH are differential forms in the space of variables (S, V) and (S, p), respectively, in mathematical terminology. It is often convenient to use other sets of variables. For example, the variable set (T, V) may be used because they are more convenient than, for example, the variable pair (S, V). To express the fundamental relation for the set (T, V), we introduce the Legendre transformation

$$A = E - TS. \tag{4.97}$$

Taking differential of this relation and using fundamental relation (4.43), we find the fundamental relation for Helmholtz free energy[20] A:

$$dA = -SdT - pdV. \tag{4.98}$$

This is clearly a differential form in the space of variables (T, V). This is a measure of energy available for useful work for the task of interest.

4.8.2. *Gibbs Free Energy*

It is possible to have a Legendre transformation of H as follows:

$$G = H - TS. \tag{4.99}$$

On substitution of H from this relation into the fundamental relation for H in Eq. (4.88), there follows the fundamental relation for Gibbs free energy:

$$dG = -SdT + Vdp, \tag{4.100}$$

which is a differential form in the space of variables (T, p). This variable set is the most convenient of the variable sets considered so far from the thermodynamic viewpoint, since temperature and pressure are easily amenable to measurement and controllable. This is also one of the reasons that the Gibbs free energy is one of the most convenient and commonly used among thermodynamic functions. It is also a measure of available energy for a given task.

---

[20]Sometimes, this Helmholtz free energy is simply called the free energy in contrast to the Gibbs free energy.

4.9. Maxwell's Relations

The various forms of energy introduced in the previous section are all state functions and their differential forms are exact. They are combined expressions for the first and second laws in the case of a closed system undergoing a reversible process. We will collect them in one place:

$$dE = TdS - pdV, \tag{4.101}$$

$$dH = TdS + Vdp, \tag{4.102}$$

$$dA = -SdT - pdV, \tag{4.103}$$

$$dG = -SdT + Vdp. \tag{4.104}$$

Since these are all exact differentials, they satisfy the exactness conditions discussed in the mathematical notes in Chapter 2.

Equation (4.101) enables us to compute two derivatives of E:

$$\left(\frac{\partial E}{\partial S}\right)_V = T, \quad \left(\frac{\partial E}{\partial V}\right)_S = -p. \tag{4.105}$$

Differentiating the derivatives once again with V and S, respectively, and remembering that

$$\frac{\partial^2 E}{\partial S \partial V} = \frac{\partial^2 E}{\partial V \partial S},$$

we find the relation

$$\left(\frac{\partial T}{\partial V}\right)_S = -\left(\frac{\partial p}{\partial S}\right)_V. \tag{4.106}$$

Performing similar calculations with Eqs. (4.102)–(4.104), we derive the following relations:

$$\left(\frac{\partial T}{\partial p}\right)_S = \left(\frac{\partial V}{\partial S}\right)_p, \tag{4.107}$$

$$\left(\frac{\partial S}{\partial V}\right)_T = \left(\frac{\partial p}{\partial T}\right)_V, \tag{4.108}$$

$$\left(\frac{\partial S}{\partial p}\right)_T = -\left(\frac{\partial V}{\partial T}\right)_p. \tag{4.109}$$

These derivatives involving T, S, p, and V are called Maxwell's relations. They are equivalent to the integrability conditions for the differential forms (4.101)–(4.104) and are extremely powerful and useful relations for various calculations in thermodynamics.

Mnemonics: It is very useful for computational purposes to devise a mnemonic scheme to construct the Maxwell relations, particularly, since their similarity creates considerable difficulty to remember them, yet one does not wish to compute them from the fundamental relations while encountering them. The following rules for constructing them are found to be effective.

First, take the quadruplet of S, T, p, and V of variables and group them into conjugate pairs, (S, T) and (p, V). Then the products of differentials are balanced in the form[21]

$$dSdT = (-1)^\alpha \, dpdV. \tag{4.110}$$

Note that the conjugate variable pairs (S, T) and (p, V) are each made up of an intensive and an extensive variable. The value of exponent α will be determined by the rules stated below — it is the essence of the mnemonics presented here.

The Maxwell relations are of partial derivatives which are formed with intensive and extensive variables, one each, from the two conjugate variable pairs of the quadruplet. For example, to form relation (4.109), we divide the equation with $dpdT$ in which p and T are the variables used to differentiate S on the left and V on the right. Thus, we obtain

$$\left(\frac{\partial S}{\partial p} \right)_l = (-1)^\alpha \left(\frac{\partial V}{\partial T} \right)_r.$$

---

[21] The mathematical basis is as follows: Under the transformation $(S, T) \to (p, V)$,

$$dSdT = \frac{\partial (S, T)}{\partial (p, V)} dpdV,$$

and

$$\frac{\partial (S, T)}{\partial (p, V)} = \pm 1.$$

The Jacobian of transformation is usually taken as 1, but here it is relaxed to take -1 as well.

Then the value of l on the left is $l = T$, the conjugate variable of S of the pair (S, T), whereas the value of r on the right is $r = p$, the conjugate variable of the pair (p, V). Thus, we obtain

$$\left(\frac{\partial S}{\partial p}\right)_T = (-1)^\alpha \left(\frac{\partial V}{\partial T}\right)_p.$$

The value of α is determined by the following rule:

$\alpha = 1$ if both intensive variables (e.g., p and T) occur on both sides of the equation,

$\alpha = 0$ if both intensive variables occur on one side only, left or right, of the equation.

Since $\alpha = 1$ for the present example, it follows that

$$\left(\frac{\partial S}{\partial p}\right)_T = -\left(\frac{\partial V}{\partial T}\right)_p,$$

which is relation (4.109).

Using these rules, other Maxwell relations can be easily constructed. For example, by dividing Eq. (4.110) with $dV\,dT$, we find the relation

$$\left(\frac{\partial S}{\partial V}\right)_l = (-1)^\alpha \left(\frac{\partial p}{\partial T}\right)_r,$$

where by the rules stated earlier, $l = T$, $r = V$, and $\alpha = 0$. Hence,

$$\left(\frac{\partial S}{\partial V}\right)_T = \left(\frac{\partial p}{\partial T}\right)_V,$$

which is relation (4.108). If we divide Eq. (4.110) with the product $dS\,dp$ and use the rules stated, we find

$$\left(\frac{\partial T}{\partial p}\right)_S = \left(\frac{\partial V}{\partial S}\right)_p,$$

and if we divide with $dV\,dS$, there follows relation

$$\left(\frac{\partial T}{\partial V}\right)_S = -\left(\frac{\partial p}{\partial S}\right)_V,$$

which is relation (4.106).

We have earlier utilized calorimetric data to compute the internal energy and the enthalpy as well as the entropy. In order to reinforce the method of computation, we observe the following relations:

$$C_v = \left(\frac{\partial E}{\partial T}\right)_V = T\left(\frac{\partial S}{\partial T}\right)_V, \tag{4.111}$$

$$C_p = \left(\frac{\partial H}{\partial T}\right)_p = T\left(\frac{\partial S}{\partial T}\right)_p. \tag{4.112}$$

Note in these relations that the entropy has the same dimension as the specific heats.

We illustrate applications of the Maxwell relations with a few examples in the following.

Example 1. $(\partial T/\partial p)_E$ may be expressed in terms of more easily measurable quantities. By the chain relation introduced in the mathematical notes in Chapter 3,

$$\left(\frac{\partial T}{\partial p}\right)_E = -\frac{\left(\frac{\partial E}{\partial p}\right)_T}{\left(\frac{\partial E}{\partial T}\right)_p}.$$

Since

$$\left(\frac{\partial E}{\partial T}\right)_p = \left[\frac{\partial}{\partial T}(H - pV)\right]_p$$

$$= C_p - p\left(\frac{\partial V}{\partial T}\right)_p, \tag{4.113}$$

$$\left(\frac{\partial E}{\partial p}\right)_T = \left[\frac{\partial}{\partial p}(G - pV + TS)\right]_T$$

$$= -p\left(\frac{\partial V}{\partial p}\right)_T - T\left(\frac{\partial S}{\partial p}\right)_T$$

$$= -p\left(\frac{\partial V}{\partial p}\right)_T + T\left(\frac{\partial V}{\partial T}\right)_p, \tag{4.114}$$

by putting these equations together, we find

$$\left(\frac{\partial T}{\partial p}\right)_E = -\frac{V\left(p\kappa - T\alpha\right)}{C_p - pV\alpha},\tag{4.115}$$

where α and κ are, respectively, the isobaric expansion coefficient and the isothermal compressibility:

$$\alpha = \frac{1}{V}\left(\frac{\partial V}{\partial T}\right)_p,$$

$$\kappa = -\frac{1}{V}\left(\frac{\partial V}{\partial p}\right)_T.\tag{4.116}$$

Example 2. We show that for ideal gases,

$$\left(\frac{\partial E}{\partial V}\right)_T = 0, \qquad \left(\frac{\partial E}{\partial p}\right)_T = 0.$$

Proof. (a)

$$\left(\frac{\partial E}{\partial V}\right)_T = \left[\frac{\partial}{\partial V}(A + TS)\right]_T$$

$$= -p + T\left(\frac{\partial S}{\partial V}\right)_T$$

$$= -p + T\left(\frac{\partial p}{\partial T}\right)_V.$$

Since for ideal gases

$$\left(\frac{\partial p}{\partial T}\right)_V = \frac{R}{V},$$

it follows that

$$\left(\frac{\partial E}{\partial V}\right)_T = 0.$$

(b)

$$\left(\frac{\partial E}{\partial p}\right)_T = \left[\frac{\partial}{\partial p}(G - pV + TS)\right]_T$$

$$= -p\left(\frac{\partial V}{\partial p}\right)_T - T\left(\frac{\partial V}{\partial T}\right)_p.$$

For ideal gases, the right-hand side is equal to zero and consequently,

$$\left(\frac{\partial E}{\partial p}\right)_T = 0.$$

Example 3. We show that for any substance,

$$C_p = C_v + \frac{TV\alpha^2}{\kappa}.$$

To show it, the following calculation is performed:

$$C_p - C_v = T\left[\left(\frac{\partial S}{\partial T}\right)_p - \left(\frac{\partial S}{\partial T}\right)_V\right] = T\left(\frac{\partial S}{\partial V}\right)_T\left(\frac{\partial V}{\partial T}\right)_p$$

$$= T\left(\frac{\partial p}{\partial T}\right)_V\left(\frac{\partial V}{\partial T}\right)_p = -\frac{T\left(\frac{\partial V}{\partial T}\right)_p^2}{\left(\frac{\partial V}{\partial p}\right)_T}$$

$$= \frac{TV\alpha^2}{\kappa}.$$

Problems

(1) Show that the efficiency of the Carnot cycle of a mole of a gas obeying the equation of state

$$p(V - b) = RT$$

is given by

$$\eta = 1 - \frac{T_2}{T_1}.$$

Assume that the specific heats of the gas are constants.

(2) Show that the efficiency of the Carnot cycle of a mole of a van der Waals gas obeying the equation of state

$$\left(p + \frac{a}{V^2}\right)(V - b) = RT$$

is given by

$$\eta = 1 - \frac{T_2}{T_1}.$$

Assume that the specific heats of the gas are constants.

(3) By using Eq. (4.87), show that Eq. (4.86) still holds true.

(4) Prove the following relations:

(a) $\left(\dfrac{\partial T}{\partial V}\right)_S = -\dfrac{T}{C_v}\left(\dfrac{\partial p}{\partial T}\right)_V$,

(b) $\left(\dfrac{\partial T}{\partial p}\right)_S = \dfrac{T}{C_p}\left(\dfrac{\partial V}{\partial T}\right)_p$,

(c) $\left(\dfrac{\partial V}{\partial p}\right)_S = \dfrac{C_v}{C_p}\left(\dfrac{\partial V}{\partial p}\right)_T$,

(d) $\left(\dfrac{\partial T}{\partial p}\right)_H = \dfrac{-V + T\left(\frac{\partial V}{\partial T}\right)_p}{C_p}$,

(e) $\left(\dfrac{\partial T}{\partial V}\right)_E = \dfrac{p - T\left(\frac{\partial p}{\partial T}\right)_V}{C_v}$,

(f) $\left(\dfrac{\partial H}{\partial G}\right)_E = \dfrac{C_p(\kappa p - \alpha T) - (1 - \alpha T)(C_p - \alpha p V)}{-S(\kappa p - \alpha T) - (C_p - \alpha p V)}$.

(5) Show the following:

(a) $\left(\dfrac{\partial E}{\partial S}\right)_p = T - p\left(\dfrac{\partial T}{\partial p}\right)_S = T\left[1 - \dfrac{p}{C_p}\left(\dfrac{\partial V}{\partial T}\right)_p\right]$,

(b) $\left(\dfrac{\partial G}{\partial T}\right)_V = -S + V\left(\dfrac{\partial p}{\partial T}\right)_V = -S + \dfrac{V\alpha}{\kappa}$,

where α and κ are defined in Example 1.

(6) The following change is maintained isobaric and isothermal:

$$C_6H_6(l) + \tfrac{7}{2}O_2(g) \rightarrow 6CO_2(g) + 3H_2O(l).$$

It is found that $\Delta G = -3174.9\,\text{kJ}\,\text{mol}^{-1}$. Find ΔA at the same condition at $p = 1\,\text{atm}$ and $T = 298.2\,\text{K}$.

(7) Show that dQ/T is an exact differential by using the fundamental relation for E. Also show that for a Carnot cycle working with an ideal gas, its cyclic integral vanishes.

(8) An extended strip of rubber has a length l when subjected to a tensile force f. If the volume change on extension may be neglected,

show that

$$\left(\frac{\partial E}{\partial l}\right)_T = f - T\left(\frac{\partial f}{\partial T}\right)_l .$$

Show that the small temperature rise, ΔT, which takes place in a slow adiabatic stretching, is given by

$$\frac{\Delta T}{T} = \int_{l_0}^{l} \frac{1}{C_l}\left(\frac{\partial f}{\partial T}\right)_l dl,$$

where l_0 and l refer to the initial and final lengths, respectively, and C_l is the heat capacity of the rubber at constant extension. [Note that the work is equal to $f\,dl$ so that $dE = T\,dS + f\,dl$. For the second part of the problem, calculate $(\partial T/\partial l)_S$ in terms of C_l and $(\partial f/\partial T)_l$. New "Maxwell's relations" are necessary for the present situation.]

(9) Between 0 and 60 K, the molar heat capacity of Ag(s) is given approximately by the following expression:

$$c_p = \left(0.23\,T + 2.5 \times 10^{-3}\,T^2 - 1.9 \times 10^{-5}\,T^3\right) \text{ J mol}^{-1}.$$

Calculate the entropy of Ag(s) at 60 K. Note $S(T = 0) = 0$.

(10) Find ΔA for O_2 $(g,\ 300.15\ \text{K},\ 2\,\text{atm}) \to O_2$ $(g,\ 300.15\ \text{K},\ 1\,\text{atm})$, assuming that the gas is ideal.

(11) In the temperature range of 298–2000 K, the molar heat capacity of NH_3 (g) obeys the equation,

$$c_p = \left(29.75 + 2.51 \times 10^{-2}\,T - 1.55 \times 10^{5}\,T^{-2}\right) \text{ J K}^{-1}\,\text{mol}^{-1}.$$

Find the entropy change at constant pressure arising from the temperature change from $T = 300$ to 1000 K.

(12) At $0°C$, aluminum has the following properties: atomic weight $= 2.70 \times 10^{-2}$ kg mol^{-1}, density $= 2.70$ kg M^{-3}, $c_p = 0.930 \times 10^3$ J kg^{-1}K^{-1}, $\alpha = 71.4 \times 10^{-6}$ K^{-1}, $\kappa = 1.34 \times 10^{-21}$ m^{2}N^{-1}. Calculate at $0°C$ the molar heat capacity at constant volume and the ratio of heat capacities γ, i.e., the polytropic ratio.

(13) Find the entropy change for the reversible process of taking n moles of an ideal gas with constant heat capacity from T_1 and p_1 to T_2 and p_2.

(14) A gas obeys the following molar heat capacity equation and equation of state:

$$c_p = a + bT + cT^2,$$

and

$$pv = RT + B_2(T)p + B_3(T)p^2.$$

Calculate the enthalpy and entropy changes for the reversible process from (T_1, p_1) to (T, p).

(15) Show Eq. (4.72).

(16) Show that for the Carnot cycle using an ideal gas as the working substance

$$\oint dE = 0, \quad \oint dH = 0, \quad \oint dA = 0, \quad \oint dG = 0, \quad \oint dS = 0.$$

Assume constant specific heats.

(17) Calculate the efficiency of the Rankine cycle and show

$$\oint dS = 0$$

for the cycle.

(18) Obtain the efficiency of the Otto cycle. Compare the efficiency of this cycle with the efficiency of the Carnot cycle. Also show that for the cycle

$$\oint dS = 0.$$

(19) Calculate the efficiency of the Diesel cycle and compare it with the efficiency of the Carnot cycle. Show that the efficiency of the Diesel cycle is greater than the efficiency of an Otto cycle at the same peak pressure. Also show that for the cycle

$$\oint dS = 0.$$

(20) The van der Waals constants for nitrogen are as follows:

$$a = 1.39 \text{ L}^2 \text{ atm mol}^{-2}, \quad b = 3.92 \times 10^{-2} \text{ L mol}^{-1}.$$

Taking the virial form for the van der Waals equation of state

$$\frac{pv}{RT} = 1 + \left(b - \frac{a}{RT} \right) \frac{1}{V} + \cdots,$$

calculate the enthalpy change $\Delta h = h(T, p) - h(T, 0)$ arising from compression of the gas from $p = 0$ to $p = 100$ atm at constant $T = 300$ K.

(21) By using the van der Waals equation of state for real gas, calculate the reversible work done on the system due to a volume change from V_1 to V_2 at temperature T.

(22) Mercury vapor at 630.15 K and 1 atm pressure is heated to 823.15 K and its pressure is increased from 1 atm to 5 atm. Calculate the entropy change in the entropy units, the vapor being treated as an ideal monatomic gas.

(23) Calculate the nonideality corrections for enthalpy, internal energy and entropy by using the van der Waals equation of state.

(24) Calculate the molar heat capacity difference $c_p - c_v$ for nitrogen at 298.15 K and 200 atm pressure, to first order in p, using the van der Waals equation of state, the van der Waals constants a and b being 1.39 L^2 atm mol^{-2} and 3.92×10^{-2} L mol^{-1}.

(25) Calculate for nitrogen the change of molar heat capacity c_p when the pressure is increased from 1 atm to 100 atm at 298.15 K using the van der Waals equation of state. The constants are given in Problem 21.

(26) The molar heat capacity of solid iodine at temperature between 298.15 K to the melting point (386.75 K) at 1 atm pressure is given by the relation

$$c_p = \left[54.64 + 1.34 \times 10^{-3}(T - 298.15)\right] \text{ J K}^{-1} \text{ mol}^{-1}.$$

Determine the increase of entropy accompanying the change of 1 mole of iodine from 298.15 K to 386.75 K.

Chapter 5

Equilibrium Conditions and Thermodynamic Stability

We have seen in Chapter 4 that the second law of thermodynamics supplies various fundamental equations involving thermodynamic functions associated with the entropy for reversible processes or when the system is in equilibrium. These fundamental equations provide means for us to analyze thermophysical data and correlate them to each other. In addition to this aspect, the laws of thermodynamics also enable us to understand under what conditions the systems reach equilibrium and how they behave in the neighborhood of the equilibrium state. This aspect owes to the fact that, associated with the second law of thermodynamics, there are inequalities descending from the Clausius inequality, which may be used to study the stability of the thermodynamic state of the system away from equilibrium. Together with the fundamental equations, the Clausius inequality and related inequalities enable us to study and comprehend under what conditions the systems reach equilibrium and how they behave in the neighborhood of the equilibrium state. This aspect of thermodynamics is richer than the fundamental equations provided by themselves. The former, in fact, describe the system at equilibrium condition, whereas the latter provide conditions for the thermodynamic stability of the system at equilibrium. This aspect of thermodynamics is physically richer than the former.

5.1. Inequalities

Since the second law of thermodynamics gives rise to inequalities involving (compensated) heat and entropy (or calortropy) as we have seen, we make use of them to examine the behavior of the systems in the neighborhood of equilibrium. From the second law of thermodynamics, we have been able to deduce the differential form for the calortropy

$$d\Psi = \frac{dQ}{T} + d\text{N}. \tag{5.1}$$

Owing to the fact that $d\text{N} \geq 0$, for an infinitesimal process, we deduce the inequality

$$d\Psi \geq \frac{dQ}{T}. \tag{5.2}$$

Since the calortropy Ψ reaches a maximum as the system tends toward equilibrium, if we imagine a virtual variation of the state from equilibrium that arises from variation in heat δQ, then since $\max(\Psi) = S$ at equilibrium, it follows

$$\delta\Psi \leq \frac{\delta Q}{T}. \tag{5.3}$$

This can be made the starting point of a thermodynamic stability theory. This, however, is in contrast to the equilibrium thermodynamics practised to this day, since the conventional thermodynamic stability theory is formulated in terms of the entropy obeying the Clausius–Duhem inequality, Eq. (4.59). In the traditional approach, we, in fact, begin with the observation that the entropy is maximum at equilibrium, that is, the entropy always increases toward its equilibrium value if the system is not in equilibrium. Therefore, a variation of entropy from equilibrium

$$\delta S = S - S_{\text{eq}} \tag{5.4}$$

must be such that it satisfies the inequality

$$\delta S \leq \frac{\delta Q}{T} \tag{5.5}$$

for all arbitrary virtual variations if the system is in equilibrium. This reasoning is troublesome and illogical because the entropy S is not defined for irreversible processes, but only for reversible processes or for a system at equilibrium only. After all, if the system is varied from the state of equilibrium by any means, it must be in a nonequilibrium state characterized by the calortropy. Such a difficulty arises when the calortropy is confused with the entropy as discussed in Chapter 4. The troublesome feature vanishes if the calortropy is employed instead of S. Here, we would like to add a point not discussed in Chapter 4.

The change in calortropy accompanying an irreversible process from a state to another in a system should obey the inequality

$$\Delta\Psi \geq \int_{(irr)} \frac{dQ}{T}, \tag{5.6}$$

where the equality holds at equilibrium. If we construct a cycle that returns the system from the final state to the initial state through a reversible process in interaction with the surroundings (see Fig. 4.6), inequality (5.6) can be expressed in terms of the entropy change complementary to the irreversible process in question and, as shown in Chapter 4, we obtain the Clausius–Duhem inequality

$$\Delta S \geq \int_{(irr)} \frac{dQ}{T}. \tag{5.7}$$

This deduction for ΔS, however, is for a finite process only, not for an infinitesimal process for which one might infer the inequality

$$dS \geq dQ/T, \tag{5.8}$$

which in turn would imply inequality (5.5). We now would like to show that while inequality (5.2) is a natural consequence of the second law of thermodynamics expressed in terms of the calortropy differential,

$$d\Psi = \frac{dQ}{T} + dN, \tag{5.9}$$

inequality (5.7) does not imply inequality (5.5). This can be seen as shown below.

In the case of a cycle depicted in Fig. 4.6, it is possible to write

$$\Delta S = \int_{(rev)} dS = \int_{(irr)} \frac{dQ}{T}. \tag{5.10}$$

However, since the line integral along the reversible process in Eq. (5.10) is not the same as the line integral along the irreversible process on the right-hand side of inequality (5.7), an inequality $dS \geq dQ/T$ cannot be inferred from the integral inequality (5.7). Put in another way, inequality (5.7) can be written as

$$\int_{(rev)} dS - \int_{(irr)} \frac{dQ}{T} \geq 0, \tag{5.11}$$

but for this to be written as a differential form holding everywhere in the reversible or irreversible segment of the cycle, it is necessary to cast it either in the form

$$\int_{(\text{rev})} \left(dS - \frac{\text{d}Q}{T} \right) \geq 0, \tag{5.12}$$

or in the form

$$\int_{(\text{irr})} \left(dS - \frac{\text{d}Q}{T} \right) \geq 0, \tag{5.13}$$

so that, under the continuity assumption, there follows the conclusion that the integrand is positive everywhere along the path of integration, that is, the inequality

$$dS \geq \frac{\text{d}Q}{T}.$$

From this, we may infer the inequality for δS given in inequality (5.5). However, it is not possible to obtain inequalities (5.12) and (5.13) from inequality (5.11) since

$$\int_{(\text{irr})} \frac{\text{d}Q}{T} \neq \int_{(\text{rev})} \frac{\text{d}Q}{T}. \tag{5.14}$$

Thus, we see that inequality (5.5) is not necessarily true for an infinitesimal process. It should be replaced by inequality (5.3) if we wish to examine the question of thermodynamic stability of a system near equilibrium. In retrospect, this conclusion should come as natural because the calortropy is a nonequilibrium extension of entropy defined for equilibrium only.

5.2. Equilibrium Conditions

Thermodynamic equilibrium is a singular state in the manifold of nonequilibrium thermophysical states of a thermodynamic system — the thermodynamic manifold, and the thermodynamic second law shows that this singular state is marked by inequalities, such as the Clausius inequality or inequality (5.3). The thermodynamic inequality (5.3) is expected to give rise to equilibrium conditions.

Inequality (5.3) and the identity of $\Psi_e := \Psi(\text{equil.})$ with the Clausius entropy S — namely, the entropy of a reversible process — at equilibrium suggests that as the parameters $(x) := (x_1, x_2, \ldots)$ characterizing

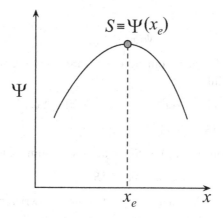

Fig. 5.1. The maximum value of the calortropy is S. Here, x is a property characterizing the state of the system which is not in equilibrium. As x is varied and the system has reached equilibrium, the calortropy acquires a maximum value and the corresponding value of x is x_e, the equilibrium value of x, where $S = \Psi(x_e)$.

the state of the system are varied toward those, (x_e), corresponding to equilibrium, the calortropy increases to a maximum where it becomes identical with S. This behavior is schematically depicted in Fig. 5.1. We take advantage of this fact to formulate a thermodynamic stability theory.

Let $\{x\}$ denote a set of macroscopic variables by which the thermodynamic state of the system is described. This set will be called the thermodynamic variable manifold. For example, the set $\{x\}$ may be considered to be the variables $(E, v, c_i, \Phi_{qi} : r \leq i \leq 1; q \geq 1)$ where c_i denotes the mass fraction of species i defined by the ratio of the mass density ρ_i of species i to the total mass density ρ, and Φ_{qi} the variables characterizing the nonequilibrium state of the substance, such as the stress, heat flux, diffusion fluxes, and so on. The value of the set $\{x\}$ at equilibrium is denoted by $\{x_e\}$ at which Ψ acquires a maximum. In the case of the example taken for $\{x\}$, the set $\{x_e\}$ simply consists of $(E, v, c_i, : r \leq i \leq 1)$ because $\Phi_{qi} = 0$ at equilibrium and $\{x_e\}$ obey the equilibrium Gibbs relation. The calortropy Ψ is defined in the space of the full set $\{x\}$ of variables, whereas the (Clausius) entropy S lives in the space of the subset $\{x_e\}$ that corresponds to the equilibrium manifold of states, for example, $(E, v, c_i : r \leq i \leq 1)$. With this understanding, we may simply write $\Psi(x)$ and $S(x)$ to indicate their dependence on the thermodynamic variables. The results of the discussions in Chapter 4 and the previous paragraph suggest that the following inequality holds[1]:

$$\Psi(x) \leq S(x). \tag{5.15}$$

The system is now varied from its equilibrium state. For an infinitesimal variation from equilibrium,

$$\delta\Psi = \Psi(x) - \Psi(x_e) = \Psi(x) - S(x_e),$$

$$\delta S = S(x) - S(x_e),$$

and furthermore, since $\Psi(x_e) = S(x_e)$ at equilibrium, it follows that

$$\delta\Psi \leq \delta S. \tag{5.16}$$

Since $\delta S = \delta Q_{\text{rev}}/T$, the inequality (5.16) may be written as

$$\delta\Psi \leq \frac{\delta Q_{\text{rev}}}{T} \tag{5.17}$$

if a variation from equilibrium is performed on the system at equilibrium. Since the first law may be written in the case of a reversible process as

$$\delta E = \delta Q_{\text{rev}} + \delta W, \tag{5.18}$$

where δW stands for the reversible work, we may write Eq. (5.17) in the form

$$\delta E + p\delta V - T\delta\Psi \geq \delta W_n, \tag{5.19}$$

where

$$\delta W_n = \delta W + p\delta V, \tag{5.20}$$

which denotes works other than the pressure–volume work. In the expression for the first law of thermodynamics, we have replaced the derivative symbol d with the variation symbol δ. It must be noted that δQ_{rev} appears instead of δQ as in inequality (5.2) since the entropy S is defined for a reversible process only. This also means that δW should be a reversible work that does not include the internal work associated with an irreversible process (see Appendix C for the meaning of dissipative internal work). Therefore, if a closed system at a constant energy and volume is to be in equilibrium, it must satisfy the equilibrium condition

$$(\delta\Psi)_{E,V} \leq -\delta W_n. \tag{5.21}$$

---

[1] The difference $S(x) - \Psi(x) := S_{\text{rel}}$ is called the relative Boltzmann entropy in kinetic theory of fluids. See B. C. Eu, *J. Chem. Phys.* **106**, 2388 (1997) and B. C. Eu, *Nonequilibrium Statistical Mechanics* (Kluwer, Dordrecht, 1998).

If there is no internal work and the pressure–volume work is the only work, then $\delta W_n = 0$, and the equilibrium condition (5.19) becomes

$$\delta E + p\delta V - T\delta\Psi \geq 0. \tag{5.22}$$

It is possible to deduce that

$$(\delta\Psi)_{E,V} \leq 0. \tag{5.23}$$

Henceforth, we will confine our discussion to the case of $\delta W_n = 0$. For a closed system at constant energy and volume, the equilibrium condition becomes

$$(\delta E)_{\Psi,V} \geq 0. \tag{5.24}$$

This implies that the internal energy is at minimum at equilibrium if Ψ and V are kept constant. This is schematically shown in Fig. 5.2. If T and V are kept constant, condition (5.22) becomes

$$(\delta E)_{T,V} - T(\delta\Psi)_{T,V} \geq 0. \tag{5.25}$$

It must be emphasized that the calortropy replaces the entropy in the aforementioned inequalities in contrast to the traditional formulation of thermodynamic stability theory, where the entropy S is used. We have earlier pointed out why the calortropy Ψ should replace the Clausius entropy S in the inequalities mentioned.

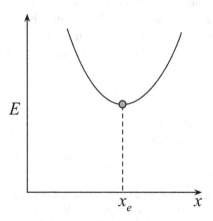

Fig. 5.2. The internal energy near equilibrium. The behavior of E near $x = x_e$ is deducible from inequality (5.17) or (5.19).

Since at equilibrium $\Psi_e = S$, the inequalities mentioned become equalities. For example, we obtain from inequality (5.22) the equation

$$\delta E = T\delta S - p\delta V. \qquad (5.26)$$

This relation suggests that E may be regarded as a function of S and V at equilibrium and in general as a function of Ψ and V.

It is often useful to work with thermodynamic functions which are functions of variables other than Ψ. To find such a function, we introduce a Legendre transformation

$$A = E - T\Psi. \qquad (5.27)$$

The new thermodynamic function A is called the Helmholtz free energy or the work function.[2] Performing a variation of A and using Eq. (5.25) yields the inequality

$$\delta A + p\delta V + \Psi\delta T \geq 0, \qquad (5.28)$$

and it yields the equilibrium condition at constant temperature and volume

$$(\delta A)_{T,V} \geq 0. \qquad (5.29)$$

In other words, the Helmholtz free energy is minimum at equilibrium if T and V are kept constant (see Fig. 5.3). At equilibrium where nonequilibrium fluxes vanish, the calortropy becomes equal to the entropy and the inequality becomes the equality holding for δA

$$\delta A = - S\delta T - p\delta V, \qquad (5.30)$$

which implies that A at equilibrium is a function of T and V. However, the variables Ψ, E, and A considered so far in this chapter are not the most convenient variables for laboratory experimental purposes. We may introduce a more suitable function by taking another Legendre transformation as follows:

$$G = A + pV = H - T\Psi. \qquad (5.31)$$

This new function is called the Gibbs free energy. By taking a variation of G and making use of Eq. (5.28), we find

$$\delta G \geq V\delta p - \Psi\delta T \qquad (5.32)$$

---

[2]Although the same terminology is used as for the equilibrium case, it must be noted that the Helmholtz free energy defined in terms of Ψ is a generalization of the equilibrium Helmholtz free energy $A = E - TS$. It may be called nonequilibrium work function or nonequilibrium Helmholtz free energy. As the system approaches the equilibrium state, it tends to the equilibrium Helmholtz free energy or work function.

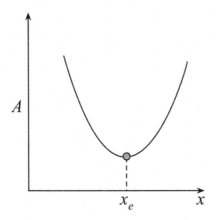

Fig. 5.3. The Helmholtz free energy near equilibrium.

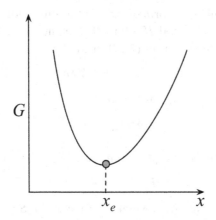

Fig. 5.4. The Gibbs free energy near equilibrium.

as another form of the equilibrium condition. At constant T and p,

$$(\delta G)_{T,p} \geq 0. \tag{5.33}$$

That is, the Gibbs free energy G has a minimum at $x = x_e$ and at equilibrium the following form holds:

$$\delta G = -S\delta T + V\delta p, \tag{5.34}$$

which means that G at equilibrium is a function of T and p. If p and T are kept constant, $\delta G = 0$ at the minimum at equilibrium as shown in Fig. 5.4.

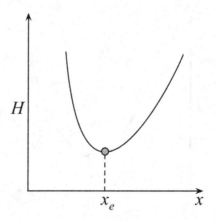

Fig. 5.5.   The enthalpy near equilibrium.

Still another equivalent form of the equilibrium condition can be derived in terms of enthalpy denoted H. With the definition (3.12) for H, we may write the equilibrium condition (5.22) in the form

$$\delta H \geq T\delta\Psi + V\delta p. \tag{5.35}$$

If Ψ and p are kept constant, it becomes

$$(\delta H)_{p,\Psi} \geq 0. \tag{5.36}$$

Since the equality holds at equilibrium and $\Psi_e = S$, there follows

$$\delta H = T\delta S + V\delta p, \tag{5.37}$$

which implies that H at equilibrium is a function of S and p (see Fig. 5.5). Equilibrium thermodynamics is based on equivalent fundamental relations (5.26), (5.30), (5.34), and (5.37) whose power and utility have already been demonstrated in the Chapter 4.

5.3.   Stability of Equilibrium

The equilibrium condition was determined by the vanishing first variation of the calortropy from equilibrium

$$(\delta\Psi)_{E,V} = 0,$$

but it does not say whether equilibrium is stable or not. The stability of the equilibrium state requires investigation of higher order variations of Ψ

or equivalent thermodynamic potentials, for example, the second variation, $\delta^2\Psi$. If the system at an equilibrium state is displaced to an arbitrary state, then the calortropy is changed by $\Delta\Psi$, and if the equilibrium is stable, then

$$\Delta\Psi < 0. \tag{5.38}$$

If the equilibrium is of neutral stability, then

$$\Delta\Psi = 0, \tag{5.39}$$

and if it is unstable, then

$$\Delta\Psi > 0. \tag{5.40}$$

These criteria can be put in the context of virtual variation since the calortropy change $\Delta\Psi$ may be expressed in a sum of variations of different orders:

$$\Delta\Psi = \delta\Psi + \delta^2\Psi + \delta^3\Psi + \cdots . \tag{5.41}$$

Here, $\delta\Psi$ is called the first variation, $\delta^2\Psi$ the second variation, $\delta^3\Psi$ the third variation, etc. If the complete set of thermodynamic variables is denoted by $\{x_i : i \geq 1\}$, then various variations are expressible in terms of derivatives of corresponding order:

$$\delta\Psi = \sum_i \frac{\partial\Psi}{\partial x_i}\delta x_i, \tag{5.42}$$

$$\delta^2\Psi = \frac{1}{2}\sum_{i,j} \frac{\partial^2\Psi}{\partial x_i\partial x_j}\delta x_i\delta x_j, \tag{5.43}$$

$$\delta^3\Psi = \frac{1}{3!}\sum_{i,j,k} \frac{\partial^3\Psi}{\partial x_i\partial x_j\partial x_k}\delta x_i\delta x_j\delta x_k, \tag{5.44}$$

and so on. Here, we are interested in continuous variation in $\{x_i\}$. If $\Delta\Psi < 0$ is to be satisfied, the first nonvanishing variation in Ψ must be of an even order for the same variation from a stable equilibrium state. Therefore, we obtain from conditions (5.38)–(5.40) the criteria

$$\delta^2\Psi < 0 \quad \text{for stable equilibrium,} \tag{5.45}$$

$$\delta^2\Psi = 0 \quad \text{for neutral stability,} \tag{5.46}$$

$$\delta^2\Psi > 0 \quad \text{for unstable equilibrium.} \tag{5.47}$$

For further discussion on the stability of thermodynamic systems on the basis of these criteria, the reader is referred to the literature.[3]

Since we are only interested in the stability of the system in equilibrium submanifold, the nonequilibrium variables $\{\Phi_q : l < q \leq n\}$ in the manifold $\{x_i : 1 \leq i \leq n\}$ may be set as equal to zero. We may then consider Ψ in the equilibrium submanifold $\{y_j : 1 \leq j \leq l\}$ of the manifold $\{x_i : 1 \leq i \leq n\}$, where $l < n$. In this case, the calortropy Ψ and its derivatives are evaluated in the submanifold $\{y_j\}$.

To make discussion as simple as possible, let us confine the discussion to a single-component fluid. When explicitly written out, $\delta^2\Psi$ is given by

$$\delta^2\Psi = \frac{1}{2}\left[\frac{\partial^2\Psi}{\partial E^2}(\delta E)^2 + \left(\frac{\partial^2\Psi}{\partial E\partial V} + \frac{\partial^2\Psi}{\partial V\partial E}\right)\delta E\delta V\right.$$
$$\left. + \frac{\partial^2\Psi}{\partial V^2}(\delta V)^2 + \cdots\right], \tag{5.48}$$

where the ellipsis represents the second derivatives of Ψ also with respect to the nonequilibrium fluxes. This expression is evaluated at the equilibrium submanifold

$$\{y_j\} = \{x_1, \ldots, x_l; x_i = 0, l < i \leq n\},$$

in which the nonequilibrium fluxes x_i $(l < i < n)$ are set equal to zero:

$$\delta^2\Psi|_{\{x_j\}=\{y_j\}} = \frac{1}{2}\left[\frac{\partial^2\Psi}{\partial E^2}(\delta E)^2 + \left(\frac{\partial^2\Psi}{\partial E\partial V} + \frac{\partial^2\Psi}{\partial V\partial E}\right)\delta E\delta V\right.$$
$$\left. + \frac{\partial^2\Psi}{\partial V^2}(\delta V)^2 + \cdots\right]_{\{x_j\}=\{y_j\}}. \tag{5.49}$$

It is convenient to remove the cross-terms in this equation. First, by using the extended Gibbs relation, Eq. (5.48) may be cast in the form

$$\delta^2\Psi|_{\{x_j\}=\{y_j\}} = \frac{1}{2}\left[\delta E\delta\left(\frac{1}{T}\right) + \delta V\delta\left(\frac{p}{T}\right)\right]_{\{x_j\}=\{y_j\}}, \tag{5.50}$$

[3]B. C. Eu, *Generalized Thermodynamics: The Thermodynamics of Irreversible Processes and Generalized Hydrodynamics* (Kluwer, Dordrecht, 2002), pp. 73–78.

which then can be put in the form

$$\delta^2 \Psi|_{\{x_j\}=\{y_j\}} = \frac{1}{2}\left[-\frac{C_v}{T^2}(\delta T)^2 - \frac{1}{T\kappa V}(\delta V)^2\right], \tag{5.51}$$

where

$$C_v = \left(\frac{\partial E}{\partial T}\right)_{V,\{\Phi_q\}}\bigg|_{\{\Phi_q\}=0}, \tag{5.52}$$

$$\kappa = -\frac{1}{V}\left(\frac{\partial V}{\partial p}\right)_{T,\{\Phi_q\}}\bigg|_{\{\Phi_q\}=0}. \tag{5.53}$$

Here, we have used the extended Gibbs relation (The extended Gibbs relation is a generalized form of the equilibrium Gibbs relation for nonequilibrium processes. See Chapter 19.), which reduces to the equilibrium Gibbs relation at equilibrium. C_v is the isochoric heat capacity and κ is the isothermal compressibility at $\{\Phi_q\} = 0$, that is, at equilibrium. The quadratic form (5.51) is negative if the following conditions are satisfied:

$$C_v > 0, \quad \kappa > 0. \tag{5.54}$$

That is, if the conditions in Eq. (5.54) are satisfied, then the equilibrium state is stable. These conditions also imply that

$$C_p > C_v \tag{5.55}$$

as a consequence of the stability condition of equilibrium. The consideration made here can be easily generalized to the case of open systems and also of multiphase systems.

Chapter 6

The Third Law of Thermodynamics

Since we are from now on concerned with reversible processes only in the rest of chapters in Part I, the calortropy coincides with the Clausius entropy. Therefore, in this and subsequent chapters except for chapters in Part II on irreversible processes, we study the thermodynamics of reversible processes or systems at equilibrium by using only the concept of entropy. In this chapter, for brevity, only pure (single-component) systems are considered in connection with the third law.

We have seen that the entropy of a system at temperature T can be calculated within a constant of integration. Thus, if the volume is constant, the entropy at temperature T is given by the formula

$$S(T) = S_0 + \int_0^T dT' \frac{C_v(T')}{T'}, \tag{6.1}$$

where S_0 is the value of the entropy at the absolute zero. If the value of S_0 is known, the absolute entropy can be obtained at any temperature. It is indeed possible to find the absolute entropy for perfect crystals according to the third law of thermodynamics.

In 1902, T. W. Richards experimentally observed that ΔS vanishes as the temperature tends to the absolute zero. This observation enabled W. Nernst to formulate his heat theorem. For an isothermal reversible process, a Helmholtz free energy change ΔA may be written as

$$\Delta A = \Delta E - T\Delta S. \tag{6.2}$$

This may be cast in the form

$$\Delta A - \Delta E = T \left(\frac{\partial \Delta A}{\partial T} \right)_V. \tag{6.3}$$

Based on the experimental data available to him, Nernst observed that the difference on the left-hand side of Eq. (6.3) always vanishes as T tends to zero and thus concluded that the difference not only vanishes, but ΔA and ΔE also have the same tangent asymptotically as T tends to zero. Therefore, the following limits should hold:

$$\lim_{T \to 0} \frac{\partial \Delta A}{\partial T} = \lim_{T \to 0} \frac{\partial \Delta E}{\partial T} = 0. \tag{6.4}$$

Consequently, by differentiating Eq. (6.3) with T, we find

$$\lim_{T \to 0} \frac{\partial \Delta A}{\partial T} = \lim_{T \to 0} \left(\frac{\partial \Delta E}{\partial T} - T \frac{\partial \Delta S}{\partial T} - \Delta S \right).$$

The limits in Eq. (6.4) suggest the following limit:

$$\lim_{T \to 0} \left(T \frac{\partial \Delta S}{\partial T} + \Delta S \right) = 0$$

or equivalently,

$$\lim_{T \to 0} \Delta S = 0. \tag{6.5}$$

This is called Nernst's heat theorem. Based on this experimental observation, Max Planck stated the third law of thermodynamics as follows.

The Third Law of Thermodynamics. *The entropy of perfect crystals is equal to zero at the absolute zero of temperature independently of pressure and volume.*

That is, independently of pressure and volume,

$$\lim_{T \to 0} S(T) = 0. \tag{6.6}$$

Since the entropy in a reversible temperature change near the absolute zero at constant volume is given by the formula

$$S(T) = S_0 + \int_0^T dT' \frac{C_v(T')}{T'}, \tag{6.7}$$

the third law means $S_0 = 0$, for $S(T) \to 0$ independently of p and V as $T \to 0$. The undetermined constant S_0 is thereby removed. The third law not only allows one to determine the entropy absolutely but also has a number of thermodynamic implications descending therefrom. We discuss some of them in the following cases.

(1) *Specific heats tend to zero as T approaches the absolute zero.*

Since

$$S(T) = \int_0^T dT' \, \frac{C_v(T')}{T'}, \tag{6.8}$$

and $S(T) \to 0$ as $T \to 0$ by the third law, the integral must be a decreasing function of T in the neighborhood of the absolute zero of temperature. That is, near $T = 0$

$$S(T) = cT^\alpha, \tag{6.9}$$

where $\alpha \geq 0$ and c is a constant. This implies that $C_v(T)$ cannot be a constant in the neighborhood of $T = 0$. Indeed, according to Debye's theory of simple solids, the exponent α is equal to 3. That is,

$$C_v(T) = aT^3, \tag{6.10}$$

and consequently, the entropy near absolute zero behaves like

$$S(T) = \frac{1}{3}aT^3. \tag{6.11}$$

Experiments show that the specific heats of amorphous glasses, however, are approximately linear in T near the absolute zero of temperature:

$$C_v(T) = aT. \tag{6.12}$$

The entropy is then $S(T) = aT$ near $T = 0$.

(2) $\left(\frac{\partial V}{\partial T}\right)_p = 0$ *and* $\left(\frac{\partial p}{\partial T}\right)_V = 0$ *as* $T \to 0$.

To show it, we recall one of Maxwell's relations and the third law:

$$\lim_{T \to 0} \left(\frac{\partial V}{\partial T}\right)_p = \lim_{T \to 0} -\left(\frac{\partial S}{\partial p}\right)_T = 0,$$

independently of p at $T = 0$ by the third law. The second identity may be shown similarly:

$$\lim_{T \to 0} \left(\frac{\partial p}{\partial T}\right)_V = \lim_{T \to 0} \left(\frac{\partial S}{\partial V}\right)_T = 0,$$

independently of V at $T = 0$ by the third law. The first identity also implies that the isobaric expansion coefficient tends to zero as $T \to 0$. Since we may write the second identity in terms of the isobaric expansion coefficient and the isothermal compressibility

$$\left(\frac{\partial p}{\partial T}\right)_V = -\frac{\left(\frac{\partial V}{\partial T}\right)_p}{\left(\frac{\partial V}{\partial p}\right)_T} = \frac{\alpha}{\kappa},$$

it also implies that α tends to zero faster than κ as $T \to 0$.

(3) *The absolute zero of temperature cannot be reached by any finite number of processes; it may be only reached asymptotically.*

In 1926, Giauque and Debye independently showed that low temperature can be achieved by adiabatic demagnetization. In this technique, a paramagnetic salt (gadolinium sulfate, for example) is repeatedly magnetized and then demagnetized. Demagnetization is accompanied by a decrease in temperature and it is found to be possible to reach $0.0014\,\mathrm{K}$ by the procedure. The thermodynamic basis of this method is described in Chapter 18, Sec. 18.3.2.

As a magnetic field H is turned on a paramagnetic salt, the entropy decreases from its value at $H = 0$ because of the paramagnetic ordering effect of the magnetic field. Let us set

$$S_0 = \lim_{T \to 0} S\,(H = 0),$$

and assume that $S_0 \neq 0$ contrary to the third law. Then it would be possible to attain $T = 0$ in a single step of adiabatic demagnetization by starting the procedure at an appropriate temperature. This then would allow the system to reach $T = 0$ in a finite number of steps. But since $S_0 = 0$, it is practically impossible to attain $T = 0$ by a finite number of steps of adiabatic demagnetization since the temperature decrease achieved in each step gets smaller and smaller as the absolute zero is approached.

As a corollary to this theorem, we may state that *it is practically impossible to attain the unit efficiency for reversible Carnot cycles.* Since, by the Carnot theorem and equivalently the second law of thermodynamics, the efficiency of reversible cycles is always larger than those of irreversible cycles, the present corollary may be regarded as a consequence of the second law of thermodynamics. Indeed, this should be so since the second law gives rise to entropy for reversible processes, and hence the corollary must be an attribute of the second law (Fig. 6.1).

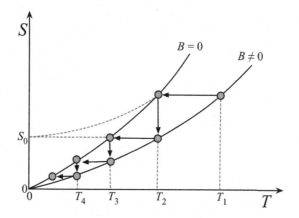

Fig. 6.1. Cooling by the Debye–Giauque adiabatic demagnetization method.

Problems

(1) Show that according to the third law of thermodynamics,

$$\lim_{T \to 0} (C_p - C_v)/TV = 0.$$

(2) Assuming that a simple monatomic liquid forms a simple crystal obeying the Debye law of specific heat $C_v = cT^3$ when it crystallizes, calculate the entropy of the liquid at $T > T_m$ where T_m is the melting point of the crystal. The heat of melting of the crystal is ΔH_m. The pressure remains fixed at p throughout the process.

Chapter 7

Thermodynamics of Mixtures and Open Systems

The thermodynamics of mixtures requires a new concept unfamiliar to the thermodynamics of pure substances. It is embodied by chemical potentials first introduced by J. W. Gibbs. The thermodynamics of mixtures and open systems is formulated by making use of the concept. It would be most appropriate from the viewpoint of the thermodynamic laws that the concept is incorporated into the representations of the thermodynamic laws. However, since Gibbs takes an axiomatic approach, in deference to his method, we will first follow his formulation. Then the former approach will be followed.

To discuss a theory of mixtures, the composition variables must be specified. The mole numbers of components are used for the purpose. Thus, the thermodynamic variable space for a mixture requires an extension to account for the composition of the mixture; it is now spanned by the variable set $(S, E, V, n_1, n_2, \ldots, n_r)$, where n_i denote the mole numbers of the species in the mixture. The present chapter also prepares us for the studies of open systems in the subsequent chapters, where mixtures of various kinds are treated thermodynamically.

7.1. Chemical Potentials

7.1.1. *The Gibbs Theory*

Let us consider a mixture consisting of $n_1, n_2, \ldots, n_r$ moles of components, $1, 2, \ldots, r$. The internal energy of the mixture then may be regarded as a function of $S, V, n_1, n_2, \ldots, n_r$:

$$E = E(S, V, n_1, n_2, \ldots, n_r). \tag{7.1}$$

It must be noted here that this assumption of a function E in the manifold of S, V, n_1, n_2, $\ldots$, and n_r presumes that there is a variable called entropy S already established on the basis of the laws of thermodynamics. This variable S, however, is fundamentally different from the rest of the variables spanning the manifold because it is not a mechanical quantity as the others are. This should be clearly kept in mind. It would have been better if the fundamental relation for E were deduced for a mixture on the basis of the laws of thermodynamics, as will be done later, but in this section, we first follow the conventional Gibbsian approach, which is axiomatic. For this reason, the connection with the thermodynamic laws is unfortunately obscured.

The differential of E is easily obtained from the functional relation in Eq. (7.1):

$$dE = \left(\frac{\partial E}{\partial S}\right)_{V,n_i} dS + \left(\frac{\partial E}{\partial V}\right)_{S,n_i} dV + \sum_{i=1}^{r} \left(\frac{\partial E}{\partial n_i}\right)_{S,V,n_{j \neq i}} dn_i. \tag{7.2}$$

If we identify the first two derivatives in Eq. (7.2), in analogy to the case of the thermodynamics of a closed system, with temperature and pressure

$$T = \left(\frac{\partial E}{\partial S}\right)_{V,n_i}, \tag{7.3}$$

$$p = -\left(\frac{\partial E}{\partial V}\right)_{T,n_i}, \tag{7.4}$$

and, in addition, denote by μ_i the derivative of E with respect to n_i

$$\mu_i = \left(\frac{\partial E}{\partial n_i}\right)_{S,V,n_{j \neq i}}, \tag{7.5}$$

the differential of E is then written in the form

$$dE = TdS - pdV + \sum_{i=1}^{r} \mu_i dn_i. \tag{7.6}$$

Here, μ_i is called the chemical potential of component i of the mixture. It was introduced by J. W. Gibbs for the first time and plays very important roles in thermodynamics. It is also useful to note that temperature and pressure are given by the same derivatives as those in Eq. (4.101) except that n_i must also be kept constant while differentiating E. The definitions (7.3)–(7.5) are mathematical and based on analogy to the mathematical structure of the thermodynamics of a closed single-component system; they are not deduced by applying the laws of thermodynamics to mixtures as has been for a closed single-component system. To gain a more physically cogent formulation that closely makes use of the thermodynamic laws, we consider an alternative approach in the following.

7.1.2. *Alternative Consideration*

From the strict standpoint of thermodynamics, the differential form (7.6) for an open system should have been deduced from the proper physical interpretation of what $đQ$ should be in the expression for the entropy difference

$$dS = \frac{đQ}{T}, \tag{7.7}$$

when the system is open with regard to matter.

When the system exchanges matter with the surroundings, energy is also carried by matter, and the heat transfer between the system and the surroundings must be corrected for the amount of energy carried by the matter transferred. This correction may be written as

$$- \sum_{i=1}^{r} \mu_i dn_i, \tag{7.8}$$

where μ_i is the amount of energy (equivalently heat) carried by a mole of species i and the sign convention on dn_i is taken such that dn_i is taken negative when matter is given up by the system to the surroundings and positive in the opposite case. Therefore, the net heat taken up by the system is[1]

$$\text{d}Q = \text{d}Q' - \sum_{i=1}^{r} \mu_i dn_i, \qquad (7.9)$$

where $\text{d}Q'$ is the heat change that would occur in the expression for the first law of thermodynamics if the system were not open. It is important to take note of the difference in the meanings of heats in the two cases of open and closed systems. With the $\text{d}Q$ in (7.9) substituted into Eq. (7.7) and with the help of the first law of thermodynamics, or equivalently (3.2), we find the differential form for dE presented for an open system in the same form as by the previous axiomatic method (Gibbs' method).

In the present manner of approaching the subject, the derivatives (7.3)–(7.5) are consequences of the differential form (7.6) derived from the second law of thermodynamics in the case of reversible processes in an open system, for which we are compelled to revise the concept of heat so as to take into account the heat energies carried by individual species. However, in the Gibbs axiomatic method, the existence of $S(E, V, n_1, \ldots, n_r)$ is implicitly assumed in analogy to the thermodynamics of closed systems and the differential form for dE is axiomatically taken. We have presented the Gibbs's derivation first in this section out of historical deference to the axiomatic method of Gibbs, which is mathematically more elegant, but obscure with regard to the physical origin of the notion of chemical potentials, especially for a mixture.

7.1.3. *Fundamental Relations for Open Systems*

Upon substitution of Eq. (7.7) and the first law of thermodynamics

$$dE = \text{d}Q' - pdV \qquad (7.10)$$

---

[1]Equivalently, one may regard the term (7.8) as a work performed when substances are transferred between the system and the surroundings which act as reservoirs of matter. In this manner of looking at the situation, the work term in the first law of thermodynamics is modified as follows:

$$dW \Rightarrow dW + \sum_{i=1}^{r} \mu_i dn_i$$

under the same sign convention for dn_i as for Eq. (7.9).

in the case of pressure–volume work, there follows the fundamental relation (7.6) for E. If there are other kinds of work present, then they must be added to the right-hand side of Eq. (7.10).

By introducing the following Legendre transformations

$$H = E + pV,$$
$$A = E - TS, \quad\quad\quad (7.11)$$
$$G = H - TS,$$

taking differential forms of H, A, and G, and making use of Eq. (7.6), we easily find the fundamental relations for thermodynamic potentials[2] H, A, and G:

$$dH = TdS + Vdp + \sum_{i=1}^{r} \mu_i dn_i, \quad\quad\quad (7.12)$$

$$dA = -SdT - pdV + \sum_{i=1}^{r} \mu_i dn_i, \quad\quad\quad (7.13)$$

$$dG = -SdT + Vdp + \sum_{i=1}^{r} \mu_i dn_i, \quad\quad\quad (7.14)$$

which are the generalizations of Eqs. (5.18), (5.20), (5.23), and (5.28) to an r-component mixture. These differential forms obviously imply that H, A, and G are functions of S, p, n_1, n_2, ..., n_r; T, V, n_1, n_2, ..., n_r; and T, p, n_1, n_2, ..., n_r, respectively:

$$H = H(S, p, n_1, n_2, \ldots, n_r),$$
$$A = A(T, V, n_1, n_2, \ldots, n_r),$$
$$G = G(T, p, n_1, n_2, \ldots, n_r).$$

The fundamental relations for H, A, and G also imply that the chemical potentials can be expressed in different ways depending on which one of E, H, A, and G is used to express chemical potential μ_i:

$$\mu_i = \left(\frac{\partial E}{\partial n_i}\right)_{S,V,n'} = \left(\frac{\partial H}{\partial n_i}\right)_{S,p,n'} = \left(\frac{\partial A}{\partial n_i}\right)_{T,V,n'} = \left(\frac{\partial G}{\partial n_i}\right)_{T,p,n'}. \quad (7.15)$$

---

[2]We call these functions thermodynamic potentials since their differential one-forms are monogenic, that is, the differential forms are integrable to scalar functions H, A, and G.

Here, n' denotes the set $\{n_j\}$ excluding n_i. Nevertheless, the last equality in Eq. (7.15) gives the most convenient interpretation of μ_i as the partial molar Gibbs free energy of species i as will be seen shortly. In this regard, it should be noted that T and p are two most readily applicable parameters to keep constant among similar constraints such as S and p, T, V, etc.

7.2. Partial Molar Properties

Let M denote an extensive thermophysical property. Extensive thermodynamic properties may be generally considered as functions of T, p, n_1, n_2, ..., n_r:

$$M = M(T, p, n_1, n_2, \ldots, n_r). \tag{7.16}$$

By definition, extensive properties depend on the masses of species or the mole numbers, and they scale linearly, being a homogeneous function of degree 1. For example, take the internal energy E. If the mass of the system is increased, then the internal energy will increase in direct proportion to the increase in mass. This implies, in mathematical terms, that extensive quantities are first degree homogeneous functions of n_1, n_2, ..., n_r of Euler. That is, if the mole numbers are increased by a factor λ for all components, then M increases by the same factor λ. This may be expressed by the relation

$$M(T, p, \lambda n_1, \lambda n_2, \ldots, \lambda n_r) = \lambda M(T, p, n_1, n_2, \ldots, n_r). \tag{7.17}$$

This is a special case of functions called Euler's homogeneous functions of degree q, which have the property

$$M(T, p, \lambda n_1, \lambda n_2, \ldots, \lambda n_r) = \lambda^q M(T, p, n_1, n_2, \ldots, n_r). \tag{7.18}$$

Therefore, for the present case, $q = 1$. We digress a little to discuss about Euler's homogeneous functions.

If a function $M(x_1, x_1, \ldots, x_r)$ has the property (7.18), namely, M is a homogeneous function of Euler of degree q, then a differentiation of the equation with respect to λ yields

$$\sum_{i=1}^{r} \left(\frac{\partial M}{\partial \lambda x_i} \right) x_i = q \lambda^{q-1} M(x_1, x_2, \ldots, x_r). \tag{7.19}$$

If λ is set as equal to 1, there follows the relation

$$M(x_1, x_2, \ldots, x_r) = q^{-1} \sum_{i=1}^{r} x_i \left(\frac{\partial M}{\partial x_i}\right). \tag{7.20}$$

This mathematical relation is useful for discussing properties of mixtures in thermodynamics. In particular, if $q = 1$ and $x_i = n_i$, we obtain

$$M(T, p, n_1, n_2, \ldots, n_r) = \sum_{i=1}^{r} \left(\frac{\partial M}{\partial \lambda n_i}\right)_{T,p,n'} \frac{d\lambda n_i}{d\lambda}$$

$$= \sum_{i=1}^{r} n_i \left(\frac{\partial M}{\partial n_i}\right)_{T,p,n'} \quad (\lambda = 1). \tag{7.21}$$

Here, the derivative is an intensive quantity and may be abbreviated by

$$\overline{M}_i = \left(\frac{\partial M}{\partial n_i}\right)_{T,p,n'}, \tag{7.22}$$

which allows to write M in the form

$$M(T, p, n_1, n_2, \ldots, n_r) = \sum_{i=1}^{r} n_i \overline{M}_i. \tag{7.23}$$

In this form, the extensive property M appears as an additive sum of contributions weighted by the mole number n_i of the components making up the mixture. It, however, should be kept in mind that $\overline{M}_i$ are generally functions of n_1, n_2, $\ldots$, n_r or more precisely, $x_i = n_i/n$ $(i = 1, \ldots, r)$, mole fractions, where $n = n_1 + n_2 + \cdots + n_r$. Note that mole fractions may serve as concentrations of species.

These mathematical properties are used in our discussion on the thermodynamics of mixtures. In particular, $\overline{M}_i$ is called the partial molar M of i. Since the chemical potential of i may be given by

$$\mu_i = \left(\frac{\partial G}{\partial n_i}\right)_{T,p,n'}, \tag{7.24}$$

we see that it is the partial molar Gibbs free energy of i in a mixture consisting of r components. Note that, according to the mathematical definition of $\overline{M}_i$ given in Eq. (7.22), μ_i is not a partial molar property of E, H, or A although it may be given by the derivative of E, H, and A as in Eq. (7.15), but qualifies for a partial molar property of G. Therefore, from Eq. (7.21) follows the relation

$$G = \sum_{i=1}^{r} n_i \mu_i. \tag{7.25}$$

In this form, the extensive property G appears as an additive sum of $n_i \mu_i$ of the components making up the mixture. Here lies one of the utilities of the notion of partial molar properties in the thermodynamics of mixtures.

We now further develop consequences and thermodynamic relations of partial molar properties and, in particular, of chemical potentials—partial molar Gibbs free energies. Since

$$\left(\frac{\partial G}{\partial T} \right)_{p,n_i} = -S, \tag{7.26}$$

$$\left(\frac{\partial G}{\partial n_i} \right)_{T,p,n_j} = \mu_i, \tag{7.27}$$

Euler's condition on exact differentials yields the Maxwell relation

$$-\left(\frac{\partial \mu_i}{\partial T} \right)_{p,n_i} = \left(\frac{\partial S}{\partial n_i} \right)_{T,p,n'}. \tag{7.28}$$

By definition, the right-hand side of Eq. (7.28) is the partial molar entropy of i:

$$\bar{s}_i = \left(\frac{\partial S}{\partial n_i} \right)_{T,p,n'}. \tag{7.29}$$

Therefore, we obtain

$$\bar{s}_i = -\left(\frac{\partial \mu_i}{\partial T} \right)_{p,n_i}. \tag{7.30}$$

Similarly, since

$$V = \left(\frac{\partial G}{\partial p} \right)_{T,n_i}, \tag{7.31}$$

by differentiating it with n_i, we obtain the relation

$$\bar{v}_i = \left(\frac{\partial \mu_i}{\partial p} \right)_{T,n_i}, \tag{7.32}$$

where $\bar{v}_i$ is the partial molar volume of component i:

$$\bar{v}_i = \left(\frac{\partial V}{\partial n_i} \right)_{T,p,n'}. \tag{7.33}$$

Since G is a function of n_i, $i = 1, 2, \ldots, r$, the Maxwell relations for single-component systems must be generalized. The additional Maxwell relations involve chemical potentials. The examples are

$$\left(\frac{\partial \mu_i}{\partial n_j}\right)_{T,p,n_k} = \left(\frac{\partial \mu_j}{\partial n_i}\right)_{T,p,n_k} \quad (k \neq i \text{ or } j). \tag{7.34}$$

Inasmuch as G is a function of T, p, n_1, $n_2, \ldots, n_r$, the chemical potentials must be functions of the same variables. If $n = \sum_{i=1}^{r} n_i$ is fixed, there are $r - 1$ independent mole numbers, and the chemical potentials may be regarded as functions of T, p, x_1, x_2, $\ldots$, x_{r-1} where x_i is the mole fraction of i:

$$x_i = \frac{n_i}{n}, \tag{7.35}$$

which normalizes to unity:

$$\sum_{i=1}^{r} x_i = 1. \tag{7.36}$$

Equations (7.30), (7.33), and (7.34) together imply the following differential form for μ_i:

$$d\mu_i = -\bar{s}_i dT + \bar{v}_i dp + \sum_{j=1}^{r-1} \mu_{ij} dx_j, \tag{7.37}$$

where

$$\mu_{ij} = \left(\frac{\partial \mu_i}{\partial x_j}\right)_{T,p,x_{k \neq j}}. \tag{7.38}$$

From Eq. (7.25) follows the differential form

$$dG = \sum_{i=1}^{r} \mu_i dn_i + \sum_{i=1}^{r} n_i d\mu_i.$$

Since we also have the fundamental relation

$$dG = -SdT + Vdp + \sum_{i=1}^{r} \mu_i dn_i, \tag{7.39}$$

by comparing these two equations, we arrive at the following important equation:

$$-SdT + Vdp = \sum_{i=1}^{r} n_i d\mu_i. \tag{7.40}$$

This is called the Gibbs–Duhem equation. It plays important roles in the studies of thermodynamics and equilibria, in particular, of multicomponent systems. We will have many occasions to use it in subsequent chapters. Incidentally, we remark that the Gibbs–Duhem equation is an integrability condition of the differential form (7.39), and Eq. (7.25) is an integral of differential form (7.39) in the space spanned by T, p, n_1, ..., n_r. The reason is that if Eqs. (7.39) and (7.40) are added side by side, there follows the integral G of the differential form (7.39) in space[3] T, p, n_1, ..., n_r. It is easy to verify that the integrability condition in the thermodynamic space for the fundamental relation

$$dS = T^{-1}\left(dE + pdV - \sum_{i=1}^{r}\mu_i dn_i\right) \qquad (7.41)$$

is then given by

$$Ed\left(\frac{1}{T}\right) + Vd\left(\frac{p}{T}\right) - \sum_{i=1}^{r} n_i d\left(\frac{\mu_i}{T}\right) = 0, \qquad (7.42)$$

which can be shown to be equivalent to the Gibbs–Duhem relation (7.40). Upon addition of Eqs. (7.41) and (7.42), there follows the integral of the differential form (7.41) within a constant of integration:

$$S = T^{-1}\left(E + pV - \sum_{i=1}^{r}\mu_i n_i\right). \qquad (7.43)$$

Equation (7.25) and the Legendre transformations (7.11) imply that Eq. (7.43) is, of course, another way of writing the Gibbs free energy.

Lastly, multiplying n_i to Eq. (7.37) and summing over i, and then making use of the Gibbs–Duhem equation (7.40) in the resulting equation, we obtain the equation

$$\sum_{i=1}^{r}\sum_{j=1}^{r-1} n_i \mu_{ij} dn_j = 0, \qquad (7.44)$$

which relates μ_{ij} to one another. This equation will be found useful for the theory of mixtures.

---

[3]More precisely, the term 'manifold' should be used in place of space.

7.3. Measurement of Partial Molar Properties

Thermodynamic functions are easily evaluated for mixtures once the partial molar properties are tabulated as a function of temperature, pressure, and concentrations. It is then obviously important to examine how we might measure them in the laboratory. Here, we discuss some experimental procedures for measuring partial molar properties. To prepare for the discussion, we first examine another useful relation underlying such measurements.

Let M be an extensive property. Then

$$dM = dM(T, p, n_1, n_2, \ldots, n_r)$$

$$= \left(\frac{\partial M}{\partial T}\right)_{p,n_i} dT + \left(\frac{\partial M}{\partial p}\right)_{T,n_i} dp + \sum_{i=1}^{r} \left(\frac{\partial M}{\partial n_i}\right)_{T,p,n_{j\neq i}} dn_i. \quad (7.45)$$

But we know

$$M = \sum_{i=1}^{r} n_i \overline{M}_i, \quad (7.46)$$

where $\overline{M}_i$ is the partial molar properties of i. On taking the differential form of M by using Eq. (7.46), we obtain

$$dM = \sum_{i=1}^{r} \overline{M}_i dn_i + \sum_{i=1}^{r} n_i d\overline{M}_i. \quad (7.47)$$

Comparing it with Eq. (7.45) yields a relation similar to the Gibbs–Duhem equation,

$$\sum_{i=1}^{r} n_i d\overline{M}_i = \left(\frac{\partial M}{\partial p}\right)_{p,n_i} dp + \left(\frac{\partial M}{\partial T}\right)_{T,n_i} dT. \quad (7.48)$$

If T and p are kept constant, it reduces to

$$\sum_{i=1}^{r} n_i d\overline{M}_i = 0. \quad (7.49)$$

Dividing it with dn_1, we obtain

$$\sum_{i=1}^{r} n_i \left(\frac{\partial \overline{M}_i}{\partial n_1}\right)_{T,p,n_2,\ldots,n_r} = 0. \quad (7.50)$$

In terms of mole fractions, this equation is expressible as

$$\sum_{i=1}^{r} x_i \left(\frac{\partial \overline{M}_i}{\partial x_1}\right)_{T,p,x_2,\ldots,x_{r-1}} = 0. \quad (7.51)$$

This relation will be useful for devising methods of measuring partial molar properties discussed below.

7.3.1. Method of Intercepts

We wish to devise methods of measuring partial molar property M for a two-component mixture composed of n_1 and n_2 moles of components 1 and 2, respectively. Setting $r = 2$ and dividing Eq. (7.46) with $n = n_1 + n_2$, we obtain the mean molar value of M:

$$M_m = x_1 \overline{M}_1 + x_2 \overline{M}_2. \tag{7.52}$$

A differentiation of Eq. (7.52) with respect to x_1 yields

$$\left(\frac{\partial M_m}{\partial x_1} \right)_{T,p} = \overline{M}_1 - \overline{M}_2, \tag{7.53}$$

for which we have made use of Eq. (7.51) for $r = 2$:

$$x_1 \left(\frac{\partial \overline{M}_1}{\partial x_1} \right)_{T,p} + x_2 \left(\frac{\partial \overline{M}_2}{\partial x_1} \right)_{T,p} = 0.$$

Since $x_1 + x_2 = 1$, we may eliminate x_1 or x_2 from Eq. (7.52) and, by making use of Eq. (7.53), obtain the equation in terms of x_2

$$
\begin{aligned}
M_m(x_2) &= \overline{M}_1(x_2) + \left[\overline{M}_2(x_2) - \overline{M}_1(x_2) \right] x_2 \\
&= \overline{M}_1(x_2) - \left(\frac{\partial M_m}{\partial x_2} \right)_{T,p} x_2,
\end{aligned} \tag{7.54}
$$

or the equation in terms of x_1

$$
\begin{aligned}
M_m(x_1) &= \overline{M}_2(x_1) + \left[\overline{M}_1(x_1) - \overline{M}_2(x_1) \right] x_1 \\
&= \overline{M}_2(x_1) + \left(\frac{\partial M_m}{\partial x_1} \right)_{T,p} x_1.
\end{aligned} \tag{7.55}
$$

If M_m is measured and plotted against x_2, say, a curve in Fig. 7.1 is obtained. The tangent line to $M_m(x_2)$ at $x_2 = x$ is

$$
\begin{aligned}
y(x_2) &= A - \left(\frac{\partial M_m}{\partial x_2} \right)_{T,p} |_{x_2 = x} x_2 \\
&= A + \left[\overline{M}_2(x) - \overline{M}_1(x) \right] x_2,
\end{aligned} \tag{7.56}
$$

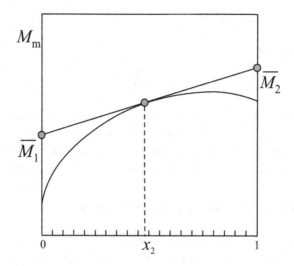

Fig. 7.1.   Method of intercepts. The partial molar properties $\overline{M}_1$ and $\overline{M}_2$ are the intercepts of the tangent to the curve at x_2.

where A is the intercept at $x_2 = 0$. Since at $x_2 = x$,

$$M_m(x) = \overline{M}_1(x) + \left[\overline{M}_2(x) - \overline{M}_1(x)\right] x,$$

and

$$y(x) = M_m(x),$$

it follows

$$A = \overline{M}_1(x).$$

Therefore, the tangent line to the curve at $x_2 = x$ is given by

$$y(x_2) = \overline{M}_1(x) + \left[\overline{M}_2(x) - \overline{M}_1(x)\right] x_2. \tag{7.57}$$

Hence $\overline{M}_1(x)$ at $x_2 = x$ is given by the intercept of this tangent line, and varying the position of x continuously between $0 \le x_2 \le 1$, the intercept $\overline{M}_1(x)$, the partial molar value of M, can be measured as a function of x for component 1. Similarly, by plotting M_m against x_1 and using the tangent equation

$$y(x_1) = \overline{M}_2(x) + \left[\overline{M}_1(x) - \overline{M}_2(x)\right] x_1, \tag{7.58}$$

$\overline{M}_2(x)$ at $x_1 = x$ is obtained from the intercept of the tangent line.

7.3.2.   Direct Method

Since by definition

$$\overline{M}_i = \left(\frac{\partial M}{\partial n_i}\right)_{T,p,n_{j\neq i}},$$

we see that partial molar properties can be measured by simply following the change in M as the mole number of a component is changed while the mole numbers of other components are kept constant. The slope of the curve for M vs. the mole number can then be obtained graphically.

7.3.3.   Method of Apparent Molar Property

We illustrate this method with the example of two components. In this method, we measure apparent molar property Ω_M defined by

$$\Omega_M = \frac{(M - n_1 M_1)}{n_2}, \tag{7.59}$$

where M_1 is the molar property of pure component 1. By rearranging the terms in the form

$$M - n_1 M_1 = n_2 \Omega_M$$

and differentiating it with n_2 at constant p and T, we obtain

$$\Omega_M = \overline{M}_2 - n_2 \left(\frac{\partial \Omega_M}{\partial n_2}\right)_{T,p,n_1}. \tag{7.60}$$

This equation suggests that $\overline{M}_2(a)$ at $n_2 = a$ can be determined by measuring the position of the intercept of the tangent line to Ω_M at $n_2 = a$ as schematically shown in Fig. 7.2.

7.3.4.   Density Dependence of $\overline{M}_i$

Partial molar properties thus measured can be summarized with a power series in mole numbers. For example, in the case of two-component mixtures, we may express them in the series of mole fractions

$$\overline{M}_1 = M_{10} + M_{11}x_2 + M_{12}x_2^2 + \cdots, \tag{7.61}$$

$$\overline{M}_2 = M_{20} + M_{21}x_1 + M_{22}x_1^2 + \cdots. \tag{7.62}$$

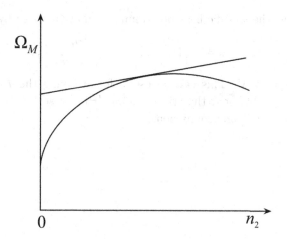

Fig. 7.2.   The method of apparent molar property. The intercept of the tangent is $\overline{M}_2(a)$
at $n_2 = a$.

In this mode of representation, M_{10} and M_{20} are respectively the molar
property of pure substance 1 and that of pure substance 2, and M_{11}, M_{21},
and so on are parameters determined experimentally. These parameters
as well as M_{10} and M_{20} are, of course, dependent on temperature and
pressure.

Problems

(1) The volume of aqueous solution of NaCl at 298.15 K and 1 atm depends
   on the molality m of NaCl as follows: in the units of L,

   $$V = 1.00294 + 0.01640\, m + 0.00214\, m^{3/2} + 2.7 \times 10^{-6}\, m^{5/2}.$$

   Find molar volumes $\overline{V}_{NaCl}$ and $\overline{V}_{H_2O}$.
(2) When 1.158 moles of water are dissolved in 0.842 moles of ethanol,
   the volume of the solution is 0.06816 L at 298.15 K. If $\overline{V}_{H_2O} = 0.01698$ L mol^{-1} in this solution, how large is $\overline{V}_{ethanol}$?
(3) Compare the partial molar volumes of the components with their
   molar volumes if $H_2O(l)$ and $C_2H_5OH(l)$ have molecular weights of
   1.802×10^{-2} and 4.602×10^{-2} kg mol^{-1}, and densities of 0.9970 and
   0.7852 kg L^{-1}, respectively, at the same temperature.
(4) Show the equivalence of Eq. (7.40) and Eq. (7.42).

(5) Show that the specific heat per volume c_v may be given by

$$c_v = T \left[\left(\frac{\partial^2 p}{\partial T^2} \right)_V - n \left(\frac{\partial^2 \mu}{\partial T^2} \right)_V \right],$$

where $n = N/V$. This can be useful for studying the T-dependence of c_v, especially, near the critical point. [Hint. Use the Gibbs–Duhem equation for a single component.]

Chapter 8

Heterogeneous Equilibria

We have defined the term *heterogeneous* as the state of a system in which the intensive properties vary with position in space. For example, if the density changes discontinuously across the boundary of two homogeneous regions, then the overall system is heterogeneous. Such homogeneous regions of a heterogeneous system are called the *phases* of the system. The thermodynamic theory developed up to this point in the previous chapters assumes that the system is homogeneous. However, since thermodynamic systems are often heterogeneous, it is necessary to generalize the theory so as to treat them thermodynamically. The desired generalization was originally achieved in a complete form by J. W. Gibbs in his monumental work,[1] and we follow his formulation of theory.

8.1. Equilibrium Conditions for a Multiphase System

Let us imagine a system of ν phases composed of r chemically inert components. The entire system is isolated from the surroundings. Suppose that these phases of different thermodynamic states, each of which nevertheless is in internal equilibrium by itself, are artificially separated by a device of walls, which are thermal insulators, rigid, and impermeable to matter (molecules), so that the whole system is kept from reaching equilibrium.

---

[1] J. W. Gibbs, *Trans. Connecticut Acad. II*, pp. 382–404 (1873); reprinted in *The Scientific Papers of J. Willard Gibbs* (Dover, New York, 1961).

The thermal insulators keep the phases from reaching thermal equilibrium and the rigid walls prevent them from reaching mechanical equilibrium, while impermeable membranes forbid them from attaining material equilibrium. Let us assume that there are ν phases and the states of the ν phases are specified, respectively, by T_α, p_α, $n_1^{(\alpha)}$, $n_2^{(\alpha)}$, ..., $n_r^{(\alpha)}$, $\alpha = 1, 2$, ..., ν, where T_α means the temperature, p_α the pressure, and $n_i^{(\alpha)}$ the mole numbers of species i in phase α, which add up to the total mole number n_i of species i in the system:

$$n^{(\alpha)} = \sum_{i=1}^{r} n_i^{(\alpha)}. \tag{8.1}$$

Therefore, there are $(r-1)$ independent mole fractions, say, $x_1^{(\alpha)}$, $x_2^{(\alpha)}$, ..., $x_{r-1}^{(\alpha)}$ ($x_i^{(\alpha)} = n_i^{(\alpha)}/n^{(\alpha)}$), and the thermodynamic state of each phase may be specified by $(r+1)$ variables, T_α, p_α, $x_1^{(\alpha)}$, $x_2^{(\alpha)}$, ..., $x_{r-1}^{(\alpha)}$ ($\alpha = 1, 2, ..., r$). As the constraints of thermal insulators, rigid walls, and impermeable membranes are removed in part or all, the phases will eventually reach equilibrium to the extent that the constraints are removed. We now wish to find the necessary and sufficient conditions for equilibria between different phases.

The equilibrium condition previously established for a single-phase system can be easily generalized to a multiphase system. In order for the system to be in equilibrium, the internal energy variation must be such that

$$(\delta E)_{\Psi, V, n_1, ..., n_r} \geq 0. \tag{8.2}$$

We further assume that the phases are sufficiently large in mass and size, so that the surface (interfacial) contributions to thermodynamic extensive quantities are negligible compared with their bulk values. Since the phases are internally in equilibrium, $\Psi^{(\alpha)} = S^{(\alpha)}$ for each phase. This does not necessarily mean that the total calortropy Ψ of the entire system is equal to the total entropy of the system since it may not be in equilibrium. Nevertheless, there does not arise the need to know Ψ for the discussion of equilibrium conditions under the assumption of internal equilibrium of phases since we are not interested here in how the whole system reaches equilibrium in time and space, but in the equilibrium conditions only. The extensive variables for the whole system under the assumption of no interfacial contribution then may be given simply as sums of the contributions from the ν separate phases:

$$E = \sum_{\alpha=1}^{\nu} E^{(\alpha)}, \quad S = \sum_{\alpha=1}^{\nu} S^{(\alpha)}, \quad V = \sum_{\alpha=1}^{\nu} V^{(\alpha)}, \quad n_i = \sum_{\alpha=1}^{\nu} n_i^{(\alpha)}. \quad (8.3)$$

Since the individual phases are assumed to be internally in equilibrium, the variation in $E^{(\alpha)}$ is given by

$$\delta E^{(\alpha)} = T_\alpha \delta S^{(\alpha)} - p_\alpha \delta V^{(\alpha)} + \sum_{i=1}^{r} \mu_i^{(\alpha)} \delta n_i^{(\alpha)} \quad (\alpha = 1, 2, \dots, \nu), \quad (8.4)$$

where $\mu_i^{(\alpha)}$ are chemical potentials of species i in phase α and $\Psi^{(\alpha)}$ is replaced by $S^{(\alpha)}$ because the phases are internally in equilibrium.

Substitution of this formula into Eq. (8.2) yields the inequality

$$\sum_{\alpha=1}^{\nu} \left(T_\alpha \delta S^{(\alpha)} - p_\alpha \delta V^{(\alpha)} + \sum_{i=1}^{r} \mu_i^{(\alpha)} \delta n_i^{(\alpha)} \right) \geq 0. \quad (8.5)$$

Since S, V, and n_i are kept constant, this inequality is subject to the conditions

$$\delta S = \sum_{\alpha=1}^{\nu} \delta S^{(\alpha)} = 0, \quad (8.6)$$

$$\delta V = \sum_{\alpha=1}^{\nu} \delta V^{(\alpha)} = 0, \quad (8.7)$$

$$\delta n_i = \sum_{\alpha=1}^{\nu} \delta n_i^{(\alpha)} = 0 \ (i = 1, 2, \dots, r). \quad (8.8)$$

Inequality (8.5) is the general equilibrium condition for the heterogeneous system we are considering here. To better understand its significance in more detail, we consider special cases by removing the constraints initially imposed on the nature of phase boundaries, namely, by making them movable, diathermal, or permeable.

8.1.1. *Mechanical Equilibrium*

Let us assume that the boundary (i.e., wall) between phases 1 and 2 is made movable, so that phases 1 and 2 are allowed to attain mechanical equilibrium, while all other constraints are still kept imposed. This means that

$$\delta S^{(\alpha)} = 0, \quad \delta n_i^{(\alpha)} = 0$$

for all α and

$$\delta V^{(\alpha)} = 0$$

for all α except for $\alpha = 1, 2$. The physical meanings of these conditions are that there are no heat and material exchanges between subsystems (i.e., phases) and the subsystems for $\alpha \neq 1, 2$ do not change their volume. Such conditions can be achieved if the subsystems are not allowed to reach thermal, material, and mechanical equilibrium, or the walls between subsystems are kept adiabatic, impermeable to matter, and mechanically rigid. Then, since by Eq. (8.7),

$$\delta V^{(1)} = -\delta V^{(2)},$$

the equilibrium condition (8.5) takes the inequality

$$(p_2 - p_1)\delta V^{(1)} \geq 0. \tag{8.9}$$

If $\delta V^{(1)} \geq 0$, then $p_2 \geq p_1$. If $\delta V^{(1)} \leq 0$, then $p_2 \leq p_1$. But $\delta V^{(1)}$ is arbitrary (see Fig. 8.1). Therefore, condition (8.9) can be satisfied only if

$$p_1 = p_2. \tag{8.10}$$

This is the condition for mechanical equilibrium between phases 1 and 2. Obviously, if the wall is movable, the pressure on both phases must be equalized for mechanical equilibrium to be established.

However, if $\delta V^{(1)} \geq 0$ is the only possible variation, then the equilibrium condition is

$$p_2 \geq p_1.$$

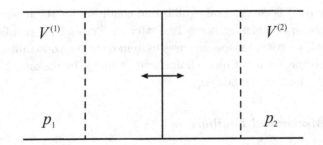

Fig. 8.1.   The wall is movable in either direction in such a way that pressures are equalized on both sides.

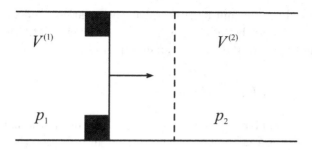

Fig. 8.2.   The wall can move only toward right from the catches.

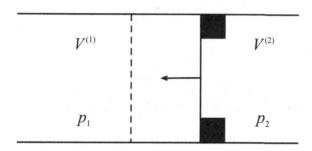

Fig. 8.3.   The wall can move only toward left from the catches.

This situation is schematically illustrated in Fig. 8.2, where there are catches on one side of the wall, so that the wall can move to the right, but not to the left. In the opposite case, that is, if $\delta V^{(1)} \leq 0$ is the only possible variation, then the equilibrium condition is

$$p_2 \leq p_1.$$

This situation is illustrated in Fig. 8.3, where the wall can move only to the left. If the wall is rigid, then

$$\delta V^{(1)} = \delta V^{(2)} = 0,$$

and there can be no condition imposed on p_1 and p_2; there is therefore no question of mechanical equilibrium.

One can repeat the same argument with different pairs of phases and finally obtain the conditions

$$p_1 = p_2 = \cdots = p_\nu \tag{8.11}$$

for mechanical equilibrium if all the phase boundaries are deformable. In this case, the phase boundaries will deform or move until the pressure becomes uniform over the whole system of phases.

8.1.2. Thermal Equilibrium

Suppose the boundary of phases 1 and 2 is made diathermal, so that phases 1 and 2 are allowed to exchange heat in order for them to reach thermal equilibrium, but all other constraints are still maintained intact. In this case,

$$\delta V^{(\alpha)} = 0, \quad \delta n_i^{(\alpha)} = 0 \tag{8.12}$$

for all α, and

$$\delta S^{(\alpha)} = 0 \tag{8.13}$$

for all α except for $\alpha = 1, 2$. But from Eq. (8.6), we obtain

$$\delta S^{(1)} = -\delta S^{(2)} \tag{8.14}$$

and the equilibrium condition becomes

$$(T_1 - T_2)\delta S^{(1)} \geq 0. \tag{8.15}$$

Since the wall is diathermal, $T_1 \geq T_2$ if $\delta S^{(1)} > 0$, but $T_1 \leq T_2$ if $\delta S^{(1)} < 0$. Since $\delta S^{(1)}$ is arbitrary, the only way for Eq. (8.15) to be satisfied is

$$T_1 = T_2. \tag{8.16}$$

This is the condition for thermal equilibrium. That is, the temperatures of phases 1 and 2 must be equal at thermal equilibrium. Repeating the same argument for other pairs of phases, we find the thermal equilibrium conditions for ν phases:

$$T_1 = T_2 = \cdots = T_\nu, \tag{8.17}$$

when all the walls are diathermal. If the walls are adiabatic, Eq. (8.17) is not required. Figure 8.4 schematically illustrates the case for Eq. (8.16). At equilibrium, the temperature thus becomes uniform throughout the phases if the walls are all diathermal.

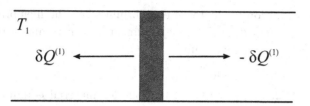

Fig. 8.4. The wall is diathermal, allowing thermal equilibrium to be reached.

8.1.3. *Material Equilibrium*

Suppose that the boundary between phases 1 and 2 is made permeable to species i so that phases 1 and 2 are allowed to reach material equilibrium with respect to species i while other constraints imposed are unchanged. Consequently, the following conditions hold:

$$\delta V^{(\alpha)} = 0, \quad \delta S^{(\alpha)} = 0 \tag{8.18}$$

for all α, but

$$\delta n_j^{(\alpha)} = 0 \tag{8.19}$$

for all j and α except for $j = i$ and $\alpha = 1, 2$. In that case, by Eq. (8.8),

$$\delta n_i^{(1)} = -\delta n_i^{(2)} \tag{8.20}$$

and the material equilibrium condition (8.5) takes the form

$$(\mu_i^{(1)} - \mu_i^{(2)})\delta n_i^{(1)} \geq 0. \tag{8.21}$$

Since the membrane is permeable to i, $\delta n_i^{(1)}$ is arbitrary. Therefore, if $\delta n_i^{(1)} > 0$, then

$$\mu_i^{(1)} \geq \mu_i^{(2)},$$

and if $\delta n_i^{(1)} < 0$, then

$$\mu_i^{(1)} \leq \mu_i^{(2)}.$$

Hence, the only way for Eq. (8.21) to be satisfied is for the chemical potentials of i in the two phases to be equal:

$$\mu_i^{(1)} = \mu_i^{(2)}. \tag{8.22}$$

This is the condition for material equilibrium when the membrane is permeable to i. By repeating the same argument for other pairs of phases and also for other species, we obtain

$$\mu_i^{(1)} = \mu_i^{(2)} = \cdots = \mu_i^{(\nu)} \tag{8.23}$$

for all $i = 1, 2, \ldots, r$ for the conditions for material equilibrium when the membranes are permeable to all species. These conditions mean that chemical potentials become uniform throughout the phases at material equilibrium if the membranes are permeable to all species.

By collecting the results (8.11), (8.17), and (8.23) obtained earlier, we reach the conclusion that the necessary conditions for equilibrium of a heterogeneous systems are

$$p_1 = p_2 = \cdots = p_\nu,$$

$$T_1 = T_2 = \cdots = T_\nu,$$

$$\mu_1^{(1)} = \mu_1^{(2)} = \cdots = \mu_1^{(\nu)}, \tag{8.24}$$

$$\vdots$$

$$\mu_r^{(1)} = \mu_r^{(2)} = \cdots = \mu_r^{(\nu)}.$$

These are also the sufficient conditions. It can be shown as follows: If $T_\alpha = T$, $p_\alpha = p$, and $\mu_i^{(\alpha)} = \mu_i$ for all α and i, that is, if Eq. (8.24) holds, then

$$(\delta E)_{\Psi, V, n_1, \ldots, n_r} = \sum_{\alpha=1}^{\nu} \left(T_\alpha \delta S^{(\alpha)} - p_\alpha \delta V^{(\alpha)} + \sum_{i=1}^{r} \mu_i^{(\alpha)} \delta n_i^{(\alpha)} \right)$$

$$= T \sum_{\alpha=1}^{\nu} \delta S^{(\alpha)} - p \sum_{\alpha=1}^{\nu} \delta V^{(\alpha)} + \sum_{i=1}^{r} \mu_i \sum_{\alpha=1}^{\nu} \delta n_i^{(\alpha)},$$

but this must be identically equal to zero

$$(\delta E)_{\Psi, V, n_1, \ldots, n_r} = 0, \tag{8.25}$$

since Eqs. (8.6)–(8.8) must hold. This is the condition for equilibrium when the calortropy (entropy if in equilibrium), volume, and mole numbers are fixed. Therefore, Eq. (8.24) is the necessary and sufficient conditions for ν heterogeneous phases to be at equilibrium when the phase boundaries are free from constraints mentioned earlier. By this, the sufficiency of conditions (8.24) is proved.

The same conclusion as for the equilibrium conditions in Eq. (8.24) can be arrived at if the uncompensated heat is calculated by means of

Eqs. (4.38) and (8.4) subject to the constraints (8.6)–(8.8) and it is set equal to zero at equilibrium. We elaborate on this statement. For the ν phase system, the following general relation holds:

$$\sum_{\alpha=1}^{\nu} \left(T_\alpha \delta \Psi^{(\alpha)} - \delta E^{(\alpha)} - p_\alpha \delta V^{(\alpha)} + \sum_{i=1}^{r} \mu_i^{(\alpha)} \delta n_i^{(\alpha)} \right) = \delta N, \qquad (8.26)$$

for which we have used Eqs. (7.9) and (7.10). Here, N is the uncompensated heat. Since $\Psi^{(\alpha)} = S^{(\alpha)}$ for all α on account of the assumption of internal equilibrium of phases and, furthermore, since

$$E = \sum_{\alpha=1}^{\nu} E^{(\alpha)} = \text{constant}$$

on account of the assumption that the entire system is isolated and hence

$$\sum_{\alpha=1}^{\nu} \delta E^{(\alpha)} = 0,$$

Eq. (8.26) can be written as

$$\sum_{\alpha=1}^{\nu} \left(T_\alpha \delta S^{(\alpha)} - p_\alpha \delta V^{(\alpha)} + \sum_{i=1}^{r} \mu_i^{(\alpha)} \delta n_i^{(\alpha)} \right) = \delta N. \qquad (8.27)$$

This variation is subject to the constraints of Eqs. (8.6)–(8.8). On applying the Lagrange multiplier method to this variational problem, we obtain

$$\sum_{\alpha=1}^{\nu} \left[(T_\alpha - T) \delta S^{(\alpha)} - (p_\alpha - p) \delta V^{(\alpha)} + \sum_{i=1}^{r} (\mu_i^{(\alpha)} - \mu_i) \delta n_i^{(\alpha)} \right] = \delta N$$
$$\geq 0,$$
$$(8.28)$$

where T, p, and μ_i are the Lagrange multipliers. Since at equilibrium or for a reversible process,

$$\delta N = 0,$$

and variations $\delta S^{(\alpha)}$, $\delta V^{(\alpha)}$, and $\delta n_i^{(\alpha)}$ are arbitrary, there follow the equilibrium conditions

$$T = T_\alpha, \quad p = p_\alpha, \quad \mu_i = \mu_i^{(\alpha)},$$

which are the same as conditions in Eq. (8.24).

8.2. Gibbs Phase Rule

For phases composed of r nonreacting components, there are $(r+1)$ intensive variables, T_α, p_α, $x_1^{(\alpha)}$, $x_2^{(\alpha)}$, ..., $x_{r-1}^{(\alpha)}$, where $x_i^{(\alpha)}$ is the mole fraction of species i in phase α. Consequently, there are $\nu(r+1)$ intensive variables for a heterogeneous system with ν phases. Since there are $(2+r)(\nu-1)$ conditions to be satisfied by these variables according to the equilibrium conditions just established, the number of free variables is

$$f = \nu(r+1) - (2+r)(\nu-1)$$

$$= r + 2 - \nu. \tag{8.29}$$

This is the number of thermodynamic degrees of freedom and is called the Gibbs phase rule. This rule holds subject to the validity of the assumptions and approximations made to derive equilibrium conditions (8.24). Let us recall that equilibrium conditions (8.24) are derived under the assumptions of deformable, diathermal, and permeable boundaries (or interfaces). Other assumptions are that there are negligible contributions to E and S from the interfaces, so that *E and S are sums of bulk phase contributions only, that there are uniform normal pressures on phases, and that there should be no chemical reactions in the system.*

Consider, for example, a single-component liquid in equilibrium with its own vapor. Since $r = 1$ and $\nu = 2$ in this case, the Gibbs phase rule gives $f = 1$. Therefore, the temperature or the pressure can be the variable in this case, and when either one of them is given, the state of the system is completely determined. For instance, if the temperature is given, the vapor pressure is uniquely determined. If three phases of a substance, for example, water, ice, and its vapor, are in equilibrium, then $f = 0$. This means that the triple point has no thermodynamic degree of freedom and therefore, it is uniquely given and invariant for the substance. This invariance was one of the reasons for choosing the triple point of water as the reference point of temperature scale discussed in Chapter 2. We shall consider some simple cases to illustrate applications of the Gibbs theory of heterogeneous phase equilibria in subsequent chapters.

8.3. Single-Component, Two-Phase Systems

Let us assume that the system consists of two phases of a single species, for example, vapor and liquid, or liquid and solid, or vapor and solid of a

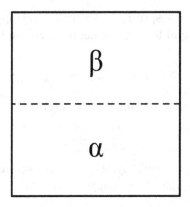

Fig. 8.5. Two-phase equilibrium: phases α and β are in equilibrium across the boundary.

substance. As is clear from the Gibbs phase rule, $f = 1$ in this case, and it implies that p may be regarded as a function of T, and vice versa. Let us then find out the relationship between the two variables.

Specializing equilibrium conditions (8.24) to the case of two phases, we obtain for the equilibrium conditions

$$\mu^{(\alpha)} = \mu^{(\beta)},$$

$$T_\alpha = T_\beta = T, \tag{8.30}$$

$$p_\alpha = p_\beta = p,$$

where α and β denote the two phases (see Fig. 8.5). Since there is now no need for a subscript to distinguish species, we have dropped the species subscript from chemical potentials. Differentiating the first equation in Eq. (8.30) yields

$$d\mu^{(\alpha)} = d\mu^{(\beta)}. \tag{8.31}$$

Since T and p are the variables, we may write this equation in the form

$$\left(\frac{\partial \mu^{(\alpha)}}{\partial T}\right)_p dT + \left(\frac{\partial \mu^{(\alpha)}}{\partial p}\right)_T dp = \left(\frac{\partial \mu^{(\alpha)}}{\partial T}\right)_p dT + \left(\frac{\partial \mu^{(\alpha)}}{\partial p}\right)_T dp. \tag{8.32}$$

Recalling that

$$\left(\frac{\partial \mu^{(i)}}{\partial T}\right)_p = -s^{(i)}, \quad \left(\frac{\partial \mu^{(i)}}{\partial p}\right)_T = v^{(i)} \ (i = \alpha, \beta), \tag{8.33}$$

where $s^{(i)}$ and $v^{(i)}$ are, respectively, the molar entropy and volume of the substance in phase i, and by rearranging the terms in Eq. (8.32), we obtain the equation

$$(v^{(\beta)} - v^{(\alpha)})dp = (s^{(\beta)} - s^{(\alpha)})dT. \qquad (8.34)$$

This may be written in the form of a differential equation

$$\frac{dp}{dT} = \frac{\Delta s}{\Delta v}, \qquad (8.35)$$

where Δs and Δv are, respectively, the molar entropy and the molar volume change accompanying the phase change from α to β:

$$\Delta s = s^{(\beta)} - s^{(\alpha)}, \quad \Delta v = v^{(\beta)} - v^{(\alpha)}. \qquad (8.36)$$

If we denote the molar enthalpies of phases α and β by $h^{(\alpha)}$ and $h^{(\beta)}$, the first equation in Eq. (8.36) may be written

$$h^{(\alpha)} - Ts^{(\alpha)} = h^{(\beta)} - Ts^{(\beta)},$$

where the temperature is the same on both sides since the two phases are in equilibrium. By using this relation, we may cast Eq. (8.35) in the form

$$\frac{dp}{dT} = \frac{\Delta h}{T\Delta v}, \qquad (8.37)$$

where Δh is the molar heat of phase transformation, namely, the molar heat of melting, vaporization, and so on:

$$\Delta h = h^{(\beta)} - h^{(\alpha)}. \qquad (8.38)$$

Equation (8.37) is called the Clapeyron equation. It describes the temperature dependence of p and vice versa.

If one of the two phases is in a condensed state, such as solid or liquid, and the other is vapor, there is a large difference in molar volumes of phases α and β. In that case, since the volume of the condensed phase may be neglected, compared with that of the vapor phase, we may approximate

$$\Delta v \approx v^{(\beta)}, \qquad (8.39)$$

where the phase β is designated as the vapor phase. Then follows from Eq. (8.37) the approximate equation

$$\frac{dp}{dT} = \frac{\Delta h}{Tv^{(\beta)}}. \qquad (8.40)$$

Furthermore, if the vapor pressure of phase β is low, the vapor may be regarded as an ideal gas, and hence the ideal gas equation of state may be used to a good approximation. Eliminating $v^{(\beta)}$ by using the ideal gas equation of state, we then obtain from Eq. (8.41) the equation

$$\frac{d\ln p}{dT} = \frac{\Delta h}{RT^2}. \tag{8.41}$$

This is called the Clapeyron–Clausius equation. It is often employed for discussing liquid–vapor or solid–vapor equilibria.

In the case of solid–liquid equilibria, it is not generally possible to approximate Δv with $v^{(\beta)}$, since the volumes of the two phases are comparable. In fact, the sign of Δv is crucial for determining the slope of the p–T curve. Thus, in the convention taking the heat of fusion Δh positive, the slope has two different signs:

$$\frac{dp}{dT} > 0 \quad \text{if } \Delta v > 0, \qquad \frac{dp}{dT} < 0 \quad \text{if } \Delta v > 0. \tag{8.42}$$

In the case of $\Delta v > 0$, the p–T curve has a positive slope, whereas in the case of $\Delta v < 0$, it has a negative slope. The former corresponds to an expansion of volume on melting whereas the latter corresponds to a contraction of volume. A typical example for a volume contraction on melting is the ice–water phase transition (Fig. 8.6).

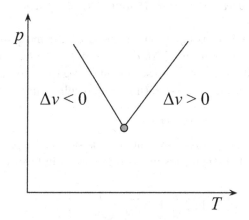

Fig. 8.6. Two different cases of the slope of the p vs. T curves at the triple point (the dot on the figure). The solid–vapor and liquid–vapor coexistence curves are omitted from the figure for brevity.

It has a significance in the recreational sport of ice skating.[2] Since ice melts under pressure according to the prediction of Eq. (8.37), the water generated serves the role of a lubricant, thus making skating facilitated. It is possible to estimate the temperature change arising from a pressure difference Δp in the case of ice–water equilibrium. Since Δh and Δv do not change much if the temperature is not widely different from the melting point, they may be regarded as independent of temperature. The Clapeyron equation may then be easily integrated to yield

$$T \exp(-p\Delta v/\Delta h) = C, \tag{8.43}$$

where C is the integration constant. Take T_0 for the melting point at pressure p_0, for example, 1 atm. Eliminating the integration constant C with this set of variables (T_0, p_0), we obtain from Eq. (8.43)

$$T \exp(-p\Delta v/\Delta h) = T_0 \exp(-p_0\Delta v/\Delta h). \tag{8.44}$$

Setting $T = T_0 + \Delta T$, we finally obtain from Eq. (8.44) the formula for the temperature variation:

$$\Delta T = T_0[\exp(\Delta p\Delta v/\Delta h) - 1]. \tag{8.45}$$

Since $\Delta v \leq 0$ for the ice–water transition, the right-hand side is negative and as the pressure is increased from p_0, the melting point decreases below the melting point of ice at p_0. This results in the melting of ice under pressure p if the ambient temperature is maintained at T_0.

8.3.1.   *Vapor Pressure and Measurement of Δh*

The Clapeyron–Clausius equation can be used for measuring the heat of vaporization or sublimation. If Δh does not depend on T too strongly over the temperature range of interest, we may regard it as a constant and integrate Eq. (8.41) over temperature. The result of integration is

$$\ln p = -\frac{\Delta h}{RT} + C, \tag{8.46}$$

where C is the integration constant. Let us denote by p_0 the pressure at temperature T_0. It is then possible to eliminate C in favor of T_0 and p_0 and obtain

$$\ln\left(\frac{p}{p_0}\right) = \frac{\Delta h}{R}\left(\frac{1}{T_0} - \frac{1}{T}\right). \tag{8.47}$$

---

[2]For a recent account of physics underlying this phenomenon, see R. Rosenberg, *Phys. Today*, **58**(12), 50 (2005).

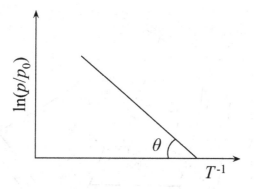

Fig. 8.7. Schematic drawing of the logarithm of vapor pressure against $1/T$. The slope gives Δh.

This equation suggests that it is possible to obtain Δh by plotting $\ln(p/p_0)$ against T^{-1} to measure the slope of the line. This procedure is illustrated in Fig. 8.7. If the line intersects with the T^{-1} axis at an angle θ, the heat of vaporization is given by the formula

$$\Delta h = R \tan \theta.$$

The assumption that Δh is constant with respect to T can be removed and the procedure described earlier can be refined with T-dependent Δh. For the purpose of integration, it would require fitting Δh to a function of T and experimentally determining the parameters in the fitting function for Δh. Such functions are often available in the literature on thermophysical data, for example, the National Institute of Science and Technology (NIST) reference data base (NIST webbook).

8.3.2. *Phase Diagrams*

When the Clapeyron–Clausius equation or the Clapeyron equation is integrated for various two-phase equilibria and the results are assembled together, we obtain a surface in the three-dimensional space of p, V, and T as schematically shown in Fig. 8.8. It is often convenient to cut this three-dimensional surface in a plane in order to display the cross-section, which shows schematically the essential features of phase equilibria of a single-component system — a phase diagram. Figure 8.9, which is an example for phase diagrams, perhaps the simplest and most common, schematically shows the essential features of phase equilibria of a single-component

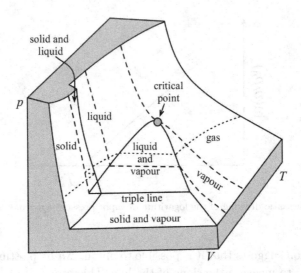

Fig. 8.8.   Three-dimensional phase surface of a simple material.

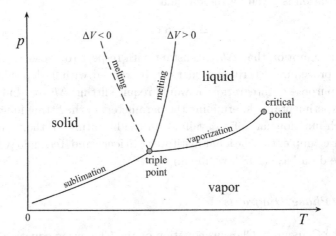

Fig. 8.9.   Phase diagram for a single-component system in the $(p-T)$ plane. This is a
cut of the surface shown in Fig. 8.8 such that it includes the critical and triple points.
Two alternative cases are shown for the sign of ΔV.

system in $p-T$ plane. It shows phases around the triple point in two different cases of volume change involved in solid–vapor equilibrium, since upon
melting, the accompanying volume change can be positive or negative. One
can also construct other examples of projected phase diagrams in $p-V$
plane or $T-V$ plane, which reveal more details of two-phase regions. They

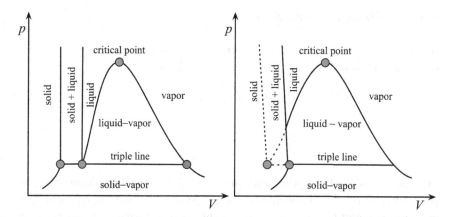

Fig. 8.10. Schematic projections of the three-dimensional phase diagram onto the $p-V$ plane in the neighborhood of the triple line.

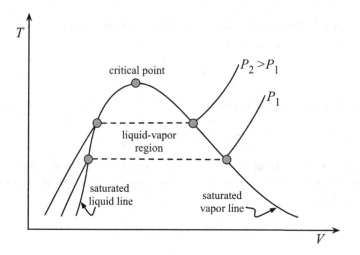

Fig. 8.11. A schematic projection of the three-dimensional phase diagram onto the $T-V$ plane below the critical point in the liquid–vapor phase region.

are shown in the following two figures: In Fig. 8.10 the solid–liquid–vapor phase diagrams are constructed around the triple line. In Fig. 8.11, the three-dimensional phase diagram is projected on the $T-V$ plane such that the liquid–vapor region below the critical point is displayed at two different pressures.

8.3.3. Ramsay–Young Rule

If the heat of vaporization is assumed to be constant, the Clapeyron–Clausius equation is easily integrated to the form

$$\ln p = \frac{A}{T} + C, \tag{8.48}$$

where

$$A = -\frac{\Delta h_v}{R},$$

Δh_v is the heat of vaporization, and C is the integration constant. The parameter depends on the nature of the substance. Let us consider two substances a and b at temperature T_a and T_b, respectively. Assuming that the approximation implied by (8.48) holds for substances a and b, we obtain

$$\ln p_a = \frac{A_a}{T_a} + C_a, \quad \ln p_b = \frac{A_b}{T_b} + C_b. \tag{8.49}$$

If the two substances have the same vapor pressure at T_a and T_b, respectively, there holds the relation

$$\frac{A_a}{T_a} + C_a = \frac{A_b}{T_b} + C_b. \tag{8.50}$$

Rearranging this equation yields

$$\frac{T_a}{T_b} = \frac{T_a}{A_b}(C_a - C_b) + \frac{A_a}{A_b}. \tag{8.51}$$

Let (T_a', T_b') be the temperatures at which the vapor pressures are equal to p', which is different from the vapor pressure $p_a = p_b$ at which (8.51) holds. Then we have

$$\frac{T_a'}{T_b'} = \frac{T_a'}{A_b}(C_a - C_b) + \frac{A_a}{A_b}. \tag{8.52}$$

Subtracting (8.52) from (8.51), there follows the equation

$$\frac{T_a}{T_b} - \frac{T_a'}{T_b'} = C_1(T_a - T_a'), \tag{8.53}$$

where

$$C_1 = \frac{(C_a - C_b)}{A_b}. \tag{8.54}$$

This result was empirically obtained by W. Ramsay and S. Young. They found that C_1 is small if a and b are chemically similar. In that case, $C_1 \simeq 0$

and, consequently, there holds the relation

$$\frac{T_a}{T_b} = \frac{T'_a}{T'_b}.$$ (8.55)

This means that the ratio of the boiling points of two similar liquids should have the same value at all pressures. This is known as the Ramsay–Young rule. For example, in the case of H_2O and ethanol,

$$p_{H_2O} = 0.0122 \text{ m Hg at } T = 287.5 \text{ K},$$

$$p_{ethanol} = 0.122 \text{ m Hg at } T = 273.2 \text{ K}.$$

Thus, the temperature ratio is

$$\frac{287.5}{273.2} = 1.052.$$

On the other hand, the normal boiling points are

$$T = 373.2 \text{ K for } H_2O,$$

$$T = 351.5 \text{ K for ethanol},$$

in which case the temperature ratio is

$$\frac{373.2}{351.5} = 1.062.$$

This value has about 1% deviation from 1.052 and indicates that the rule holds within the experimental error of measurement. This rule allows a quick approximate determination of the boiling point of a substance from the boiling point of another similar substance if the boiling point of the latter is known at a pressure and the boiling point of the former is desired at the same pressure.

8.4. Two-Component Systems

For a two-component system, the Gibbs phase rule is

$$f = 4 - \nu,$$

which, depending on the number of phases, will be

$$f = 0 \quad \text{for } \nu = 4$$
$$= 1 \quad \text{for } \nu = 3$$

$$= 2 \quad \text{for } \nu = 2$$
$$= 3 \quad \text{for } \nu = 1.$$

The last case is a homogeneous two-component system which does not interest us here. The case of $f = 0$ corresponds to a point, which is invariant, in a three-dimensional space of intensive thermodynamic variables, for example, T, p, and x_1. The case of $f = 1$ can be described by the Clapeyron–Clausius equation whereas the case of $f = 2$ is described by a differential form of two independent variables. Since we have already examined the Clapeyron–Clausius equation, we shall consider the latter case ($f = 2$) in this section.

Since there are two phases and two components in this case, the equilibrium conditions are

$$T_1 = T_2 = T,$$
$$p_1 = p_2 = p, \tag{8.56}$$
$$\mu_i^{(1)} = \mu_i^{(2)} = \mu_i \quad (i = 1, 2).$$

Taking differentials, we obtain the last condition in the form

$$d\mu_i^{(1)} = d\mu_i^{(2)} \quad (i = 1, 2). \tag{8.57}$$

Let us take T, p, $x_2^{(\alpha)}$ ($\alpha = 1, 2$) for independent variables. The chemical potentials then are functions thereof. Applying Eq. (7.37) to the present case, we obtain differential forms for chemical potentials

$$d\mu_i^{(\alpha)} = -\bar{s}_i^{(\alpha)} dT + \bar{v}_i^{(\alpha)} dp + \phi_i^{(\alpha)} dx_2^{(\alpha)} \quad (i = 1, 2; \; \alpha = 1, 2), \tag{8.58}$$

where $\bar{s}_i^{(\alpha)}$ and $\bar{v}_i^{(\alpha)}$ are, respectively, the partial molar entropy and partial molar volume of component i in phase α, and

$$\phi_i^{(\alpha)} = \left(\frac{\partial \mu_i^{(\alpha)}}{\partial x_2^{(\alpha)}} \right)_{T,p}. \tag{8.59}$$

Since in general,

$$\frac{S^{(\alpha)}}{n} = \sum_{i=1}^{r} x_i^{(\alpha)} \bar{s}_i^{(\alpha)},$$

and

$$\frac{V^{(\alpha)}}{n} = \sum_{i=1}^{r} x_i^{(\alpha)} \bar{v}_i^{(\alpha)} \quad \left(n = \sum_{i=1}^{r} n_i \right),$$

the Gibbs–Duhem equation for the present case may be written in the form

$$\sum_{i=1}^{2} x_i^{(\alpha)} d\mu_i^{(\alpha)} + \sum_{i=1}^{2} x_i^{(\alpha)} (\bar{s}_i^{(\alpha)} dT - \bar{v}_i^{(\alpha)} dp) = 0 \quad (\alpha = 1, 2). \tag{8.60}$$

This equation leads us to a pair of equations which are comparable to the Clapeyron equation for two-phase equilibria of a single-component system. In order to obtain one of them, let us take $\alpha = 1$ in Eq. (8.60):

$$\sum_{i=1}^{2} x_i^{(1)} d\mu_i^{(1)} + \sum_{i=1}^{2} x_i^{(1)} (\bar{s}_i^{(1)} dT - \bar{v}_i^{(1)} dp) = 0. \tag{8.61}$$

Upon elimination of $d\mu_i^{(1)}$ of this equation by using Eqs. (8.57) and (8.58), there follows the differential form

$$\sum_{i=1}^{2} x_i^{(1)} \phi_i^{(2)} dx_2^{(2)} + \Delta \bar{s}^{(1)} dT - \Delta \bar{v}^{(1)} dp = 0, \tag{8.62}$$

where

$$\Delta \bar{v}^{(1)} = \sum_{i=1}^{2} x_i^{(1)} \Delta \bar{v}_i, \qquad \Delta \bar{s}^{(1)} = \sum_{i=1}^{2} x_i^{(1)} \Delta \bar{s}_i, \tag{8.63}$$

$$\Delta \bar{v}_i = \bar{v}_i^{(1)} - \bar{v}_i^{(2)}, \qquad \Delta \bar{s}_i = \bar{s}_i^{(1)} - \bar{s}_i^{(2)}. \tag{8.64}$$

Since at equilibrium

$$\Delta \bar{s}_i = \frac{\Delta \bar{h}_i}{T}, \tag{8.65}$$

where $\Delta \bar{h}_i$ is the enthalpy change for component i owing to the phase change

$$\Delta \bar{h}_i = \bar{h}_i^{(1)} - \bar{h}_i^{(2)}, \tag{8.66}$$

we may recast Eq. (8.62) in a more suitable form

$$-\frac{\Delta \bar{h}^{(1)}}{T} dT = -\Delta \bar{v}_1^{(1)} dp + \left(\sum_{i=1}^{2} x_i^{(1)} \phi_i^{(2)} \right) dx_2^{(2)}, \tag{8.67}$$

with $\Delta \bar{h}^{(1)}$ defined by

$$\Delta \bar{h}^{(1)} = \sum_{i=1}^{2} x_i^{(1)} \Delta \bar{h}_i. \tag{8.68}$$

A similar analysis can be made by taking $\alpha = 2$ in Eq. (8.60). By defining

$$\Delta \bar{v}^{(2)} = \sum_{i=1}^{2} x_i^{(2)} \Delta \bar{v}_i, \quad \Delta \bar{h}^{(2)} = -\sum_{i=1}^{2} x_i^{(2)} \Delta \bar{h}_i, \tag{8.69}$$

we also obtain a differential form equivalent to Eq. (8.67):

$$-\frac{\Delta \overline{h}^{(2)}}{T} dT = -\Delta \overline{v}^{(2)} dp + \left(\sum_{i=1}^{2} x_i^{(2)} \phi_i^{(1)} \right) dx_2^{(1)}. \qquad (8.70)$$

Note that Eq. (8.67) is in reference to $x_2^{(2)}$ in phase 2 whereas Eq. (8.70) is in reference to $x_2^{(1)}$ in phase 1. The solutions of these two equations will give relations for T and p in reference to the concentration of component 2 either in phase 1 or in phase 2.

Since it is not easy to integrate the differential forms in the three-dimensional space of T, p, $x_2^{(1)}$ or T, p, $x_2^{(2)}$, we shall consider a couple of special cases. These special cases mathematically correspond to cutting the space with a plane and studying the relations between two variables in the plane.

If the pressure is kept constant, we obtain first-order differential equations from Eqs. (8.67) and (8.70):

$$\left(\frac{d \ln T}{dx_2^{(2)}} \right)_p = -\frac{1}{\Delta \overline{h}^{(1)}} \sum_{i=1}^{2} \phi_i^{(2)} x_i^{(1)},$$

$$\left(\frac{d \ln T}{dx_2^{(1)}} \right)_p = -\frac{1}{\Delta \overline{h}^{(1)}} \sum_{i=1}^{2} \phi_i^{(2)} x_i^{(2)}. \qquad (8.71)$$

For the purpose of integrating these equations, it is necessary to know the $x_2^{(2)}$-dependence of $\Delta \overline{h}^{(1)}$, $\phi_i^{(2)}$, and $x_2^{(1)}$ or the $x_2^{(1)}$-dependence of $\Delta \overline{h}^{(1)}$, $\phi_i^{(2)}$, and $x_2^{(2)}$. Assuming they are known, when these two differential equations are integrated, there arise relations between T and $x_2^{(1)}$ or T and $x_2^{(2)}$. Schematically, these relations may be represented by the curves in Figs. 8.12 and 8.13. Note that the two curves are not necessarily of the same shape. This is generally the case. By combining the two figures into one, we usually represent the phase diagrams in a single figure, and in this particular case obtain the $T-x_2$ curves for a two-phase system shown in Fig. 8.14.

On the other hand, if the temperature is kept constant, we obtain from Eqs. (8.67) and (8.70) the following pair of equations:

$$\left(\frac{dp}{dx_2^{(2)}} \right)_T = \frac{1}{\Delta \overline{v}^{(1)}} \sum_{i=1}^{2} \phi_i^{(2)} x_i^{(1)},$$

$$\left(\frac{dp}{dx_2^{(1)}} \right)_T = \frac{1}{\Delta \overline{v}_i^{(2)}} \sum_{i=1}^{2} \phi_i^{(1)} x_i^{(2)}. \qquad (8.72)$$

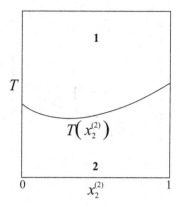

Fig. 8.12. Phase diagram for a two-component system in $T-x_2^{(2)}$ plane. The superscript refers to the phase whereas the subscript refers to the component. Numerals 1 and 2 refer to the phases.

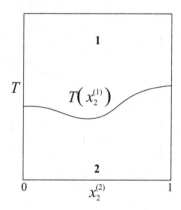

Fig. 8.13. Phase diagram for a two-component system in $T-x_2^{(1)}$ plane. The meanings of the symbols are similar to Fig. 8.12.

By integrating these equations, it is possible to find the pressure–concentration relations for the system at a given T. For example, they may look as shown in Figs. 8.15 and 8.16. In Figs. 8.14 and 8.16, two curves coincide at a value of x. At that temperature and pressure, two phases acquire the same composition. Such a mixture is called *azeotropic*. When a mixture becomes azeotropic, fractional distillation is no longer possible unless the variable kept fixed, namely, T or p, is appropriately changed. Water–alcohol mixtures or benzene–ethanol mixtures can be azeotropic at certain conditions. If x_2 is kept constant, then there results from Eqs. (8.67) and (8.70)

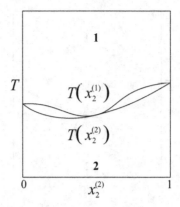

Fig. 8.14.   Phase diagram for a two-component system in T–x_2 plane. This is a combined figure of Figs. 8.12 and 8.13.

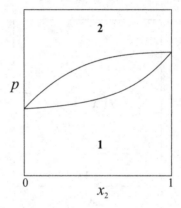

Fig. 8.15.   Phase diagram for a two-component system in p–x_2 plane.

Clapeyron equations, which we have already studied. Therefore, we do not consider this case here. We will study phase equilibria of two-component solutions in more detail in later chapters, where thermodynamics of solutions is dealt with.

It is, of course, possible to study multicomponent mixtures, but the thermodynamic principles required are the same and the general theory developed can be followed, except that the phase diagrams become more complex and the equations involved are more complicated; for example, triangular phase diagrams for three-component systems. These topics are

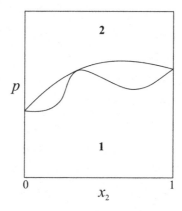

Fig. 8.16. Phase diagram for a two-component system in p–x_2 plane.

referred to the monographs[3] specializing on the subjects, such as material sciences and metallurgy in which phase diagrams help design alloys for specific purposes, for example.

Problems

(1) The vapor pressure of a liquid is represented by the expression

$$\ln p = -\frac{A}{T} + B - CT + DT^2.$$

Find the temperature dependence of the heat of vaporization under the assumption that the vapor is ideal.

(2) Assume that the chemical potentials for a binary mixture are given by

$$\mu_i^{(\alpha)} = \mu_i^{0(\alpha)} + RT \ln x_i^{(\alpha)} \quad (1,2; \ i = 1,2),$$

when two phases (liquid and vapor) are in equilibrium, and obtain from Eq. (8.71) the differential equations for T with regard to variables $x_i^{(\alpha)}$. Integrate these differential equations to obtain the relations to $x_i^{(\alpha)}$ $(\alpha = 1,2)$.

---

[3] For example, see A. Findlay, *The Phase Rule and Its Applications*, 9th edition revised by A. N. Campbell and O. S. Smith (Dover, New York, 1951); L. S. Palatnik and A. I. Landau, *Phase Equilibria in Multicomponent Systems* (Holt, Rinehart, and Winston, New York, 1964).

Chapter 9

Thermodynamics of Real Fluids

We have discussed the basic thermodynamic concepts and principles in the previous chapters and indicated some calculational procedures for thermodynamic quantities of ideal gases. In this and following chapters, we apply those concepts and principles to study the thermodynamic properties of real gases and liquids. Since the results for ideal gases can be obtained at zero pressure or low density limits from the thermodynamic properties of real fluids, we will not separately consider the ideal gas thermodynamics, but only indicate the ideal gas results whenever necessary and possible. The thermodynamics of real fluids requires a suitable generalization of an equation of state as well as the caloric equation of state and the constitutive relations for chemical potentials — generally called constitutive equations — beyond the ideal gas equation of state.

9.1. Constitutive Equations

The thermodynamic state of a system at equilibrium is described by the equilibrium Gibbs relation (fundamental relation), which is a differential form for the entropy of an r component mixture regardless of the state of aggregation of the system,

$$TdS = dE + pdV - \sum_{i=1}^{r} \mu_i dn_i, \qquad (9.1)$$

or by the differential form in the energy representation

$$dE = TdS - pdV + \sum_{i=1}^{r} \mu_i dn_i. \tag{9.2}$$

This differential form gives in principle the internal energy E in terms of independent variables S, V and n_i, $i = 1, 2, \ldots, r$, or in the entropy mode, the entropy S in terms of E, V and n_i, $i = 1, 2, \ldots, r$, if the differential form is integrated. For a closed system, $dn_i = 0$, and we have

$$dE = TdS - pdV. \tag{9.3}$$

These equilibrium Gibbs relations are not complete for the description of thermodynamic states of a system since the derivatives are not specified in terms of the independent variables chosen. That is, for example, in the case of the energy mode, the derivatives of the differential form are

$$T = \left(\frac{\partial E}{\partial S}\right)_{V,n_i}, \tag{9.4}$$

$$p = \left(\frac{\partial E}{\partial V}\right)_{S,n_i}, \tag{9.5}$$

$$\mu_i = \left(\frac{\partial E}{\partial n_i}\right)_{S,V,n_{j\neq i}}, \tag{9.6}$$

which evidently must be functions of $s = S/n$, $v = V/n$, $n_i = 1, 2, \ldots, r$, but their precise functional forms in terms of the independent variables are not revealed by the differential form (9.1) or (9.2) itself. In thermodynamics, the functional forms of such derivatives must be supplied on the empirical grounds. They are called the constitutive relations. They are characteristics of the substance of interest, indicating their thermophysical properties in terms of material parameters.

It is common to take the caloric equation of state—i.e., c_v or c_p as a function of T and v or p—in place of $T = T(s, v, n_i, \ldots, n_r)$. Thus, we choose the following set of constitutive equations:

$$c_v = c_v(T, v, x_1, \ldots, x_{r-1}), \tag{9.7}$$

or

$$c_p = c_p(T, p, x_1, \ldots, x_{r-1}), \tag{9.8}$$

and

$$p = f(T, v, x_1, \ldots, x_{r-1}), \tag{9.9}$$

$$\mu_i = \mu_i(T, p, x_1, \ldots, x_{r-1}) \quad (i = 1, 2, \ldots, r), \tag{9.10}$$

where it is assumed that there are r components in the system. These relations reflect the properties of the substance in question and hence the name constitutive equations (or relations). Equation (9.9) is, especially, called the equation of state, and it must be empirically determined through experiment or theoretically derived by means of statistical mechanics. There are numerous empirical equations of state known in thermodynamics. Different substances also have different equations of state, although the same class of substances may have the same mathematical form for the equation of state with different parameters which reflect the characteristics of the substances.

9.1.1. *Ideal Gas Equation of State*

For ideal gases, as shown earlier, the equation of state takes the universal form

$$pV = nRT, \tag{9.11}$$

where n is the mole number of the gas in volume V. This equation of state holds only when the pressure is sufficiently low. More rigorously, we may write it in the form

$$\lim_{p \to 0} \frac{pV}{nRT} = 1. \tag{9.12}$$

Thus, the equation of state as written in Eq. (9.11) must be understood by the limiting form, Eq. (9.12). Since by definition

$$\left(\frac{\partial E}{\partial V} \right)_T = 0,$$

or

$$\left(\frac{\partial E}{\partial p} \right)_T = 0$$

for ideal gases, the internal energy of ideal gases depends on the temperature only.

9.1.2. Caloric Equation of State

Since specific heats of monatomic ideal gases are known to be independent
of T, we will specifically denote them by $C_p^* = nc_p^*$ and $C_v^* = nc_v^*$:

$$c_p^* = \lim_{p \to 0} c_p, \quad c_v^* = \lim_{p \to 0} c_v. \tag{9.13}$$

If the temperature is not so high as to excite electrons in atoms into the
excited states, the specific heats of monatomic gases are those of the translational degrees of freedom and have the following values: $c_p^* = \frac{5}{2}nR$ and
$c_v^* = \frac{3}{2}nR$ per mole. Specific heats of ideal diatomic and polyatomic gases
depend on temperature because the excitation of internal motions depends
on temperature, and they are sometimes a rather strong function of T. It is
generally possible to split specific heats into the translational and internal
contributions:

$$c_p(T) = c_p^* + c_p^{\text{int}}(T), \qquad c_v(T) = c_v^* + c_v^{\text{int}}(T). \tag{9.14}$$

Note that for ideal gases, the translational part is the same for all molecules
regardless of whether they are monatomic, diatomic, or polyatomic. It is
possible to derive explicitly the contributions of the internal degrees of freedom to the specific heats by using the method of statistical mechanics, but
it is sufficient for our purpose here to express them in a power series of T as
indicated in Eq. (3.39), Chapter 3. In this work, we are consciously avoid-
ing blending thermodynamics with statistical mechanics — the molecular
theory—as much as possible.

9.1.3. Ratio of Specific Heats and Compressibility

The ratio of specific heats C_p and C_v can be measured from compressibility
data. To see this, we calculate the polytropic ratio

$$\gamma = \frac{C_p}{C_v} = \frac{c_p}{c_v}. \tag{9.15}$$

This can be easily expressed as

$$\gamma = \frac{\left(\frac{\partial V}{\partial p}\right)_T}{\left(\frac{\partial V}{\partial p}\right)_S}, \tag{9.16}$$

which can be shown to be true as follows:

$$\gamma = \frac{\left(\frac{\partial S}{\partial T}\right)_p}{\left(\frac{\partial S}{\partial T}\right)_V} = \frac{\partial(S,p)}{\partial(T,p)} \cdot \frac{\partial(T,V)}{\partial(S,V)}$$

$$= \frac{\partial(T,V)}{\partial(T,p)} \cdot \frac{\partial(S,p)}{\partial(T,V)} \cdot \frac{\partial(T,V)}{\partial(S,V)} = \frac{\partial(T,V)}{\partial(T,p)} \cdot \frac{\partial(S,p)}{\partial(S,V)}$$

$$= \frac{\left(\frac{\partial V}{\partial p}\right)_T}{\left(\frac{\partial V}{\partial p}\right)_S}.$$

This implies that it is simply the ratio of isothermal compressibility κ_T to adiabatic compressibility κ_S defined by

$$\kappa_S = -v^{-1}\left(\frac{\partial v}{\partial p}\right)_S, \tag{9.17}$$

i.e.,

$$\gamma = \frac{\kappa_T}{\kappa_S}. \tag{9.18}$$

This shows that γ can be measured if isothermal and adiabatic compressibilities are determined. We will discuss a way of determining the polytropic ratio γ in the following section.

9.1.4. Sound Wave Velocity and Polytropic Ratio

A small amplitude, long wavelength compressional oscillatory motion propagating in a compressible fluid is called a sound wave. At each point of the fluid, compression and rarefaction are alternately caused by the sound wave. The velocity of such a sound wave is related to the polytropic ratio γ as will be shown. Therefore, the former can be used for measuring the latter.

As preparation for the discussion, we examine the thermodynamics of compression and rarefaction related to such a wave in an ideal gas. We assume that the gas obeys the ideal constitutive relations and is also non-dissipative, that is, there is no viscosity and thermal conductivity of the fluid. This latter assumption is not generally true, but simplifies the discussion in making the point.

A sound wave in an ideal monatomic gas is an adiabatic process and hence isentropic: For we find that the equilibrium Gibbs relation can be written as

$$dS = T^{-1}(dE + pdV)$$

$$= nRd[\ln(T^{3/2}V)]. \tag{9.19}$$

For an adiabatic process, the argument of the logarithmic function is indeed a constant as shown in a previous chapter, and hence,

$$dS = 0. \tag{9.20}$$

Therefore, a sound wave is an isentropic process.

If we denote by p_0 and ρ_0 the equilibrium pressure and mass density, and by p' and ρ' their variations caused by the sound wave, then p and ρ are, to the first order, given by the formulas

$$p = p_0 + p', \quad \rho = \rho_0 + \rho', \tag{9.21}$$

where

$$p' \ll p_0, \quad \rho' \ll \rho_0.$$

In view of this disparities in magnitude of the variables, the equation of continuity for ρ and the momentum balance equation may be linearized as follows:

$$\frac{\partial \rho'}{\partial t} = -\rho_0 \nabla \cdot \mathbf{u}, \tag{9.22}$$

$$\rho_0 \frac{\partial \mathbf{u}}{\partial t} = -\nabla p', \tag{9.23}$$

where $\mathbf{u}$ is the velocity of the fluid. Since the flow process in the sound wave is adiabatic, or isentropic, as shown earlier and the pressure varies together with the density, p may be expanded in ρ'. So, to the first order in ρ', we obtain

$$p' = \left(\frac{\partial p}{\partial \rho_0}\right)_S \rho'. \tag{9.24}$$

Furthermore, by the definition of adiabatic compressibility, we obtain

$$\rho_0 \left(\frac{\partial p}{\partial \rho_0}\right)_S = -v \left(\frac{\partial p}{\partial v}\right)_S = \kappa_S^{-1}, \tag{9.25}$$

where $v = 1/\rho_0$, the specific volume. From Eqs. (9.22) and (9.24) follows the equation

$$\frac{\partial p'}{\partial t} + \rho_0 \left(\frac{\partial p}{\partial \rho_0}\right)_S \nabla \cdot \mathbf{u} = 0. \tag{9.26}$$

It is convenient to introduce a scalar potential $\phi(\mathbf{r}, t)$ such that

$$\mathbf{u} = \nabla\phi. \tag{9.27}$$

Then we find, by combining Eqs. (9.23) and (9.27), the relation between p' and ϕ:

$$p' = -\rho_0 \frac{\partial\phi}{\partial t}. \tag{9.28}$$

Upon substitution of Eqs. (9.27) and (9.28) into Eq. (9.26), we obtain the wave equation for ϕ:

$$\frac{\partial^2\phi}{\partial t^2} = c^2\nabla^2\phi, \tag{9.29}$$

where c is the speed of sound wave defined by

$$c = \sqrt{\left(\frac{\partial p}{\partial \rho_0}\right)_S}. \tag{9.30}$$

Since for an ideal gas,

$$\left(\frac{\partial p}{\partial \rho_0}\right)_S = \frac{1}{\kappa_S \rho_0} = \frac{\gamma}{\kappa_T \rho_0} = \frac{\gamma RT}{m}, \tag{9.31}$$

where m is the molecular mass of the gas, we obtain the sound wave speed in the form

$$c = \sqrt{\frac{\gamma RT}{m}}. \tag{9.32}$$

This is a desired relation we are looking for.

The general solution of the wave equation, Eq. (9.29), is given by

$$\phi(\mathbf{r}, t) = f_1(\mathbf{r} - ct) + f_2(\mathbf{r} + ct). \tag{9.33}$$

Here, we will confine the discussion to a one-dimensional sound wave in the direction of x-axis. Since the x-component of the velocity u_x is given by

$$u_x = \frac{\partial}{\partial x} f_1(x - ct), \tag{9.34}$$

but

$$p' = \rho_0 c \frac{\partial}{\partial x} f_1(x - ct), \tag{9.35}$$

we find

$$u_x = \frac{p'}{\rho_0 c} = \frac{\rho'}{\rho_0} c, \tag{9.36}$$

for which we have used Eq. (9.24), $p' = c^2\rho'$. The discussion given here and, especially, formula (9.32) shows that the polytropic ratio γ can be determined by measuring the sound wave speed u_x.

The discussion given in this section is not of equilibrium thermodynamics, but, as a matter of fact, of hydrodynamics in essence. It shows the role played by equilibrium thermodynamics in the derivation of the wave equation. It underscores the importance of the concepts in equilibrium thermodynamics even for investigation of dynamic problems, although such concepts require some subtle assumptions on time and spatial scales of the processes involved. For more appropriate description of sound wave propagations, a method of irreversible thermodynamics must be employed. An example is discussed later in Chapter 20.

It is useful to mention an empirical relation relating the sound velocity, density, and surface tension of a liquid, known as the Auerbach relation[1]:

$$c = \left(\sigma/6.33 \times 10^{-10}\rho\right)^{2/3}, \tag{9.37}$$

where σ is the surface tension and ρ is the density of the liquid. According to Blairs,[2] this relation may be written in the form

$$c = Av^{1/6}\left(\gamma\sigma/\rho\right)^{1/2}, \tag{9.38}$$

where A is a constant, which remains empirical.

9.2. Virial Equation of State

One of the most widely used equations of state is the virial equation of state, which is often called the virial expansion[3] for pressure. This equation of state has a firm statistical mechanical theory to support it, although it does not describe the liquid regime as well as the van der Waals equation of state does. In fact, historically, the latter preceded the former. It is common to express the virial expansion, in the case of a single-component system, in the form

$$\frac{pv}{RT} = 1 + B(T)v^{-1} + C(T)v^{-2} + \cdots, \tag{9.39}$$

---

[1]See N. Auerbach, *Experientia* **4**, 473 (1948).

[2]B. S. Blairs, J. *Colloid Interface Sci.* **302**, 312 (2006).

[3]M. Kammerlingh-Onnes for the first time proposed the virial expansion as an alternative to the van der Waals equation of state.

where the coefficients B, C, and so on, are called the second, third virial coefficient, and so on. They are functions of T alone. Alternatively, the virial expansion may be written as a power series of pv

$$pv = RT + B'(T)p + C'(T)p^2 + \cdots . \qquad (9.40)$$

Since the two forms of the virial expansion must be equivalent, the virial coefficients in the two modes of expansion are related to each other. By substituting Eq. (9.39) into Eq. (9.40) and comparing the resulting series in inverse power of v with Eq. (9.39), we find their relationships:

$$B' = B,$$

$$C' = \frac{1}{RT}(C - B^2),$$

and so on.

Second virial coefficients vary with temperature as shown in Fig. 9.1 and evident from an approximate van der Waals form, Eq. (9.49), given below. The temperature at which $B(T)$ vanishes is called the Boyle temperature: $B(T_B) = 0$. The van der Waals equation predicts the Boyle temperature

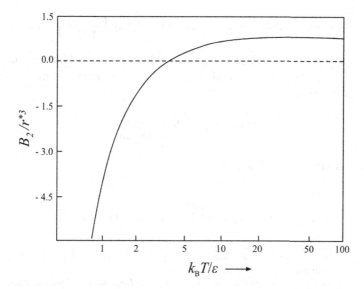

Fig. 9.1. The temperature dependence of the second virial coefficient. It has a zero at $T = T_B$, the Boyle point. $r^* \equiv \sigma$, and ε is the well depth of the potential.

at $T_B = a/Rb$. Compared with the experimental value, the van der Waals equation predicts the Boyle temperature too high. If the pressure is moderate, the virial expansion for pressure may be truncated at the v^{-1} term inclusive:

$$\frac{pv}{RT} = 1 + B(T)v^{-1}. \tag{9.41}$$

Then at the Boyle temperature, at which the second virial coefficient vanishes, the gas represented by this equation of state behaves as if it is ideal, since at the Boyle temperature T_B the second virial coefficient vanishes:

$$B(T_B) = 0. \tag{9.42}$$

Before closing this section, it is interesting to note that by using the statistical mechanical formula for $B(T)$, it is possible to derive the temperature dependence of the second virial coefficient of the Lennard-Jones (LJ) fluid obeying the potential

$$U(r) = 4\varepsilon \left[\left(\frac{r}{\sigma} \right)^{12} - \left(\frac{r}{\sigma} \right)^{6} \right].$$

It is given by the formula[4]

$$B_2 = -\sqrt{2}\Gamma \left(\frac{1}{4} \right) v_0 t^{1/4} \left[6tM \left(\frac{7}{4}, \frac{3}{2}, t \right) - (1 + 4t) M \left(\frac{3}{4}, \frac{1}{2}, t \right) \right.$$
$$\left. + \frac{20}{3} t^{3/2} M \left(\frac{9}{4}, \frac{5}{2}, t \right) + 2\sqrt{t} \, (1 - 4t) M \left(\frac{5}{4}, \frac{3}{2}, t \right) \right]. \tag{9.43}$$

Here, $t = \varepsilon\beta$ ($\beta = 1/k_B T$), $v_0 = \pi\sigma^3/6$, $\Gamma(x)$ is a Gamma function $[\Gamma(1/4) = 3.6256099082\ldots]$, and $M(a, b, t)$ is Kummer's confluent hypergeometric function[5]

$$M(a, b, t) = \sum_{n=0}^{\infty} \frac{(a)_n}{(b)_n} \frac{t^n}{n!} \tag{9.44}$$

with the coefficients defined by

$$(a)_0 = 1,$$
$$(a)_n = a(a+1)(a+2)\cdots(a+n-1) \quad (n \geq 1).$$

---

[4]See Appendix A of this book in which the derivation of the second virial coefficient for Lennard-Jones fluids is described in detail.
[5]See, for example, M. Abramowitz and I. A. Stegun, *Handbook of Mathematical Functions* (National Bureau of Standards, Washington, D.C., 1964), pp. 504–535.

This formula for $B_2(\beta)$ predicts its asymptotic behaviors—the limiting laws—as follows: as $T \to \infty$ or $\varepsilon\beta \to 0$,

$$B_2 = 4\sqrt{2}\Gamma\left(\frac{3}{4}\right) v_0 \left(\varepsilon\beta\right)^{1/4} \left[1 + O(\varepsilon\beta)\right],$$

(9.45)

$$\Gamma\left(\frac{3}{4}\right) = 1.2254167024\ldots$$

and thus $B_2 \to +0$ as $T \to \infty$ and, on the other hand, as $T \to 0$ or $\varepsilon\beta \to \infty$,

$$B_2(T) = -16\sqrt{2\pi}v_0 e^{\varepsilon\beta} \left(\varepsilon\beta\right)^{\frac{3}{2}} \left[1 + \frac{19}{16\varepsilon\beta} + \frac{105}{512\left(\varepsilon\beta\right)^2} + \cdots\right], \quad (9.46)$$

and thus $B_2 \to -\infty$ as $T \to \infty$. These asymptotic formulas are consistent with the experimental behavior of B shown in Fig. 9.1. As a matter of fact, with an appropriate choice of potential parameters, formula (9.41) predicts experimental data of simple fluids in excellent accuracy. The exact value of the Boyle point is also computable from formula (9.43).

9.3. van der Waals Equation of State

As the density of gas increases, the ideal gas equation of state begins to show its inadequacy. To remedy it van der Waals, based on an insightful argument on dense fluids, proposed an equation of state, which impressively displays the behavior of dense fluids in the liquid–vapor coexistence regime of density and temperature:

$$\left(p + \frac{a}{v^2}\right)(v - b) = RT.$$

(9.47)

Here, a and b are parameters related to the attractive dispersion force between molecules and the size of the molecule, respectively. Note that $v = V/n$ is the specific volume.

The van der Waals equation of state may be cast into a virial expansion if it is expanded in a power series of b/v:

$$\frac{pv}{RT} = 1 + \left(b - \frac{a}{RT}\right)v^{-1} + b^2 v^{-2} + \cdots.$$

(9.48)

By comparing it with the virial expansion, Eq. (9.39), we therefore obtain the second virial coefficient in terms of the van der Waals parameters a and b:

$$B(T) = b - \frac{a}{RT}. \qquad (9.49)$$

Since b and a are known to be related to the repulsive and the attractive force, respectively, the aforementioned relation shows that the second virial coefficient is related to the intermolecular forces as shown already by Eq. (9.43). In fact, the second virial coefficient can be a source for information on intermolecular forces. We note in passing that intermolecular forces can also be deduced from data on transport coefficients such as viscosities and from molecular beam experiments. In the van der Waals theory, the Boyle temperature is given by

$$T_B = \frac{a}{Rb}. \qquad (9.50)$$

The van der Waals equation of state describes continuous pressure–volume isotherms even in the subcritical regime, although it is known that they are discontinuous in the liquid–vapor coexistence regime. Despite this difficulty, if the Maxwell construction is employed as shown in the following, it impressively displays the behavior of dense fluids in the liquid–vapor coexistence regime covering both liquid and gas phases. It is capable of describing the phase transition unlike the virial equation of state earlier. In Fig. 9.2, pressure is plotted against volume for various temperatures by using Eq. (9.47). The van der Waals equation is continuous and gives rise to loops for temperatures below the critical temperature T_c. The states corresponding to the loops between (v_{min}, p_{min}) and (v_{max}, p_{max}) are not thermodynamically stable since in the interval

$$\left(\frac{\partial p}{\partial v} \right)_T > 0, \qquad (9.51)$$

and therefore they are not realizable. The actual transition may be regarded to occur along a horizontal line between points l and g. J. C. Maxwell suggested to construct the horizontal lines such that the areas of the shaded regions are equal. This procedure is called the Maxwell construction. It can be rationalized as follows.

We calculate the area under the curve between points l and g. Since the area under the curve is an integral of p over an interval of v,

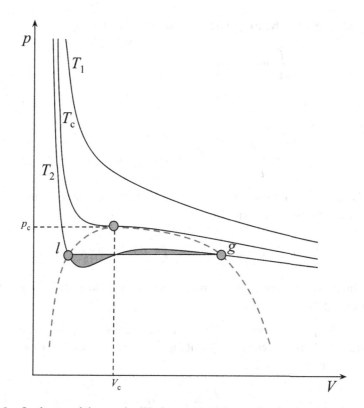

Fig. 9.2. Isotherms of the van der Waals equation of state. l and g stand for the coexist-ing liquid and vapor at equilibrium, respectively, determined by the Maxwell construc-tion. The broken curve represents the liquid–vapor coexistence curve.

we have

$$\int_l^g dv\, p = -\int_l^g [dg - d(pv) + s dT],\qquad (9.52)$$

where g in the integrand is the molar Gibbs free energy, and s the molar entropy. Since the temperature is fixed and

$$g = \mu,$$

we find

$$\text{area} = g_g - g_l + p\,(v_g - v_l)\,.$$

Since by the equilibrium condition for the two-phase single-component sys-tem $\mu_g = \mu_l$, where μ_g and μ_l are the chemical potential of the gas and of the liquid, respectively, Eq. (9.52) can be written as

$$\int_l^g dv\, p = p\,(v_g - v_l)$$

$$= p\,(v_g - v_m) + p\,(v_m - v_l)\,. \tag{9.53}$$

On the other hand,

$$\int_l^g dv\, p = \int_l^m dv\, p + \int_m^g dv\, p. \tag{9.54}$$

Therefore, by substituting this equation into the previous equation and rearranging the resulting equation, we obtain

$$\int_m^g dv\, p - p\,(v_g - v_m) = -\int_l^m dv\, p + p\,(v_m - v_l), \tag{9.55}$$

which implies that the shaded areas marked by points (l, m) and points (m, g) in Fig. 9.2 are equal.

Since the $p - v$ curve at $T = T_c$ has an inflection point, the critical temperature T_c is determined by putting the first two derivatives equal to zero at the inflection point:

$$\left(\frac{\partial p}{\partial v}\right)_{T_c} = 0, \tag{9.56}$$

$$\left(\frac{\partial^2 p}{\partial v^2}\right)_{T_c} = 0. \tag{9.57}$$

Twice differentiating the van der Waals equation of state (9.47) and setting the derivatives equal to zero at $T = T_c$, $p = p_c$, and $v = v_c$ yields two equations corresponding to those in Eqs. (9.56) and (9.57):

$$p_c + \frac{a}{v_c^2} - \frac{2a}{v_c^3}(v_c - b) = 0, \tag{9.58}$$

$$-\frac{4a}{v_c^3} + \frac{6a}{v_c^4}(v_c - b) = 0. \tag{9.59}$$

At the critical point the van der Waals equation is

$$\left(p_c + \frac{a}{v_c^2}\right)(v_c - b) = RT_c. \tag{9.60}$$

Table 9.1. van der Waals Constants for Various Fluids

Fluids	$a\,(L^2\,\text{atm}\,\text{mol}^{-2})$	$b\,(\text{L}\,\text{mol}^{-1}) \times 10^2$
Ar	1.35	3.23
He	0.034	2.38
H_2	0.245	2.67
O_2	1.36	3.19
N_2	1.39	3.92
CO_2	3.60	4.28
NH_3	4.17	3.72
H_2O	5.46	3.30

When these three equations are solved for p_c, v_c and T_c, there follow three expressions for v_c, p_c, and T_c given in terms of parameters a and b only:

$$v_c = 3b,$$
$$p_c = \frac{a}{27b^2}, \tag{9.61}$$
$$T_c = \frac{8a}{27Rb}.$$

It is amusing to note that elimination of a and b from three equations in Eq. (9.61) yields an equation

$$p_c v_c = \frac{3}{8}RT_c,$$

which looks like an ideal gas equation of state at the critical point except for the factor $3/8$. Equation (9.61) may be used to obtain the parameters a and b from the critical data. Some values of a and b so determined are listed in Table 9.1. It is possible to calculate various thermodynamic properties of the system by using the van der Waals equation of state once the van der Waals constants are determined. The results thus obtained are not always completely satisfactory, but give useful approximations for thermodynamic quantities when no other information is available at high pressures where a nonideality correction must be made in order to properly account for thermodynamic data.

9.4. Law of Corresponding States

It is possible to define reduced temperature τ, reduced pressure π, and reduced volume ω by scaling T, p, and v with their critical values:

$$\tau = T/T_c, \quad \omega = v/v_c, \quad \pi = p/p_c. \tag{9.62}$$

In the case of the van der Waals equation of state, since the critical data
are given by Eq. (9.61), upon elimination of the van der Waals constants a
and b from the equation of state, there follows a reduced equation of state

$$\left(\pi + \frac{3}{\omega^2}\right)(3\omega - 1) = 8\tau. \tag{9.63}$$

An interesting feature of this equation is that it does not contain parameters indicative of the substance. It is a universal relationship for the reduced
variables involved and therefore is expected to hold for a class of substances.
If equimolar amounts of two gases, whose $p - v - T$ relation may be represented by a van der Waals equation of state, are at the same reduced
pressure and reduced volume, then they are expected to be at the same
reduced temperature. They are then said to be in the corresponding states.
Equation (9.63) is an example of the law of corresponding states. In fact,
the van der Waals equation is not the only equation of state for which the
law of corresponding states holds. For example, the virial equation of state
can be put into a reduced form if the reduced variables are suitably defined.
Thus, we may write the equation of state in a reduced form valid for a wide
class of substances, independent of material parameters

$$\pi = f(\tau, \omega), \tag{9.64}$$

and compactly summarize the equation of state data in a single universal
equation. As Fig. 9.3 shows, there are many fluids which obey the law of
corresponding states to a good approximation. Note that data for various
gases follow the same reduced curve at a given reduced temperature.

9.5.   Thermodynamic Functions

In the following sections, we shall calculate various thermodynamic quantities by using the equations of state for real gases given in the previous
section. Although there are many more equations of state known in the
literature, we first will use either the van der Waals or the virial equation
of state, since they have statistical mechanical foundations[6] and sufficiently
well illustrate applications of the thermodynamic principles and methods

---

[6]It must be noted that the van der Waals equation of state can be derived in statistical
mechanics only if some intuitive approximations are made on the basis of physical arguments, whereas the virial equation of state can be justified rigorously by expanding the
partition function in a density series by means of the cluster expansion. However, the
cluster expansion has a limited radius of convergence.

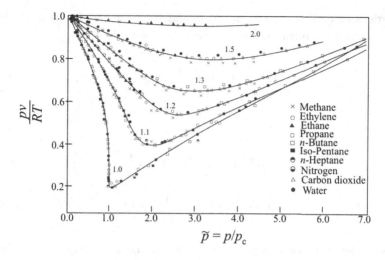

$$\tilde{p} = p/p_c$$

Fig. 9.3. Law of corresponding states. The compressibility factor pv/RT for several gases are plotted as a function of the reduced pressure p/p_c at various values of the reduced temperature T/T_c. Reproduced with permission from G. J. Su, *Ind. Eng. Chem.* **38**, 803 (1946) © 1946 American Chemical Society.

developed so far. Then in the next chapter, we will discuss the canonical equation of state, which can be given a full statistical mechanical support unlike the original van der Waals equation of state for which only an approximate derivation can be given. The thermodynamic equations developed are valid for any aggregate of matter up to a certain extent. It is only when we choose a particular equation of state that the calculations must be specialized to gases, liquids, or solids.

9.5.1. *Reversible Work*

Let us consider a pressure–volume work performed reversibly on the system at constant temperature. Assume that the volume is changed from v_1 to v_2. For a van der Waals gas, the work done on the system is easily found:

$$W = -\int_{v_1}^{v_2} p\, dv$$

$$= -RT \ln\left(\frac{v_2 - b}{v_1 - b}\right) - a\left(\frac{1}{v_2} - \frac{1}{v_1}\right). \tag{9.65}$$

In order to identify the contribution arising from the nonideality of the gas, we rearrange the terms and write this equation in the form

$$W = -RT \ln \left(\frac{v_2}{v_1} \right) + W_{\text{vdw}}, \tag{9.66}$$

where

$$W_{\text{vdw}} = -RT \ln \left(\frac{1 - bv_2^{-1}}{1 - bv_1^{-1}} \right) - a \left(\frac{1}{v_2} - \frac{1}{v_1} \right)$$
$$= (bRT - a) \left(\frac{1}{v_2} - \frac{1}{v_1} \right) + O(v^{-2}). \tag{9.67}$$

This is the correction to the work done on the ideal gas as a result of molecular interactions in the gas. The sign of the first-order correction depends on the temperature and the relative magnitudes of a and b. If the temperature is such that $b \geq a/RT$, then W_{vdw} is positive, otherwise it is negative for the process considered. If the virial equation of state (9.39) is used, then the work is given by the formula

$$W = -RT \ln \left(\frac{v_2}{v_1} \right) + W_{\text{vir}}, \tag{9.68}$$

where

$$W_{\text{vir}} = -BRT \left(\frac{1}{v_1} - \frac{1}{v_2} \right) - \tfrac{1}{2} CRT \left(\frac{1}{v_1^2} - \frac{1}{v_2^2} \right) + \cdots . \tag{9.69}$$

The first-order term agrees with the first-order term in Eq. (9.67) by the van der Waals equation of state in view of Eq. (9.49). Therefore, to this order of approximation, the van der Waals and the virial equation of state have the same value for work.

9.5.2. *Heat Change in Isothermal Expansion*

We now calculate the heat change accompanying a reversible isothermal expansion for a van der Waals gas. Similar calculations can be, of course, performed for real gases obeying other kinds of equation of state. Since in the case of an isothermal expansion of a mole of a substance,

$$dQ = Tds = T\left(\frac{\partial s}{\partial V}\right)_T dv = T\left(\frac{\partial p}{\partial T}\right)_v dv,$$

the reversible heat change accompanying the volume change from v_1 to v is given by the formula

$$Q = \int_{v_1}^{v} dv' \, T\left(\frac{\partial p}{\partial T}\right)_{v'}. \tag{9.70}$$

The van der Waals equation of state

$$\left(p + \frac{a}{v^2}\right)(v - b) = RT$$

may be arranged to the form

$$p = \frac{RT}{v - b} - \frac{a}{v^2},$$

which upon differentiation with T yields

$$\left(\frac{\partial p}{\partial T}\right)_v = \frac{R}{v - b} - \frac{1}{v^2}\frac{da}{dT}.$$

It must be noted that the parameter a is generally dependent on T and the van der Waals attraction force constant. Upon substitution of this derivative into Eq. (9.70) in the case of a mole of a van der Waals gas and performing the integration, we easily find

$$Q = \int_{v_1}^{v} dv' \left[\frac{RT}{v' - b} - \frac{T}{v'^2}\left(\frac{da}{dT}\right)\right]$$

$$= RT \ln\left(\frac{v - b}{v_1 - b}\right) + \left(\frac{1}{v} - \frac{1}{v_1}\right) T\frac{da}{dT}. \tag{9.71}$$

This may be recast in a form better exhibiting the nonideal part by using a procedure similar to that leading to Eq. (9.66):

$$Q = RT \ln\left(\frac{v}{v_1}\right) + Q_{\text{vdw}}, \tag{9.72}$$

where

$$Q_{\text{vdw}} = RT \ln\left(\frac{1 - bv^{-1}}{1 - bv_1^{-1}}\right) + \left(\frac{1}{v} - \frac{1}{v_1}\right) T\frac{da}{dT}. \tag{9.73}$$

The first term on the right in Eq. (9.72) is obviously the heat change due to the volume change for the ideal gas and Q_{vdw} is the correction to it which originates from the molecular interaction.

9.5.3. *Standard States*

Since in thermodynamics it is possible to determine only the changes in thermodynamic functions accompanying a reversible process from a state to another, we are not able to determine the absolute values of thermodynamic state functions. In order to tabulate them, it is therefore desirable to choose a reference state and fix thermodynamic state functions relative to the reference state chosen. It was indicated in Chapter 3 that specific heats are usually tabulated for the ideal gas in the case of gases and at 1 atm (1 atm $:= 1.01325 \times 10^5$ Pa) in the case of liquids and solids. The ideal gas state is chosen for gases because specific heats are independent of pressure and volume for ideal gases.

In the case of enthalpy, the reference state is the ideal gas state or the limit for the real gas as the pressure approaches the zero pressure, and the standard state is then the hypothetical ideal gas at $T = 298.15$ K and $p = 1$ atm in the case of real gases. For liquids or solids of elements, the standard state is their most stable state at the pressure of 1 atm. At the standard state at $T = 298.15$ K, the enthalpy is set equal to zero for liquids and solids of elements:

$$H(p = 1\,\text{atm};\ T = 298.15\ \text{K}) = 0, \tag{9.74}$$

and the entropy is also taken equal to zero at $T = 0$:

$$S(p = 1\,\text{atm};\ T = 0) = 0. \tag{9.75}$$

Since the entropy is a logarithmic function of pressure, the standard state entropy for a gas cannot be determined simply at $p = 0$ as in the case of enthalpy. The standard state entropy for a gas is defined as that of the hypothetical ideal gas at 1 atm pressure in the following manner:

$$S^0 = \lim_{p \to 0}[(S(T, p) + R \ln p]. \tag{9.76}$$

With this preparation, we are now able to calculate various thermodynamic functions at arbitrary temperature and pressure.

9.5.4. *Enthalpy*

Let us denote by h the molar enthalpy of a pure substance. Since it is convenient to take the experimentally most suitable variables as independent variables, we shall assume that h is a function of T and p: $h = h(T, p)$. Then the differential of h is given by

$$dh(T, p) = \left(\frac{\partial h}{\partial T} \right)_p dT + \left(\frac{\partial h}{\partial p} \right)_T dp. \tag{9.77}$$

This differential form is then integrated from state (p_1, T_1) to state (p, T). Since dh is an exact differential in the space of T and p, the process in question can be arbitrarily and conveniently distorted into two steps: $(p_1, T_1) \rightarrow (p_1, T)$ (isobaric path) and $(p_1, T) \rightarrow (p, T)$ (isothermal path), as shown in Fig. 9.4. This distortion of the path is possible owing to the fact that dh is a total (exact) differential. First, we consider the isobaric process along the vertical path from (p_1, T_1) to (p_1, T). The enthalpy change accompanying the isobaric process is

$$h(T, p_1) - h(T_1, p_1) = \int_{T_1}^{T} dT' \left(\frac{\partial h}{\partial T'} \right)_p (p_1) = \int_{T_1}^{T} dT' \, c_p(T', p_1), \quad (9.78)$$

where c_p is the molar specific heat at constant pressure. Since

$$\left(\frac{\partial h}{\partial p} \right)_T = v - T \left(\frac{\partial v}{\partial T} \right)_p, \tag{9.79}$$

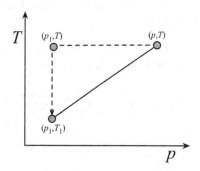

Fig. 9.4. Path of integration for h in $p - T$ plane.

the enthalpy change along the isothermal (horizontal) path is given by the expression

$$h(T, p) - h(T, p_1) = \int_{p_1}^{p} dp' \left[v - T \left(\frac{\partial v}{\partial T} \right)_{p'} \right] (T, p'). \qquad (9.80)$$

By combining Eqs. (9.78) and (9.80), we find the enthalpy change for the process from (p_1, T_1) to (p, T)

$$h(T, p) = h(T_1, p_1) + \int_{T_1}^{T} dT' \, c_p(T', p_1)$$

$$+ \int_{p_1}^{p} dp' \left[v - T \left(\frac{\partial v}{\partial T} \right)_{p'} \right] (T, p'). \qquad (9.81)$$

If the initial pressure is taken as equal to zero, namely, if $p_1 = 0$, we obtain

$$h(T, p) = h^*(T) + \int_{0}^{p} dp' \left[v - T \left(\frac{\partial v}{\partial T} \right)_{p'} \right] (T, p'), \qquad (9.82)$$

where

$$h^*(T) = h(T_1, 0) + \int_{T_1}^{T} dT' \, c_p^*(T') \qquad (9.83)$$

with $c_p^* = \lim_{p \to 0} c_p$, then $h(T_1, 0)$ is the enthalpy of the substance at $T = T_1$, where we may take $T_1 = 298.15$ K. We thus see that

$$h^*(T) = \lim_{p \to 0} h(T, p),$$

that is, $h^*(T)$ is the molar enthalpy of a hypothetical ideal gas. Therefore, the second term on the right-hand side of Eq. (9.82) is the nonideality correction for the enthalpy.

In the case of liquids and solids of elements, if we take $p_1 = 1$ atm and $T_1 = 298.15$ K, we obtain $h(T, p)$ as given below:

$$h(T, p) = h^0(T) + \int_{1}^{p} dp' \left[v - T \left(\frac{\partial v}{\partial T} \right)_{p'} \right] (T, p'), \qquad (9.84)$$

where

$$h^0(T) = \int_{T_1}^{T} dT \, c_p(T, p = 1), \qquad (9.85)$$

for which we have prescribed the standard enthalpy by the convention introduced in the previous section:

$$h(T_1 = 298.15 \text{ K}, \, p = 1 \text{ atm}) = 0.$$

Let us further examine Eq. (9.82). Since for ideal gases,

$$v - T\left(\frac{\partial v}{\partial T}\right)_p = 0,$$

the integral in Eq. (9.82) vanishes and the enthalpy becomes that of an ideal gas:

$$h(T, p) = h^*(T).$$

The nonideality correction given by the integral

$$h_{\text{real}} = \int_0^p dp' \left[v - T\left(\frac{\partial v}{\partial T}\right)_{p'} \right] (T, p') \tag{9.86}$$

can be calculated explicitly with the equation of state for the system. If we assume the virial equation of state (9.40), the integrand is calculated to be

$$v - T\left(\frac{\partial v}{\partial T}\right)_p = \left(B - T\frac{dB}{dT}\right) + \left(C' - T\frac{dC'}{dT}\right)p + \cdots .$$

Upon putting this result into the integral, we obtain the nonideality correction in terms of virial coefficients

$$h_{\text{real}} = \left(B - T\frac{dB}{dT}\right)p + \frac{1}{2}\left(C' - T\frac{dC'}{dT}\right)p^2 + \cdots . \tag{9.87}$$

If the second virial coefficient is estimated with the van der Waals theory form

$$B = b - \frac{a}{RT},$$

we obtain to first order in p the real gas part of h:

$$h_{\text{real}} = \left(b - \frac{2a}{RT}\right)p + O(p^2). \tag{9.88}$$

Observe that the real gas correction h_{real}, to first order in p, is positive if

$$b - \frac{2a}{RT} \geq 0,$$

but negative otherwise. Since parameter a is representative of the attractive part of the intermolecular force whereas b is a measure of the repulsive part, the enthalpy becomes smaller than $h^*(T)$ if the temperature is sufficiently low so that the attractive part of the intermolecular force becomes dominant over the repulsive part.

Since specific heats can be calculated with $h(T, p)$, we can obtain the pressure dependence of the nonideality correction for $c_p(T, p)$ as follows:

$$c_p(T, p) = c_p^*(T) + c_{p\,\text{real}}(T, p), \qquad (9.89)$$

where

$$c_p^* = \left(\frac{\partial h^*}{\partial T} \right)_p,$$

$$(9.90)$$

$$c_{p\,\text{real}} = \left(\frac{\partial h_{\text{real}}}{\partial T} \right)_p = -T \frac{d^2 B}{dT^2} p + O(p^2).$$

In the van der Waals theory, $c_{p\,\text{real}}$ is found to be as follows:

$$c_{p\,\text{real}} = \frac{2a}{RT^2} p + O(p^2). \qquad (9.91)$$

This gives a rough idea of how c_p depends on T and p for real gases. We thus see that data for c_p^* and the equation of state would be sufficient for determining c_p at $p \neq 0$.

If the process $(p_1, T_1) \to (p, T)$ involves a phase transition at $T = T_m$, it is necessary to take the latent heat into account. For the purpose of calculating the enthalpy change for the process, we return to Eq. (9.81) and modify the integral. The range of integral must be split into two parts: (T_1, T_m) and (T_m, T); see Fig. 9.6 for the integration path superimposed on the phase diagram. Thus, we find

$$h(T, p) - h(T_1, p_1) = \Delta h_m(T_m, p_1) + \int_{T_1}^{T_m} dT \, c_p^{(1)}(T, p_1)$$

$$+ \int_{T_m}^{T} dT' \, c_p^{(2)}(T', p_1), \qquad (9.92)$$

where Δh_m is the latent heat at $T = T_m$ and $p = p_1$, and $c_p^{(1)}$ and $c_p^{(2)}$ are, respectively, the specific heat of the low temperature phase 1 and of the

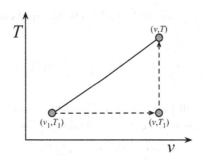

Fig. 9.5.   Path of integration for $\mathcal{E}$ in $T - v$ plane.

high temperature phase 2. The value of $h(T_1, p_1)$ may be fixed by choosing an appropriate standard state.

9.5.5.   *Internal Energy*

It is possible to calculate the volume dependence of the internal energy by using a method similar to that for the enthalpy since it is permissible to distort the path of integration into isochoric and isothermal paths because $d\mathcal{E}$ is an exact differential. Let us denote the molar internal energy by $\mathcal{E} = E/n$, where n is the mole number. It is convenient to regard $\mathcal{E}$ as a function of v and T. To investigate the T and v dependence of $\mathcal{E}$, we derive the following differential form for $\mathcal{E}$:

$$
d\mathcal{E}(T, v) = \left(\frac{\partial \mathcal{E}}{\partial T}\right)_v dT + \left(\frac{\partial \mathcal{E}}{\partial v}\right)_T dv
$$

$$
= c_v dT + \left[T\left(\frac{\partial p}{\partial T}\right)_v - p\right] dv, \tag{9.93}
$$

where c_v is the molar specific heat at constant volume. For this formula, we have made use of $\mathcal{E} = \mathcal{A} + Ts$, where $\mathcal{A}$ is the molar work function and s the molar entropy, and one of Maxwell's relations. By integrating this differential form along the path $(T_1, v_1) \to (T, v_1)$ as shown in Fig. 9.5, we obtain

$$
\mathcal{E}(T, v_1) - \mathcal{E}(T_1, v_1) = \int_{T_1}^{T} dT' \, c_v(T', v_1), \tag{9.94}
$$

and, integrating along the path $(T, v_1) \to (T, v)$,

$$
\mathcal{E}(T, v) = \mathcal{E}(T, v_1) + \int_{v_1}^{v} dv' \left[T\left(\frac{\partial p}{\partial T}\right)_v - p\right](v'). \tag{9.95}
$$

Combining Eqs. (9.94) and (9.95) yields the internal energy change for the process $(T_1, v_1) \rightarrow (T, v)$

$$\mathcal{E}(T, v) = \mathcal{E}(T_1, v_1) + \int_{T_1}^{T} dT' \, c_v(T', v_1) - \int_{v_1}^{v} dv' \left[p - T \left(\frac{\partial p}{\partial T} \right)_{v'} \right] (T, v').$$

(9.96)

Now, let us take the limit $v_1 \rightarrow \infty$. This limit is equivalent to the limit $p \rightarrow 0$ at which the gas behaves as an ideal gas. Therefore,

$$\lim_{v_1 \rightarrow \infty} c_v(T, v_1) = c_v^*(T).$$

(9.97)

Furthermore, we may choose the same standard state for the internal energy as for the enthalpy and set

$$\mathcal{E}(T_1 = 298.15 \text{ K}, \, p_1 = 1 \text{ atm}, \, v_1 = \infty) = 0$$

(9.98)

to remain consistent with the standard state for the enthalpy of the gas. Therefore, if we take the initial volume to be infinite at a fixed T, so that the gas becomes ideal, the internal energy may be given in the form

$$\mathcal{E}(T, v) = \mathcal{E}^*(T) + \mathcal{E}_{\text{real}}(T, v),$$

(9.99)

where

$$\mathcal{E}^*(T) = \int_{T_1}^{T} dT' \, c_v^*(T') + \mathcal{E}(T_1, \infty)$$

$$= \int_{T_1}^{T} dT' \, c_v^*(T') \qquad \text{if } T_1 = 298.15 \text{ K},$$

(9.100)

$$\mathcal{E}_{\text{real}} = \int_{v}^{v_1} dv' \left[p - T \left(\frac{\partial p}{\partial T} \right)_{v'} \right] (T, v').$$

(9.101)

Note that as $v \rightarrow \infty$,

$$\lim_{v \rightarrow \infty} \frac{pv}{RT} = 1$$

and, therefore,

$$p - T \left(\frac{\partial p}{\partial T} \right)_v \rightarrow 0,$$

which implies that

$$\lim_{v \rightarrow \infty} \mathcal{E}_{\text{real}}(T, v) = 0,$$

and, consequently,

$$\lim_{v \rightarrow \infty} \mathcal{E}(T, v) = \mathcal{E}^*(T).$$

(9.102)

We thus see that $\mathcal{E}^*(T)$ is the internal energy of a mole of the gas and $\mathcal{E}_{\text{real}}(T, v)$ is the nonideality correction. Note that this result is consistent with the definition of ideal gases

$$\left(\frac{\partial \mathcal{E}}{\partial v}\right)_T = 0,$$

which means that the internal energy of ideal gases is independent of v.

We may calculate the nonideality correction by using the virial equation of state (9.39). Since the virial equation of state gives

$$T\left(\frac{\partial p}{\partial T}\right)_v - p = \frac{RT^2}{v^2}\frac{dB}{dT} + \frac{RT^2}{v^3}\frac{dC}{dT} + \cdots,$$

substitution of this equation into the integral for $\mathcal{E}_{\text{real}}$ in Eq. (9.101) yields

$$\mathcal{E}_{\text{real}}(T, v) = -\frac{RT^2}{v}\frac{dB}{dT} - \frac{1}{2}\frac{RT^2}{v^2}\frac{dC}{dT} - \cdots \tag{9.103}$$

and, therefore,

$$\mathcal{E}(T, v) = \mathcal{E}^*(T) - \frac{RT^2}{v}\frac{dB}{dT} - \frac{1}{2}\frac{RT^2}{v^2}\frac{dC}{dT} - \cdots. \tag{9.104}$$

If the second virial coefficient is estimated from the van der Waals equation of state, we find

$$\mathcal{E}_{\text{real}} = -\frac{a}{v} + O(v^{-2}), \tag{9.105}$$

which implies that the internal energy is smaller than the ideal gas value because of the intermolecular attraction, at least, to first order in v^{-2} in the van der Waals theory.

It is also possible to calculate the nonideality correction for c_v by using Eq. (9.105):

$$c_v(T, v) = c_v^*(T) + c_{v\,\text{real}}(T, v), \tag{9.106}$$

where

$$c_{v\,\text{real}} = -\frac{R}{v}\frac{d}{dT}\left(T^2\frac{dB}{dT}\right) - \frac{R}{2v^2}\frac{d}{dT}\left(T^2\frac{dC}{dT}\right) - \cdots. \tag{9.107}$$

If Eq. (9.105) is used for calculating the specific heat, there follows

$$c_v(T, v) = c_v^*(T) + O(v^{-2}).$$

The specific heat in this case is equal to the ideal gas value c_v^* to second order in v^{-2}.

9.5.6. *Entropy*

Entropy may be calculated in basically the same way as for other thermodynamic functions. Let us denote the molar entropy by $s = S/n$ where n
is the mole number of the substance. It may be considered a function of T
and p. To find its dependence on T and p, we integrate its differential form

$$ds(T,p) = \left(\frac{\partial s}{\partial T}\right)_p dT + \left(\frac{\partial s}{\partial p}\right)_T dp$$

$$= \frac{c_p}{T} dT - \left(\frac{\partial v}{\partial T}\right)_p dp. \tag{9.108}$$

The second line is due to the definition of heat capacity and the use of a
Maxwell relation. This differential form is now integrated from (T_1, p_1) to
(T, p). Since s is a state function, the path may be distorted to the dotted
line in Fig. 9.4. By integrating Eq. (9.108) along the isobaric path, we obtain

$$s(T, p_1) = s(T_1, p_1) + \int_{T_1}^{T} dT \, \frac{c_p(T)}{T}. \tag{9.109}$$

The integration along the isothermal path requires a little more careful
consideration. Since along the isothermal path,

$$ds(T, p) = -\left(\frac{\partial v}{\partial T}\right)_p dp$$

and the p-dependence of $(\partial v/\partial T)_p$ is roughly

$$\left(\frac{\partial v}{\partial T}\right)_p \sim \frac{R}{p}$$

and therefore is singular with respect to p, the differential form must be
cast into a more suitable form as follows:

$$s(T, p) = \left[\frac{R}{p} - \left(\frac{\partial v}{\partial T}\right)_p\right] dp - \left(\frac{R}{p}\right) dp. \tag{9.110}$$

Upon integration of this equation along the isothermal path, there follows

$$s(T, p) = s(T, p_1) - R \ln\left(\frac{p}{p_1}\right) + \int_{p_1}^{p} dp' \left[\frac{R}{p'} - \left(\frac{\partial v}{\partial T}\right)_{p'}\right] (T, p'). \tag{9.111}$$

When Eqs. (9.109) and (9.111) are combined, there results the entropy change for the process under consideration

$$s(T,p) = s(T_1,p_1) + \int_{T_1}^{T} dT' \frac{c_p(T',p_1)}{T'} - R \ln\left(\frac{p}{p_1}\right)$$

$$+ \int_{p_1}^{p} dp' \left[\frac{R}{p'} - \left(\frac{\partial v}{\partial T}\right)_{p'}\right](T,p'). \tag{9.112}$$

Note that the specific heat c_p in Eq. (9.112) depends on the pressure p_1 at which it is measured. We consider this expression for systems in different states of aggregation

9.5.6.1. *Real Gases*

If the substance is a real gas, the thermodynamic functions, such as the enthalpy, the free energy, and so on, are calculated in reference to the standard state which is taken as the state of a hypothetical ideal gas at $p = 1$ atm. For this purpose, it is convenient to cast the entropy formula, Eq. (9.112), in reference to such standard state. This aim is easily achieved if we rearrange the terms in Eq. (9.112) and take the limit $p_1 \to 0$ as follows:

$$\lim_{p_1 \to 0} s(T,p) = \lim_{p_1 \to 0} \left\{ s(T_1,p_1) + R \ln p_1 - \int_{0}^{p_1} dp' \left[\frac{R}{p'} - \left(\frac{\partial v}{\partial T}\right)_{p'}\right](p') \right.$$

$$\left. + \int_{T_1}^{T} dT' \frac{c_p(T',p_1)}{T'} \right\}$$

$$- R \ln p + \int_{0}^{p} dp' \left[\frac{R}{p'} - \left(\frac{\partial v}{\partial T}\right)_{p'}\right](p'). \tag{9.113}$$

Note that the following limits hold:

$$\lim_{p_1 \to 0} c_p(T,p_1) = c_p^*(T),$$

and

$$\lim_{p_1 \to 0} \int_{0}^{p_1} dp' \left[\frac{R}{p'} - \left(\frac{\partial v}{\partial T}\right)_{p'}\right](p') = 0$$

because the equation of state becomes that of the ideal gas as $p_1 \to 0$ where the gas obeys the ideal gas equation of state. Therefore, Eq. (9.113) becomes

$$\lim_{p_1 \to 0} s(T, p) = \lim_{p_1 \to 0} [s(T_1, p_1) + R \ln p_1] + \int_{T_1}^{T} dT' \, \frac{c_p^*(T')}{T'}$$

$$-R \ln p + \int_{0}^{p} dp' \left[\frac{R}{p'} - \left(\frac{\partial v}{\partial T} \right)_{p'} \right] (p'). \qquad (9.114)$$

According to the standard entropy of gas introduced in Eq. (9.76), we have

$$s^0 = \lim_{p_1 \to 0} [s(T_1, p_1) + R \ln p_1]. \qquad (9.115)$$

Note that s^0 depends on T_1. We thus finally obtain the entropy in the form

$$s(T, p) = s^*(T) - R \ln p + s_{\text{real}}, \qquad (9.116)$$

where

$$s^*(T) = s^0 + \int_{T_1}^{T} dT \, \frac{c_p^*(T)}{T}, \qquad (9.117)$$

and

$$s_{\text{real}}(T, p) = \int_{0}^{p} dp' \left[\frac{R}{p'} - \left(\frac{\partial v}{\partial T} \right)_{p'} \right] (T, p').$$

Since in the case of an ideal gas,

$$s_{\text{real}} = 0,$$

$s^*(T)$ may be interpreted as the molar entropy of the hypothetical ideal gas at 1 atm pressure. Then s_{real} is the nonideality correction.

It is easy to calculate the nonideality correction by using the virial equation of state (9.40). The result is

$$s_{\text{real}} = -\frac{dB}{dT} p - \frac{1}{2} \frac{dC'}{dT} p^2 - \cdots . \qquad (9.118)$$

If the van der Waals estimate (9.49) is used for the second virial coefficient, s_{real} to first order in p is

$$s_{\text{real}} = -\frac{a}{RT^2} p + O(p^2). \qquad (9.119)$$

The entropy of a substance in a condensed phase may be calculated with Eq. (9.112) in reference to the standard state prescribed in the previous section. Thus, one may choose $p_1 = 1$ atm. In that case, $s(T_1, p_1 = 1$ atm) is the entropy of the substance at T_1 in reference to the standard state at which the entropy is prescribed equal to zero.

9.5.6.2. Substances in Condensed Phase

If the substance of interest is in a condensed phase, the reference state is taken differently from that of a gas. Therefore, there is no need for taking the limit of $p_1 \to 0$. To be general, we consider the case of the pressure range involving two condensed phases (see Fig. 9.6). The path of integration $(p_1, T_0) \to (p, T)$ in the case may be distorted into two components $(p_1, T_0) \to (p_1, T)$ and $(p_1, T) \to (p, T)$ as indicated by the broken lines in Fig. 9.6, the first of which traverses the solid–liquid phase equilibrium curve at $T = T_m$ and the second of which is an isothermal process within the liquid phase. Upon application of Eq. (9.108) along the isobaric path $(p_1, T_0) \to (p_1, T)$ we find

$$s(T, p_1) = \int_0^{T_m} dT \, \frac{c_p^{(s)}(T, p_1)}{T} + \Delta s_m (T_m, p_1) + \int_{T_m}^{T} dT' \, \frac{c_p^{(l)}(T', p_1)}{T'},$$

where $c_p^{(s)}$ and $c_p^{(l)}$ are the molar specific heat of the solid and of the liquid at $p = p_1$, respectively, and T_m is the melting point at $p = p_1$, and $\Delta s_m (T_m, p_1)$ is the discontinuity in entropy at $T = T_m$ and $p = p_1$

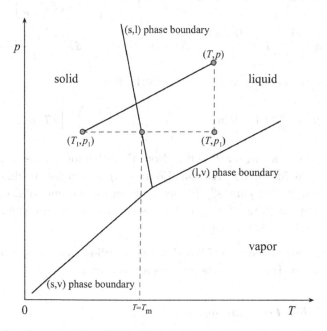

Fig. 9.6. A path crossing the solid–liquid phase boundary in $p-T$ plane. The integration path $(p_0, T_0) \to (p, T)$ can be distorted into path $(p_0, T_0) \to (p_0, T) \to (p_0, T) \to (p, T)$. The melting point is $T = T_m$.

owing to the solid–liquid phase transition. The third law of thermodynamics is made use of for the first integral on the right. Along the isothermal path $(p_1, T) \to (p, T)$, we find

$$s(T, p) = s(T, p_1) - R \ln \left(\frac{p}{p_1} \right) + \int_{p_1}^{p} dp' \left[\frac{R}{p'} - \left(\frac{\partial v}{\partial T} \right)_{p'} \right] (T, p').$$

Combining these two equations yields

$$s(T, p) = \int_0^{T_m} dT \frac{c_p^{(s)}(T, p_1)}{T} + \Delta s_m (T_m, p_1) + \int_{T_m}^{T} dT' \frac{c_p^{(l)}(T', p_1)}{T'}$$

$$- R \ln \left(\frac{p}{p_1} \right) + \int_{p_1}^{p} dp' \left[\frac{R}{p'} - \left(\frac{\partial v}{\partial T} \right)_{p'} \right] (T, p'). \qquad (9.120)$$

The last term can be computed from the knowledge of the equation of state of the liquid. Defining the reference entropy at $p_1 = 1$ atm

$$s_m^0(T) = \lim_{p_1 \to 1} \left[s(T, p) - R \ln \left(\frac{p}{p_1} \right) + \int_{p_1}^{1} dp' \left[\frac{R}{p'} - \left(\frac{\partial v}{\partial T} \right)_{p'} \right] (T, p') \right]$$

$$= \int_0^{T_m} dT \frac{c_p^{(s)}(T, 1)}{T} + \Delta s_m (T_m, 1) + \int_{T_m}^{T} dT' \frac{c_p^{(l)}(T', 1)}{T'}, \qquad (9.121)$$

we express the entropy at $p_1 = 1$ atm in the form

$$s(T, p) = s_m^0(T) - R \ln p + \int_1^{p} dp' \left[\frac{R}{p'} - \left(\frac{\partial v}{\partial T} \right)_{p'} \right] (T, p') \qquad (9.122)$$

for the liquid at arbitrary T and p. Note that this formula is in the same form as for the formula for real gas, Eq. (9.116), except for the difference in the meanings of $s^*(T)$ and $s_m^0(T)$, arising from the presence of $\Delta s_m (T_m, 1)$ and the split integrals of $c_p^{(\alpha)}/T$ ($\alpha = s, l$) for the entropy contributions from the solid and liquid phases.

The entropy in the case of liquid–vapor or solid–vapor transition can be found similarly. It is left to the readers as an exercise.

9.5.7. *Gibbs Free Energy*

Since the molar enthalpy and entropy of a substance are already calculated, the molar Gibbs free energy is easily obtained from them by using the formula

$$g = h - Ts.$$

In the gas phase excluding a phase transition, substitution of Eqs. (9.81) and (9.112) yields the molar free energy of a gas

$$g(T, p) = g^*(T) + RT \ln p + \int_0^p dp \left(v - \frac{RT}{p} \right), \qquad (9.123)$$

where

$$g^*(T) = h^*(T) - Ts^*(T). \qquad (9.124)$$

The meaning of $g^*(T)$ is clear from the meanings of h^* and s^*: it is the molar Gibbs free energy of the hypothetical ideal gas at 1 atm pressure. Therefore, the last term in Eq. (9.123) is the nonideality correction. It vanishes for ideal gases:

$$\lim_{p \to 0} \int_0^p dp' \left(v - \frac{RT}{p'} \right) = 0.$$

This is easily verified. In this case, we have for the molar Gibbs free energy

$$g(T, p) = g^*(T) + RT \ln p, \qquad (9.125)$$

which is the molar Gibbs free energy we have calculated previously for ideal gas.

Now, the question arises as to what would be the free energy for the case of a process in which a phase transition is involved, for example, the process described in Fig. 9.6. Since the free energy is continuous across two phases in equilibrium for the isobaric process in Fig. 9.6, if we denote the free energies of the solid and liquid phases by $g^{(s)}$ and $g^{(l)}$, then

$$g^{(s)}(T_m, p) = g^{(l)}(T_m, p), \qquad (9.126)$$

the free energy can still be written as formula (9.123). This is left to the readers as an exercise.

9.6. Fugacity

In analogy to the ideal gas formula for $g(T, p)$ we write the molar Gibbs free energy of a real gas in the form

$$g(T, p) = g^*(T) + RT \ln p_f. \qquad (9.127)$$

Comparing it with the formula in Eq. (9.123) yields

$$p_f = p \exp \int_0^p dp' \left(\frac{v}{RT} - \frac{1}{p'} \right). \tag{9.128}$$

This was introduced by G. N. Lewis who called p_f the fugacity of the real gas at T and p. Obviously, as the pressure tends to zero, the equation of state becomes that of the ideal gas and, consequently, the fugacity p_f approaches the pressure p. Therefore, the fugacity is a measure of nonideality of the substance. The nonideality correction is sometimes expressed as excess-free energy:

$$g_{ex} = RT \ln \left[\exp \int_0^p dp' \left(\frac{v}{RT} - \frac{1}{p'} \right) \right]. \tag{9.129}$$

The fugacity then may be written as

$$p_f = p \exp \left(\frac{g_{ex}}{RT} \right) := fp, \tag{9.130}$$

where f is called the fugacity coefficient:

$$f = \exp \int_0^p dp' \left(\frac{v}{RT} - \frac{1}{p'} \right). \tag{9.131}$$

The fugacity coefficient f tends to 1 as p decreases to zero, namely, as the gas becomes ideal. Having obtained the fugacity explicitly as in Eqs. (9.130) and (9.131), we see that the fugacity is nothing but a quantity that introduces the equation of state of the fluid into the chemical potential of the substance and therefore its thermodynamic description.

Performing the integration in Eq. (9.128), for example, with the virial equation of state (9.39) for pressure, we obtain p_f as follows:

$$p_f = p \exp \left(\frac{B}{RT} p + \frac{C'}{2RT} p^2 + \cdots \right). \tag{9.132}$$

If the pressure is not too high, the exponential factor may be expanded to yield a series in p:

$$p_f = p + \frac{B}{RT} p^2 + O(p^3). \tag{9.133}$$

In order to gain more insight into the concept of fugacity, let us use the van der Waals estimate for B:

$$B = b - \frac{a}{RT}.$$

To the first order in p, Eq. (9.133) becomes

$$p_f = p\left[1 + \frac{1}{RT}\left(b - \frac{a}{RT}\right)p\right].$$
(9.134)

In the domain of T where the repulsive force is dominant over the attractive force, the fugacity is larger than the pressure predicted by the ideal gas equation of state, whereas, in the opposite case, the fugacity is smaller than the pressure value of the ideal gas value because of the attractive pull by the molecules. This is precisely the picture provided by the van der Waals theory, and the fugacity reflects this intermolecular interaction effect on thermodynamic functions and the Gibbs free energy in particular.

Since the molar Gibbs free energy for a pure substance is simply the chemical potential

$$g = \mu(T, p),$$

the chemical potential of the substance may be written as

$$\mu(T, p) = \mu^*(T) + RT\ln p + \int_0^p dp'\left(v - \frac{RT}{p'}\right).$$
(9.135)

This can be put in a form analogous to the formula for an ideal gas if the fugacity is used:

$$\mu(T, p) = \mu^*(T) + RT\ln p_f.$$
(9.136)

It is instructive to note that we could have obtained it if integration was performed on the differential form

$$d\mu(T, p) = -sdT + vdp$$
(9.137)

along the path indicated by the dotted line in Fig. 9.4. The meaning of $\mu^*(T)$ can be inferred from that of $g^*(T)$: that is, *it is the chemical potential of the hypothetical ideal gas at 1 atm pressure. It is called the standard chemical potential of the substance*. It may be evaluated from the standard enthalpy and the standard entropy. By definition,

$$\mu^*(T) = h^*(T) - Ts^*(T),$$
(9.138)

where

$$h^*(T) = h^0 + \int_0^T dT'\, c_p^*(T'),$$

$$s^*(T) = s^0 + \int_0^T dT'\, \frac{c_p^*(T')}{T'},$$
(9.139)

with h^0 and s^0 denoting the integration constants that may be determined from the standard enthalpy and the standard entropy, respectively.

As is often the case for gases, if the specific heat c_p may be expanded in a temperature series

$$c_p^*(T) = c_{p0}^* + c_{p1}^* T + c_{p2}^* T^2 + \cdots , \tag{9.140}$$

the integrals in Eq. (9.139) can be easily calculated to obtain

$$h^*(T) = h^0 + c_{p0}^* T + \tfrac{1}{2} c_{p1}^* T^2 + \cdots , $$
$$s^*(T) = s^0 + c_{p0}^* \ln T + c_{p1}^* T + \cdots . \tag{9.141}$$

When these results are combined into Eq. (9.138), there arises a series for $\mu^*(T)$:

$$\mu^*(T) = \left(h^0 - s^0 T \right) + c_{p0}^* T (1 - \ln T) - \tfrac{1}{2} c_{p1}^* T^2 - \cdots . \tag{9.142}$$

An alternative way of calculating $\mu^*(T)$ is as follows.

We first observe the identity

$$\left(\frac{\partial}{\partial T} \frac{\mu}{T} \right)_p = - \frac{h}{T^2} . \tag{9.143}$$

In the limit of zero pressure, this equation takes the ideal gas form

$$\left[\frac{\partial}{\partial T} \frac{\mu}{T} \right]_{p=0} = - \frac{h^*(T)}{T^2} . \tag{9.144}$$

Integrating it over T at $p = 0$ yields

$$\mu^* = - T \int^T dT \, \frac{h^*(T)}{T^2} - IT, \tag{9.145}$$

where I is the integration constant. Substituting Eq. (9.141) and performing the integration, we obtain

$$\mu^*(T) = h^0 - c_{p0}^* T \ln T - \tfrac{1}{2} c_{p1}^* T^2 - \cdots - IT. \tag{9.146}$$

The integration constant I in Eq. (9.145) is identified by comparing Eq. (9.145) with Eq. (9.142):

$$I = s^0 - c_{p0}^* . \tag{9.147}$$

It is called Nernst's chemical constant. This constant may be experimentally determined from calorimetry data with the help of the third law of thermodynamics. These chemical constants are useful for a number of thermodynamic calculations and, especially, for chemical equilibrium constants.

9.7. Joule–Thomson Experiment

When the volume of a gas is expanded reversibly and adiabatically, it is accompanied by a change in temperature. The change can be studied by the Joule–Thomson experiment.

Let us consider a gas contained in a cylinder AB whose wall is thermally insulated (see Fig. 9.7). The gas in the cylinder is infinitesimally slowly compressed at pressure p by the piston on the left. The gas then effuses through the porous plug (the shaded part within the cylinder) into the right-hand side of the cylinder. Pressure p' is applied to the piston on the right. At the end of the process, the piston on the left is brought to position A' while the piston at B is pushed to position B'. Let the area of the piston by $\mathcal{A}$. Then the force $p\mathcal{A}$ is acted on the gas from the left. The opposing force is $p'\mathcal{A}$ on the gas on the right-hand side of the porous plug. If the volume of the gas on the left is V, the path traversed by the gas is $V/\mathcal{A}$ on the left of the plug and $V'/\mathcal{A}$ on the right. The total work performed on the gas is then

$$\int_{in}^{fin} dW = pV - p'V'.$$

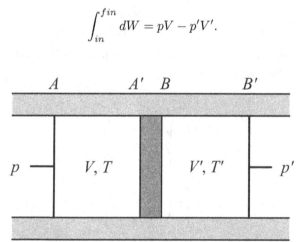

Fig. 9.7. Joule–Thompson experiment. A gas is adiabatically effused through the porous plug between A' and B under a difference in pressure p and p'. The (shaded) wall of the cylinder is thermally insulated.

Since the experiment is done adiabatically and very slowly so that the irreversibility is minimized, we have

$$Q = \int_{in}^{fin} dQ = 0,$$

and, consequently, by the first law of thermodynamics

$$E' - E = pV - p'V'.$$

By rearranging it, we obtain

$$E + pV = E' + p'V'.$$

This relation means that the process is isoenthalpic. Therefore, the molar enthalpy of the gas remains unchanged during the process:

$$h = h'. \tag{9.148}$$

This condition must hold because the heat content of the gas remains unchanged owing to the particular setup of the experiment adiabatically performed.

Let us now examine the accompanying temperature change. This temperature change arising from the pressure change at an adiabatic condition is given by the Joule–Thomson coefficient

$$\mu_{JT} = \left(\frac{\partial T}{\partial p} \right)_h. \tag{9.149}$$

This coefficient may be expressed in terms of more easily measurable quantities in the following manner:

$$\begin{aligned}
\mu_{JT} &= -\frac{\left(\frac{\partial h}{\partial p} \right)_T}{\left(\frac{\partial h}{\partial T} \right)_p} \\
&= -c_p^{-1} \left[\left(\frac{\partial g}{\partial p} \right)_T + T \left(\frac{\partial s}{\partial p} \right)_T \right] \\
&= c_p^{-1} \left[T \left(\frac{\partial v}{\partial T} \right)_p - v \right] \\
&= v c_p^{-1} (T\alpha - 1), \tag{9.150}
\end{aligned}$$

where α is the isothermal expansion coefficient. The first line follows upon use of the chain relation (3.68). Since $v \geq 0$ and $c_p \geq 0$, the Joule–Thomson coefficient is positive if $T\alpha \geq 1$ and it is negative if $T\alpha \leq 1$. Put in other words, the temperature decreases with decreasing pressure if

$$-\left(\frac{\partial h}{\partial p}\right)_T > 0,$$

and the temperature increases with decreasing pressure if

$$-\left(\frac{\partial h}{\partial p}\right)_T < 0.$$

As a matter of fact, if temperature is plotted against pressure at constant values of h, the isoenthalpic curves are obtained and they go through a maximum. When the locus of the maximum is plotted in the $T - p$ plane, we obtain the inversion curve as shown in Fig. 9.8. The temperature at which

$$\mu_{JT} = 0 \tag{9.151}$$

is called the inversion temperature T_i which is determined from the equation

$$T_i \alpha(T_i) = 1, \tag{9.152}$$

the zero of μ_{JT} (see Eq. (9.150)). Here,

$$\alpha(T_i) = \left[v^{-1}\left(\frac{\partial v}{\partial T}\right)_p\right]_{T=T_i}. \tag{9.153}$$

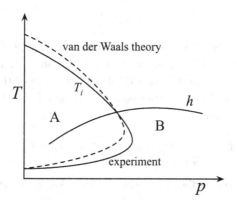

Fig. 9.8. Inversion curve, which is the locus of $\mu_{JT} = 0$. The solid curve denotes the experimental curve and the broken, the van der Waals theory prediction. In the inside region of the inversion curve, $\mu_{JT} > 0$ whereas $\mu_{JT} < 0$ in the outer region of the curve.

In the domain A in Fig. 9.8, the Joule–Thomson coefficient is positive since the temperature along an isenthalpic curve increases with the pressure, whereas in the domain B it is negative since the temperature decreases with increasing pressure. Thus, in the former domain the gas cools as it expands, whereas in the latter it heats up on expansion. This cooling effect has an important technological application since it forms the thermodynamic foundation of the procedure for liquefying gases.

In order to better understand the inversion temperature and the inversion curve, let us calculate the Joule–Thomson coefficient by using the virial equation of state. When the virial expansion is substituted into Eq. (9.150), we obtain to the lowest order in p

$$\mu_{JT} = c_p^{-1} \left(T \frac{dB}{dT} - B \right) + O(p). \tag{9.154}$$

If the van der Waals theory estimate is used for the second virial coefficient B, the Joule–Thomson coefficient takes a more transparent form

$$\mu_{JT} = c_p^{-1} \left(\frac{2a}{RT} - b \right), \tag{9.155}$$

which gives the inversion temperature in the form

$$T_i = \frac{2a}{Rb}.$$

Since the second virial coefficient is equal to zero at the Boyle temperature, we also see that

$$T_i = 2T_B$$

in the case of the van der Waals approximation, Eq. (9.155), for μ_{JT} (see Eq. (9.49)). It is possible to obtain a more precise formula for the Joule–Thomson coefficient than Eq. (9.155) if we make use of the van der Waals equation of state without an approximation. Since from the van der Waals equation state follows

$$\left(\frac{\partial v}{\partial T} \right)_p = R[p + av^{-2} - 2av^{-3}(v - b)]^{-1}, \tag{9.156}$$

the Joule–Thomson coefficient in the van der Waals theory is given by the formula

$$\mu_{JT} = c_p^{-1} \frac{2av^{-2}(v - b) - b(p + av^{-2})}{p + av^{-2} - 2av^{-3}(v - b)}. \tag{9.157}$$

Therefore, the inversion temperature is determined by the solution of the equation

$$2av^{-2}(v - b) - b(p + av^{-2}) = 0. \tag{9.158}$$

Note that v must be regarded as a function of T and p in this equation. This equation can be cast into a relation for T_i and p.

Upon use of the van der Waals equation of state, it can be put into the form

$$2av^{-2}(v - b)^2 - bRT_i = 0, \tag{9.159}$$

which is easily solved for b/v:

$$\frac{b}{v} = 1 \pm \sqrt{\frac{bRT_i}{2a}}. \tag{9.160}$$

Using this in Eq. (9.153), we find a quadratic equation for T_i, which is easily solved for T_i:

$$T_i = \frac{10a}{9bR} \left[1 - \frac{3b^2 p}{5a} \pm \frac{4}{5}\sqrt{1 - \frac{3b^2 p}{a}} \right]. \tag{9.161}$$

The solutions are real for all p satisfying the condition

$$p \le \frac{a}{3b^2}. \tag{9.162}$$

This result for the inversion temperature is correct only qualitatively because of the approximate nature of the van der Waals equation of state, but still gives a fairly good idea of the pressure–temperature domain where the Joule–Thomson coefficient is negative. The inversion curve calculated with Eq. (9.155) is schematically plotted in Fig. 9.8 (the broken curve).

Equation (9.155) gives us a useful insight into what happens in the gas as the pressure is changed. As the gas is compressed, the intermolecular distances will decrease and the molecules will feel stronger intermolecular attractions. In such a range of distance, the interaction energy is dominated by the attractive potential. This energy is converted into the kinetic energy through collisions and the gas will be heated up as it is compressed. This is easily explained by formula (9.155), which shows that the Joule–Thomson coefficient will be positive as long as the strength of the attractive forces is sufficiently large so as to make $2a/RT$ larger than b which is a measure of the repulsive force. As the pressure is further increased, the intermolecular distances decrease further and the repulsive potential becomes predominant.

The effect of this exchange in predominance of the attractive and repulsive potentials can be understood with Eq. (9.155) if we note that to the lowest order of approximation, it is possible to write

$$\frac{2a}{RT} - b = \frac{2a}{pv} - b.$$

Therefore, at a constant volume, there will be a value of pressure at which $2a/pv - b$ changes its sign, resulting in a negative Joule–Thomson coefficient. This interpretation is in accordance with the condition on pressure (9.162) given above.

9.8.   Liquefaction of Gases

Examination of the Joule–Thomson coefficient of a gas provides the information on the state of the gas from which to start its liquefaction successfully. If the temperature and pressure are in the domain bounded by the inversion curve (see the domain A in Fig. 9.8) at a given value of enthalpy, a simple adiabatic expansion of the gas will lower the temperature and, by repeating the process, it will be possible to lower the temperature below the critical temperature. Then mere compression of the cooled gas will be sufficient to put it into the liquid form.

Figure 9.9 shows a schematic diagram of the experimental setup. A compressed gas at p and T is adiabatically expanded through the nozzle to attain pressure p_b and eventually temperature T_b, the boiling point of the liquid. The portion of the gas unliquefied attains thermal equilibrium with the gas on the verge of expansion at T. This gas is compressed to p and the cycle is repeated. Since the whole system is thermally insulated, the process is isoenthalpic.

Let us calculate the fraction ϵ of the gas liquefied. Since the process is isoenthalpic, the enthalpy of the substance before and after throttling must be equal to the sum of the enthalpies of the fraction liquefied and the remainder of the gas unliquefied. Therefore, for a mole of the gas,

$$h_g(T, p) = (1 - \epsilon) h_g(T, p_b) + \epsilon h_l(T_b, p_b), \qquad (9.163)$$

where h_g and h_l denote the molar enthalpy of the gas and of the liquid, respectively, at the state indicated. Solving this equation for ϵ, we obtain

$$\epsilon = \frac{h_g(T, p_b) - h_g(T, p)}{h_g(T, p_b) - h_l(T_b, p_b)}. \qquad (9.164)$$

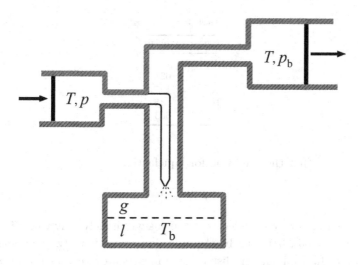

Fig. 9.9. Liquefaction of gas: a schematic illustration.

The enthalpy changes in Eq. (9.164) may be easily calculated from Eq. (9.81):

$$h_g(T, p_b) - h_g(T, p) = \int_p^{p_b} dp \left[v_g - T \left(\frac{\partial v_g}{\partial T} \right)_p \right],$$

$$h_g(T, p_b) - h_l(T_b, p_b) = \Delta h_v(T_b, p_b) + \int_{T_b}^T dT\, c_p^{(g)}(T, p_b),$$

where Δh_v is the heat of vaporization at T_b and p_b,

$$\Delta h_v(T_b, p_b) = h_g(T_b, p_b) - h_l(T_b, p_b),$$

$c_p^{(g)}$ is the heat capacity of the gas, and v_g denotes the molar volume of the gas. Substituting these results into Eq. (9.164) yields

$$\epsilon = \frac{-\int_p^{p_b} dp'\, \mu_{JT}(p')c_p^{(g)}(T, p')}{\Delta h_v(T_b, p_b) + \int_{T_b}^T dT'\, c_p^{(g)}(T', p_b)}. \tag{9.165}$$

Since $\Delta h_v \geq 0$ and $c_p^{(g)} \geq 0$, the denominator is always positive and hence it follows that

$$\epsilon > 0,$$

that is, the gas will liquefy only if

$$\int_{p_b}^p dp\, \mu_{JT}(p)c_p^{(g)}(T, p) > 0,$$

Table 9.2. T_i^*.

CO_2	1500 K
Ar	723
N_2	621
H_2	195
He	35
air	603

which implies that the condition for liquefaction is

$$\mu_{JT} > 0. \tag{9.166}$$

Therefore, it can be concluded that the gas will liquefy only when T and p are in the domain bounded by the inversion curve, namely, the domain A in Fig. 9.8. In Table 9.2 are listed some maximum zero pressure inversion temperatures T_i^* for various gases. The maximum efficiency of liquefaction at a given operating temperature is achieved at the pressure determined by the equation

$$\left(\frac{\partial \epsilon}{\partial p}\right)_T = c_p^{(g)}(T, p)\mu_{JT}(T, p)\left(\Delta h_v + \int_{T_b}^T dT' \, c_p^{(g)}\right)^{-1} = 0. \tag{9.167}$$

Since the specific heat is not equal to zero, the condition is satisfied only if p is such that

$$\mu_{JT}(T, p) = 0. \tag{9.168}$$

That is, the temperature and pressure must be those corresponding to the inversion point, T_i and p_i, for the maximum efficiency to be achieved in liquefaction.

If the virial equation of state is used to the first order in p, we obtain the fraction ϵ in a rather simple form

$$\epsilon = \frac{(T\frac{dB}{dT} - B)(p - p_b)}{\Delta h_v + \int_{T_b}^T c_p^{(g)} dT'} \approx \frac{(T\frac{dB}{dT} - B)(p - p_b)}{\Delta h_v + c_p^{(g)}(T_p - T_b)}, \tag{9.169}$$

where $c_{p0}^{(g)}$ is the temperature, independent part of $c_p^{(g)}$. The second approximation will hold if the specific heat of the gas does not vary strongly with temperature in the range of temperature of interest.

9.9. Entropy Surface

The Gibbs relation

$$TdS = dE + pdV - \sum_{i=1}^{r} \mu_i dn_i$$

implies that the entropy is a surface in the space of $(S, E, V, n_1, n_2, \ldots, n_r)$ in the case of a mixture, or, in the case of a closed system, the Gibbs relation

$$TdS = dE + pdV$$

implies that the entropy is a surface in the space of (S, E, V). The calculations which we have performed for thermodynamic quantities in Section 9.6 are in effect attempts at constructing such surfaces although the results obtained have not been cast in such forms. Such a surface contains all the necessary constitutive information on the system of which the entropy surface is representative, and, as we shall see, such surfaces can indeed generate all the thermodynamic constitutive relations when appropriate derivatives are taken from them. For this reason, the entropy is said to be a storage of the macroscopic information on the material in equilibrium.

Since constructing entropy surfaces in the general case is as difficult as integrating an arbitrary differential form, we obviously have to specialize to particular cases if we wish to explicitly demonstrate the method. Here, we shall consider the cases of an ideal gas and a van der Waals gas under the assumption that the heat capacities are independent of temperature.

It is worthwhile noting that in the axiomatic theory[7] of thermodynamics, the existence of entropy is taken as an axiom following from the thermodynamic laws, and a theory of thermodynamics is formulated axiomatically. Such an approach is mathematically appealing, but it tends to remove us from the physical aspects important to the true understanding of thermodynamics, especially, of irreversible processes.

9.9.1. *Ideal Gas*

Let us collect the results for the internal energy and the entropy which we have calculated for the ideal gas:

$$S = C_v \ln T + R \ln V + S_0 - (C_v \ln T_0 + R \ln V_0), \qquad (9.170)$$

---

[7] For example, see H. B. Callen, *Thermodynamics* (Wiley, New York, 1960).

$$E = C_v T + E_0 - C_v T_0. \tag{9.171}$$

If S and E are determined such that

$$S_0 = C_v \ln T_0 + R \ln V_0, \tag{9.172}$$

$$E_0 = C_v T_0, \tag{9.173}$$

then S and E are given by

$$S = C_v \ln T + R \ln V, \tag{9.174}$$

$$E = C_v T. \tag{9.175}$$

By eliminating T from (9.174) and (9.175), we find

$$S = C_v \ln \left(\frac{E V^{R/C_v}}{C_v} \right) \tag{9.176}$$

for the entropy surface in the space spanned by (S, E, V). This entropy contains all the information on the system (i.e., ideal gas) since it can yield all the thermodynamic quantities for the ideal gas.

In thermodynamics, the temperature may be expressed such that

$$\frac{1}{T} = \left(\frac{\partial S}{\partial E} \right)_v. \tag{9.177}$$

The derivative on the right-hand side can be calculated from (9.176):

$$\left(\frac{\partial S}{\partial E} \right)_v = \frac{C_v}{E}.$$

Combining it with (9.177) yields Eq. (9.175). Analogously, the pressure may be expressed by the relation

$$\frac{p}{T} = \left(\frac{\partial S}{\partial V} \right)_E. \tag{9.178}$$

Since from (9.174),

$$\left(\frac{\partial S}{\partial V} \right)_E = \frac{R}{V},$$

we recover the equation of state for a mole of an ideal gas from (9.178):

$$pV = RT. \tag{9.179}$$

This is another constitutive equation postulated for calculating (9.174). These simple calculations imply the following: $1/T$ and p/T — the constitutive relations — are simply the tangents to the entropy surface.

Nevertheless, however clean and appealing mathematically this point may be, the physical meanings of these tangents and their historical and experimental origins are much deeper than their mathematical meanings.

By combining the aforementioned results, we also obtain from (9.174)

$$dS = \frac{1}{T}dE + \frac{p}{T}dV, \tag{9.180}$$

namely, the equilibrium Gibbs relation. This is the differential form which, with the help of the constitutive equations (9.175) and (9.179), has yielded the entropy surface (9.174) in the first place. Other thermodynamic quantities follow from either the set (9.174), (9.177), and (9.178) or the set (9.175), (9.179), and (9.180). These two sets represent two different viewpoints to the same thermodynamic quantities.

9.9.2. *van der Waals Gas*

If the heat capacity C_v is independent of T, then the molal energy E and entropy S of a van der Waals gas are given by

$$E = -\frac{a}{V} + C_v T + (E_0 - C_v T_0) + \frac{a}{V_0}, \tag{9.181}$$

$$S = C_v \ln T + R \ln(V - b) + [S_0 - C_v \ln T_0 - R \ln (V_0 - b)], \tag{9.182}$$

where E_0 and S_0 are the molal energy and entropy of the gas in the hypothetical state of an ideal gas at 1 atm and T_0. We select the integration constants E_0 and S_0 such that

$$E_0 = C_v T_0, \quad S_0 = C_v \ln T_0 + R \ln (V_0 - b). \tag{9.183}$$

Then E and S are given by

$$E = -\frac{a}{V} + C_v T, \quad S = C_v \ln T + R \ln(V - b). \tag{9.184}$$

Elimination of T from the two equations in (9.184) yields the entropy surface for the van der Waals gas:

$$S = C_v \ln \left[\frac{(V - b)^{R/C_v}}{C_v} \left(E + \frac{a}{V} \right) \right]. \tag{9.185}$$

By following the same procedure as for the ideal gas considered earlier, this surface can be shown to yield the constitutive equations

$$E = -\frac{a}{V} + C_v T, \quad p = \frac{RT}{V-b} - \frac{a}{V^2}. \tag{9.186}$$

The verification of these results is left to the readers as an exercise.

If the specific heat is not assumed to be independent of T, then the entropy surface can be a more complicated function of E. In any event, the entropy surface (9.185) contains in it all the macroscopic constitutive information on the system under the assumption of constant C_v. In this sense, the entropy is a storage of the constitutive information on the system in hand. For this reason and its statistical interpretation, the concept of entropy holds a considerable importance in information science. However, it should be kept in mind that the entropy in the present work is defined only for reversible processes or systems in equilibrium. For irreversible processes, the calortropy may be regarded as a storage of information in lieu of the entropy for systems at equilibrium.

Problems

(1) A mole of a liquid is reversibly heated from T_m to its boiling point T_b at 1 atm pressure and then the vapor is reversibly compressed and heated at the same time to pressure p and temperature T. Calculate the enthalpy change for the reversible process. The following data are available for the substance:

$$c_p(l) = a + bT + cT^2,$$
$$c_p(v) = c_p^* + c_{p1}T + c_{p2}T^2,$$
$$pv = RT + BP \text{ for the vapor,}$$
$$\Delta h_v = \text{heat of vaporization at } T = T_b \text{ and } p = 1 \text{ atm.}$$

(2) Verify the results in Eq. (9.186) for the van der Waals fluid.

(3) Utilize the following data at $0°$ C to calculate the fugacity of carbon monoxide at 50, 100, 400, and 1000 atm by graphical integration based on the expression for fugacity.

p atm	pv/RT
25	0.9890
50	0.9792
100	0.9741
200	1.0196
400	1.2482
800	1.8075
1000	2.0819

(4) Calculate the efficiency of liquefaction by using the van der Waals model for the second virial coefficient $B = b - a/RT$.

(5) Using the pv/RT values for 50 and 400 atm given in Problem 3, calculate the van der Waals constants a and b for carbon monoxide. From these, determine the fugacities at various pressures and compare the results with those obtained in Problem 3.

(6) By using the approximation $v \simeq RT/p$ in Eq. (9.158) and solving the resulting quadratic equation for T, show that the inversion temperature is given by

$$T_i = \frac{a}{Rb}\left[1 \pm \left(1 - \frac{3b^2}{a}p\right)^{\frac{1}{2}}\right],$$

which is real for all p satisfying the condition

$$p \leq \frac{a}{3b^2},$$

the same condition as Eq. (9.162).

(7) Derive the entropy formula similar to Eq. (9.122) for change in T and p that involves (a) liquid–vapor phase transition and also (b) solid–vapor phase transition.

(8) Show that the formula for free energy at (T, p), Eq. (9.123), remains the same even though there exists a phase transition in the path of change of the system.

Chapter 10

Canonical Equation of State

In the previous Chapter 9 on real fluids, we have shown that the van der Waals equation of state is capable of qualitatively describing the thermodynamic properties of fluids, although its description of critical and subcritical properties is inadequate in accuracy and in its ability to describe the details if examined in depth. Nevertheless, it is qualitatively better than the virial equation of state in its capability. It thus indicates that it somehow contains the basic elements required for correct description of fluid properties over the whole range of density and temperature. This attractiveness of the van der Waals equation of state has been enduring motivations for numerous attempts[1] at improving it in the literature.

In this chapter,[2] we present the canonical (i.e., generic van der Waals) equation of state that fulfills the desired aim of satisfactorily extending[3] the van der Waals equation of state. Here, we develop the algorithms and formulas for the critical properties of fluids by means of which we can phenomenologically relate the critical characteristics of fluids to the parameters in the canonical equation of state. The critical and subcritical thermodynamic data, such as the critical parameters and the

---

[1]See, for example, J. J. van Laar, *Proc. Sci. Sec. Kon. ned. Akad. Wiesensch.* **14**, 278, 428, 563, 711 (1912); R. Planck, *Forsch. Geb. Ingenieures* **7**, 161 (1936); R. Planck and L. Riedel, *Ing. Arch.* **16**, 255 (1947/48); H. D. Baehr, *Forsch. Geb. Ingenieures.* **29**, 143 (1963).
[2]This chapter is based on an unpublished research note on the subject by B. C. Eu.
[3]See B. C. Eu and K. Rah, *Phys. Rev. E* **63**, 031303 (2001); B. C. *Eu, J. Chem. Phys.* **114**, 10899 (2001); K. Rah and B. C. *Eu, J. Phys. Chem. B* **107**, 4382 (2003).

critical pressure–density data, spinodal and coexistence curves, isothermal compressibility, and specific heat in the subcritical neighborhood of the critical point, are shown closely related to the generic van der Waals parameters in the canonical equation of state — a phenomenological model. Each of these characteristic properties makes it possible to determine the empirical parameters in the model. It is thus shown that with suitable choices of parameter values, the model would be capable of describing the thermodynamics of critical and subcritical fluids, appropriately generalizing the original van der Waals equation of state theory. The present chapter therefore shows how to fashion the equation of state in a van der Waals-like form, so that the experimental thermodynamic properties of fluids are described better than that by the van der Waals equation of state, which is found to be defective in a number of aspects.

10.1. Canonical Equation of State

The genesis of the canonical equation of state is in a statistical mechanics study[4] of equation of state of real fluids and, in particular, of the Lennard-Jones (LJ) fluid, whose interaction potential energy consists of repulsive and attractive contributions. For such fluids, it can be shown that the statistical mechanical virial equation of state can be split into the repulsive and attractive contributions, which can then be rearranged to a form reminiscent of the van der Waals equation of state:

$$\left[p + A\left(\rho, \beta\right)\rho^2\right]\left[1 - B\left(\rho, \beta\right)\rho\right] = \rho\beta^{-1}, \tag{10.1}$$

where p, ρ, and β are, respectively, the pressure, density, and inverse temperature $\beta = 1/k_B T$ with T denoting the temperature and k_B the Boltzmann constant, and $A(\rho, \beta)$ and $B(\rho, \beta)$ are constants independent of p, which we call the generic van der Waals (GvdW) parameters. Equation (10.1) not only retains the useful features of the van der Waals equation of state but also removes the defects of the latter in describing the thermodynamic properties of real fluids, particularly, in the critical and subcritical regimes.

According to the statistical mechanics study of Eq. (10.1), the parameters $A(\rho, \beta)$ and $B(\rho, \beta)$ are generally functions of ρ and β, but as the density ρ and the inverse temperature β diminish to zero, they become the

---

[4]See footnotes 2 and 3, and also K. Rah and B. C. Eu, *J. Chem. Phys.* **115**, 2634 (2003).

van der Waals constants a' and b' of the original van der Waals equation of state

$$\left(p + a'\rho^2\right)\left(1 - b'\rho\right) = \rho\beta^{-1}, \tag{10.2}$$

i.e.,

$$\lim_{\rho \to 0, T \to \infty} A\left(\rho, \beta\right) = a', \qquad \lim_{\rho \to 0, T \to \infty} B\left(\rho, \beta\right) = b'. \tag{10.3}$$

Because of this limiting behavior of A and B, the equation of state, (10.1), was initially called the generic van der Waals equation of state. Since the form of equation of state (10.1) is generic to any fluid characterized by an interaction potential energy consisting of attractive and repulsive potential energies, the terminology canonical equation of state is quite appropriate for fluids obeying such potentials of interaction. In fact, the van der Waals equation of state may be regarded as a special case of the canonical form.

By using phenomenological forms for $A(\rho, \beta)$ and $B(\rho, \beta)$ and calculating the critical isotherm, it will be shown that the canonical equation of state is capable of describing the critical isotherm correctly. Furthermore, its utility was also shown for developing a molecular theory of transport coefficients in the liquid density regime with the help of the modified free volume (MFV) theory,[5] since the canonical form (10.1) provides a well-defined formula for the mean free volume v_f in terms of generic van der Waals parameter $B(\rho, \beta)$:

$$v_f = \rho^{-1}\left[1 - B(\rho, \beta)\,\rho\right], \tag{10.4}$$

and A and B can be rigorously calculated from their statistical mechanical representations by using the pair correlation function of the fluid. Such an expression for the mean free volume has played a crucial role in the calculation of diffusion coefficients of liquids in the MFV theory of diffusion in liquids.

The statistical mechanics of the GvdW (i.e., canonical) equation of state indicates that A and B are not only dependent on density and temperature but also discontinuous in the subcritical regime, displaying a nonanalytic behavior characteristic of the equation of state for a real fluid in the subcritical regime. Relying on the nonanalytic behavior of A and B

---

[5]See K. Rah and B. C. Eu, *J. Chem. Phys.* **115**, 2634 (2001); K. Rah and B. C. Eu, *Phys. Rev. Lett.* **88**, 065901(2002); K. Rah and B. C. Eu, *J. Chem. Phys.* **116**, 7967 (2002). See also B. C. Eu, *Transport Coefficients of Fluids* (Springer, Heidelberg, 2006).

exhibited in the statistical mechanical study, we may phenomenologically assume empirical nonanalytic formulas — irrational functions of density and temperature — for them and then demonstrate that the equation of state thus constructed can be indeed capable of yielding the critical isotherm for pressure of the fluid.

Turning this procedure around, we now formulate algorithms to construct a canonical equation of state so as to account for various thermodynamic properties and thus relate them to some characteristic aspects of the GvdW parameters, such as the virial coefficients, Boyle temperature, Joule–Thomson coefficient and inversion temperature, various critical properties of fluids — the critical isotherm, liquid–vapor coexistence curve, spinodal curve, isothermal compressibility, and specific heat in the neighborhood of the critical point. In this manner, the canonical equation of state is assured of summarizing as broadly as possible the fluid properties around the critical state and beyond.

The critical properties we are discussing in this chapter have been considered in the literature on the basis of the renormalization group theory,[6] but the present canonical-equation-of-state approach is empirical and, moreover, mathematically quite low-browed and algebraic, so that it can be easily grasped, although it appears algebraically complicated at first glance. It does not require anything other than a pair of constitutive equations for pressure and chemical potential based on the canonical equation of state. But it involves only simple calculus and algebraic equations.

10.2. Reduced Variables

For the purpose of carrying out the desired analysis, it is convenient to use nondimensionalized variables. We define the following reduced (dimensionless) variables:

$$p^* = \frac{pv_0}{\varepsilon}, \qquad T^* = \frac{k_B T}{\varepsilon}, \qquad \eta = v_0 \rho,$$

$$A^* = \frac{A}{\varepsilon v_0}, \qquad B^* = \frac{B}{v_0}, \tag{10.5}$$

---

[6]References to renormalization group theory.

where ε is the potential well depth and σ is the size parameter, which may be identified with the LJ potential parameters, and $v_0 := \pi\sigma^3/6$, the volume of a sphere of diameter σ. The critical parameters are denoted, with subscript c, by p_c, η_c, T_c. These reduced variables will be used when we examine the thermodynamic properties away from the critical region.

For studying the critical and subcritical behavior of the fluid, it is convenient to scale p^*, η, and T^* with the reduced critical parameters p_c^*, η_c, and T_c^*. With the definitions of scaled reduced variables

$$\psi = \frac{p^*}{p_c^*} = \frac{p}{p_c}, \quad y = \frac{\eta}{\eta_c}, \quad \theta = \frac{T^*}{T_c^*} = \frac{T}{T_c}, \tag{10.6}$$

we define relative reduced variables

$$\phi := \psi - 1, \quad x := y - 1, \quad t := \theta - 1. \tag{10.7}$$

These relative reduced variables vanish at the critical point (p_c, η_c, T_c). The reduced critical parameters p_c^*, η_c, and T_c^* will be expressed in terms of parameters making up A^* and B^*. We also define the dimensionless parameters

$$\nu := B_c^* \eta_c, \quad \zeta := \frac{A_c^* \eta_c^2}{p_c^*}, \quad \tau := \frac{\eta_c T_c^*}{p_c^*}, \tag{10.8}$$

where A_c^* and B_c^* are the reduced values of A^* and B^* at the critical point. They are constants independent of density and temperature and may be related to the van der Waals parameters a' and b' in the van der Waals equation of state. The critical parameters can then be expressed in terms of these dimensionless parameters and A_c^* and B_c^*:

$$\eta_c = \frac{\nu}{B_c^*}, \quad p_c^* = \frac{\nu^2 A_c^*}{\zeta B_c^{*2}}, \quad T_c^* = \frac{\tau \nu A_c^*}{\zeta B_c^*}. \tag{10.9}$$

The dimensionless parameter τ, being independent of A_c^* and B_c^*, is an important quantity that enables us to gauge the quality of models for the GvdW parameters A^* and B^*, as will be shown later with more explicit, but simpler models for them — the quadratic model discussed later. The critical parameters given in Eq. (10.9) should be compared with the corresponding results in the van der Waals theory:

$$(\eta_c)_{\text{vdw}} = \frac{1}{3b}, \quad (p_c^*)_{\text{vdw}} = \frac{a}{27b^2}, \quad (T_c^*)_{\text{vdw}} = \frac{8a}{27b}, \tag{10.10}$$

where

$$a = a'/\varepsilon v_0, \quad b = b'/v_0.$$

Note that these parameters a and b are dimensionless in the present scheme of reducing variables and parameters. Since A_c^* and B_c^* are qualitatively comparable, respectively, to the reduced van der Waals constants a and b, we see that in the low density limit,

$$\nu \to \frac{1}{3}, \quad \frac{\nu^2}{\zeta} \to \frac{1}{27}, \quad \frac{\tau\nu}{\zeta} \to \frac{8}{27},$$

or

$$\nu \to \frac{1}{3}, \quad \zeta \to 3, \quad \tau \to \frac{8}{3}. \tag{10.11}$$

The structural similarities of the critical parameters in the van der Waals theory and the present canonical equation of state theory is one of the attractiveness of the latter, firstly, because with relatively simple models for A^* and B^*, the canonical equation of state can be shown capable of accounting phenomenologically for the critical phenomena with a relatively unsophisticated mathematical method as does the van der Waals theory, and, secondly, because it can be made use of formulating the thermodynamics of fluids as the thermodynamics of fluids has been built on the van der Waals theory, which is capable of qualitatively describing most features of real fluids in the entire density range, as demonstrated in Chapter 9 on real fluids.

10.3. Reduced Canonical Equation of State

Upon using the aforementioned reduced variables and parameters ζ, ν, and τ, the canonical equation of state (10.1) can be reduced with respect to the critical parameters and expressed in the nondimensional form

$$[\phi + 1 + \zeta(1 + x)^2 \mathcal{A}(x, t)][1 - \nu(1 + x)\mathcal{B}(x, t)] = \tau(1 + t)(1 + x),$$
$$\tag{10.12}$$

where

$$\mathcal{A}(x, t) = \frac{A^*}{A_c^*}, \quad \mathcal{B}(x, t) = \frac{B^*}{B_c^*}. \tag{10.13}$$

In the limit where Eq. (10.11) holds, $\mathcal{A} = 1$ and $\mathcal{B} = 1$, then Eq. (10.12) becomes the nondimensionalized (in fact, corresponding state) van der Waals equation of state. If $\mathcal{A}$ and $\mathcal{B}$ are functions of x and t that are independent of material parameters, then Eq. (10.12) would also represent the law of corresponding states. However, that will not be generally the case because $\mathcal{A}$ and $\mathcal{B}$ are not free from (reduced) material parameters, which then may not be universal.

As mentioned earlier, in the statistical mechanics study of $\mathcal{A}$ and $\mathcal{B}$, it was found that $\mathcal{A}$ and $\mathcal{B}$ become discontinuous in the subcritical regime and there is a region of density in which they do not exist. The breakdown point[7] of $\mathcal{A}$ and $\mathcal{B}$ will be denoted by z_{sl} and z_{sv}, where $z_{sl} = \eta_1/\eta_c$ is the reduced breakdown density η_1 on the liquid branch, whereas $z_{sv} = \eta_2/\eta_c$ is the reduced breakdown density η_2 on the vapor branch $(\eta_1 > \eta_2)$. Note that η_1 and η_2 are different from the liquid–vapor coexistence densities η_l and η_v, respectively. As the critical point is approached, η_1 and η_2 tend to the critical density. Therefore,

$$\lim_{t \to 0} z_{sk} = 1 \quad (k = l, v). \tag{10.14}$$

That is, $z_{sk}(t = 0) = 1$ coincides with the density at the critical point — the critical density. These reduced breakdown densities are functions of temperature only. It is convenient to define deviations from the breakdown densities z_{sk}:

$$x_{sk} = z_{sk} - 1 \quad (k = l, v),$$

which then has the limit

$$\lim_{t \to 0} x_{sk} = 0. \tag{10.15}$$

In Fig. 10.1, the points x_{sl} and x_{sv} represent points of limit of stability of a subcritical isotherm and their locus is called the spinodal curve. The system on the interval of curve between x_{sl} and x_{sv} is thermodynamically unstable, whereas it is metastable between l and x_{sl} and also g and x_{sv}. Therefore, the darker shaded area represents an unstable regions whereas the lighter shaded regions between the binodal (i.e., liquid–vapor coexistence) and the branches of the spinodal curve are metastable regions. The breakdown densities of stability will, in fact, turn out to be the spinodal densities at t, and in the present theory such points will be mathematically defined as the points at which the first four density derivatives of pressure vanish in a manner similar to the density derivatives of pressure at the critical point.

---

[7]The locus of the breakdown points will turn out to be the spinodal curve in the case of liquid–vapor transition, and the spinodal curve is the limit of stability.

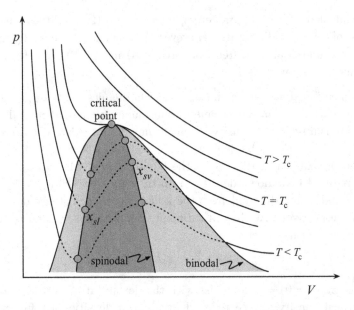

Fig. 10.1. Schematic illustration of the meanings of x_{sl} and x_{sv}. The points l and g respectively represent the coexisting liquid and vapor at equilibrium.

10.4. Models for the GvdW Parameters

10.4.1. *Subcritical Regime*

We have mentioned earlier that the GvdW parameters are discontinuous with respect to density for $t < 0$ (the subcritical regime of temperature) and do not exist in the interval $x_{sv} < x < x_{sl}$. Outside this interval, they are continuous functions of density and temperature. In the subcritical regime of temperature, they can also be irrational functions of t. Any phenomenological forms for the GvdW parameters should display these features so as to correctly describe the experimentally observed data on the thermophysical properties, especially, in the vicinity of the critical state. The following forms consisting of series of integer and fractional powers in $(x - x_{sk})$ meet the requirements mentioned:

$$\mathcal{A} = \sum_{i \geq 0} a_i^{(k)}(t)(x - x_{sk})^i + \sum_{i \geq 0} \alpha_{nai}^{(k)}(t)(x - x_{sk})^{f+i}|x - x_{sk}|^{1+\delta}, \quad (10.16)$$

$$\mathcal{B} = \sum_{i \geq 0} b_i^{(k)}(t)(x - x_{sl})^i + \sum_{i \geq 0} \beta_{nai}^{(k)}(t)(x - x_{sk})^{f+i}|x - x_{sk}|^{1+\delta}, \quad (10.17)$$

where the coefficients $a_i^{(k)}(t)$, $\alpha_{nai}^{(k)}(t)$, $b_i^{(k)}(t)$, and $\beta_{nai}^{(k)}(t)$ $(k = l, v)$ are temperature-dependent and generally nonanalytic with respect to t and the exponent f is an integer whereas δ is fractional $(\delta < 1)$. In the quadratic model we will later use, $f = 3$ and the value of i in the first terms on the right is $i = 2$ whereas it is $i = 0$ in the nonanalytic terms — the second terms on the right. The fractional number δ will be fixed from the empirical information on the critical isotherm. In this model, the spinodes (points on the spinodal curve) x_{sk} and the spinodal curve (locus of spinodes) are integral parts of the equation of state.

The coefficients are chosen such that the liquid and vapor branches of the coefficients coincide with each other at the critical temperature, namely,

$$a_i^{(l)}(0) = a_i^{(v)}(0) := a_i \neq 0, \quad \alpha_{nai}^{(l)}(0) = \alpha_{nai}^{(v)}(0) := \alpha_{ani} \neq 0,$$

$$b_i^{(l)}(0) = b_i^{(v)}(0) := b_i \neq 0, \quad \beta_{nai}^{(l)}(0) = \beta_{nai}^{(v)}(0) := \beta_{nai} \neq 0, \qquad (10.18)$$

$$\lim_{t \to 0} a_0^{(k)}(t) = 1, \quad \lim_{t \to 0} b_0^{(k)}(t) = 1 \quad (k = l, v),$$

and the fractional exponent $(4 + \delta)$ $(\delta = 0.3 - 0.5)$ is chosen in anticipation of the critical pressure–density isotherm discussed later. These properties suggest that in the subcritical regime near the critical point, we may look for $a_i^{(k)}(t)$ and $b_i^{(k)}(t)$ $(k = l, v; i = 0, 1, 2)$ in the forms

$$a_i^{(k)}(t) = a_i + \hat{a}_i^{(k)}(t) = a_i + \mathfrak{a}_i^{(k)} |t|^{\varepsilon_i} [1 + O(|t|^{1+\varepsilon_i})], \qquad (10.19)$$

$$b_i^{(k)}(t) = b_i + \hat{b}_i^{(k)}(t) = b_i + \mathfrak{b}_i^{(k)} |t|^{\varepsilon_i} [1 + O(|t|^{1+\varepsilon_i})] \quad (i = 0, 1, 2),$$

where $\mathfrak{a}_i^{(k)}$ and $\mathfrak{b}_i^{(k)}$ $(k = l, v)$ are constants independent of x and t, and ε_i are fractional numbers, probably less than unity. Their values will be fixed empirically. Therefore,

$$\hat{a}_i^{(k)}(0) = \hat{b}_i^{(k)}(0) = 0 \quad (i \geq 0). \qquad (10.20)$$

Similarly, the coefficients of the irrational nonanalytic terms must also be in the forms

$$\alpha_{na}^{(k)}(t) = \alpha_{na} + \hat{\alpha}_{na}^{(k)}(t) = \alpha_{na} + \mathfrak{a}_{na}^{(k)} |t|^{\varepsilon_n},$$

$$\beta_{na}^{(k)}(t) = \beta_{na} + \hat{\beta}_{na}^{(k)}(t) = \beta_{na} + \mathfrak{b}_{na}^{(k)} |t|^{\varepsilon_n}, \qquad (10.21)$$

where ε_n is a fractional number, which may be less than unity. Its value will also be determined empirically. Therefore,

$$\widehat{\alpha}_{na}^{(k)}(t) = \mathfrak{a}_{na}^{(k)} |t|^{\varepsilon_n} \to 0, \quad \widehat{\beta}_{na}^{(k)}(t) = \mathfrak{b}_{na}^{(k)} |t|^{\varepsilon_n} \to 0 \quad \text{as } t \to 0. \quad (10.22)$$

With this model for the t-dependence of the parameters and with the exponents ε_i [$\varepsilon = \min(\varepsilon_0, \varepsilon_1, \ldots)$] and ε_n appropriately chosen in comparison of the theory with experiment, the equation of state can be used to calculate the critical properties of fluids in the subcritical regime, as will be shown later. We have taken the same exponents for both $\mathcal{A}$ and $\mathcal{B}$ because their statistical mechanical formulas are given in terms of an identical pair correlation function for the liquid (l) and vapor (v) branches — in which the temperature dependence is entirely vested — because they are symmetric in the neighborhood of the critical point. As will be shown later, the parameters $\mathfrak{a}_i^{(k)}$, $\mathfrak{b}_i^{(k)}$, $\mathfrak{a}_{na}^{(k)}$, and $\mathfrak{b}_{na}^{(k)}$ ($i = 0, 1$; $k = l, v$) are not arbitrary, but constrained by the compatibility with experiment and by the stability conditions. *Our basic point in this chapter is that the GvdW parameters are directly related to certain specific and characteristic features of thermophysical data of the fluid under consideration, especially, in the neighborhood of the critical point.*

10.4.2. *Supercritical Regime*

Owing to the aforementioned properties of $a_i^{(k)}(t)$, $b_i^{(k)}(t)$, $\alpha_{nai}^{(k)}(t)$, and $\beta_{nai}^{(k)}(t)$ and the fact that at the critical point,

$$x_{sl} = x_{sv} = 0,$$

the expansions for $\mathcal{A}$ and $\mathcal{B}$ in Eqs. (10.16) and (10.17) reduce, on the critical isotherm, to the forms

$$\mathcal{A} = 1 + \sum_{i \geq 1} a_i x^i + \sum_{i \geq 0} \alpha_{nai} x^{3+i} |x|^{1+\delta} \quad (t = 0), \quad (10.23)$$

$$\mathcal{B} = 1 + \sum_{i \geq 1} b_i x^i + \sum_{i \geq 0} \beta_{nai} x^{3+i} |x|^{1+\delta} \quad (t = 0). \quad (10.24)$$

These models can be extended into the supercritical regime of $t > 0$, in which case a_i, b_i, α_{nai}, and β_{nai} may have to be made t-dependent if we want to discuss the thermophysical properties of the fluid in the supercritical regime. It should be noted that $\mathcal{A}$ and $\mathcal{B}$ are analytic for $t > 0$ and $x \neq 0$ despite the presence of the irrational terms.

10.5. Reduced Chemical Potential

The chemical potential μ is reduced as follows:

$$\varpi\left(t,x\right) = \frac{\zeta B_c^*}{\nu A_c^* \varepsilon} \mu\left(x,t\right) = \frac{\tau \mu\left(x,t\right)}{T_c^* \varepsilon}. \tag{10.25}$$

This reduced chemical potential is described by the reduced equilibrium Gibbs relation for ϖ

$$d\varpi = -\widehat{s}dt + \frac{1}{(1+x)}d\psi \quad [v = 1/y = 1/\left(1+x\right)], \tag{10.26}$$

where the molar entropy $\overline{s}$ is scaled as follows:

$$\widehat{s} = \frac{\tau}{k_B}\overline{s}. \tag{10.27}$$

Now, consider an isothermal process $(t,\psi_0) = (t,\phi_0+1) \rightarrow (t,\psi) = (t,\phi+1)$. Then upon integration of Eq. (10.26), we obtain

$$\varpi\left(t,\phi\right) = \varpi\left(t,\phi_0\right) + \int_{\psi_0}^{\psi} d\psi \frac{1}{1+x}. \tag{10.28}$$

The reference reduced pressure ψ_0 will be more explicitly specified later. The integral on the right of Eq. (10.28) is now calculated with the reduced equation of state (10.12).

By using Eq. (10.12), we express the integral for the chemical potential in terms of the GvdW parameters $\mathcal{A}$ and $\mathcal{B}$. First, the integral in Eq. (10.28) may be recast as follows:

$$K_k = \int_{\psi_0}^{\psi} d\psi \frac{1}{1+x} = \int_{x_0}^{x} dx \frac{1}{1+x}\left(\frac{\partial \psi}{\partial x}\right)_t. \tag{10.29}$$

Performing integration by parts and using the reduced canonical equation of state (10.12), we obtain

$$K_k = \frac{1+\phi\left(x,t\right)}{1+x} - \frac{1+\phi\left(x_0,t\right)}{1+x_0} - \zeta \int_{x_0}^{x} dx \mathcal{A}_k\left(x,t\right)$$

$$+ \tau\left(1+t\right) \int_{x_0}^{x} \frac{dx}{(1+x)\left[1-\nu\left(1+x\right)\mathcal{B}_k\left(x,t\right)\right]}. \tag{10.30}$$

Here, the subscript k refers to the liquid (l) or vapor (v) branch; the liquid and vapor branches of the equation of state are different in the subcritical regime, but in the supercritical regime, the equation of state involves a single set of continuous $\mathcal{A}$ and $\mathcal{B}$ representative of the gaseous states of the fluid. The expression for K_k is rearranged to the form which we find most convenient for the questions addressed in this chapter:

$$K_k = \frac{\phi(x,t)}{1+x} - \frac{\phi(x_0,t)}{1+x_0} + \int_{x_0}^{x} dx \frac{\Pi(x,t)}{(1+x)^2 [1 - \nu(1+x)\mathcal{B}_k]}, \quad (10.31)$$

with the definition of $\Pi(x,t)$

$$\Pi(x,t) = \tau(1+x)(1+t)$$
$$- [1 + \zeta(1+x)^2 \mathcal{A}(x,t)][1 - \nu(1+x)\mathcal{B}(x,t)], \quad (10.32)$$

which is related to the reduced pressure as will be evident from Eq. (10.60). Then we will be able to take advantage of the mathematical results developed for the equation of state when we deduce the mathematical properties of the chemical potential. The reduced chemical potential now is expressible in the form

$$\varpi(t,\phi) = \varpi_0(t) + \varpi_{\text{ex}}(x,t), \quad (10.33)$$

where the excess chemical potential is written as

$$\varpi_{\text{ex}}(x,t) = \frac{\phi(x,t)}{1+x} + \int_{x_0}^{x} dx \frac{\Pi(x,t)}{(1+x)^2 [1 - \nu(1+x)\mathcal{B}]}. \quad (10.34)$$

With $\phi_0 = \phi(x_0,t)$, the reference chemical potential is defined by

$$\varpi_0(t) = \lim_{\phi \to \phi_0, x \to x_0} \left[\varpi(t,\phi) - \frac{\phi(x,t)}{1+x} \right]. \quad (10.35)$$

The chemical potential $\varpi(t,\phi)$ presented is formal in the sense that the integral is not performed as yet. We may specify the reference state more explicitly: we find it convenient to take $x_0 := x_c = 0$ for both $k = l, v$ or $x_0 = x_{sk}$ $(k = l, v)$, especially, when the excess chemical potential $\varpi_{ex}(x,t)$ is expressed in series of $(x - x_{sk})$. We will return to this question after we have discussed the definition of critical point.

10.6. Specific Heat

The isochoric specific heat $\hat{c}_v$ may be calculated from the internal energy $\mathcal{E}$, which may be calculated in the same manner as for the chemical potential by following the well-known procedure in thermodynamics, namely, by varying the internal energy from state (x_0, t_0) to state (x, t). Then the specific heat is calculated with the derivative

$$\hat{c}_v(x, t) = \left(\frac{\partial \mathcal{E}}{\partial t}\right)_v, \tag{10.36}$$

which can be eventually expressible in terms of the equation of state. It is also possible to obtain $\hat{c}_v(x, t)$ by integrating the derivative

$$\left(\frac{\partial \hat{c}_v}{\partial x}\right)_t = -\frac{(1+t)}{(1+x)^2}\left(\frac{\partial^2 \psi}{\partial t^2}\right)_x \tag{10.37}$$

over x, given the equation of state; see (10.6) and (10.7) for the definition of ψ. Yet another way of calculating $\hat{c}_v$ employs the reduced work function (Helmholtz free energy) F. It gives rise to the formula

$$\hat{c}_v(x, t) = -(1+t)\left(\frac{\partial^2 F}{\partial t^2}\right)_x$$

$$= (1+t)\left[\left(\frac{\partial^2 \psi}{\partial t^2}\right)_x - (1+x)\left(\frac{\partial^2 \varpi}{\partial t^2}\right)_x\right]. \tag{10.38}$$

It should be remarked that this formula can also be obtained by using the Gibbs–Duhem equation; see Problem 4 of Chapter 7. All these relations can be shown to yield eventually the same formula expressed in terms of the equation of state. However, we will find relation (10.38) more convenient to use for our purpose of calculating the asymptotic behavior of the anomalous specific heat in the subcritical neighborhood of the critical point because the temperature derivative is taken after integration performed over x.

By using Eq. (10.33), it is easy to show that the excess part $\hat{c}_{vk}^{(ex)}$ of the specific heat above and beyond the specific heat corresponding to $\varpi_0(t)$ is given by the formula

$$\hat{c}_{vk}^{(ex)} = -(1+t)(1+x)\left(\frac{\partial^2 I_k}{\partial t^2}\right)_x, \tag{10.39}$$

where $I_k(x_k, t)$ is defined by the integral

$$I_k(x_k, t) = \int_{x_0}^{x_k} dx \frac{\Pi_k(x, t)}{(1+x)^2\left[1 - \nu\mathcal{B}_k(x+1)\right]} \qquad (k = l, v) \tag{10.40}$$

with x_0 chosen appropriately and $\Pi_k(x, t)$ is $\Pi(x, t)$ with the GvdW parameters for the liquid or vapor branch: for $k = l$ or v,

$$\Pi_k(x, t) = \tau(1+x)(1+t)$$

$$-[1 + \zeta(1+x)^2\mathcal{A}_k(x, t)][1 - \nu(1+x)\mathcal{B}_k(x, t)]. \tag{10.41}$$

Here, $\mathcal{A}_k$ and $\mathcal{B}_k$ denote the k branch of $\mathcal{A}$ and $\mathcal{B}$, respectively. As will be shown later, the behavior of $\widehat{c}_{vk}^{(ex)}(k=l,v)$ put further restriction on the t-dependence of parameters $\mathcal{A}(t,x)$ and $\mathcal{B}(t,x)$ because of their characteristic singular behavior in the neighborhood of the critical point.

10.7. Virial and Joule–Thomson Coefficients

The virial coefficients provide useful information on fluid behaviors and their experimental data are available for many fluids in the literature. So do the Joule–Thomson coefficients. Since these coefficients can be related to the GvdW parameters, the experimental information on the virial and Joule–Thomson coefficients provide important and valuable information on the GvdW parameters, particularly, in the supercritical regime. We relate them to each other in the following. The canonical equation of state reduced with respect to the potential parameters, but not with respect to the critical parameters, is used in this section.

10.7.1. *Virial Coefficients*

The virial coefficients C_i $(i \geq 2)$ of the virial equation of state in the supercritical regime

$$\frac{p^*\beta^*}{\eta} = 1 + C_2\eta + C_3\eta^2 + \cdots + C_i\eta^{i-1} + \cdots \tag{10.42}$$

are given by the formulas in terms of A^* and B^* and their derivatives:

$$C_2 = B^*(T^*,0) - A^*(T^*,0)\beta^*, \tag{10.43}$$

$$C_3 = [B^*(T^*,0)]^2 + \left(\frac{dB^*}{d\eta}\right)_{\eta=0} - \beta^*\left(\frac{dA^*}{d\eta}\right)_{\eta=0}, \tag{10.44}$$

$$\vdots$$

Here, C_j are reduced virial coefficients $(j \geq 2)$. Since the second virial coefficient vanishes at the Boyle temperature, it follows that at the Boyle temperature $T^* = T_B^*$,

$$B^*(T_B^*,0) = \beta^* A^*(T_B^*,0). \tag{10.45}$$

Since C_2 tends to negative infinity as $T^* \to 0$, the parameters A and B should be such that

$$\lim_{T^*\to 0}[B^*(T^*,0) - \beta^* A^*(T^*,0)] = -\infty. \tag{10.46}$$

On the other hand,

$$\lim_{T^* \to \infty} C_2 = C_2^{(\infty)} > 0. \tag{10.47}$$

Moreover,

$$\lim_{T^* \to T_c^*} C_2 = C_2^{(c)}. \tag{10.48}$$

The third virial coefficient also behaves similar to the second virial coefficient with respect to T^*. In particular, it is known experimentally that there exists temperature T_3^* at which

$$C_3(T_3^*) = 0, \tag{10.49}$$

and

$$\lim_{T^* \to 0} C_3(T^*) = -\infty. \tag{10.50}$$

The Boyle temperature T_B^* does not coincide with T_3^* at which the third virial coefficient vanishes. If by any chance $T_3^* = T_B^*$, then at that temperature, the equation of state assumes an ideal gas equation of state to the order of η^3.

We may conjecture that such a temperature exists for the canonical equation of state. Denote it by T_{can}^*. Then at $T^* = T_{\text{can}}^*$, the following relation holds:

$$\frac{B^*(T_{\text{can}}^*, \eta)}{1 - B^*(T_{\text{can}}^*, \eta)\eta} = \beta^* A^*(T_{\text{can}}^*, \eta). \tag{10.51}$$

At temperature T_{can}^*, the equation of state then takes the form for an ideal gas. However, at present, it is not known that such a temperature is experimentally identified. If it was, it should be possible to define the standard state of the fluid with it instead of the hypothetical ideal gas that is conventionally defined as the standard state. It is the state at which $T = T_{\text{can}}^*$ and the equation of state is that of an ideal gas.

10.7.2. *Joule–Thomson Coefficient*

Another important property of fluids is the Joule–Thomson coefficient (9.150), which in terms of the canonical equation of state may be expressed as

$$\mu_{JT} = (\eta c_p)^{-1} \left[\frac{\eta T^{*2} \left(B^* + B^{*\prime} \eta \right)}{p^* \left(1 - B^* \eta \right)^2} - \frac{\eta T^*}{p^*} (A^* + A^{*\prime} \eta) - 1 \right]. \qquad (10.52)$$

Here, the prime on A^* and B^* denotes the temperature derivative. The Joule–Thomson coefficient vanishes at the inversion temperature. Therefore, at $T^* = T_{in}^*$, inversion temperature, the following equation holds:

$$p_{in}^* = \frac{\eta T_{in}^{*2} \left(B_{in}^* + B_{in}^{*\prime} \eta \right)}{\left(1 - B_i^* \eta \right)^2} - \eta T_{in}^* (A_{in}^* + A_{in}^{*\prime} \eta), \qquad (10.53)$$

where $A_{in}^* = A^*(T_{in}^*, \eta)$, $A_{in}^{*\prime} = (\partial A^* / \partial T^*)_{p^*} |_{T^* = T_{in}^*}$, $B_{in}^* = B^*(T_{in}^*, \eta)$, and $B_{in}^{*\prime} = (\partial B^* / \partial T^*)_{p^*} |_{T^* = T_{in}^*}$. If p_{in}^* on the left is replaced with the reduced canonical equation of state at $T^* = T_{in}^*$, Eq. (10.53) becomes an algebraic equation for T_{in}^* and η. The relation such as Eq. (10.53) puts constraints on the mathematical expressions for the GvdW parameters such that the parameters are empirically determined to reproduce the inversion curve.

10.7.3. *Asymptotic Behavior of the Second Virial Coefficient*

The asymptotic behaviors of the virial coefficients also put additional constraints on the GvdW parameters. For example, since the second virial coefficient for the LJ fluid asymptotically behaves as shown in Chapter 9 and derived in Appendix A

$$C_2 = \sqrt{2} \Gamma \left(\tfrac{1}{4} \right) v_0 T^{*-1/4} [1 + O(T^{*-1})] \quad \text{as } T^* \to \infty \qquad (10.54)$$

$$= -16 \sqrt{2\pi} v_0 e^{1/T^*} T^{*-\frac{3}{2}} \left[1 + O\left(T^* \right) \right] \quad \text{as } T^* \to 0, \qquad (10.55)$$

the limiting behaviors of the GvdW parameters must conform to these limiting behaviors to be consistent with the LJ fluid and experiment. Therefore, these are additional means to fix the GvdW parameters.

10.8. Stability Conditions of Mechanical and Material Equilibria

10.8.1. *Classical Definition of Critical Point*

The critical state is a singular state of the fluid, and its definition is closely associated with the thermodynamic stability of the critical state of the fluid.

In the conventional approach, the critical point is defined by the vanishing first two density derivatives of pressure or the chemical potential, and the sign of the third density derivative determines the stability of the system at the critical point so defined: it is stable if the third derivative is negative, unstable if it is positive, and neutral if it is equal to zero.

Thus, if the reduced canonical equation of state is used, the critical point $(x, t, \phi) = (0, 0, 0)$ is given by (ν, τ, ζ) (see Eq. (10.8)) obeying the three algebraic equations

$$(1 + \zeta)(1 - \nu) = \tau, \tag{10.56}$$

$$\left(\frac{\partial^i \phi}{\partial x^i}\right)_{t=0, x=0} = 0 \quad \text{for } i = 1, 2. \tag{10.57}$$

This definition is possible for van der Waals equation of state, but it is necessary to extend it. It is possible to see its necessity if we examine it with the canonical equation of state in which the GvdW parameters can depend on t as well as x in general. Then it would be conceivable to imagine that there arise some cases in which the third density derivative of ϕ happens to be equal to zero.

10.8.2.   *Extended Definition of Critical Point*

J. J. van Laar[8] considered for the first time the variability of the van der Waals constant b and its relation to the critical parameters, although van der Waals himself disfavored such an extension. As noted earlier, introduction of density- and temperature-dependent van der Waals parameters might necessitate extending the definition of the critical point since such parameters could give rise to the vanishing third density derivative. He thus required the first four density derivatives of pressure to identically vanish at the critical point.

We elaborate on this point by considering the thermodynamic stability theory of the critical point. In the classical thermodynamic stability theory of the critical state, the third density derivative provides the stability criteria for the fluid at the critical point. If the van der Waals parameters are

---

[8]See J. J. van Laar in footnote 1.

assumed to depend[9] on density and temperature, it is clearly possible to have a situation in which the third density derivative might identically vanish at the critical point. In that event, the thermodynamic stability of the fluid is determined by the fifth density derivative with the fourth density derivative vanishing as well.

Thus, the critical point is determined by the equation of state and the first four density derivatives of pressure vanishing at the critical state. *The system is then thermodynamically stable if the fifth density derivative is negative at the critical point, unstable if it is positive, and neutral if it is equal to zero.* The sign of the fifth density derivative, therefore, also puts an additional condition on the equation of state or its parameters. This kind of situation arises for the canonical equation of state as will be shown. In fact, this extension of the definition of critical point makes it possible to relate the parameters of the canonical equation of state to the critical properties of the fluid.

Since the GvdW parameters $\mathcal{A}$ and $\mathcal{B}$ are generally density- and temperature-dependent, it is natural to extend the classical definition of the critical point. As a matter of fact, the GvdW parameters can be given exact statistical mechanical representations, which can indeed be shown as density- and temperature-dependent. Although this extension necessarily gives rise to a larger and more complicated set of algebraic equations to solve for various thermodynamic properties, it widens the scope of the ideas embodied by the van der Waals theory and makes the theory more flexible, thereby enabling us to more effectively treat the critical phenomena phenomenologically than the van der Waals equation of state. But it also renders the basic mathematical structures of algebraic equations that determine some thermophysical properties simpler than otherwise. As a matter of fact, such a simplification can be more attractive in practice. Moreover, the additional conditions significantly reduce the degrees of freedom for the parameters necessarily introduced in $\mathcal{A}$ and $\mathcal{B}$ in the phenomenological approach. Therefore, we make the following proposition.

Proposition 1. *The critical point of the fluid described by the canonical (GvdW) equation of state is defined by requiring the first four density derivatives of pressure to vanish at the critical point.*

---

[9]From the molecular theory standpoint — namely, the generic van der Waals equation of state of Eu and Rah (see footnote 3) — the van der Waals parameters would generally depend on density and temperature.

This proposition is in essence about the thermodynamic stability of mechanical equilibrium of the fluid at the critical temperature and density. Away from the critical point in the subcritical regime, there exists a locus of the limit of stability in density, and this locus is the spinodal curve. There are two densities x_{sl} and x_{sv} ($x_{sv} < x_{sl}$) within the interval of which the fluid is unstable. See Fig. 10.1 for the meanings of x_{sl} and x_{sv}. According to the statistical mechanical expressions for the GvdW parameters, which have been investigated by means of an integral equation theory and simulations,[10] the GvdW parameters extend into the liquid–vapor coexistence regime to the points, where they become discontinuous. This locus of discontinuity in GvdW parameters $\mathcal{A}$ and $\mathcal{B}$ is known to coincide with the spinodal curve in liquid–vapor transition. Since $x_{sk} = 0$ at the critical point $t = 0$, it is reasonable to extend Proposition 1 to the subcritical regime and assume that it is also applicable at $x = x_{sk}$ for $t < 0$. Therefore, we make the following proposition for the spinodal curve.

Proposition 2. *Along the locus $x_{sk}(t)$ of the limit of instability, the first four density derivatives of ϕ at $t < 0$ vanish. The fluid at the spinodal state is thermodynamically stable if the fifth density derivative of ϕ is negative, unstable if it is positive, and neutral if it is equal to zero.*

These propositions not only provide additional conditions on the GvdW parameters $\mathcal{A}$ and $\mathcal{B}$ and their density derivatives but also produce, as a function of t, the spinodal curve that otherwise would appear as parameters in the model for $\mathcal{A}$ and $\mathcal{B}$ we have chosen. This point will become evident as we develop the theory for the critical characteristics of fluids in the following section.

10.9. Critical Properties of Fluids

10.9.1. *Critical Point*

According to *Proposition* 1, the nondimensional form of GvdW equation of state (10.12) at the critical point is given by

$$(1 + \zeta)(1 - \nu) = \tau, \tag{10.58}$$

---

[10]K. Rah and B. C. Eu, *J. Chem. Phys.* **115**, 9370 (2001).

and the first four density derivatives are equal to zero:

$$\left(\frac{\partial^i \phi}{\partial x^i}\right)_{t=0,x=0} = 0 \quad \text{for } i = 1, 2, \ldots, 4. \tag{10.59}$$

Note that the first equation (10.58) is equivalent to $\phi(x,t)|_{t=0,x=0} = 0$.

The reduced equation of state, Eq. (10.12), can be rearranged to the form

$$\phi\left[1 - \nu\left(1 + x\right)\mathcal{B}\left(x,t\right)\right] = \Pi(x,t). \tag{10.60}$$

Henceforth $\Pi(x,t)$ will be referred to as the effective reduced pressure.

Then, since

$$\left(\frac{\partial^i p^*}{\partial \eta^i}\right)_{T^*=T_c^*,\eta=\eta_c} = \left(\frac{\partial^i \phi}{\partial x^i}\right)_{t=0,x=0} \quad \text{for } i = 1, 2, \ldots, 4,$$

and furthermore,

$$\left(\frac{\partial^i \phi}{\partial x^i}\right)_{t=0,x=0} = \left(\frac{\partial^i \Pi}{\partial x^i}\right)_{t=0,x=0} = 0 \quad \text{for } i = 1, 2, \ldots, 4, \tag{10.61}$$

owing to the fact that $\phi = 0$ at the critical point, Eqs. (10.58) and (10.59) take the forms

$$(1 + \zeta)(1 - \nu) = \tau, \tag{10.62}$$

$$\left(\frac{\partial^i \Pi}{\partial x^i}\right)_{t=0,x=0} = 0 \quad \text{for } i = 1, 2, \ldots, 4. \tag{10.63}$$

These equations are algebraic equations for the reduced critical parameters (ν, ζ, τ) and, when solved, yield (ν, ζ, τ) in terms of parameters making up $\mathcal{A}$ and $\mathcal{B}$. They are simpler to handle than Eqs. (10.58) and (10.59). Since there are five algebraic equations for the parameters appearing in the models for $\mathcal{A}(x,t)$ and $\mathcal{B}(x,t)$, the number of degrees of freedom is reduced by 5 for the parameters making up $\mathcal{A}$ and $\mathcal{B}$.

The first equation of this set is the reduced equation of state at the critical point $(\phi, t, x) = (0, 0, 0)$. If the density derivatives (10.63) are explicitly worked out at the critical point, we have the set of algebraic equations for the parameters:

$$P_0 := \tau - (1 + \zeta)(1 - \nu) = 0, \tag{10.64}$$

$$1! P_1 := \tau - \zeta(2 + a_1)(1 - \nu) + \nu(1 + \zeta)(1 + b_1) = 0, \tag{10.65}$$

$$2! P_2 := -2\zeta(1 - \nu)(1 + 2a_1 + a_2) + 2\nu\zeta(2 + a_1)(1 + b_1)$$
$$+ 2\nu(1 + \zeta)(b_1 + b_2) = 0, \tag{10.66}$$

$$3! P_3 := -6(\zeta a_1 + 2\zeta a_2 + \zeta a_3)(1 - \nu) + 6\zeta\nu(1 + 2a_1 + a_2)(1 + b_1)$$
$$+ 6\zeta\nu(2 + a_1)(b_1 + b_2) + 6\nu(1 + \zeta)(b_2 + b_3) = 0, \tag{10.67}$$

$$4! P_4 := -24(1 - \nu)\zeta(a_2 + 2a_3 + a_4)$$
$$- 576\zeta^2\nu^2(a_1 + 2a_2 + a_3)(1 + 2a_1 + a_2)(1 + b_1)(b_1 + b_2)$$
$$+ 24\zeta\nu(2 + a_1)(b_2 + b_3) + 24\nu(1 + \zeta)(b_3 + b_4) = 0. \tag{10.68}$$

It should be noted that these algebraic equations are exact for the models chosen, Eqs. (10.16) and (10.17).

The set of algebraic equations, (10.64)–(10.68), determines the reduced parameters (τ, ζ, ν) in terms of a_i and b_i ($i \leq 4$). Since the critical parameters are available experimentally, these parameters can be determined consistently with the experimental data, thus facilitating the determination of a_i and b_i from experiment. It is important to see that a_i and b_i ($i \leq 4$) are determined from the experimental values for the reduced critical parameters (τ, ζ, ν) such that conditions (10.64)–(10.68) are satisfied. Such a set of (τ, ζ, ν) and a_i and b_i ($i \leq 4$) now assures the desired behavior for the critical isotherm for pressure as will be shown in the following section.

10.9.2. *Critical Isotherm for Pressure*

With the preparation made earlier, the critical isotherm for ϕ can be calculated if the effective reduced pressure $\Pi(x, t)$ is expanded around $x = 0$ on the critical isotherm $t = 0$. To obtain the idea of what sort of form $\Pi(x, t)$ would take, if the models for $\mathcal{A}$ and $\mathcal{B}$, for example, (10.16) and (10.17), for the GvdW parameters are employed, we may express it in the form

$$\Pi(x,t) = \sum_{i \geq 0} L^{(i)}(x_{sk},t)(x - x_{sk})^i$$

$$+ \sum_{i \geq 0} N^{(i)}(x_{sk},t)(x - x_{sk})^{3+i}|x - x_{sk}|^{1+\delta}$$

$$+ \sum_{i \geq 0} M^{(i)}(x_{sk},t)(x - x_{sk})^{6+i}|x - x_{sk}|^{2+2\delta}, \quad (10.69)$$

where $L^{(i)}(x_{sk},t)$, $N^{(i)}(x_{sk},t)$, and $M^{(i)}(x_{sk},t)$ can be explicitly found in terms of $\mathcal{A}$ and $\mathcal{B}$ by using the models (10.16) and (10.17). This in fact is a fairly general expansion arising from the required behavior of the GvdW parameters in the subcritical regime. In view of P_i defined earlier, it is convenient to express $L^{(i)}(x_{sk},t)$ in the forms

$$L^{(i)}(x_{sk},t) = P_i + \widehat{L}^{(i)}(x_{sk},t), \quad (10.70)$$

with $\widehat{L}^{(i)}(x_{sk},t)$ denoting that part of $L^{(i)}$ depending on t away from the critical temperature. Similarly, $N^{(i)}(x_{sk},t)$ can be expressed in the form

$$N^{(i)}(x_{sk},t) = Q_i + \widehat{N}^{(i)}(x_{sk},t), \quad (10.71)$$

where

$$Q_i = N^{(i)}(x_{sk},t)|_{t=0}, \quad (10.72)$$

and $\widehat{N}^{(i)}(x_{sk},t)$ is the t-dependent part. We note that

$$Q_0 = \nu(1 + \zeta)\beta_{na0} - \zeta(1 - \nu)\alpha_{na0}, \quad (10.73)$$

which will be used later. The coefficient $M^{(i)}(x_{sk},t)$ can be similarly expressed. The t-dependent parts will be proportional to t^ε if the power series models (10.19) and (10.20) are chosen for $N^{(i)}(x_{sk},t)$.

Since the first four density derivatives of $\Pi(x,t)$ vanishes at the critical point according to Proposition 1, we see that the power of the leading nonvanishing term in the expansion of $\Pi(x,t)$ near $x = 0$ should be at least larger than 4. Therefore, upon use of Eq. (10.60), we obtain

$$\phi \simeq \frac{\Pi(x,0)}{1 - \nu} = \frac{Q_0}{1 - \nu}x^{4+\delta}[1 + O(x)] \quad (\delta > 0). \quad (10.74)$$

Thus, Eq. (10.74) is one of the necessary conditions for $\mathcal{A}$ and $\mathcal{B}$ to account for the experimental data on critical phenomena. It in fact enables us to vest the critical isotherm in the nonanalytic term in $\mathcal{A}$ and $\mathcal{B}$ and choose the exponent δ on the basis of experiment, namely, $\delta = 0.3$–0.5. This indicates to us which part of the canonical equation of state is principally responsible for the critical isotherm for pressure. *This is the first important result by the canonical equation of state assumed in consideration of subcritical phenomena.* Other subcritical characteristics of fluids can be vested in the *t*-dependent parts of $\mathcal{A}$ and $\mathcal{B}$, as will be shown subsequently.

10.9.3. *Spinodal Curve*

To calculate the spinodal curve as a function of t, it is convenient to recast the conditions of Proposition 2

$$\left(\frac{\partial^i \phi}{\partial x^i}\right)_{x=x_{sk}} = 0 \quad (i = 1, \ldots, 4) \tag{10.75}$$

into another more convenient form by using Eq. (10.60). Upon use of this particular form of equation of state, we obtain in place of Eq. (10.75) the set of equations

$$-\nu[B_k^{(i-1)} + (1 + x_{sk}) B_k^{(i)}]\Pi_k(x_{sk}, t) = \Pi_k^{(i)}[1 - \nu(1 + x_{sk})B_k^{(0)}], \tag{10.76}$$

where $i = 1, \ldots, 4$, $k = l, v$, and various symbols are defined by the following equations:

$$B_k^{(i)}(t) = \left(\frac{\partial^i \mathcal{B}}{\partial x^i}\right)_{x=x_{sk}}, \tag{10.77}$$

$$\Pi^{(i)}(x_{sk}, t) = \left(\frac{\partial^i \Pi}{\partial x^i}\right)_{x=x_{sk}}, \tag{10.78}$$

$$\phi_0 = \phi(x_{sk}, t). \tag{10.79}$$

It can be shown that Eq. (10.76) is quartic with respect to x_{sk} for all i. Thus, the set can be written as

$$\varphi_{i4}^{(k)} x_{sk}^4 + \varphi_{i3}^{(k)} x_{sk}^3 + \varphi_{i2}^{(k)} x_{sk}^2 + \varphi_{i1}^{(k)} x_{sk} + \varphi_{i0}^{(k)} = 0 \tag{10.80}$$

for $k = l, v$ and $i = 1, \ldots, 4$. The coefficients $\varphi_{ij}^{(k)}$ in Eq. (10.80) can be worked out explicitly; they are presented in Appendix B. This set of four algebraic equations can be reduced to a single quadratic equation in x_{sk}

$$D_2^{(k)} x_{sk}^2 + D_1^{(k)} x_{sk} + D_0^{(k)} = 0, \tag{10.81}$$

where

$$D_2^{(k)} = \frac{1}{\Delta_1}(\varphi_{24}^{(k)}\varphi_{12}^{(k)} - \varphi_{14}^{(k)}\varphi_{22}^{(k)}) - \frac{1}{\Delta_2}(\varphi_{44}^{(k)}\varphi_{32}^{(k)} - \varphi_{34}^{(k)}\varphi_{42}^{(k)}),$$

$$D_1^{(k)} = \frac{1}{\Delta_1}(\varphi_{24}^{(k)}\varphi_{11}^{(k)} - \varphi_{14}^{(k)}\varphi_{21}^{(k)}) - \frac{1}{\Delta_2}(\varphi_{44}^{(k)}\varphi_{31}^{(k)} - \varphi_{34}^{(k)}\varphi_{41}^{(k)}), \quad (10.82)$$

$$D_0^{(k)} = \frac{1}{\Delta_1}(\varphi_{24}^{(k)}\varphi_{10}^{(k)} - \varphi_{14}^{(k)}\varphi_{20}^{(k)}) - \frac{1}{\Delta_2}(\varphi_{44}^{(k)}\varphi_{30}^{(k)} - \varphi_{34}^{(k)}\varphi_{40}^{(k)}),$$

with the definitions

$$\Delta_1 = \varphi_{24}^{(k)}\varphi_{13}^{(k)} - \varphi_{14}^{(k)}\varphi_{23}^{(k)}, \quad (10.83)$$

$$\Delta_2 = \varphi_{44}^{(k)}\varphi_{33}^{(k)} - \varphi_{34}^{(k)}\varphi_{43}^{(k)}. \quad (10.84)$$

The spinodal curve is then given by a solution of the quadratic equation (10.81):

$$x_{sk} = -\frac{D_0^{(k)}}{D_1^{(k)}} \cdot \frac{2}{\left(1 + \sqrt{1 - 4D_0^{(k)}D_2^{(k)}/D_1^{(k)2}}\right)}, \quad (10.85)$$

for which we have chosen the positive sign in the denominator. The important factor indicating the t-dependence of x_{sk} is the numerator.

It can be shown that $\varphi_{ij}^{(k)}$ can be decomposed into t-independent and t-dependent parts

$$\varphi_{ij}^{(k)}(t) = \omega_{ij} + \widehat{\varphi}_{ij}^{(k)}(t). \quad (10.86)$$

The t-independent part ω_{ij} is dependent on τ, ζ, ν, a_i, and b_i ($i \geq 0$). More specifically,

$$\omega_{i0} = 0, \quad (10.87)$$

$$\omega_{ij} \neq 0 \quad (0 \leq i \leq 4; \, 1 \leq j \leq 4). \quad (10.88)$$

Therefore, $\varphi_{i0}^{(k)}(t)$ appearing in $D_0^{(k)}$ do not contain the t-independent part. This means that in the leading order in t, the numerator $D_0^{(k)}$ is a linear combination of $\widehat{\varphi}_{i0}^{(k)}(t)$ and hence x_{sk} is a linear combination of $\widehat{a}_i^{(k)}(t)$ and $\widehat{b}_i^{(k)}(t)$ in the leading order in t:

$$x_{sk} = -\sum_{i \geq 0}[s_{ai}^{(k)}\widehat{a}_i^{(k)}(t) + s_{bi}^{(k)}\widehat{b}_i^{(k)}(t)] + O(\widehat{a}^{(k)}\widehat{b}^{(k)}), \quad (10.89)$$

where $s_{ai}^{(k)}$ and $s_{bi}^{(k)}$ are constants composed of parameters ν, τ, ζ, $a_1, \ldots, a_4, b_1, \ldots, b_4$ related to the critical point and hence known by now; parameters $\nu, \tau, \zeta, a_1, a_2, b_1, b_2$ are explicitly determined in the quadratic model discussed later. These coefficients can be worked out more explicitly from the formulas for $\varphi_{ij}^{(k)}(t)$ given in Appendix B.

In any case, we now have a relatively simple algorithm to calculate the spinodal curve in terms of the model assumed. Therefore, the spinodal curve may be written as

$$x_{sk} = |t|^\varepsilon f_k(t), \tag{10.90}$$

where the exponent ε is given by

$$\varepsilon = \min(\varepsilon_0, \varepsilon_1, \varepsilon_2, \ldots),$$

and the function $f_k(t)$ may be written as

$$f_k(t) = f_0^{(k)} \left[1 + O\left(|t|^\varepsilon, |t|^{\varepsilon_1}, \ldots\right)\right]. \tag{10.91}$$

Thus, we see that the leading t-dependence is determined by that part of the GvdW parameters corresponding to the exponent ε. This means that the t-dependent part of the GvdW parameters is directly associated with the spinodal curve that is the locus of stability of the fluid. *We have now acquired the second important result for the model.*

To work out the coefficients $\varphi_{ij}^{(k)}$, it is necessary to calculate $\Pi^{(i)}(x_{sk}, t)$, which is the ith derivative of $\Pi(x, t)$ at $x = x_{sk}$. They are found to be cubic polynomials of x_{sk}:

$$\Pi^{(i)}(x_{sk}, t) = \Pi_0^{(i)} + \Pi_1^{(i)} x_{sk} + \Pi_2^{(i)} x_{sk}^2 + \Pi_3^{(i)} x_{sk}^3 \quad (0 \le i \le 4). \tag{10.92}$$

The coefficients $\Pi_j^{(i)}$ in this polynomial consist of t-independent and t-dependent parts as follows:

$$\Pi_j^{(i)} = P_{ij} + \widehat{\Pi}_j^{(i)}(t), \tag{10.93}$$

where P_{ij} is the t-independent part made up of the parameters determining the critical point and $\widehat{\Pi}_j^{(i)}(t)$ is the t-dependent part determined by the t-dependent part of the GvdW parameters, namely, $\widehat{a}_i(t)$ and $\widehat{b}_i(t)$. In particular, we note that

$$P_{i0} = P_i = 0 \tag{10.94}$$

by virtue of Proposition 1; Eq. (10.64). This means that

$$\Pi_0^{(i)} = \widehat{\Pi}_0^{(i)}(t) \quad (0 \le i \le 4). \tag{10.95}$$

These coefficients, although somewhat complicated but straightforward to obtain from the definition of $\Pi^{(i)}(x,t)$, are listed in Appendix B. It is useful to note that by virtue of Eqs. (10.94) and (10.95), we obtain Eq. (10.87) that gives rise to the conclusion (10.89) for x_{sk} given earlier.

10.9.4. *Excess Chemical Potential*

We have proposed Proposition 2 as a subcritical extension of the stability theory at the critical point. The proposition defines the stability of the fluid along the line of limit of stability — *the spinodal curve*. Since the derivatives of ϕ can be transcribed to those of chemical potentials, the proposition can be phrased in terms of excess chemical potential ϖ_{ex}. Therefore, equivalent to Proposition 2 for the derivatives of ϕ, we have the conditions for the spinodal curve in the form

$$\left(\frac{\partial^i \varpi_{ex}}{\partial x^i}\right)_{x=x_{sk}} = 0 \quad (i = 1, \ldots, 4; \ k = l, v). \tag{10.96}$$

Along the spinodal curve, we also have the excess chemical potential $\varpi_{ex}(x,t)$ given by the formula

$$\varpi_{ex}(x_{sk},t) = \frac{\phi(x_{sk},t)}{1+x_{sk}} + \int_0^{x_{sk}} dx \frac{\Pi(x,t)}{(1+x)^2 \left[1 - \nu(1+x)\mathcal{B}\right]}, \tag{10.97}$$

where the k branch (i.e., either the liquid or vapor branch) should be taken for $\mathcal{A}$ and $\mathcal{B}$ in the quantities involved. *The spinodal state is stable if the fifth density derivative of ϖ_{ex} is negative, unstable if positive, and neutral if equal to zero.* In fact, since the fluid is unstable at the spinodal state, the fifth density derivative of ϖ_{ex} should be positive and, consequently, parameters associated with the t-dependent part of the GvdW parameters should be such that the derivative in question is positive at $x = x_{sk}$ for $t < 0$.

It is now necessary to examine the behaviors of Eq. (10.96) around x_{sk} to deduce some of the thermodynamic properties in the neighborhood of x_{sk}. First, Eq. (10.96) can be written more explicitly:

$$0 = D_k(x_{sk},t)\,\Pi(x_{sk},t) + \int_0^{x_{sk}} dx \frac{D_k(x,t)}{(1+x)}\Pi(x,t), \tag{10.98}$$

$$0 = \left(D_k' + \frac{D_k}{1+x}\right)_{x=x_{sk}} \Pi^{(0)} + D_k(x_{sk})\,\Pi^{(1)}, \tag{10.99}$$

$$0 = \left(D_k' + \frac{D_k}{1+x}\right)'_{x=x_{sk}} \Pi^{(0)} + \left(2D_k' + \frac{D_k}{1+x}\right)_{x=x_{sk}} \Pi^{(1)}$$

$$+ D_k(x_{sk})\,\widehat{\Pi}^{(2)},\tag{10.100}$$

$$0 = \left(D_k' + \frac{D_k}{1+x}\right)''_{x=x_{sk}} \Pi^{(0)} + \left(3D_k' + \frac{2D_k}{1+x}\right)'_{x=x_{sk}} \Pi^{(1)}$$

$$+ \left(3D_k' + \frac{D_k}{1+x}\right)_{x=x_{sk}} \Pi^{(2)} + D_k(x_{sk})\Pi^{(3)},\tag{10.101}$$

$$0 = \left(D_k' + \frac{D_k}{1+x}\right)'''_{x=x_{sk}} \Pi^{(0)} + \left(4D_k' + \frac{3D_k}{1+x}\right)''_{x=x_{sk}} \Pi^{(1)}$$

$$+ \left(6D_k' + \frac{3D_k}{1+x}\right)'_{x=x_{sk}} \Pi^{(2)} + \left(4D_k' + \frac{D_k}{1+x}\right)_{x=x_{sk}} \Pi^{(3)}$$

$$+ D_k(x_{sk})\Pi^{(4)},\tag{10.102}$$

where the primes and multiple primes denote the density derivatives and

$$D_k(x,t) = \frac{1}{(1+x)\,[1 - \nu\,(1+x)\,\mathcal{B}_k(x,t)]} \quad (k = l, v).\tag{10.103}$$

Since the integral in Eq. (10.97) is simply $I_k(x_{sk}, t)$ (see Eq. (10.40)), we begin the analysis with it. We are interested in the asymptotic behavior of the integral in x_{sk} in the neighborhood of $t = 0$. For this purpose, $D_k(x,t)/(1+x)$ is expanded in series of $(x - x_{sk})$:

$$\frac{D_k(x,t)}{1+x} = d_0^{(k)}[1 + d_1^{(k)}(x - x_{sk}) + \cdots],\tag{10.104}$$

where the coefficients $d_0^{(k)}$, $d_1^{(k)}$, etc. are independent of x, e.g.,

$$d_0^{(k)} = \frac{1}{(1 + x_{sk})^2\,[1 - \nu\,(1 + x_{sk})\,\mathcal{B}_k^{(0)}(t)]}.\tag{10.105}$$

Note that in fact,

$$\mathcal{B}_k^{(0)}(t) = b_0^{(k)}(t)\tag{10.106}$$

in the model taken. Then, since Π is a series in $x - x_{sk}$, the integral $I_k(x_{sk}, t)$ can be evaluated. It is given by the form

$$I_k(x_{sk}, t) = d_0^{(k)}[-\Pi^{(0)}x_{sk} + \tfrac{1}{2}(\Pi^{(1)} + d_1^{(k)}\Pi^{(0)})x_{sk}^2 + \cdots]$$

$$+ I_n^{(k)}(x_{sk}, t),\tag{10.107}$$

where $I_n^{(k)}(x_{sk}, t)$ represents the contributions to $I_k(x_{sk}, t)$ that are led by terms of $O(x_{sk}^{5+\delta})$ or higher:

$$I_n^{(k)}(x_{sk}, t) = -d_0^{(k)} \frac{N^{(0)}(x_{sk}, t)}{5+\delta} x_{sk}^{5+\delta} + \cdots . \tag{10.108}$$

For this result, we have used the following: For $x_k > x > x_{sk}$, we obtain it in the form

$$I_n^{(k)}(x, t) = \sum_{i \geq 0}^{5} \frac{[Q_i + \widehat{N}^{(i)}(x_{sk}, t)]}{5+\delta} [(x - x_{sk})^{5+i+\delta} - x_{sk}^{5+\delta+i}]$$

$$+ d_0^{(k)} \sum_{i=5}^{7} \frac{\Pi^{(i)}(x_{sk}, t)}{1+i} [(x - x_{sk})^{i+1} - (-x_{sk})^{1+i}]$$

$$+ \sum_{i \geq 0}^{3} \frac{[Q_{i+6} + \widehat{M}^{(i)}(x_{sk}, t)]}{9+2\delta} [(x_{sk} - x)^{9+2\delta+i} - x_{sk}^{9+2\delta+i}]. \tag{10.109}$$

On the other hand, for $x_k < x < x_{sk}$ the irrational terms are different, and we obtain

$$I_n^{(k)}(x, t) = d_0^{(k)} \sum_{i \geq 0}^{5} \frac{[Q_i + \widehat{N}^{(i)}(x_{sk}, t)]}{5+\delta} [(x_{sk} - x)^{5+\delta+i} - (-x_{sk})^{5+\delta+i}]$$

$$+ d_0^{(k)} \sum_{i \geq 5} \frac{\Pi^{(i)}(x_{sk}, t)}{1+i} [(x - x_{sk})^{i+1} - (-x_{sk})^{1+i}]$$

$$+ d_0^{(k)} \sum_{i \geq 0} \frac{[Q_{i+6} + \widehat{M}^{(i)}(x_{sk}, t)]}{9+2\delta}$$

$$\times [(x_{sk} - x)^{9+2\delta+i} - (-x_{sk})^{9+2\delta+i}]. \tag{10.110}$$

The spinodal density x_{sk} is, in fact, proportional to $|t|^\varepsilon$ in the leading order (see Eq. (10.90)). For calculation of the isochoric specific heat, the integral $I_k(x, t)$ is needed as will be seen.

By using the expansion (10.104), we find the leading order terms for the integral. Use of Eq. (10.107) in Eq. (10.98) provides us with an inhomogeneous linear algebraic set for $\Pi^{(i)}$ ($i = 0, 1, \ldots, 4$) with the inhomogeneous term led by $x_{sk}^{5+\delta}$, which arises from the irrational term in $\mathcal{A}$ and $\mathcal{B}$. We emphasize that $\Pi^{(i)}$ in Eq. (10.107) depend on t. With this result for I_k and upon solving the set Eqs. (10.98)–(10.102) for $\Pi^{(i)}$, we find

$$\Pi^{(i)}\left(x_{sk}, t\right) + O\left(x_{sk}^{5+\delta}\right) = 0 \ (i = 0, 1, \ldots, 4).$$ (10.111)

That is, $\Pi^{(i)}(x_{sk}, t)$ are equal to zero to $O(x_{sk}^{5+\delta})$.

Since we are interested in the asymptotic behavior near the critical point, the leading terms of the polynomial equations obtained will be of main interest. Therefore, for our purpose in this work we have the following lemma.

Lemma 1. *The leading order contribution to the integral $I_k(x, t)$ necessary in the asymptotic analysis is given by the form*

$$I_k\left(x, t\right) = \frac{d_0^{(k)} Q_0}{5 + \delta}[(x - x_{sk})^{5+\delta} - x_{sk}^{5+\delta}] + \cdots \quad \text{for } x_k > x > x_{sk},$$ (10.112)

or

$$I_k\left(x, t\right) = \frac{d_0^{(k)} Q_0}{5 + \delta}[(x_{sk} - x)^{5+\delta} - (-x_{sk})^{5+\delta}] + \cdots \quad \text{for } x_k < x < x_{sk}.$$ (10.113)

Lemma 2. *The excess chemical potential $\varpi_{ex}(x, t)$ is then given by*

$$\varpi_{ex}(x, t) = \frac{D_k(x, t)}{1 + x}\Pi(x, t) + I_k(x, t),$$ (10.114)

where $\Pi(x, t)$ is given by Eq. (10.69) and $I_k(x, t)$ by Eq. (10.112) or Eq. (10.113).

All thermodynamic properties for the subcritical regime in the neighborhood of $t = 0$ can be deduced with this chemical potential and $\Pi(x, t)$, for which the parameters are determined subject to conditions deduced by applying Propositions 1 and 2. We make use of the results above in the analysis for the coexistence curve, excess heat capacity, and so on.

10.9.5.   *Liquid–Vapor Coexistence Curve*

10.9.5.1.   *Equilibrium Conditions*

The liquid–vapor coexistence curve is determined by the pressure and material equilibrium conditions. The pressure equilibrium condition in the present notation is given by

$$\frac{\Pi(x_l, t)}{1 - \nu \mathcal{B}_l (x_l + 1)} = \frac{\Pi(x_v, t)}{1 - \nu \mathcal{B}_v (x_v + 1)}, \tag{10.115}$$

whereas the material equilibrium condition is

$$\varpi_l (x_l, t) = \varpi_v (x_v, t), \tag{10.116}$$

where the reduced chemical potential $\varpi(x, t)$ is given by Eq. (10.33) with Eq. (10.34). The GvdW parameters

$$\mathcal{A}_k = \mathcal{A}(x_k, t), \quad \mathcal{B}_k = \mathcal{B}(x_k, t) \quad (k = l, v) \tag{10.117}$$

are for the liquid (l) and vapor (v) branch, respectively. Then the material equilibrium condition may be written as

$$\frac{\Pi(x_l, t)}{(1 + x_l)[1 - \nu(x_l + 1)\mathcal{B}_l]} - \frac{\Pi(x_v, t)}{(1 + x_v)[1 - \nu(x_v + 1)\mathcal{B}_v]} = I_v(x_v, t) - I_l(x_l, t). \tag{10.118}$$

Using Eq. (10.115) in Eq. (10.118), this equation may be recast into the form

$$(x_v - x_l) \Pi(x_l, t) - [I_v(x_v, t) - I_l(x_l, t)] (1 + x_l)(1 + x_v)$$
$$\times [1 - \nu(x_l + 1)\mathcal{B}_l] = 0. \tag{10.119}$$

Lemma 1 for the integral $I_k(x_k, t)$ asymptotically evaluated in the subcritical regime in the neighborhood of $t = 0$ will be used in this equation.

10.9.5.2. Coexistence Curve

The calculation of the coexistence curve requires solving the pair of equilibrium conditions (10.115) and (10.118) for x_l and x_v. For this purpose, transform the variables:

$$x_l = R + \tfrac{1}{2}u, \quad x_v = R - \tfrac{1}{2}u, \tag{10.120}$$

and expand first $\mathcal{A}_k$ and $\mathcal{B}_k$ in series of R and u to obtain

$$\mathcal{A}_k = \alpha_0^{(k)} + \alpha_R^{(k)} R + \alpha_u^{(k)} u + O(Ru),$$
$$\mathcal{B}_k = \beta_0^{(k)} + \beta_R^{(k)} R + \beta_u^{(k)} u + O(Ru), \tag{10.121}$$

where expansion coefficients are given by the formulas

$$\alpha_0^{(k)} = a_0^{(l)} - a_1^{(l)} x_{sl} + a_2^{(l)} x_{sl}^2 - a_3^{(l)} x_{sl}^3 + \cdots,$$
$$\alpha_R^{(k)} = a_1^{(l)} - 2a_2^{(l)} x_{sl} + 3a_3^{(l)} x_{sl}^2 + \cdots,$$
$$\alpha_u^{(k)} = \tfrac{1}{2}a_1 - a_2 x_{sl} + \tfrac{3}{2}a_3^{(l)} x_{sl}^2 + \cdots, \tag{10.122}$$

$$\beta_0^{(k)} = b_0^{(k)} - b_1^{(k)} x_{sk} + b_2^{(k)} x_{sk}^2 - b_3^{(k)} x_{sk}^3 + \cdots,$$

$$\beta_R^{(k)} = b_1^{(k)} - 2b_2^{(k)} x_{sk} + 3b_3^{(k)} R x_{sk}^2 + \cdots,$$

$$\beta_u^{(k)} = \tfrac{1}{2} b_1^{(l)} - b_2^{(l)} x_{sl} + \tfrac{3}{2} b_3^{(l)} x_{sl}^2 + \cdots \qquad (k = l, v). \qquad (10.123)$$

The ellipsis represents terms of parameters $a_i^{(k)}$ and $b_i^{(k)}$ for $i \geq 4$. These series stem from the expansions for $\mathcal{A}_k$ and $\mathcal{B}_k$ — the models (10.16) and (10.17). Therefore, we find

$$\Pi_k(x_k, t) = T_0^{(k)} + T_R^{(k)} R + T_u^{(k)} u + O(Ru) \qquad (10.124)$$

with the t-dependent coefficients defined by

$$T_R^{(k)} = \tau + \nu(\beta_0^{(k)} + \beta_R^{(k)})(\zeta \alpha_0^{(k)} + 1) \qquad (10.125)$$

$$+ \zeta(\nu \beta_0^{(k)} - 1)(2\alpha_0^{(k)} + \alpha_R^{(k)}) + \tau t,$$

$$T_u^{(k)} = \tfrac{1}{2} \tau + \zeta(\nu \beta_0^{(k)} - 1)(\alpha_0^{(k)} + \alpha_u^{(k)}) \qquad (10.126)$$

$$+ \nu(\tfrac{1}{2} \beta_0^{(k)} + \beta_u^{(k)})(\zeta \alpha_0^{(k)} + 1) + \tfrac{1}{2} \tau t,$$

$$T_0^{(k)} = \tau + (\zeta \alpha_0^{(k)} + 1)(\nu \beta_0^{(k)} - 1) + \tau t. \qquad (10.127)$$

The integral $I_k(x_k, t)$ is also expanded in series of R and u. Since

$$x_{sk}^{5+\delta} g(x_k) = -(5 + \delta) x_{sk}^{4+\delta} x_k + O(x_k^2), \qquad (10.128)$$

we find the integral along the isochore $x = x_k$ is given by[11]

$$I_k(x_k) = (\tau\nu - \zeta d_{na}^{(k)}) \beta_{na0}^{(k)}(t) x_{sk}^{4+\delta} x_k + O(x_k^2) \qquad (k = l, v), \qquad (10.129)$$

where

$$d_{na}^{(k)}(t) = \frac{\alpha_{na0}^{(k)}(t)}{\beta_{na0}^{(k)}(t)}. \qquad (10.130)$$

Therefore, we find

$$I_l(x_l) - I_v(x_v) = \mu_R R + \mu_u u + \cdots, \qquad (10.131)$$

---

[11] We remark that along the isochore $x = x_{sk}$, the result for the integral $I_k(x)$ would have been simpler than for the isochore $x = x_k$. Nevertheless, the isochore $x = x_k$ is chosen here because the value of the integral along $x = x_k$ is required to calculate the specific heat along the isochore mentioned, owing to the experimental data measured along the isochore.

where

$$\mu_R = [\tau\nu - \zeta d_{na}^{(l)}(t)]\beta_{na0}^{(l)}(t)\, x_{sl}^{4+\delta}\,(1 - \vartheta_{vl}),\qquad(10.132)$$

$$\mu_u = \tfrac{1}{2}[\tau\nu - \zeta d_{na}^{(l)}(t)]\beta_{na0}^{(l)}(t)\, x_{sl}^{4+\delta}\,(1 + \vartheta_{vl}),\qquad(10.133)$$

with ϑ_{vl} defined by the ratio

$$\vartheta_{vl} = \frac{[\tau\nu - \zeta d_{na}^{(v)}(t)]\beta_{na0}^{(v)}(t)\, x_{sv}^{4+\delta}}{[\tau\nu - \zeta d_{na}^{(l)}(t)]\beta_{na0}^{(l)}(t)\, x_{sl}^{4+\delta}} \approx \frac{[\tau\nu - \zeta d_{na}^{(v)}(0)]\beta_{na0}^{(v)}(0)\, x_{sv}^{4+\delta}}{[\tau\nu - \zeta d_{na}^{(l)}(0)]\beta_{na0}^{(l)}(0)\, x_{sl}^{4+\delta}}.$$
$$(10.134)$$

It should be noted that μ_R and μ_u are made up of quantities associated with the nonanalytic terms in $\mathcal{A}$ and $\mathcal{B}$ and $\beta_{na0}^{(k)}(t) = \beta_{na}^{(k)}(t) = \beta_{na} + \widehat{\beta}_{na}(t)$ with $\widehat{\beta}_{na}(0) = 0$. Thus, we see that Eqs. (10.115) and (10.119) give rise to the following pair of algebraic equations:

$$\Lambda_{1R}R + \Lambda_{1u}u = -T_0^{(v)}(\nu\beta_0^{(l)} - 1) + O\,(Ru),\qquad(10.135)$$

$$\Lambda_{2R}R + \Lambda_{2u}u = 0 + O\,(Ru),\qquad(10.136)$$

where the coefficients are defined by

$$\Lambda_{1R} = T_R^{(v)}(\nu\beta_0^{(l)} - 1) + \nu T_0^{(v)}(\beta_0^{(l)} + \beta_R^{(l)}),\qquad(10.137)$$

$$\Lambda_{1u} = T_u^{(v)}(\nu\beta_0^{(l)} - 1) + \nu T_0^{(v)}(\tfrac{1}{2}\beta_0^{(l)} + \beta_u^{(l)}),\qquad(10.138)$$

$$\Lambda_{2R} = (\nu\beta_0^{((l)} - 1)\mu_R,\qquad(10.139)$$

$$\Lambda_{2u} = T_0^{(l)} + (\nu\beta_0^{((l)} - 1)\mu_u.\qquad(10.140)$$

Thus, the leading order of the set is linear in R and u.

This set is solved to the linear order to obtain the coexistence curve and the rectilinear diameter:

$$R := \tfrac{1}{2}\,(x_l + x_v) = -\frac{T_0^{(v)}(\nu\beta_0^{(l)} - 1)}{\det|\Lambda|}\Lambda_{2u},\qquad(10.141)$$

$$u := x_l - x_v = \frac{T_0^{(v)}(\nu\beta_0^{(l)} - 1)}{\det|\Lambda|}\Lambda_{2R}.\qquad(10.142)$$

We now observe that these solutions are intimately related to $x_{sk}^{4+\delta}$ and also to the linear combinations of $a_i^{(k)}(t)$ and $b_i^{(k)}(t)$. We also emphasize that these results are rigorous asymptotic formulas holding in the neighborhood of the critical point and that, as will be shown, the t-dependence of the coexistence curve (i.e., u) is given by $|t|^{(7+\delta)\varepsilon}$ in the leading order, where ε is the smallest of the exponents of the t-dependence of the spinodal curve. It should be remembered that the exponent $(4 + \delta)$ is that of the critical isotherm and hence the exponent $(7 + \delta)\varepsilon$ is a composite of the exponents of the critical isotherm and spinodal curve, reflecting its physical origin.

To learn more details of these results, let us explore the coefficients. Since

$$\beta_0^{(k)} = 1 - b_1 x_{sk} + b_2 x_{sk}^2 + \widehat{b}_0^{(k)} - \widehat{b}_1^{(k)} x_{sk} + \widehat{b}_2^{(k)} x_{sk}^2 + \cdots ,$$

$$\alpha_0^{(k)} = 1 + \widehat{a}_0^{(k)} + b_2 x_{sk}^2 + \widehat{b}_0^{(k)} - \widehat{b}_1^{(k)} x_{sk} + \widehat{b}_2^{(k)} x_{sk}^2 + \cdots , \qquad (10.143)$$

and

$$T_0^{(v)} = \zeta (\nu - 1) \widehat{a}_0^{(v)} + \nu (1 + \zeta) \widehat{b}_0^{(v)} + \zeta \nu \widehat{a}_0^{(v)} \widehat{b}_0^{(v)} + \tau t + \cdots , \qquad (10.144)$$

for which we have made use of Eq. (10.64), the factors in the numerators of the solutions for R and u are found in the forms

$$T_0^{(v)} (\nu \beta_0^{(l)} - 1) = \zeta \nu (\nu - 1) \widehat{a}_0^{(v)} \widehat{b}_0^{(l)} + \nu^2 (1 + \zeta) \widehat{b}_0^{(l)} \widehat{b}_0^{(v)} + \cdots , \qquad (10.145)$$

$$\Lambda_{2R} = \nu \widehat{b}_0^{(l)} (\tau \nu - \zeta d_{na}^{(l)}) \beta_{na}^{(l)} (t) x_{sl}^{4+\delta} + \cdots , \qquad (10.146)$$

$$\Lambda_{2u} = \tfrac{1}{2} [\zeta (\nu - 1) \widehat{a}_0^{(l)} + \nu (1 + \zeta) \widehat{b}_0^{(l)}]$$

$$+ \nu (1 + \zeta) (\tau \nu - \zeta d_{na}) \beta_{na} \widehat{b}_0^{(v)} x_{sl}^{4+\delta} + \cdots , \qquad (10.147)$$

where the ellipsis stands for higher order terms. This means that R and u are asymptotically

$$R \sim |t|^{3\varepsilon} , \qquad (10.148)$$

$$u \sim |t|^{(7+\delta)\varepsilon} \qquad (10.149)$$

in the regime of $|t| \ll 1$. These are, respectively, the asymptotic behavior of the rectilinear diameter and the coexistence curve in the neighborhood of the critical point in the model employed. We see that the critical exponents are composites of the exponents for the spinodal curve and the critical isotherm. The value of ε may be chosen from the information on the excess specific heat near the critical point.

10.9.6. *Excess Heat Capacity*

To calculate the excess heat capacity anomaly, the formal result presented for the excess chemical potential ϖ_{ex} given in Eqs. (10.34) and (10.33) is preferable to use. In the following, we examine the leading order excess heat capacity which exhibits a singular behavior in the critical region.

To obtain the asymptotic behavior of the excess heat capacity in the neighborhood of the critical point in the subcritical regime, we need to consider only the integral $I_k(x,t)$, which, according to *Lemma 1* (Eq. (10.112) or Eq. (10.113)), can be shown expressible in the form

$$I_k(x,t) = \frac{1}{(5+\delta)}\left[\tau\nu - \zeta\frac{\alpha_{na0}^{(k)}(t)}{\beta_{na0}^{(k)}(t)}\right]g(x)\beta_{na0}^{(k)}(t)x_{sk}^{5+\delta} + \cdots, \quad (10.150)$$

where

$$g(x) = \left(\frac{x}{x_{sl}} - 1\right)^{5+\delta} - 1 \quad \text{for liquid} \quad (10.151)$$

$$= \left(1 - \frac{x}{x_{sv}}\right)^{5+\delta} - 1 \quad \text{for vapor.} \quad (10.152)$$

Since $\alpha_{na0}^{(k)}(t)$ and $\beta_{na0}^{(k)}(t)$ remain adjustable, we are free to further assume that $\alpha_{na0}^{(k)}(t)$ and $\beta_{na0}^{(k)}(t)$ are such that

$$\frac{\alpha_{na0}^{(k)}(t)}{\beta_{na0}^{(k)}(t)} = d_{na} \text{ independent of } t + O(|t|^{\varepsilon_n}). \quad (10.153)$$

Again, the argument for this can be made based on the statistical mechanics that the GvdW parameters $\mathcal{A}$ and $\mathcal{B}$ share the same pair distribution function and hence $\alpha_{na0}^{(k)}(t)$ and $\beta_{na0}^{(k)}(t)$ should have a common t-dependence, at least, in the neighborhood of $t = 0$ and hence scale to a constant. Since we are interested in x along the isochore $x = x_k$ ($k = l, v$) and since x_k/x_{sk} is a constant independent of t, the t-dependence of the excess chemical potential in the order considered is mainly vested in $\beta_{na0}^{(k)}(t)x_{sk}^{5+\delta}$. Since we may take

$$\beta_{na0}^{(k)}(t) = \beta_{na0} + b_{na0}^{(k)}|t|^{\varepsilon_n} \quad (10.154)$$

with $0 < \varepsilon_n < 1$ and $x_{sk} \sim t^{\varepsilon}$ ($t > 0$), we find to the leading order

$$\frac{\partial^2}{\partial t^2}x_{sk}^{5+\delta} \sim x_{sk}^{(7+\delta)\varepsilon-2}.$$

This shows that in the neighborhood of $t \lesssim 0$, we have the temperature dependence

$$-(1+t)(1+x_k)\frac{\partial^2}{\partial t^2}I_k(x,t) \sim t^{-\alpha'}, \quad (10.155)$$

where
$$\alpha' = 2 - (7 + \delta)\,\epsilon, \tag{10.156}$$
and the excess heat capacity in the neighborhood of the critical point is therefore given by
$$\widehat{c}_{vk}^{(ex)} \sim |t|^{-\alpha'} \quad (k = l, v). \tag{10.157}$$
The exponent α' is made up of the exponents characterizing the critical isotherm, coexistence curve, and spinodal curve. Therefore, their values may be chosen to optimize collectively the behaviors of the properties mentioned in addition to the excess specific heat and the isothermal compressibility discussed below.

Tentatively, with the values for ϵ and δ taken earlier, respectively, for the spinodal and coexistence curves and the critical isotherm in this work, it follows
$$\alpha' = 2 - (7 + 0.3) \times 0.25 = 0.18. \tag{10.158}$$
This value of α' for $\widehat{c}_{vk}^{(ex)}$ along the isochore $x = x_k$ is consistent with the experimental observation on the behavior of the specific heat along the isochore $x = x_k$ $(k = l, v)$. Thus, we have shown that the specific heat anomaly, from the standpoint of the canonical equation of state, arises from the combination of the exponents of nonanalytic irrational terms in $\mathcal{A}_k^{(i)}(t)$ and $\mathcal{B}_k^{(i)}(t)$ making up the GvdW parameters $\mathcal{A}$ and $\mathcal{B}$. Furthermore, it also provides information on the parameters $\mathcal{A}_k^{(i)}(t)$ and $\mathcal{B}_k^{(i)}(t)$; in fact, their mutual relations provided by Propositions 1 and 2, which are algebraic equations of parameters and their relations, drastically reduce the number of free parameters to just only a few.

10.9.7. *Isothermal Compressibility*

The isothermal compressibility κ_T along the critical isochore or along the isochores x_k $(k = l, v)$ is experimentally known to behave[12]
$$\frac{\kappa_T}{\kappa_I} = Ct^{-\gamma} \quad (\rho = \rho_c, \, t > 0) \tag{10.159}$$

$$= C\,(-t)^{-\gamma'} \quad (\rho = \rho_l(T) \text{ or } \rho = \rho_g(T), \, t < 0). \tag{10.160}$$

---

[12]P. Heller, *Rept. Prog. Phys.* **30**, 731 (1967); J. V. Sengers and J. M. H. Levelt Sengers, in *Progress in Liquid Physics*, C. A. Croxton, ed. (Wiley, New York, 1978), pp. 103–174; P. A. Egelstaff and J. W. Ring, in *Physics of Simple Liquids*, eds. H. N. V. Temperley, J. S. Rowlinson and G. S. Rushbrooke (North-Holland, Amsterdam, 1968), pp. 255–297.

By using the equation of state near $t = 0$, we find that along the isochore at $x = x_k$

$$\kappa_T \sim x_k^{-2} \left| x_k \right|^{-1-\delta} + \cdots . \tag{10.161}$$

To determine the t-dependence of κ_T, it is necessary to know the t-dependence of x_k, namely, the liquid–vapor coexistence curve, which has been already determined in the previous section. Using the result for x_k presented earlier, the t-dependence of the isothermal compressibility in the neighborhood of $t = 0$ is deduced to be

$$\kappa_T \sim \left| t \right|^{-\gamma'} , \tag{10.162}$$

where $\gamma' = 3(3 + \delta)\varepsilon$. Therefore, if we choose $\varepsilon = 0.25$ as we have for the coexistence curve, we find γ' within the range of experimental observation, namely, $\gamma' = 2.48$. This value is about two times too large, but it should be noted that the estimate of the exponent for the coexistence curve itself given here is too rough; a more accurate value, however, would require numerical evaluations of various quantities involved that we have avoided to see how the exponents arise within the range of experimental values for them without resorting to numerical methods. Moreover, as mentioned earlier, the values of the exponents should be optimized for the whole set of properties together. *The point we would like to emphasize here is that the exponents stem from the t-dependence of the GvdW parameters A and B.* Note that the exponent γ' is a composite number stemming from the critical isotherm, the spinodal curve, and the coexistence curve, which ultimately originates from the t-dependence of A and B away from the critical temperature. This is the principal point we would like to make with the model for the canonical equation of state studied here.

10.10. Quadratic Model

In this section, we implement the algorithm developed to calculate the critical behavior of fluids by using particular models for the GvdW parameters of the canonical equation of state. Specifically, we assume a quadratic model in which the number of terms in A and B is limited to $i \le 2$ for the terms regular in x and to $i = 0$ for the nonanalytic terms. This model, obviously renders various coefficients simpler. To be more precise, by the quadratic model, we mean the pair of the formulas for A and B in the following two sections:

10.10.1. *Subcritical Regime*

In the subcritical regime, the parameters $\mathcal{A}$ and $\mathcal{B}$ collectively denoted by $\mathcal{C} = (\mathcal{A}, \mathcal{B})$ may be expressed

$$\mathcal{C} = \sum_{i=0}^{2} \mathcal{C}_i^{(k)}(t) (x - x_{sk})^i + \mathcal{C}_{na}^{(k)}(t) (x - x_{sk})^3 |x - x_{sk}|^{1+\delta}, \quad t \leq 0,$$

(10.163)

where the coefficients are as follows:

$$\mathcal{C}_i^{(k)}(t) = [a_i^{(k)}(t), b_i^{(k)}(t); \ 0 \leq i \leq 2],$$

$$\mathcal{C}_{na}^{(k)}(t) = [\alpha_{na}^{(k)}(t), \beta_{na}^{(k)}(t)] \quad (k = l, v).$$

The exponent $\delta < 1$ (e.g., $\delta = 0.3 \sim 0.5$) is a parameter already familiar to us here.

These coefficients are temperature-dependent and nonanalytic (discontinuous) with respect to t. They are assumed to take the form

$$\mathcal{C}_i^{(k)}(t) = \mathcal{C}_i + \widehat{\mathcal{C}}_i^{(k)}(t) = \mathcal{C}_i + \mathfrak{c}_i^{(k)} |t|^{\varepsilon_i} \quad (i = 0, 1, 2),$$

(10.164)

where $\mathcal{C}_i = (a_i, b_i)$, $\widehat{\mathcal{C}}_i^{(k)} = (\widehat{a}_i^{(k)}, \widehat{b}_i^{(k)})$ and $\mathfrak{c}_i^{(k)} = (\mathfrak{a}_i^{(k)}, \mathfrak{b}_i^{(k)})$ $(k = l, v)$ are constants, and ε_i $(\varepsilon_0 \leq \varepsilon_1 \leq \varepsilon_2)$ are fractional numbers less than unity. Therefore, at the critical temperature (i.e., at $t = 0$),

$$\widehat{\mathcal{C}}_i^{(k)}(0) = 0 \quad (i = 0, 1, 2).$$

(10.165)

The liquid and vapor branches of the coefficients are sought to coincide with each other at the critical temperature, namely,

$$\mathcal{C}_i^{(l)}(0) = \mathcal{C}_i^{(v)}(0) := \mathcal{C}_i \neq 0 \ (i = 1, 2), \quad \lim_{t \to 0} \mathcal{C}_0^{(k)}(t) = 1.$$

(10.166)

Similarly, the coefficients of the irrational nonanalytic terms must also be in the forms

$$\mathcal{C}_{na}^{(k)}(t) = \mathcal{C}_{na} + \widehat{\mathcal{C}}_{na}^{(k)}(t) = \mathcal{C}_{na} + \mathfrak{c}_{na}^{(k)} |t|^{\varepsilon_n},$$

(10.167)

where $\mathcal{C}_{na} = (\alpha_{na}, \beta_{na})$ and $\mathfrak{c}_{na}^{(k)} = (\mathfrak{a}_{na}^{(k)}, \mathfrak{b}_{na}^{(k)})$, and it is assumed that $\varepsilon_n \leq \varepsilon_2$ and is also a fractional number less than unity. Therefore,

$$\widehat{\mathcal{C}}_{na}^{(k)}(t) \to 0 \quad \text{as } t \to 0.$$

(10.168)

The irrational terms multiplied by $\alpha_{na}^{(k)}(t)$ and $\beta_{na}^{(k)}(t)$ $(k = l, v)$ are the nonanalytic parts, which give rise to the nonanalytic (in fact, discontinuous) behavior of the equation of state with respect to density, and the nonclassical critical exponent for the pressure–density relation different from the mean field (van der Waals) theory values. This is the principal reason for assuming the nonanalytic term with a fractional exponent for the density dependence. The parameters $\alpha_{na}^{(k)}$ and $\beta_{na}^{(k)}$ may depend on temperature t only, but are constants on the critical isotherm in the same sense as for $a_i^{(k)}$ and $b_i^{(k)}$ in Eq. (10.166), namely,

$$\alpha_{na}^{(l)}(0) = \alpha_{na}^{(v)}(0) := \alpha_{na}, \quad \beta_{na}^{(l)}(0) = \beta_{na}^{(v)}(0) := \beta_{na}. \qquad (10.169)$$

With this model for the t-dependence of the parameters and with the exponents ε_i ($\epsilon := \varepsilon_0$) and ε_n appropriately chosen in comparison of the theory with experiment (principally on the critical exponents), the equation of state can be used to calculate the critical properties of fluids in the subcritical regime. We have taken the same exponents ε_i and ε_n for both vapor (v) and liquid (l) branches as an approximation near the critical point because the vapor and liquid branches of the experimental coexistence curve is practically symmetric in the close neighborhood of the critical point. As will be explicitly ascertained, the parameters $c_i^{(k)} = (a_i^{(k)}, b_i^{(k)} : i = 0, 1, 2; \ k = l, v)$ and $c_{na}^{(k)} = (a_{na}^{(k)}, b_{na}^{(k)} : k = l, v)$ are not arbitrary, but constrained by some conditions given by Propositions 1 and 2, which remove the degree of freedom for the parameters except for only a few. Therefore, there are not so many free parameters left, as one might get such an impression at first glance. Put differently, the quadratic model is an empirical model constructed to ensure to reproduce the critical characteristics of fluids with a least number of adjustable parameters by imposing Propositions 1 and 2.

10.10.2. *Supercritical Regime*

Owing to the aforementioned properties of $\mathcal{C}_i^{(k)}$ and $\mathcal{C}_{na}^{(k)}$ and the fact that the spinodal curve coincides with the coexistence curve at the critical point, i.e.,

$$x_{sl} = x_{sv} = 0,$$

the expansions for $\mathcal{C}$ in (10.167) reduce, on the critical isotherm or in the supercritical regime, to the form

$$\mathcal{C} = 1 + \sum_{i=1}^{2} \mathcal{C}_i x^i + \mathcal{C}_{na} x^3 |x|^{1+\delta} \quad (t = 0). \qquad (10.170)$$

The parameters $\mathcal{C}_c = (A_c^*, B_c^*)$ of the quadratic model are related to the van der Waals parameters a and b in the following sense: since $x \to -1$ as

$\eta \to 0$, it follows that, if T_c is such that $\exp(\varepsilon/k_B T_c) - 1 \ll 1$ — in other words, $T_c^* \gg 1$, then we obtain

$$a = A_c^* \left[1 + \sum_{i=1}^{2} (-1)^i a_i - \alpha_{na} \right], \quad b = B_c^* \left[1 + \sum_{i=1}^{2} (-1)^i b_i - \beta_{na} \right].$$

(10.171)

This means that A_c^* and B_c^* do not coincide with the reduced van der Waals parameters a and b, respectively, unless a_i, b_i, α_{na}, and β_{na} all vanish.

10.10.3. Critical Point

In the quadratic model, five of the seven parameters a_1, a_2, b_1, b_2, τ, ζ, and ν are determined in terms of two free parameters from the information on the critical isotherm ($t = 0$) with the help of Proposition 1. As a consequence, with an appropriate choice for the exponent δ so that the critical ϕ vs. x relation (reduced critical isotherm) agrees with experiment, the parameters a_1, b_1, τ, ζ, and ν are determined by suitably choosing a_2 and b_2 such that the critical parameters agree with experiment. The remaining two parameters are in fact constrained by the stability condition — namely, the fifth derivative — that should be either positive or negative or equal to zero, depending on the stability of the critical state. Therefore, they are not completely arbitrary. The so-determined parameters and the parameters necessary for the critical isotherm are summarized in Table 10.1. The critical isotherm calculated in the quadratic model is found in excellent agreement with the experiment as shown in Fig. 10.2.

It is interesting to note that the critical parameters are given by the parameters of the regular, rational power terms in the model whereas the critical isotherm is determined by the parameters making up the irrational part of the model. Thus, we see that in this model, the irrational part is essential to account for the characteristic behavior of the critical isotherm of the fluid. It has already been shown that the same irrational term, in coupling with the (irrational) t-dependence of $a_i^{(k)}(t)$ and $b_i^{(k)}(t)$ ($k = l, v$), is closely associated with the critical behaviors of other thermodynamic quantities, such as the isothermal compressibility, heat capacity, etc. This feature is seen unchanged in the case of the quadratic model. It is remarkable that the quadratic model works rather well for the equation of state for the fluid in the neighborhood of the critical point. The important point

Table 10.1. Various parameter values at $t = 0$ in the vdW theory and the GvdW theory in the quadratic model.

a_i	vdW	GvdW	Parameters	vdW	GvdW
a_0	1	1	ν	1/3	0.535
b_0	1	1	ζ	3	6.424
a_1	0	−0.336	τ	8/3	3.448
b_1	0	−0.618	α_{na}	0	0.01
a_2	0	−0.360	β_{na}	0	−0.015
b_2	0	0.0436			

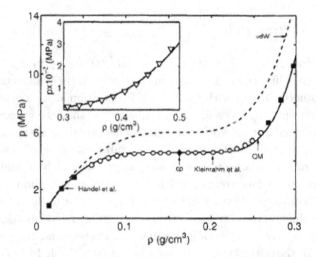

Fig. 10.2. Critical isotherms for methane predicted by the quadratic model and the van der Waals theory in comparison with experiment. The solid curve is for the quadratic model (QM) with the nonanalytic terms included, for which $\alpha_{na} = 0.01$ and $\beta_{na} = -0.015$ have been used, and the dashed curve (vdW) is for the van der Waals theory. The symbols are for the experimental data: ■ by Händel *et al.* and ○ by Kleinrahm *et al.* The critical point is indicated by the filled diamond(◆), and the triangles in the inset (▽) (for the high-density regime) have been calculated by employing the empirical equation of state for methane proposed by Setzmann and Wagner. Reproduced with permission from K. Rah and B. C. Eu, *J. Phys. Chem. B* **107**, 4388 (2003) © American Chemical Society.

we would like to make with the canonical equation of state postulated is that specific parts of the equation of state are associated with characteristic properties of thermodynamic properties, subcritical and supercritical, of the fluid of interest and we now have an equation of state in the neighborhood of critical point with a reliable accuracy.

10.10.4. *Critical Isotherms*

It is easy to show that the critical isotherm is in the same form as before:

$$\phi(x,0) = \left[\frac{\nu(1+\zeta)\beta_{na} - \zeta(1-\nu)\alpha_{na}}{1-\nu b_0}\right] x^{4+\delta}[1+O(x)]. \quad (10.172)$$

With the values of the parameters given in Table 10.1, the critical isotherm is in excellent agreement with the experiment.

10.10.5. *Spinodal Curve*

In the quadratic model, since to the leading order in $|t|^\varepsilon$ and so on

$$\varphi_{24}^{(k)}\varphi_{13}^{(k)} - \varphi_{14}^{(k)}\varphi_{23}^{(k)} = \Delta_1 + O\left(|t|^\varepsilon, \ldots\right),$$

$$\varphi_{44}^{(k)}\varphi_{33}^{(k)} - \varphi_{34}^{(k)}\varphi_{43}^{(k)} = \Delta_2 + O\left(|t|^\varepsilon, \ldots\right), \quad (10.173)$$

$$\Delta_1 = \omega_{24}\omega_{13} - \omega_{14}\omega_{23},$$

$$\Delta_2 = \omega_{44}\omega_{33} - \omega_{34}\omega_{43},$$

we find

$$D_0^{(k)} = \frac{\omega_{24}}{\Delta_1}\widehat{\varphi}_{10}^{(k)}(t) - \frac{\omega_{14}}{\Delta_1}\widehat{\varphi}_{20}^{(k)}(t) - \frac{\omega_{44}}{\Delta_2}\widehat{\varphi}_{30}^{(k)}(t) + \frac{\omega_{34}}{\Delta_2}\widehat{\varphi}_{40}^{(k)}(t)$$

$$= \sum_{i=0}^{2} s_{ai}\widehat{a}_i^{(k)}(t) + \sum_{i=0}^{2} s_{bi}\widehat{b}_i^{(k)}(t) + O(|t|^{2\varepsilon}, \ldots), \quad (10.174)$$

$$D_1^{(k)} = D_{10}^{(k)} + O\left(|t|^\varepsilon, \ldots\right), \quad (10.175)$$

$$D_2^{(k)} = D_{20}^{(k)} + O\left(|t|^\varepsilon, \ldots\right), \quad (10.176)$$

where s_{ai}, s_{bi}, $D_{10}^{(k)}$, and $D_{20}^{(k)}$ are constants consisting of ω_{ij}; they can be calculated explicitly in terms of the parameters determined and hence known from the critical point data by now. For $D_2^{(k)}$, $D_1^{(k)}$, and $D_0^{(k)}$, see Section 10.9.3 for the general discussion on the spinodal curve. Then to the leading order in t, the spinodal curve is given by

$$x_{sk} = -\sum_{i=0}^{2}\frac{2}{D_{10}^{(k)}}[s_{ai}\widehat{a}_i^{(k)}(t) + s_{bi}\widehat{b}_i^{(k)}(t)] + \text{higher order terms.} \quad (10.177)$$

Together with the information on the critical point data, the critical isotherm and coexistence curve can be calculated by using the procedure described earlier. Therefore, the information on the spinodal curve facilitates the determination of the quadratic model. We remark that if a more accurate result is desired of x_{sk}, one should simply calculate x_{sk} in the quadratic model without approximations. With the spinodal curve thus calculated and GvdW parameters determined, we can calculate the other thermodynamic quantities as already described in Section 10.9. The expressions for other critical properties do not get simpler than those already presented in the quadratic model taken here. For the reason of space, we do not discuss them here. We leave them to the readers to work out for exercise.

10.11. Concluding Remarks

The van der Waals theory provides a fairly simple form of equation of state which accounts for most of the fluid behaviors in a qualitatively correct manner. It, however, sometimes has glaring defects that should be removed so as to acquire an improvement if quantitatively correct results are desired, especially, with regard to the critical properties. Because of the attractive features of the van der Waals theory, we have taken an approach in which an improved theory is built around the van der Waals theory, incorporating the virtuous parts of the van der Waals theory into the improved theory. The canonical equation of state is designed to achieve this objective. In the canonical equation of state approach, we phenomenologically elucidate the manner in which the critical properties and other thermodynamic properties of fluids are connected to the GvdW model with regard to its density and temperature dependences. Thus, a set of thermodynamic properties of a fluid is specifically determined in association with the GvdW parameters. Once such a GvdW equation of state is constructed for a class of fluids, we can use it to study the thermodynamics of all fluids of the class and expect to correctly reproduce, at least, various aspects of critical phenomena phenomenologically. Because of the statistical mechanical representations of GvdW parameters $\mathcal{A}$ and $\mathcal{B}$, the statistical mechanical basis of the parameters so determined can be studied and elucidated systematically phenomenological behaviors of matter from the molecular theory viewpoint especially around the critical state. We believe that therein lies the promising potential for the canonical equation of state.

The model for $\mathcal{A}$ and $\mathcal{B}$ consists of analytical and nonanalytical (irrational) parts, the former being made up of a finite-order polynomial of density, and the latter factor proportional to density with fractional exponents — thus nonanalytic. The temperature-dependent coefficients are also nonanalytic with respect to t and closely related to various critical exponents for the coexistence and spinodal curves, isothermal compressibility, excess heat capacity, and critical isotherm. Since the quadratic model yields an excellent result for the critical isotherm for reduced pressure, an empirical model for $\mathcal{A}$ and $\mathcal{B}$ is constructed to account for the critical behavior of fluids in the subcritical regime near $t = 0$ on the basis of the model that correctly yields the critical isotherm. We have therewith shown that the generic van der Waals parameters $\mathcal{A}$ and $\mathcal{B}$ indeed can determine the temperature dependence of the spinodal and coexistence curves through coefficients $a_i^{(k)}(t)$ and $b_i^{(k)}(t)$ ($k = l, v$; $i = 0, 1, 2$) (see Eqs. (10.81), (10.85), (10.141), and (10.142)). These are the exact results in the model assumed. However, the determination of the liquid–vapor coexistence curve and other related thermodynamic properties cannot be obtained exactly in a simple form for the model. If their behaviors are desired in the neighborhood of the critical point, asymptotic formulas that exhibit a qualitatively correct critical behavior of the fluid near the critical point can be extracted only as approximations, provided that the exponents are suitably chosen. This way, we have also shown that the isothermal compressibility and the specific heat diverge at the critical point in a manner consistent with the experimental observation (see Eq. (10.162)).

The divergences of isothermal compressibility and isochoric excess heat capacity essentially owe their origin to the nonanalytic fractional power term of density as well as the irrational t-dependences of $a_i^{(k)}(t)$ and $b_i^{(k)}(t)$. Moreover, their divergence is also closely related to the behavior of the coexistence curve or the spinodal curve. Therefore, it, in fact, indicates the internal consistency of the model (10.16) and (10.17) as well as the quadratic model in accounting for the critical properties. An interesting conclusion of the present study of the canonical equation of state is that the critical properties of fluids provide not only a way to construct a model for a canonical form of equation of state but also a set of algebraic relations among the parameters of the model in a manner reminiscent of the scaling theory relations.

The molecular theoretic origin of the nonanalytic behaviors of the GvdW parameters must be sought through the statistical mechanics theory, but that is beyond the scope of a phenomenological model for the equation of state described in this chapter and in the book. The present phenomenological equation of state is, in fact, built upon the salient points of the statistical mechanical studies and empirical observations of the critical phenomena.

For a global thermodynamic description over a wider range of density and temperature of subcritical fluids, the present asymptotic analysis in the subcritical neighborhood of the critical point would not be sufficient, but a more detailed numerical analysis with the full models assumed for $a_i^{(k)}(t)$, $b_i^{(k)}(t)$, $\alpha_{na}^{(k)}(t)$, and $\beta_{na}^{(k)}(t)$ would be necessary. The algorithm given already presents itself for such a numerical analysis.

Chapter 11

Thermodynamics of Real Gas Mixtures

The foregoing formalisms for calculating thermodynamic functions hold for pure substances and, in particular, for pure real fluids. The most important result among the host of thermodynamic functions is the formula for chemical potentials, since we can calculate other thermodynamic functions from them by either taking suitable derivatives or combining the derivatives thereof. The formalisms for calculating thermodynamic functions for mixtures proceed similarly regardless of the states of aggregation of the substance involved. However, we shall specialize on the case of real gas mixtures in this chapter in order to equip us with a formalism, so that gas phase chemical equilibria can be treated in the subsequent chapter.

11.1. Chemical Potentials for Mixtures

We have seen that the chemical potential of a pure gas may be given in a form similar to that for the ideal gas, if the fugacity is introduced, namely, it is written in a form reminiscent of the ideal gas form

$$\mu(T,p) = \mu^*(T) + RT \ln p_f.$$

See Eq. (9.136). Therefore, in the case of a real gas mixture, it is convenient to look for the formula for chemical potentials in a mathematically similar form

$$\mu_i(T,p) = \mu_i^*(T) + RT \ln p_{fi}, \tag{11.1}$$

where p_{fi} is the fugacity for component i in the mixture and $\mu_i^*(T)$ is the chemical potential of i when $p_{fi} = 1$. Let us now find the precise meaning of the fugacity p_{fi} and its mathematical expression.

Since the partial molar volume may be written as

$$\bar{v}_i = \left(\frac{\partial \mu_i}{\partial p} \right)_{T,x_i} \tag{11.2}$$

(see Eq. (7.32)), if we look for μ_i in the form of Eq. (11.1), we must have the relation

$$\bar{v}_i = RT \left(\frac{\partial \ln p_{fi}}{\partial p} \right)_{T,x_i}, \tag{11.3}$$

where p is the total pressure. Since chemical potentials are logarithmically singular at $p = 0$ as we have found out in the case of a single-component system, it is convenient to subtract the singular part from $\bar{v}_i$. Therefore, we cast Eq. (11.3) in a slightly different, but more suitable, form as below:

$$RT \left[\frac{\partial}{\partial p} \ln \left(\frac{p_{fi}}{x_i p} \right) \right]_{T,x_i} = \bar{v}_i - \frac{RT}{p}. \tag{11.4}$$

Upon integrating it at constant T, we obtain

$$RT \ln p_{fi} = RT \left[\ln \left(\frac{p_{fi}}{x_i p} \right) \right]_{p=0} + RT \ln (x_i p) + \int_0^p dp \left(\bar{v}_i - \frac{RT}{p} \right). \tag{11.5}$$

It is now necessary to calculate the first term on the right-hand side of Eq. (11.5). It can be rewritten as

$$\left[\ln \left(\frac{p_{fi}}{x_i p} \right) \right]_{p=0} = \lim_{p \to 0} \left[\ln \left(\frac{p_{fi}}{p_i} \right) + \ln \left(\frac{p_i}{x_i p} \right) \right]. \tag{11.6}$$

Since the Gibbs–Dalton law holds for the ideal gas, we have the limit

$$\lim_{p \to 0} \left(\frac{p_i}{x_i p} \right) = 1. \tag{11.7}$$

The existence of the limit of (p_{fi}/p_i) can be established[1] if the mixture is put in equilibrium with component i across a semipermeable membrane, which is permeable to component i only. Let us consider the following situation. A mixture of real gases is in equilibrium with a pure gas of the ith

---

[1] See J. G. Kirkwood and I. Oppenheim, *Chemical Thermodynamics* (McGraw-Hill, New York, 1960).

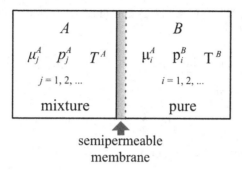

Fig. 11.1. Equilibrium between a mixture and pure species i across a semipermeable membrane.

species. The membrane is also assumed to be diathermal and deformable, so that the two systems are in thermal and mechanical equilibrium as well as in material equilibrium with respect to component i. We designate the mixture phase A, and the pure-component system phase B, and affix a superscript A or B to thermodynamic quantities of the two phases in order to distinguish them (Fig. 11.1). It then follows from the equilibrium conditions established for heterogeneous systems in Chapter 8 that

$$\mu_i^A = \mu_i^B := \mu_i,$$

$$T^A = T^B := T, \qquad (11.8)$$

$$p_i^A = p_i^B := p_i.$$

According to the analysis of pure (single-component) systems in Chapter 9, we have for the chemical potential in phase B the expression

$$\mu_i^B(T, p_i^B) = \mu_i^*(T) + RT \ln p_{fi}(T, p_i^B)$$

$$= \mu_i^*(T) + RT \ln \left[p_i^B \exp \int_0^{p_i^B} dp \left(\frac{v}{RT} - p^{-1} \right) \right], \quad (11.9)$$

which, in view of the equilibrium conditions listed in Eq. (11.8), may be written as

$$\mu_i(T, p_i) = \mu_i^*(T) + RT \ln \left[p_i \exp \int_0^{p_i} dp \left(\frac{v}{RT} - p^{-1} \right) \right]. \qquad (11.10)$$

We have made use of the fugacity expression for the pure component i. Note that $\mu_i^*(T)$ in Eq. (11.9) is the chemical potential of i when $p_{fi} = 1$ or the chemical potential of the hypothetical pure ideal gas of component i at 1 atm pressure. Therefore, if we determine the fugacity p_{fi} such that the values for $\mu_i^*(T)$ in Eqs. (11.1) and (11.9) coincide, then comparison of Eqs. (11.10) and (11.11) yields

$$\lim_{p \to 0} \left(\frac{p_{fi}}{p_i} \right) = \lim_{p \to 0} \exp \left[\int_0^{p_i} dp \left(\frac{v}{RT} - p^{-1} \right) \right] = 1. \qquad (11.11)$$

This establishes that

$$\left[\ln \left(\frac{p_{fi}}{x_i p} \right) \right]_{p=0} = 0,$$

and we finally find from Eq. (11.5)

$$p_{fi}(T, p) = x_i p \exp \left[\int_0^p dp' \left(\frac{\bar{v}_i}{RT} - \frac{1}{p'} \right) \right] \qquad (11.12)$$

for fugacities of the components of the mixture. We clearly see that it is a generalization of the fugacity formula (9.128) for a single component to which it reduces as x_i tends to unity.

Chemical potentials are sometimes given in terms of the ideal gas and excess chemical potentials:

$$\mu_i(T, p) = \mu_i^*(T) + RT \ln(x_i p) + \mu_i^{(ex)}, \qquad (11.13)$$

where

$$\mu_i^{(ex)} = \int_0^p dp' \left(\bar{v}_i - \frac{RT}{p'} \right). \qquad (11.14)$$

Since $p\bar{v}_i = RT$ for the ideal gas, for which the excess chemical potential obviously vanishes, we recover from Eq. (11.14) the chemical potential of the ideal gas i from Eq. (11.14):

$$\mu_i(T, p) = \mu_i^*(T) + RT \ln(x_i p).$$

For many common gases, the volume change on mixing is negligible. In this case, the partial molar volume $\bar{v}_i$ of gas i in the mixture is almost the same as the molar volume of the pure gas, i.e.,

$$\Delta v_i = \bar{v}_i - v_i \simeq 0. \qquad (11.15)$$

When this approximation is applicable, the partial molar volume in Eq. (11.12) may be replaced with the molar volume and we obtain

$$p_{fi} \simeq x_i p_{fi}^0, \qquad (11.16)$$

where

$$p_{fi}^0 = p \exp \left[\int_0^p dp' \left(\frac{v_i}{RT} - p'^{-1} \right) \right], \qquad (11.17)$$

which is the fugacity of pure gas i. In this approximation, the fugacity of component i in a mixture is the product of the mole fraction and the fugacity of pure gas i. This approximate relation holds with a fair accuracy up to about 100 atm pressure for common gases. Equation (11.16) represents the Lewis–Randall rule.[2]

According to statistical mechanics, the virial equation of state for a mixture may be written in the form

$$pv = RT + B_2(T)p + B_3(T)p^2 + \cdots , \qquad (11.18)$$

where

$$v = \frac{V}{n} = \frac{V}{n_1 + n_2 + \cdots + n_c},$$

$$B_2 = \sum_{i=1}^{c} \sum_{j=1}^{c} B_{ij}(T)x_i x_j, \qquad (11.19)$$

$$B_3 = \sum_{i=1}^{c} \sum_{j=1}^{c} \sum_{k=1}^{c} B_{ijk}(T)x_i x_j x_k,$$

and so on, with x_i, x, and x_k denoting mole fractions. The second virial coefficients[3] are generally symmetric with respect to the indices:

$$B_{ij} = B_{ji}. \qquad (11.20)$$

Using the equation of state (11.18), we may calculate the fugacity of a mixture. For the purpose, it is first necessary to calculate the partial molar

---

[2] G. N. Lewis and Randall, *Thermodynamics*, pp. 225–227; *J. Am. Chem. Soc.* **50**, 1522 (1928); *ibid.* **59**, 2733 (1937).

[3] According to the statistical mechanics of simple fluids, the second virial coefficient B_{ij} is given in terms of the intermolecular potential energy $u_{ij}(r)$ between two molecules of species i and j:

$$B_{ij}(T) = -2\pi \int_0^\infty dr\, r^2 \left\{ \exp\left[-\frac{u_{ij}(r)}{k_B T} \right] - 1 \right\},$$

where k_B is the Boltzmann constant. Since $u_{ij} = u_{ji}$, the symmetry property of B_{ij} in Eq. (11.20) is a consequence of the symmetry of the potential energies with respect to interchange of particle indices. For the analytic formula for the $B_{ij}(T)$ in the case of the Lennard-Jones (12-6) potential, see Appendix A of this book.

volume $\bar{v}_i$. Since from the virial equation of state (11.18),

$$V = \sum_{i=1}^{c} n_i \frac{RT}{p} + \sum_{i=1}^{c}\sum_{j=1}^{c} n_i n_j B_{ij}(T) n^{-1}$$

$$+ p \sum_{i=1}^{c}\sum_{j=1}^{c}\sum_{k=1}^{c} n_i n_j n_k B_{ijk}(T) n^{-2} + \cdots, \qquad (11.21)$$

we obtain on differentiation with respect to n_i

$$\bar{v}_i = \frac{RT}{p} + \sum_{j=1}^{c}\sum_{k=1}^{c} x_j x_k (2B_{ij} - B_{jk})$$

$$+ p \sum_{j=1}^{c}\sum_{k=1}^{c}\sum_{l=1}^{c} x_j x_k x_l (B_{ijk} + B_{jil} + B_{jki} - 2B_{jkl}) + \cdots. \qquad (11.22)$$

When this is substituted into Eq. (11.12), there follows an approximate fugacity

$$p_{fi} = x_i p \exp\left[\frac{p}{RT} \sum_{j=1}^{c}\sum_{k=1}^{c} x_j x_k (2B_{ij} - B_{jk}) \right], \qquad (11.23)$$

for which we have ignored the term linear in p in Eq. (11.22). If the van der Waals theory is applied to the mixture, the second virial coefficients are found to be in the form

$$B_{ij} = b_{ij} - \frac{a_{ij}}{RT}, \qquad (11.24)$$

where b_{ij} are known to be related to the sum of molecular radii of components i and j, and a_{ij} to the strength of the van der Waals attraction between the molecules of components i and j.

It is sometimes a fair approximation to calculate B_{ij} as the arithmetic mean[4] of the second virial coefficients of pure gases i and j, namely,

$$B_{ij} = \frac{1}{2}(B_{ii} + B_{jj}), \qquad (11.25)$$

where B_{ii} and B_{jj} are, respectively, the second virial coefficient of pure gases i and j. In this approximation, the fugacity takes a rather simple

---

[4]This approximation is called the combination rule.

form:

$$p_{fi} = x_i p \exp\left(\frac{B_{ii}p}{RT}\right). \tag{11.26}$$

Since we may write the virial equation of state for pure gas i in the form

$$v_i = \frac{RT}{p} + B_{ii}(T) + O(p),$$

where v_i is the molar volume of the gas, by combining it with Eq. (11.22), we obtain the volume change on mixing

$$\bar{v}_i - v_i = \sum_{j=1}^{c}\sum_{k=1}^{c} x_j x_k (2B_{ij} - B_{jk} - B_{ii}) + O(p). \tag{11.27}$$

When the approximation (11.25) is used for the second virial coefficients in this equation, there follows

$$\bar{v}_i - v_i = \sum_{j=1}^{c}\sum_{k=1}^{c} \tfrac{1}{2}x_j x_k (B_{jj} - B_{kk})$$
$$= 0, \tag{11.28}$$

and therefore, to the same order of approximation, the volume is unchanged on mixing

$$\Delta v = \sum_{i=1}^{c} n_i(\bar{v}_i - v_i) = 0. \tag{11.29}$$

It is useful to recall the Lewis–Randall rule which assumes the validity of a negligible change in volume on mixing. The rule therefore is seen to be related to the combination rule (11.25) and the virial equation of state truncated at the first order in p.

11.2. Entropy of Mixing

Once the chemical potentials are calculated for a mixture, other thermodynamic functions may be easily calculated from the chemical potentials. We describe the procedure here in the case of a gas. Since the partial molar entropy of component i may be written as

$$\bar{s}_i = -\left(\frac{\partial \mu_i}{\partial T}\right)_{p,x_i}, \tag{11.30}$$

differentiation of Eq. (11.13) yields the partial molar entropy

$$\bar{s}_i = \bar{s}_i^*(T) - R\ln(x_i p) - \int_0^p dp' \left[\left(\frac{\partial \bar{v}_i}{\partial T}\right)_{p',x_i} - \frac{R}{p'}\right], \tag{11.31}$$

where

$$\bar{s}_i^* = -\left(\frac{\partial \mu_i^*}{\partial T}\right)_{p,x_i}. \tag{11.32}$$

Its meaning is similar to the entropy $s^*(T)$ of a pure gas; it is the partial molar entropy of a hypothetical pure ideal gas at 1 atm pressure or alternatively

$$\bar{s}_i(T) = \lim_{p\to 0}[\bar{s}_i(T,p) + R\ln(x_i p)]$$

$$= s_i^0 + \int_{T_1}^T dT\, c_{pi}^*(T)$$

$$= s_i^*(T). \tag{11.33}$$

If the system is composed of component i alone, the partial molar entropy of component i is equal to its molar entropy which may be written as

$$s_i(T,p) = s_i^*(T) - R\ln p - \int_0^p dp' \left[\left(\frac{\partial v}{\partial T}\right)_p - \frac{R}{p'}\right]. \tag{11.34}$$

This is the molar entropy of pure real gas i.

Let us imagine the following process: $n_1, n_2, \ldots, n_c$ moles of real gases are reversibly[5] combined at constant pressure and temperature. The partial molar entropies are then expected to be different from the molar entropies and therefore there will be a change in the overall entropy of the system on mixing. This change in entropy is called the entropy of mixing. It is calculated as follows:

$$\Delta s = \sum_{i=1}^c x_i(\bar{s}_i - s_i)$$

$$= -R\sum_{i=1}^c x_i \ln x_i + \sum_{i=1}^c x_i \int_0^p dp' \left[\left(\frac{\partial \bar{v}_i}{\partial T}\right)_{p'} - \left(\frac{\partial \bar{v}_i}{\partial T}\right)_{p',x_i}\right](p'). \tag{11.35}$$

---

[5]Strictly speaking, mixing various components cannot be regarded as a reversible process because it cannot be reversed in practice. Here, it means equilibrium is maintained over the course of the process.

Equation (11.33) is made use of in this calculation. Since for ideal gases,

$$\bar{v}_i = v_i,$$

it follows that

$$\left(\frac{\partial \bar{v}_i}{\partial T}\right)_{p,x_i} = \left(\frac{\partial v_i}{\partial T}\right)_p,$$

and therefore, the entropy of mixing takes the form

$$\Delta s = -R \sum_{i=1}^{c} x_i \ln x_i. \tag{11.36}$$

This is the entropy of mixing for ideal gases. As is obvious from the formula, the entropy of mixing is always positive for ideal gases.

We see that the second term on the right-hand side of Eq. (11.35) is the nonideality correction to the ideal gas entropy of mixing. The nonideality correction to the entropy of mixing can be estimated by making use of Eq. (11.27). Differentiating Eq. (11.27) with respect to T and substituting the result into the integral in Eq. (11.35), we obtain the nonideality correction

$$\Delta s_{\text{real}} = p \sum_{i=1}^{c} \sum_{j=1}^{c} x_i x_j \left(\frac{dB_{ii}}{dT} - \frac{dB_{ij}}{dT}\right). \tag{11.37}$$

Note that Δs_{real} vanishes when the combination rule (11.25) is used for evaluating the derivatives of the second virial coefficients in Eq. (11.37). The Δs_{real} generally vanishes if the Lewis–Randall rule holds. If the van der Waals model is used for B_{ii} and B_{ij}, then

$$\Delta s_{\text{real}} = \frac{p}{RT^2} \sum_{i=1}^{c} \sum_{j=1}^{c} x_i x_j \left(a_{ii} - a_{ij}\right). \tag{11.38}$$

If the combination rule is used for a_{ij}, then still

$$\Delta s_{\text{real}} = 0,$$

and the entropy of mixing again has the same form as an ideal mixture. The entropy of mixing obtained in this section, for example, Eqs. (11.36)–(11.38), permits the following remark. In physical chemistry textbooks, it is often mentioned that the entropy of mixing is consistent with the requirement of the second law of thermodynamics since $\Delta s \geq 0$. This is obviously true in the case of an ideal gas mixture, but not assured if the gas mixture is real since, for example, if $a_{ij} \geq a_{ii}$ and if p and temperature are such that Δs_{real} is larger than Δs_{ideal} defined by Eqs. (11.36), then it is possible that

$$\Delta s < 0.$$

The ideal gas formula Δs_{ideal} for the entropy of mixing is isomorphic to the statistical mechanical entropy formula if x_i is taken to mean the distribution function of finding species i, but such an interpretation is overextending the notion of probability. Recall also the remarks made in Chapter 4 with regard to the second law of thermodynamics and $\Delta s \geq 0$ for an isolated system.

11.3.　Heat of Mixing

The partial molar enthalpy may be calculated in terms of μ_i and $\bar{s}_i$ just calculated if we make use of the formula

$$\bar{h}_i = \mu_i + T\bar{s}_i.$$

Upon substitution of the expressions for the chemical potential and the partial molar entropy, we easily find the partial molar enthalpies of a mixture

$$\bar{h}_i(T, p) = h_i^*(T) + \int_0^p dp' \left[\bar{v}_i - T \left(\frac{\partial \bar{v}_i}{\partial T} \right) \right]_{p', x_i}, \qquad (11.39)$$

where

$$h_i^*(T) = \mu_i^*(T) + Ts_i^*(T). \qquad (11.40)$$

Therefore, we again see that

$$h_i^*(T) = \lim_{p \to 0} \bar{h}_i(T, p) \qquad (11.41)$$

and hence the molar enthalpy of pure ideal gas i. As a consequence, we see that the second term on the right-hand side of Eq. (11.39) is the nonideality correction for the partial molar enthalpies. It is the generalization of the nonideality correction for the molar enthalpy of a pure gas.

The molar enthalpy was calculated for pure gases, and for pure gas i it may be written in the form

$$h_i = h_i^*(T) + \int_0^p dp' \left[v_i - T \left(\frac{\partial v_i}{\partial T} \right)_{p'} \right], \qquad (11.42)$$

where v_i is the molar volume of gas i.

The heat of mixing is defined as the sum of all enthalpy changes for all substances upon formation of a mixture of $x_1, x_2, \ldots, x_c$ mole fractions from the c pure constituents of the mixture:

$$\Delta h = \sum_{i=1}^{c} x_i (\bar{h}_i - h_i). \tag{11.43}$$

It is then easily calculated with Eqs. (11.39) and (11.42). We obtain

$$\Delta h = \sum_{i=1}^{c} x_i \int_0^p dp' \left[\Delta \bar{v}_i - T \left(\frac{\partial \Delta \bar{v}_i}{\partial T} \right)_{T,x_i} \right], \tag{11.44}$$

where $\Delta v_i = \bar{v}_i - v_i$. Since the molar volumes do not change on mixing if the gases are ideal, $\Delta v_i = 0$ as was already discussed previously, and we find that there is no heat of mixing for ideal gases:

$$\Delta h = 0.$$

For real gases, if we make use of Eq. (11.27), there follows to the first order in p the expression

$$\Delta h = p \sum_{i=1}^{c} \sum_{j=1}^{c} x_i x_j \left[B_{ij} - B_{ii} - T \left(\frac{dB_{ij}}{dT} - \frac{dB_{ii}}{dT} \right) \right]. \tag{11.45}$$

This equation indicates that the heat of mixing will be equal to zero if the virial equation of state is truncated at the first order in p and if the combination rule (11.25) holds for the second virial coefficients. There will be generally no heat of mixing for gases for which the Lewis–Randall rule holds, but it will be present for real gases not obeying the Lewis–Randall rule.

11.4. Activity and Activity Coefficient

Let us assume that we have determined the change in the chemical potential of a gaseous component by some method, when the concentrations are varied at constant temperature and pressure. This determination amounts to measuring the ratio of the fugacities since

$$\Delta \mu_i = \mu_i - \mu_i^0 = RT \ln \left(\frac{p_{fi}}{p_{fi}^{(0)}} \right), \tag{11.46}$$

where $p_{fi}^{(0)}$ is the fugacity of i at the state corresponding to μ_i^0. It is common practice in thermodynamics to determine and use such a ratio of fugacities when it is difficult to obtain the numerical value of either one of the fugacities. It is proven to be convenient to work with the ratio of the fugacity p_{fi} of a substance in a given state to its fugacity $p_{fi}^{(0)}$ in some other state chosen as a standard state. We call this relative fugacity the activity and will denote it by a_i:

$$a_i = \frac{p_{fi}}{p_{fi}^{(0)}}. \tag{11.47}$$

Therefore, the activity is equal to unity at the standard state. The chemical potential is then written

$$\mu_i = \mu_i^0 + RT \ln a_i, \tag{11.48}$$

where μ_i^0 is the chemical potential of i at the standard state chosen.

There are a number of conventions adopted for the standard state. For gases, the standard state is chosen with the one in which the fugacity at a given temperature is unity. It is also the state at which the heat capacity, enthalpy, and entropy are those of the gas at infinitely low pressure:

$$a_i = p_{fi}, \tag{11.49}$$

i.e., in such a limit

$$\lim_{p \to 0} \frac{a_i}{x_i p} = 1. \tag{11.50}$$

This choice of standard state for gases makes the standard chemical potential μ_i^0 coincide with $\mu_i^*(T)$:

$$\mu_i^0 = \mu_i^*(T),$$

where $\mu_i^*(T)$ is given in Eq. (11.1) and is the chemical potential of the hypothetical ideal gas i at 1 atm pressure.

If the activity coefficient f_i for gas i is introduced by the relation

$$a_i = f_i p_i = x_i f_i p, \tag{11.51}$$

and if we define

$$\mu_i^0(T, p) = \mu_i^*(T) + RT \ln p, \tag{11.52}$$

where it must be noted that $\mu_i^0(T, p)$ is not the same as μ_i^0 in Eq. (11.48), then the chemical potential is given in the form

$$\mu_i = \mu_i^0(T, p) + RT \ln(x_i f_i). \tag{11.53}$$

For liquids and solids, the standard chemical potentials are chosen such that

$$\mu_i^0 = \mu_i^l(T, p) \quad \text{for liquids } (i = 1, 2, \ldots, c)$$

$$= \mu_i^s(T, p) \quad \text{for solids } (i = 1, 2, \ldots, c), \tag{11.54}$$

where $\mu_i^l(T, p)$ and $\mu_i^s(T, p)$ are the chemical potential of pure liquid i and pure solid i, respectively. This will be referred to as Convention I on activity.

Sometimes, the following convention is adopted by distinguishing the solvent designated as component 1 and the solutes designated as components $2, 3, \ldots, c$:

$$\mu_1^0 = \mu_1^l(T, p) \quad \text{for the solvent,}$$
$$\mu_i^0 = \lim_{x_1 \to 1} (\mu_i - RT \ln x_i) \quad \text{for the solutes.} \tag{11.55}$$

This will be referred to as Convention II. Note that μ_i^0 for solutes $i = 2, 3, \ldots, c$ are defined as the chemical potentials in the limit of infinite dilution in this convention. One can choose one of the aforementioned conventions, depending on the situation and convenience. In any case, by introducing the activity coefficient f_i by the relation

$$a_i = x_i f_i, \tag{11.56}$$

we can express the chemical potentials of components in a solution in the form

$$\mu_i = \mu_i^0(T, p) + RT \ln(x_i f_i). \tag{11.57}$$

In this manner, we can put the formalisms for gaseous mixtures and liquid or solid solutions on the same footing. The conventions adopted for the standard state give rise to the following conventions for the activity coefficients. For gases,

$$\lim_{p \to 0} f_i = 1. \tag{11.58}$$

For liquids and solids,

Convention I

$$\lim_{x_i \to 1} f_i = 1 \ (i = 1, 2, \ldots, c); \tag{11.59}$$

Convention II

$$\lim_{x_1 \to 1} f_i = 1 \ (i = 1, 2, \ldots, c). \tag{11.60}$$

Activities and activity coefficients introduced here will be used for discussions of thermodynamics of real substances in the chapters for real gas mixtures and real solutions in this work where various theories of determining activity and activity coefficients are developed.

11.5.　Canonical Equation of State for a Mixture

Instead of the virial equation of state, the theory can be developed by means of the canonical equation of state suitably generalized for mixtures. It is just necessary to use the canonical equation of state for the equation of state in the formalism presented earlier in this chapter. The canonical equation of state for a c-component mixture may be written as

$$[p + A(n,T)n^2][1 - nB(n,T)] = nk_BT, \tag{11.61}$$

where n is the number density of molecules and

$$A(n,T) = \sum_{i,j=1}^{c} A_{ij}(n,T) x_i x_j, \tag{11.62}$$

$$B(n,T) = \sum_{i,j=1}^{c} B_{ij}(n,T) x_i x_j, \tag{11.63}$$

with x_i denoting the mole fraction of species i

$$x_i = \frac{n_i}{n} = \frac{n_i}{\sum_{i=1}^{c} n_i}.$$

The coefficients $A_{ij}(n,T)$ and $B_{ij}(n,T)$ are phenomenological functions of n_i, n_j, and T. The statistical mechanical representations of the GvdW parameters A_{ij} and B_{ij} are

$$A_{ij}(n,T) = \frac{2\pi}{3} \int_{r_{ij}^{\ddagger}}^{\infty} dr r^3 u'_{ij}(r) \exp\left[-\beta u_{ij}(r)\right] g_{ij}(r), \tag{11.64}$$

$$B_{ij}(n,T) = -\frac{2\pi}{3}\beta \int_{r_{ij}^0}^{r_{ij}^{\ddagger}} dr r^3 u'_{ij}(r) \exp\left[-\beta u_{ij}(r)\right] g_{ij}(r)$$

$$\times \left[1 - \frac{2\pi}{3}\beta n \sum_{i,j=1}^{c} x_i x_j \int_{r_{ij}^0}^{r_{ij}^{\ddagger}} dr r^3 u'_{ij}(r) \exp[-\beta u_{ij}(r)] g_{ij}(r)\right]^{-1}, \tag{11.65}$$

where $\beta = 1/k_BT$, $u_{ij}(r)$ is the potential energy of the pair (i,j), $u'_{ij}(r)$ is the derivative of $u_{ij}(r)$, $g_{ij}(r)$ is the radial distribution function of the pair (i,j), $r_{ij}^{\ddagger}$ is the position at which $-r^3 u'_{ij}(r) \exp[-\beta u_{ij}(r)] g_{ij}(r)$ is a maximum, and r_{ij}^0 is the hard core or the point below which the radial distribution function vanishes practically to zero; generally, $r_{ij}^0 = 0$ for $u_{ij} \to \infty$ as $r \to 0$. It should be noted that $B_{ij}(n,T)$ in the canonical equation of state are, of course, different from the second virial coefficients appearing in the virial expansion used earlier in this chapter. In phenomenological thermodynamics, these statistical mechanical representations, of course, have no place in the theory, but they are presented to help us cast the thermodynamic formulas in terms of the GvdW parameters introduced earlier, since the thermodynamic formulas so obtained can guide the statistical thermodynamic investigation of real fluid mixtures.

We calculate partial molar volumes which we have found to be a building block of the thermodynamic solution theory. For this purpose and also for insight and completeness, we make use of the virial equation of state expressed, for notational simplicity, in the form

$$Z = \beta p - n = \sum_{i,j=1}^{c} n_i n_j \Im_{ij}, \tag{11.66}$$

where the statistical mechanical form for $\Im_{ij}$ is given by the formula

$$\Im_{ij} = -\frac{2\pi}{3}\beta \int_0^{\infty} dr r^3 u'_{ij}(r) g_{ij}(r; \{n_k\}). \tag{11.67}$$

Here, $u'_{ij}(r) = \partial u_{ij}/\partial r$. Since the intermolecular potential energy is symmetric with respect to the interchange of indices i and i, it follows that $\Im_{ij} = \Im_{ji}$. The partial molar volume is then calculated from this form of compressibility factor Z. Differentiating with the specific volume $v = 1/n$ and using

$$\frac{\partial}{\partial n_k} = \frac{\partial x_k}{\partial n_k}\frac{\partial}{\partial x_k} = \frac{1}{n}(1 - x_k)\frac{\partial}{\partial x_k},$$

we obtain

$$\frac{\bar{v}_k}{v} = \sum_{i,j=1}^{c}\left[2\delta_{ik}x_j\Im_{ij} + x_i x_j (1 - x_k)\left(\frac{\partial\Im_{ij}}{\partial x_k}\right)_{T,p,n'}\right]. \tag{11.68}$$

We now observe that if we split $\Im_{ij}$ into two parts, attractive and repulsive, as in

$$\Im_{ij} = \Im_{ij}^{(A)} + \Im_{ij}^{(B)},$$

where

$$\Im_{ij}^{(A)} = -\frac{2\pi}{3}\beta \int_{r_{ij}^{\ddagger}}^{\infty} dr r^3 u'_{ij}(r)\exp\left[-\beta u_{ij}(r)\right]g_{ij}(r), \tag{11.69}$$

$$\Im_{ij}^{(B)} = -\frac{2\pi}{3}\beta \int_{r_{ij}^0}^{r_{ij}^{\ddagger}} dr r^3 u'_{ij}(r)\exp\left[-\beta u_{ij}(r)\right]g_{ij}(r). \tag{11.70}$$

The partial molar volume in Eq. (11.68) may then be expressible in terms of the GvdW parameters for the mixture defined by

$$A_{ij} = -\beta^{-1} \mathfrak{Z}_{ij}^{(A)}, \tag{11.71}$$

$$B_{ij} = \frac{\mathfrak{Z}_{ij}^{(B)}}{1 + n \sum_{k,l=1}^{c} x_i x_j \mathfrak{Z}_{kl}^{(B)}}. \tag{11.72}$$

These definitions put the equation of state in the canonical form given in Eq. (11.61). Inverting the relation B_{ij} and $\mathfrak{Z}_{ij}^{(B)}$ in Eq. (11.72), we find

$$\mathfrak{Z}_{kl}^{(B)} = \frac{B_{kl}}{1 - n \sum_{ij}^{c} x_i x_j B_{ij}}. \tag{11.73}$$

With these identifications of $\mathfrak{Z}_{ij}^{(A)}$ and $\mathfrak{Z}_{kl}^{(B)}$, the partial molar volume can be expressed in terms of GvdW parameters:

$$\frac{\bar{v}_k}{v} = \sum_{j=1}^{c} 2x_j \left(\beta A_{kj} - \frac{B_{kj}}{1 - n \sum_{l,m}^{c} x_l x_m B_{lm}} \right) + \sum_{i,j=1}^{c} x_i x_j (1 - x_k)$$

$$\times \left[\beta \left(\frac{\partial A_{ij}}{\partial x_k} \right)_{T,p,x'} - \frac{\partial}{\partial x_k} \left(\frac{B_{ij}}{1 - n \sum_{l,m}^{c} x_l x_m B_{lm}} \right)_{T,p,x'} \right]. \tag{11.74}$$

From the statistical mechanical viewpoint, there is little gain, but since we can model the GvdW parameters by using the phenomenological canonical equation of state as shown in Chapter 10, the formula presented for $\bar{v}_k$ provides an approach alternative to that of the virial expansion for solutions, which is not reliable for liquids. We note that for a study of subcritical fluids of a single component, a quadratic model for the GvdW parameters were used, which were nonanalytic and discontinuous functions of density and temperature (see Section 10.10, Chapter 10). It is expected that their mathematical properties would be similar even for mixtures.

Expanding the rational fraction term in Eq. (11.74) yields a virial expansion-like series

$$\frac{\bar{v}_k}{v} = \sum_{i,j=1}^{c} 2\delta_{ik} x_j \left(\beta A_{ij} - B_{ij} - n \sum_{lm}^{c} x_l x_m B_{ij} B_{lm} + \cdots \right) + \sum_{i,j=1}^{c} x_i x_j$$

$$\times (1 - x_k) \left[\frac{\partial}{\partial x_k} \left(\beta A_{ij} - B_{ij} - n \sum_{lm}^{c} x_l x_m B_{ij} B_{lm} + \cdots \right) \right]_{T,p,x'}. \tag{11.75}$$

If A_{ij} and B_{ij} are expanded in series of mole fractions

$$A_{ij} = A_{ij}^{(0)}(T) + \sum_{k=1}^{c} A_{ijk}^{(1)}(T)x_k + \sum_{k,l=1}^{c} A_{ijkl}^{(2)}(T)x_k x_l + \cdots, \quad (11.76)$$

$$B_{ij} = B_{ij}^{(0)}(T) + \sum_{k=1}^{r} B_{ijk}^{(1)}(T)x_k + \sum_{k,l=1}^{c} B_{ijkl}^{(2)}(T)x_k x_l + \cdots, \quad (11.77)$$

then to the leading order mole fractions in the supercritical regime the partial molar volume is given by the series

$$\frac{\overline{v}_k}{v} = \sum_{j=1}^{c} 2(\beta A_{kj}^{(0)} - B_{kj}^{(0)})x_j - 2n \sum_{j,l,m=1}^{c} B_{kj}^{(0)} B_{lm}^{(0)} x_j x_l x_m + \cdots. \quad (11.78)$$

The partial molar volumes[6] obtained can be made use of to deduce other thermodynamic quantities for a mixture such as the entropy and heat of mixing. The thermodynamics of critical phenomena can be similar to the formalism for the case of pure substance presented in the previous chapter, but because of the space limitation, this subject is left to the readers as an exercise.

Problems

(1) Calculate the entropy of mixing for a c-component mixture obeying the canonical equation of state.

(2) Calculate the heat of mixing for a c-component mixture obeying the canonical equation of state.

(3) Calculate the fugacity of a c-component mixture obeying the canonical equation of state to the leading order expansion (11.78) for the partial molar volume.

---

[6]We remark that molar and partial molar volumes discussed in this chapter are those of a uniform equilibrium fluid, but they are not necessarily equal to their nonequilibrium counterpart; for example, if the fluid is spatially nonuniform, then the local partial molar volumes are not conserved over space and time owing to the nonequilibrium evolution of the fluid structure in space–time. Therefore, since dynamic molar volume is not necessarily the same as the equilibrium molar volume which is conserved, care must be exercised when molar and partial molar volumes are treated for nonequilibrium fluids. See B. C. Eu, *J. Chem. Phys.* **129**, 094502, 134509 (2008).

Chapter 12

Chemical Equilibria

In most of the discussions up to now, we have excluded chemical reactions among the substances in a mixture. This assumption should be removed for chemically reacting fluids. In this chapter, we consider chemical equilibria. Nonequilibrium phenomena in reacting systems will be deferred to the last chapter in Part II of this book. Since chemical reactions transform one or more species into other species through molecular interactions of reactants or products, there are additional relations required of the chemical potentials that are absent when chemical reactions are precluded, and equilibrium is accordingly modified. We first find these additional conditions on equilibrium. The discussion can be made more general by considering a heterogeneous system of ν phases where a number of coupled chemical reactions may occur as in the cases of many biological systems and catalysis. Since a general formulation also invites complications and often cumbersome notations without necessarily increasing our understanding of the fundamental points in question, we may first consider a simple case and then extend the result to a more general situation removing the restriction. With the general results so obtained, we will also be able to consider a generalized phase rule as well.

12.1. A Single Reaction

We are interested in obtaining the equilibrium condition for a chemical reaction occurring in a closed homogeneous system composed of c components of $n_1, n_2, \ldots, n_c$ moles in a single phase, for example, the gas phase. Let us assume that there is a reversible chemical reaction occurring in the system:

$$r_1 \mathcal{R}_1 + r_2 \mathcal{R}_2 + r_3 \mathcal{R}_3 + \cdots \rightleftharpoons q_1 \mathcal{Q}_1 + q_2 \mathcal{Q}_2 + q_3 \mathcal{Q}_3 + \cdots . \tag{12.1}$$

The symbols $\mathcal{R}_i$ ($i \geq 1$) denote the reactants, and $\mathcal{Q}_i$ ($i \geq 1$) the products. The small letters r_i and q_i denote the stoichiometric coefficients of the reaction. Reaction (12.1) collectively represents a unimolecular, bimolecular, or trimolecular reaction, and so on. It is convenient to write the equation in the form

$$\sum_{i=1}^{c} \sigma_i X_i = 0, \tag{12.2}$$

where X_i stand for the species and σ_i the stoichiometric coefficients *counted positive for the products and negative for the reactants*. Equation (12.2) is obtained from Eq. (12.1) by transferring the left-hand side to the right.

If the reactants are predominant in the concentration relative to the products or if the conditions are favorable for the forward reaction, the reaction will proceed toward the right in Eq. (12.1) and vice versa. Eventually, the forward and reverse reactions will be precisely balanced, when the system has reached chemical equilibrium. The condition for chemical equilibrium is the object we wish to investigate here.

For this purpose, it is useful to introduce the progress variable λ indicating the degree of the chemical reaction progressed. Let us denote by $dn_1, dn_2, \ldots, dn_c$ the molar changes for species arising from the chemical reaction. The stoichiometry of the reaction then demands that

$$\frac{dn_1}{\sigma_1} = \frac{dn_2}{\sigma_2} = \cdots = \frac{dn_c}{\sigma_c} = d\lambda. \tag{12.3}$$

This equation basically defines the progress variable and in this definition we are counting the reaction as progressing to the right if $d\lambda > 0$, and to the left if $d\lambda < 0$. We will denote the molecular weights of compounds by M_i, $i = 1, 2, \ldots, c$. Since the total mass must be balanced before and after the reaction, as is demanded by the conservation of mass, the following relation must hold:

$$\sum_{i=1}^{c} \sigma_i M_i = 0. \tag{12.4}$$

This is the mass balance equation for the reaction. With this preparation, we are now ready to consider the chemical equilibrium condition. Since the system is closed, for an arbitrary virtual variation of the internal energy

$$(\delta E)_{\Psi, V} \geq 0 \tag{12.5}$$

in accordance with the demand made by the second law of thermodynamics. Since for reversible processes, $S = \Psi_e$ and thus the fundamental relation may be written as

$$dE = TdS - pdV + \sum_{i=1}^{c} \mu_i dn_i, \tag{12.6}$$

where n_i changes because of the chemical reaction, the equilibrium condition (12.5) now takes the form

$$\sum_{i=1}^{c} \mu_i \delta n_i \geq 0 \tag{12.7}$$

for arbitrary virtual variations in n_i, $i = 1, 2, \ldots, c$. Substitution of equations in Eq. (12.3) puts this inequality in the form

$$\delta \lambda \sum_{i=1}^{c} \mu_i \sigma_i \geq 0. \tag{12.8}$$

If $\delta \lambda > 0$, then

$$\sum_{i=1}^{c} \mu_i \sigma_i \geq 0,$$

but if $\delta \lambda < 0$, then

$$\sum_{i=1}^{c} \mu_i \sigma_i \leq 0.$$

However, since the variation in progress variable is arbitrary, its sign is also arbitrary. The only way to satisfy the aforementioned two conditions simultaneously is then

$$\sum_{i=1}^{c} \mu_i \sigma_i = 0. \tag{12.9}$$

This is the chemical equilibrium condition we are looking for. Th. de Donder[1] introduced the affinity defined by

$$\mathcal{A} = -\sum_{i=1}^{c} \mu_i \sigma_i. \tag{12.10}$$

---

[1]Th. de Donder, *Bull. Acad. Roy. Belg.* (*Cl. Sc.*) (5) **7**, 197, 205 (1922).

The chemical equilibrium condition (12.9) then implies that the affinity is equal to zero at chemical equilibrium. The concept of affinity plays an important role in irreversible phenomena of chemical reactions. Its meaning can be better understood quite easily if we simply write out Eq. (12.10) for the chemical reaction (12.1). By using the convention we have adopted for σ_i, we find

$$\mathcal{A} = \sum_{j \in \text{reac}} r_j \mu_j - \sum_{i \in \text{prod}} q_i \mu_i, \tag{12.11}$$

which shows that the affinity is the difference between the Gibbs free energies of the products and reactants of a chemical reaction. When the two Gibbs free energies are precisely balanced, the chemical equilibrium is reached and there is no longer a macroscopically discernible change in the concentrations of the reactants and the products at the given temperature and pressure.

12.2. Coupled Chemical Reactions

The equilibrium condition just obtained for a single chemical reaction occurring in a phase can be generalized to the case of a system of coupled chemical reactions occurring in a phase. Let us denote the coupled chemical reactions by the equations

$$\sum_{i=1}^{c} \sigma_{i\alpha} X_i = 0 \quad (\alpha = 1, 2, \ldots, m), \tag{12.12}$$

where $\sigma_{i\alpha}$ stands for the stoichiometric coefficient of the compound X_i in the αth chemical reaction. The same convention on the stoichiometric coefficients is used as for the single reaction considered earlier. Examples for coupled chemical reactions are as follows:

$$NaH_2PO_4 = Na^+ + H_2PO_4^-,$$

$$H_2PO_4^- = H^+ + HPO_4^{2-},$$

$$HPO_4^{2-} = H^+ + PO_4^{3-}.$$

Each chemical reaction may be followed in terms of progress variables λ_α $(\alpha = 1, 2, \ldots, m)$. We may then write

$$dn_{i\alpha} = \sigma_{i\alpha} d\lambda_{i\alpha}. \tag{12.13}$$

Since the total change in the number of moles for species i is

$$dn_i = \sum_{i=1}^{c} dn_{i\alpha} = \sum_{\alpha=1}^{m} \sigma_{i\alpha} d\lambda_{\alpha}, \qquad (12.14)$$

the Gibbs relation may be written in the form

$$dE = TdS - pdV + \sum_{i=1}^{c} \sum_{\alpha=1}^{m} \sigma_{i\alpha} \mu_i d\lambda_{\alpha}. \qquad (12.15)$$

The equilibrium condition for the closed reacting system is therefore

$$\sum_{\alpha=1}^{m} \left[\sum_{i=1}^{c} \sigma_{i\alpha} \mu_i \right] \delta\lambda_{\alpha} \geq 0.$$

By following the same line of argument as for Eq. (12.7), we find the equilibrium condition

$$\mathcal{A}_{\alpha} = -\sum_{i=1}^{c} \sigma_{i\alpha} \mu_i = 0 \quad (\alpha = 1, 2, \ldots, m). \qquad (12.16)$$

That is, the affinities must vanish for all chemical reactions at chemical equilibrium. This is obviously a generalization of Eq. (12.7).

12.3. Chemical Reactions in a Multiphase System

The chemical equilibrium condition (12.16) which we have obtained in Section 12.2 holds in the case of a single phase. In nature, there can arise a situation in which chemical reactions occur in different phases which are not necessarily in equilibrium with each other. One typical example would be biological cells where various chemical reactions are known to occur in cytoplasm, membranes, and vesicles. Another more physical example would be a system of a liquid solution and a gaseous mixture in which chemical reactions occur. We would like to extend to such systems the equilibrium conditions which we have obtained for nonreacting systems.

If there is a chemical reaction proceeding in a phase, there are two factors contributing to the concentration changes for the components involved: one stems from the chemical transformation of species, and the other arises from the permeation (diffusion) of species through the interfacial boundaries. These two factors will be denoted $d_i n_k$ and $d_c n_k$, the former standing for the transfer of matter across the interfacial boundaries, and the latter for the concentration change arising from the chemical reactions. We shall consider for the sake of simplicity the case where the same chemical reactions occur in every phase involved and each phase is in internal equilibrium with regard to temperature and pressure, apart from chemical reactions which are assumed to occur uniformly over the phases. It will be assumed that there are m chemical reactions.

In this case, because of the assumption of internal equilibrium, the Gibbs relation for phase a may be written as

$$dE^{(a)} = T^{(a)}dS^{(a)} - p_a dV^{(a)} + \sum_{k=1}^{c} \mu_k^{(a)} d_i n_k^{(a)} + \sum_{k=1}^{c} \mu_k^{(a)} d_c n_k^{(a)}, \qquad (12.17)$$

where, upon use of progress variables $\lambda_\alpha^{(a)}$ ($\alpha = 1, 2, \ldots, m$; $a = 1, 2, \ldots, \nu$) for m chemical reactions, we may express the chemical reaction part of the concentration change as follows:

$$d_c n_k^{(a)} = \sum_{\alpha=1}^{m} d_c n_{k\alpha}^{(a)} = \sum_{\alpha=1}^{m} \sigma_{i\alpha} d\lambda_\alpha^{(a)}. \qquad (12.18)$$

If Eq. (12.18) is substituted into the last term on the right-hand side in Eq. (12.17), it may be given a familiar form:

$$\sum_{k=1}^{c} \mu_k^{(a)} d_c n_k^{(a)} = \sum_{\alpha=1}^{m} \left[\sum_{k=1}^{c} \sigma_{k\alpha} \mu_k^{(a)} \right] d\lambda_\alpha^{(a)}$$

$$:= -\sum_{\alpha=1}^{m} \mathcal{A}_\alpha^{(a)} d\lambda_\alpha^{(a)}, \qquad (12.19)$$

where

$$\mathcal{A}_\alpha^{(a)} = -\sum_{k=1}^{c} \sigma_{k\alpha} \mu_k^{(a)} \qquad (12.20)$$

is the affinity of reaction α in phase a.

The equilibrium conditions for the reacting heterogeneous systems can be obtained from Eq. (12.17), which must satisfy the inequality

$$\sum_{a=1}^{\nu} dE^{(a)} \geq 0 \qquad (12.21)$$

for the system to be in stable thermodynamic equilibrium. The procedure to obtain the equilibrium conditions is the same as for the nonreacting multiphases or the reacting single phase we have previously considered for nonreacting systems. The result is

$$T_1 = T_2 = \cdots = T_\nu,$$

$$p_1 = p_2 = \cdots = p_\nu,$$

$$\mu_1^{(1)} = \mu_1^{(2)} = \cdots = \mu_1^{(\nu)},$$

$$\mu_2^{(1)} = \mu_2^{(2)} = \cdots = \mu_2^{(\nu)}, \tag{12.22}$$

$$\vdots$$

$$\mu_c^{(1)} = \mu_c^{(2)} = \cdots = \mu_c^{(\nu)},$$

$$\mathcal{A}_\alpha^{(a)} = -\sum_{k=1}^{c} \sigma_{k\alpha} \mu_k^{(a)} = 0 \quad (\alpha = 1, 2, \ldots, m).$$

Because of the material equilibrium conditions that equate the chemical potentials in the ν phases, the chemical potential $\mu_k^{(a)}$ may be replaced with the equilibrium chemical potential μ_k in the chemical equilibrium conditions which now read

$$\mathcal{A}_\alpha^{(a)} = -\sum_{k=1}^{c} \sigma_{k\alpha} \mu_k = 0 \quad (\alpha = 1, 2, \ldots, m). \tag{12.23}$$

When compared with the equilibrium conditions for nonreacting multiphase systems, *there are m additional conditions related to chemical equilibrium* which supply additional constraints on chemical potentials — the last m conditions in Eq. (12.22). The phase rule for the system under consideration is then easily found to be

$$f = c + 2 - \nu - m, \tag{12.24}$$

where m is the number of chemical reactions. This generalizes the phase rule for nonreacting multiphase systems

$$f = c + 2 - \nu, \tag{12.25}$$

which was derived in Chapter 8.

12.4. Equilibrium Constant

The equilibrium conditions obtained earlier imply that the concentrations are no longer independent, but are constrained by one or more relations.

Such constraining relations can be made more explicit if chemical potentials previously calculated for a mixture in terms of fugacities or activities are made use of. By substituting the formula for chemical potentials into, for example, the equilibrium condition (12.9), we obtain

$$RT \sum_{i=1}^{c} \sigma_i \ln a_i = -\Delta G^0, \tag{12.26}$$

where

$$\Delta G^0 = \sum_{i=1}^{c} \sigma_i \mu_i^0 (T, p)$$

$$= (q_1 \mu_1^0 + q_2 \mu_2^0 + \cdots) - (r_1 \mu_1^0 + r_2 \mu_2^0 + \cdots)$$

$$= G_{\text{prod}}^0 - G_{\text{reac}}^0. \tag{12.27}$$

Equation (12.27) may be rearranged to the form

$$K(T, p) = \exp\left(-\frac{\Delta G^0}{RT}\right), \tag{12.28}$$

where $K(T, p)$ is called the equilibrium constant. It is defined by

$$K(T, p) = \prod_{i=1}^{c} a_i^{\sigma_i} = \frac{\prod_i a_i^{q_i}}{\prod_i a_i^{r}}$$

$$= \frac{a_{\mathcal{Q}_1}^{q_1} a_{\mathcal{Q}_2}^{q_2} a_{\mathcal{Q}_3}^{q_3} \cdots}{a_{\mathcal{R}_1}^{r_1} a_{\mathcal{R}_2}^{r_2} a_{\mathcal{R}_3}^{r_3} \cdots}. \tag{12.29}$$

If the system is ideal (e.g., at infinite dilution), the activity a_i is equal to mole fraction x_i and the equilibrium constant is given in terms of mole fractions:

$$K(t, p) = \prod_{i=1}^{c} x_i^{\sigma_i}. \tag{12.30}$$

Since for real gases,

$$a_i = p_{fi}$$

by choice of the standard state, the equilibrium constant for a real gas chemical reaction takes the form

$$K(T) = \prod_{i=1}^{c} p_{fi}^{\sigma_i}, \tag{12.31}$$

where the equilibrium constant may be given in terms of the standard free energy change ΔG^*

$$K(T) = \exp\left(-\frac{\Delta G^*}{RT}\right) \tag{12.32}$$

with ΔG^* defined by

$$\Delta G^* = \sum_{i=1}^{c} \sigma_i \mu_i^*(T). \tag{12.33}$$

In this case, since the standard chemical potentials are dependent on the temperature only, the equilibrium constant is a function of temperature only. Since fugacity coefficients are unity for ideal gases, the fugacities are equal to the partial pressures. Hence, we obtain for an ideal gas reaction

$$K(T) = p_i^{\Delta\sigma} \prod_{i=1}^{c} x_i^{\sigma_i} := K_0(T, p), \tag{12.34}$$

where $\Delta\sigma$ is the difference in stoichiometric coefficients for the reaction:

$$\Delta\sigma = \sum_{i=1}^{c} \sigma_i. \tag{12.35}$$

The temperature and pressure dependence of equilibrium constants is contained in ΔG^0. It is therefore possible to investigate the T- and p-dependence of $K(T)$ by studying the thermochemical data on ΔG^0. The standard free energy change ΔG^0 may be obtained from the data on ΔH^0 and ΔS^0. Since we may write for an isothermal process

$$\Delta G^0 = \Delta H^0 - T\Delta S^0, \tag{12.36}$$

the temperature dependence of $K(T)$ is often expressed in the form

$$K(T, p) = \exp\left(\frac{\Delta S^0}{R}\right) \exp\left(-\frac{\Delta H^0}{RT}\right), \tag{12.37}$$

where

$$\Delta H^0 = \sum_{i=1}^{c} \sigma_i h_i^0(T, p), \tag{12.38}$$

$$\Delta S^0 = \sum_{i=1}^{c} \sigma_i s_i^0(T, p). \tag{12.39}$$

Here, ΔH^0 is called the heat of reaction and ΔS^0 the entropy change of reaction, and h_i^0 and s_i^0 are, respectively, the standard enthalpy and standard entropy of species i defined previously.

12.5. van't Hoff Equation

Chemical equilibria are influenced by the state of the system where chemical reactions occur. A change in temperature or pressure, especially, results in a shifting of chemical equilibrium. Here, we consider the temperature dependence of equilibrium constants. To find it, let us differentiate $K(T)$ with respect to T:

$$\left[\frac{\partial \ln K(T)}{\partial T}\right]_p = -\frac{1}{R}\sum_{i=1}^{c}\sigma_i\left(\frac{\partial}{\partial T}\frac{\mu_i^0}{T}\right)_p, \tag{12.40}$$

for which we have used

$$\Delta G^0 = \sum_{i=1}^{c}\sigma_i\mu_i^0$$

in Eq. (12.28). Since

$$\left(\frac{\partial}{\partial T}\frac{\mu_i^0}{T}\right)_p = -\frac{h_i^0}{T^2},$$

by making use of Eq. (12.38), we may cast Eq. (12.40) in the form

$$\left[\frac{\partial \ln K(T)}{\partial T}\right]_p = \frac{\Delta H^0}{RT^2}. \tag{12.41}$$

This is called the van't Hoff equation for chemical equilibrium constant. Here, ΔH^0 is the heat of reaction. If the temperature dependence of ΔH^0 is known, it is possible to find the variation of $K(T)$ with respect to T. In the case of gaseous reactions, heats of reaction are measured in reference to the ideal gas state for all the components. It is the heat absorbed when the reaction proceeds in the forward direction. Note that h_i^0 is chosen for gases such that

$$h_i^0 = \mu_i^0(T,p) + Ts_i^0(T,p) = h_i^*(T), \tag{12.42}$$

which is the enthalpy of i in the dilute ideal gas state. In the case of reactions in the liquid state, ΔH^0 may be defined by either one of the following conventions:

(a) *The heat absorbed in the forward reaction when all the components are in their most stable pure state.*

(b) *The heat absorbed in the forward reaction when the solvent is pure and the other components (solutes) are in the hypothetical, infinitely dilute state.*

It is helpful to note that these conventions are rooted on the conventions taken for the standard states for thermodynamic functions, which we have adopted previously (see Section 3.7.2). The same standard states will be adopted when we develop the thermodynamics of solutions in subsequent chapters.

Let us pursue a little further the discussion of the temperature dependence of $K(T)$ in the case of gaseous reactions. We recall that

$$h_i^*(T) = h_{i0} + \int^T dT\, c_p^*(T)$$

$$= h_{i0} + c_{i0}^* T + \frac{1}{2} c_{i1}^* T^2 + \cdots$$

and

$$s_i^*(T) = s_{i0} + c_{i0}^* \ln T + c_{i1}^* T + \cdots.$$

Therefore, the heat of reaction ΔH^0 may be obtained from the calorimetric data as indicated below:

$$\Delta H^0 = \Delta h^0 + \Delta c_0 T + \frac{1}{2} \Delta c_1 T^2 + \cdots, \tag{12.43}$$

where

$$\Delta h^0 = \sum_{i=1}^c \sigma_i h_{i0},$$

$$\Delta c_0 = \sum_{i=1}^c \sigma_i c_{i0}^*, \quad \Delta c_1 = \sum_{i=1}^c \sigma_i c_{i1}^*,$$

and so on. Similarly, with definition

$$\Delta s^0 = \sum_{i=1}^c \sigma_i s_{i0},$$

the following relation holds

$$\Delta s^0 = \Delta s^0 + \Delta c_0 \ln T + \Delta c_1 T + \cdots. \tag{12.44}$$

By using these relations, we finally obtain the temperature dependence of $K(T)$:

$$\ln K(T) = -\frac{\Delta h^0}{RT} + \frac{\Delta s^0}{R} - \frac{\Delta c_0}{R} + \frac{\Delta c_0}{R}\ln T + \frac{\Delta c_1}{2R}T + \cdots . \qquad (12.45)$$

On the other hand, by inserting the expansion (12.43) for ΔH^0 into Eq. (12.41) and integrating the equation, we find

$$\ln K(T) = I - \frac{\Delta h^0}{RT} + \frac{\Delta c_0}{R}\ln T + \frac{\Delta c_1}{2R}T + \cdots , \qquad (12.46)$$

where I is the integration constant. Comparison of Eqs. (12.45) and (12.46) yields the integration constant in the form

$$I = \frac{1}{R}(\Delta s^0 - \Delta c_0). \qquad (12.47)$$

This result suggests that to obtain $K(T)$ at a temperature, all we need is Δh^0, Δs^0, Δc^0, and so on.

If the integration constant I is eliminated with the equation for $K(T)$ at another temperature, say, T_1, then we find

$$\ln\left[\frac{K(T)}{K(T_1)}\right] = \frac{\Delta h^0}{R}\left(\frac{1}{T_1} - \frac{1}{T}\right) + \frac{\Delta c_0}{R}\ln\left(\frac{T}{T_1}\right) + \frac{\Delta c_1}{2R}(T - T_1) + \cdots .$$
$$(12.48)$$

This, of course, can be obtained directly from Eqs. (12.41) and (12.43) if Eq. (12.41) is integrated from T_1 to T.

12.6. Equilibrium Constant for Real Gases

Since chemical reactions do not necessarily occur only between ideal gases or at sufficiently low pressures so that gases may be treated as ideal, it is useful to consider what roles the nonideality of gases play in chemical reactions. We now examine the nonideality correction for $K(T)$ in the case of gaseous reactions. For this purpose, we return to Eq. (12.31). Recalling that chemical potentials for real gases may be given in terms of fugacities

$$p_{fi} = f_i p_i = x_i f_i p,$$

we separate out the ideal gas part and the nonideality correction of $K(T)$:

$$K(T) = K_0(T, p)\prod_{i=1}^{c} f_i^{\sigma_i}, \qquad (12.49)$$

where f_i is the fugacity coefficient of i

$$f_i = \exp\left[\int_0^p dp'\left(\frac{\bar{v}_i}{RT} - \frac{1}{p'}\right)\right], \qquad (12.50)$$

and $K_0(T, p)$ is the ideal gas part of $K(T)$ and defined in Eq. (12.34). The nonideality correction may be more explicitly given if Eq. (12.50) is made use of:

$$K_{\mathrm{corr}}(T, p) = \prod_{i=1}^c f_i^{\sigma_i} = \exp\left[\sum_{i=1}^c \sigma_i \int_0^p dp'\left(\frac{\bar{v}_i}{RT} - \frac{1}{p'}\right)\right]. \qquad (12.51)$$

Since according to Eq. (11.22), the partial molar volume $\bar{v}_i$ may be approximated, to the first order in p, by the formula

$$\bar{v}_i = \frac{RT}{p} + \sum_{j=1}^c \sum_{k=1}^c x_j x_k (2B_{ij} - B_{jk}),$$

substitution of this formula into Eq. (12.51) yields $K_{\mathrm{corr}}(T, p)$ in the form

$$K_{\mathrm{corr}}(T, p) = \exp\left[\sum_{i=1}^c \sum_{j=1}^c \sum_{k=1}^c \sigma_i x_j x_k (2B_{ij} - B_{jk})\frac{p}{RT}\right]. \qquad (12.52)$$

If we use the combination rule (11.25) for the second virial coefficients, the exponent in Eq. (12.52) can be considerably simplified. We thereby obtain

$$\sum_{i=1}^c \sum_{j=1}^c \sum_{k=1}^c \sigma_i x_j x_k (2B_{ij} - B_{jk}) = \sum_{i=1}^c \sigma_i B_{ii} := \Delta B, \qquad (12.53)$$

and hence

$$K(T, p) = K_0(T, p)\exp\left(\frac{p\Delta B}{RT}\right). \qquad (12.54)$$

This enables us to make a first-order correction to chemical equilibrium constants, when the gases participating in the chemical reaction are not ideal. It gives chemical equilibrium constants to a fair degree of accuracy up to a few tens of atmospheric pressure, but if the pressure is higher, the higher order terms in the virial equation of state must be included, or another more suitable equation of state must be used. It is straightforward to do so if the theory of real fluids developed earlier is made use of.

Problems

(1) For chemical reaction $\frac{1}{2}N_2 + \frac{3}{2}H_2 = NH_3$, the equilibrium constant is known to be 0.0065 at 450°C. The following is available for ΔH^0 as a function of T:

$$\Delta H^0 = (-38.20 - 3.12 \times 10^{-2}\,T$$
$$+ 1.54 \times 10^{-5}T^2 - 1.97 \times 10^{-9}\,T^3)\ \text{kJ mol}^{-1}.$$

Obtain the equilibrium constant as a function of T.

(2) Use the van der Waals model for the second virial coefficients B_{ii} of species in gas phase chemical reaction and express the nonlinearity correction for the equilibrium constant in terms of the van der Waals parameters.

(3) Two liquid phases are separated by a semipermeable membrane which permeates species 1 only. One phase is a solution of three species where species 1 is the solvent and reacts with species 2 to form the third (species 3) according to the reaction

$$\sigma_1 X_1 + \sigma_2 X_2 \leftrightharpoons \sigma_3 X_3.$$

Obtain the Gibbs phase rule and discuss the thermodynamics of this system including chemical equilibrium.

Chapter 13

Thermodynamics of Solutions

The general principles developed in the previous chapters can be applied to study thermodynamic properties of liquid solutions. A liquid composed of more than one component is called a liquid solution. The predominant component is called the solvent, and the other components the solute. Henceforth, when the term 'solution' is used in this chapter, we mean a liquid solution. We assume that the solution consists of c neutral components, which do not react with each other nor do they ionize into constituent ions. This assumption is easy to remove if the theory is developed as discussed in Chapter 12.

13.1. Chemical Potentials of Solutions

As in the case of gases, the quantity of central interest is the chemical potentials, since they are the storage of thermodynamic information of the system, and other thermodynamic functions can be derived from them. For a c-component solution, chemical potentials are generally functions of T, p, and $x_1, x_2, \ldots, x_{c-1}$, but their explicit forms are not completely known *a priori*. In effect, the aim of thermodynamics of solutions is just in finding the functional forms for chemical potentials through laboratory experiments and comparison of the results of experiments with a thermodynamic theory which one might develop for solutions of interest.

We have seen that in the thermodynamics of gases, the ideal gas plays a special role and thermodynamic functions for real gases are calculated in reference to the thermodynamic functions of the ideal gas. Thus, thermodynamic functions for real gases are invariably calculated such that they consist of the ideal gas part and the nonideality correction. We develop a theory of solutions in a way parallel to that of gases. We call a solution ideal in close analogy to gases if the chemical potentials depend on the concentrations in the following manner:

$$\mu_i(T, p, x_1, \ldots, x_{c-1}) = \mu_i^0(T, p) + RT \ln x_i, \qquad (13.1)$$

where $\mu_i^0(T, p)$ is the reference chemical potential of i whose value we may fix according to the conventions introduced in Chapters 9 and 11. Our experience with the ideal gas thermodynamics shows that thermodynamics will be simple for ideal solutions, but it will be only a limiting theory of a more realistic thermodynamics of real solutions in the ideal solution limit. Since Eq. (13.1) will generally deviate from the chemical potential for species i in the real solution, a correction must be made to account for the deviation arising from nonideal behaviors[1] of the constituents. It is generally introduced in the form of excess chemical potential

$$\mu_i^{(ex)} = \mu_i - [\mu_i^0(T, p) + RT \ln x_i]. \qquad (13.2)$$

It is common practice to express the excess chemical potential in terms of activity coefficient f_i defined as follows:

$$f_i = \exp\left(\frac{\mu_i^{(ex)}}{RT}\right) = x_i^{-1} \exp\left(\frac{\mu_i - \mu_i^0}{RT}\right). \qquad (13.3)$$

The chemical potential may then be expressed in terms of the concentration and the activity coefficient f_i

$$\mu_i = \mu_i^0(T, p) + RT \ln(x_i f_i), \qquad (13.4)$$

or more simply,

$$\mu_i = \mu_i^0(T, p) + RT \ln a_i, \qquad (13.5)$$

---

[1] The ideal gas is often regarded as a gas where constituent molecules do not interact with each other. This notion of no interaction between the molecules is quite correct for ideal gases, but it is misleading in the case of ideal solutions because the molecules do interact intimately with each other in a condense phase. The definition of ideal solutions is based on the mathematical form of the constitutive equation for the chemical potential, e.g., Eq. (13.1), where the information on molecular interactions in pure solvent is contained in the reference chemical potential $\mu_i^0(T, p)$.

where a_i is the activity defined by

$$a_i = x_i f_i. \tag{13.6}$$

The reference values of activity coefficients f_i are fixed in accordance with the standard states adopted for chemical potentials. We recall that there are two modes of choosing the reference state, and the reference activity coefficients are determined by the limits

$$\lim_{x_i \to 1} f_i(x_1, x_2, \ldots, x_{c-1}) = 1 \quad (i = 1, 2, \ldots, c), \tag{13.7}$$

or by the limits

$$\lim_{x_1 \to 1} f_i(x_1, x_2, \ldots, x_{c-1}) = 1 \quad (i = 1, 2, \ldots, c), \tag{13.8}$$

depending on the convention for the standard state—a reference state—adopted. It is conventional to express concentrations by units other than mole fractions, such as molality or molarity, although it is most convenient to use mole fractions from the theoretical viewpoint. Since the actual values of activity coefficients can be different for different concentration units taken, it is necessary to distinguish them by different symbols. Since different concentration units are interrelated, the activity coefficients in different concentration units are also related to each other.

If the concentrations are expressed in molality m_i, we will write the chemical potential μ_i as

$$\mu_i = \mu_i^{(m)}(T, p) + RT \ln(\gamma_i m_i), \tag{13.9}$$

where γ_i is the activity coefficient and $\mu_i^{(m)}$ is the corresponding standard chemical potential. When chemical potentials are given in terms of molality, only Convention II applies, since the definition of molality necessarily designates the solvent. We therefore have

$$\mu_i^{(m)} = \mu_1^l(T, p),$$
$$\mu_i^{(m)} = \lim_{x_i \to 0} (\mu_i - RT \ln m_i) \quad (i = 2, \ldots, c), \tag{13.10}$$

where μ_1^l is the chemical potential of pure liquid 1, and thus,

$$\lim_{x_1 \to 1} \gamma_i = 1 \quad (i = 2, \ldots, c). \tag{13.11}$$

By definition, the molality is related to mole fraction x_i as follows:

$$x_i = \frac{m_i M_1}{1000 \left(1 + \frac{M_1}{1000} \sum_{j=2}^c m_j\right)}, \tag{13.12}$$

where M_1 is the molecular weight of the solvent and m_i is the molality of component i — namely, the mole numbers of the solutes in 1 kg of the solvent. Since the chemical potentials in the two concentration units must be the same,

$$\mu_i^0 + RT \ln(x_i f_i) = \mu_i^{(m)} + RT \ln(\gamma_i m_i), \tag{13.13}$$

if we set

$$\mu_i^{(m)} = \mu_i^0 + RT \ln\left(\frac{M_1}{1000}\right), \tag{13.14}$$

we find by making use of Eq. (13.12) the relation between two activity coefficients f_i and γ_i :

$$\gamma_i = \frac{f_i}{1 + \frac{M_1}{1000} \sum_{j=2}^c m_j}. \tag{13.15}$$

Therefore, if the solution is dilute, the activity coefficients γ_i and f_i are approximately equal:

$$\gamma_i \approx f_i.$$

If the molarity c_i is taken to represent the concentration, the chemical potential μ_i may be written as

$$\mu_i = \mu_i^{(c)}(T, p) + RT \ln(\alpha_i c_i), \tag{13.16}$$

where α_i is the activity coefficient of component i in molarity units and $\mu_i^{(c)}$ is the corresponding standard chemical potential. Since there is no necessity of distinguishing the solvent and the solutes in this case, both Conventions I and II for the standard state apply, and we therefore have

$$\mu_i^{(c)} = \mu_i^l(T, p) \quad (i = 1, 2, \ldots, c). \tag{13.17}$$

Consequently, in Convention I,

$$\lim_{x_1 \to 1} \alpha_i = 1 \quad (i = 1, 2, \ldots, c), \tag{13.18}$$

and in Convention II,

$$\mu_1^0 = \mu_1^l(T, p),$$
$$\mu_i^{(c)} = \lim_{x_1 \to 1} (\mu_i - RT \ln c_i) \quad (i = 2, \ldots, c). \tag{13.19}$$

The chemical potential must have the same value regardless of the concentration units in which it is expressed. Therefore, the following equation should hold:

$$\mu_i^{(c)} + RT\ln(\alpha_i c_i) = \mu_i^0 + RT\ln(x_i f_i).$$

Since c_i is related to x_i as follows:

$$c_i = \frac{1000 n_i}{\sum_{j=1}^{c} \bar{v}_j n_j} = \frac{1000 x_i}{\sum_{j=1}^{c} \bar{v}_j x_j}, \qquad (13.20)$$

if we set

$$\mu_i^{(c)} = \mu_i^0 + RT\ln\left(\frac{\bar{v}_1 n_1}{1000}\right), \qquad (13.21)$$

there follows the relation between α_i and f_i:

$$f_i = \frac{\alpha_i}{1 + \sum_{j=2}^{c}(\bar{v}_j n_j / \bar{v}_1 n_1)}. \qquad (13.22)$$

Therefore, f_i coincides with α_i in the limit of infinite dilution, where the second term in the denominator becomes negligible.

13.2. Ideal Solutions

As an ideal gas mixture is an idealization of the real gas mixture at low pressure, an ideal solution is an idealization of real solutions. In this case, it is a fictitious representation of real solutions in the limit of infinite dilution.[2] Nevertheless, it is mathematically easy to handle and serves as a reference state of real solutions around which a thermodynamic theory of real solutions can be constructed. This approach makes it easier for us to draw analogy from the experience we have gained through our study of real gas mixtures and develop a theory quite parallel to the thermodynamics of real gas mixtures.

If Convention I is adopted for the reference state, ideal solutions may be defined as solutions whose chemical potentials are given by the formula

$$\mu_i = \mu_i^l(T, p) + RT\ln x_i \quad (i = 1, 2, \dots, c). \qquad (13.23)$$

---

[2]This notion is a little limited. Generally, ideal solutions include solutions of liquids of structurally similar molecules interacting similarly, as will be seen later. In this sense, the notion of ideal solutions is different from that of ideal gases, which implies no interactions between the molecules.

They are sometimes defined as solutions for which Raoult's law holds. As will be seen shortly, these two definitions are equivalent. The expression for chemical potentials is similar to the formula for chemical potentials of ideal gas mixtures except for the difference in the reference state.

If the chemical potentials for $(c-1)$-components in a c-component solution are given by Eq. (13.23), the chemical potential for the cth-component must also be in the same form as Eq. (13.23). Therefore, if we have a binary solution, it is not possible to treat only one component as an ideal component obeying Eq. (13.23): both components must be treated on the same footing. To prove this statement, we use the Gibbs–Duhem equation at constant T and p

$$\sum_{i=1}^{c} x_i d\mu_i = 0. \tag{13.24}$$

Since Eq. (13.23) holds for $(c-1)$-components, we obtain from Eq. (13.23)

$$d\mu_i = \frac{RT}{x_i} dx_i \quad (i = 1, 2, \ldots, c),$$

where T and p are kept constant. Substituting it into Eq. (13.24) and making use of

$$\sum_{i=1}^{c} dx_i = 0,$$

we obtain the differential equation

$$d\mu_c = \frac{RT}{x_c} dx_c.$$

Integration of this differential equation yields

$$\mu_c = C + RT \ln x_{cr},$$

where C is the integration constant. By using Convention II for chemical potentials, we find

$$C = \mu_c^l(T, p),$$

and thus,

$$\mu_c = \mu_c^l(T, p) + RT \ln x_c,$$

which proves the statement.

Ideal solutions have a number of characteristic properties that remind us of the properties of ideal gas mixtures. These properties in practice may be used collectively as an indicator of the ideality of a solution. We list them in the following.

1. *An ideal solution has no heat of mixing.*

As in the case of gases, the heat of mixing for a solution is defined as

$$\Delta h = \sum_{i=1}^{c} x_i (\bar{h}_i - h_i^l), \tag{13.25}$$

where h_i^l is the molar enthalpy of pure component i. The partial molar enthalpy may be calculated with Eq. (13.23). Since

$$\bar{h}_i = -T^2 \left(\frac{\partial}{\partial T} \frac{\mu_i}{T} \right)_{p,x_i}, \tag{13.26}$$

substitution of Eq. (13.23) yields

$$\bar{h}_i = -T^2 \left(\frac{\partial}{\partial T} \frac{\mu_i^l}{T} \right)_p = h_v^l \tag{13.27}$$

which implies that

$$\Delta h = 0. \tag{13.28}$$

2. *There is no volume change on mixing for ideal solutions.*

The volume change of mixing is defined by

$$\Delta v = \sum_{i=1}^{c} x_i \left(\bar{v}_i - v_i^l \right), \tag{13.29}$$

where v_i^l is the molar volume of liquid component i. Since the partial molar volume is given by

$$\bar{v}_i = \left(\frac{\partial \mu_i}{\partial p} \right)_{T,x_i}, \tag{13.30}$$

substitution of Eq. (13.23) yields

$$\bar{v}_i = \left(\frac{\partial \mu_i^l}{\partial p} \right)_T = v_i^l, \tag{13.31}$$

which implies that

$$\Delta v = 0. \tag{13.32}$$

3. *The entropy of mixing is in the same form as for an r-component ideal gas mixture.*

The entropy of mixing for a solution is defined by

$$\Delta s = \sum_{i=1}^{r} x_i (\bar{s}_i - s_i^l), \tag{13.33}$$

where s_i^l is the molar entropy of pure liquid component i. Since the partial molar entropy of component i is

$$\bar{s}_i = -\left(\frac{\partial \mu_i}{\partial T}\right)_{p, x_i}, \tag{13.34}$$

substitution of Eq. (13.23) into the right-hand side yields

$$\bar{s}_i = s_i^l - R \ln x_i, \tag{13.35}$$

where

$$\bar{s}_i = -\left(\frac{\partial \mu_i^l}{\partial T}\right)_p. \tag{13.36}$$

Therefore, Eq. (13.35) implies

$$\Delta s = -R \sum_{i=1}^{c} x_i \ln x_i, \tag{13.37}$$

which is in exactly the same form as for the entropy of mixing for an r-component ideal gas mixture. Note that since $x_i \leq 1$, the entropy of mixing is always positive. That is, the entropy always increases on mixing two or more components. This conclusion can be drawn only for ideal solutions because for real solutions, the nonideality correction may be such that Δs may not be positive in the whole range of x_i. The nonpositivity, in any case, does not mean that the second law is broken by the mixing process; the reason for this statement was discussed in Chapter 4.

13.3. Raoult's Law

The vapor pressure of a solution reflects the properties of the solution. In fact, the vapor pressure is in a well-defined, simple relationship with the composition of the solution if the solution is ideal. This relation is the content of Raoult's law. If a solution is in equilibrium with its vapor at a given T and p, by the equilibrium condition (8.22) the following relation holds:

$$\mu_i = \mu_i^v, \tag{13.38}$$

where μ_i^v is the chemical potential of species i in the vapor. It is generally given by

$$\mu_i^v = \mu_i^*(T) + RT \ln p_{fi}. \tag{13.39}$$

By substituting this equation and Eq. (13.23) into Eq. (13.38) and rearranging the terms, we obtain for the fugacity the following expression:

$$p_{fi} = x_i \exp\left[\frac{\mu_i^l(T,p) - \mu_i^*(T)}{RT}\right]. \tag{13.40}$$

If x_i is put equal to unity, then the solution consists of a single component and the vapor consists of the same component only. In that case, p_{fi} becomes the fugacity p_{fi}^0 of pure component i,. That is, as x_i tends to unity, the left-hand side becomes the fugacity of pure vapor i:

$$\exp\left[\frac{\mu_i^l(T,p) - \mu_i^*(T)}{RT}\right] = p_{fi}^0. \tag{13.41}$$

Thus, when Eqs. (13.40) and (13.41) are combined, there follows the relation

$$p_{fi} = x_i p_{fi}^0. \tag{13.42}$$

This is Raoult's law. If the vapor is ideal, then the fugacity is equal to the pressure, namely,

$$p_{fi} = p_i; \quad p_{fi}^0 = p_i^0, \tag{13.43}$$

where p_i^0 is the vapor pressure of pure component i and, consequently, we obtain Raoult's law for the ideal vapor:

$$p_i = x_i p_i^0. \tag{13.44}$$

As an application of Raoult's law, we now consider the case of a binary ideal solution in equilibrium with its vapor. An example is a solution of 2-methyl propanol-1 and propanol-2. This system is almost ideal. Let us designate the solvent as component 1 and the solute as component 2. The mole fractions in the solution will be denoted by x_i, and the mole fractions in the vapor by y_i. We assume that the vapor is ideal. Then according to Raoult's law,

$$p_i = p_i^0 x_i.$$

The total pressure is

$$p = p_1 + p_2.$$

If the Raoult's law is made use of, it is given by

$$p = p_1^0 + (p_2^0 - p_1^0)x_2. \tag{13.45}$$

Since according to the Gibbs–Dalton law

$$p_i = y_i p, \tag{13.46}$$

it is possible to express the total pressure in terms of the mole fraction of the solute in the vapor, y_2, as follows:

$$(p_2^0 - p_1^0)x_2 = \frac{(p_2^0 - p_1^0)x_2 p_2^0}{p_2^0} = \frac{(p_2^0 - p_1^0)p_2}{p_2^0}$$

$$= \frac{(p_2^0 - p_1^0)y_2 p}{p_2^0}.$$

Substituting it into (13.45) and solving the resulting equation for p, we obtain

$$p = \frac{p_1^0 p_2^0}{p_2^0 + (p_1^0 - p_2^0)y_2}. \tag{13.47}$$

A phase diagram for the two-phase binary solution system is given in Fig. 13.1. The total vapor pressure p is linear with respect to x_2 according to Eq. (13.45), whereas it is concave upward with respect to y_2 as shown in Fig. 13.1. As the pressure is reduced to p_s from p, the solution attains an equilibrium with its vapor at the composition corresponding to p_v, and the concentration of component 2 in the vapor in equilibrium with the solution is y_2. As the pressure is further reduced below p', the solution is completely vaporized and the system attains a single phase consisting of vapor at that composition. This phase diagram serves as a theoretical and thermodynamic basis of fractional distillation of a liquid mixture.

Suppose there is a two-component liquid solution at pressure p and concentration x_2, as shown in Fig. 13.1. As the pressure is lowered to p_s from some point, say, p, the state of two-phase equilibrium is reached at p_s at which point the vapor phase acquires the mole fraction y_2. Collect the vapor at y_2 and condense it to the solution at another concentration x_2' which is in equilibrium with the vapor phase at concentration y_2'. This process can be iterated until the liquid phase acquires a desired composition $y_2', y_3', \ldots$, until it asymptotically reaches pure vapor of component 2. This

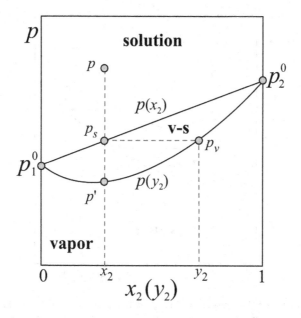

Fig. 13.1.   A phase diagram for a two-component system in $p - x(y)$ plane.

process is called fractional distillation. The process of fractional distillation is illustrated in Fig. 13.2.

Next, we consider the **lever rule** determining the decomposition in the two-phase region between the curves in phase diagram in Fig. 13.3. A point in the two-phase region of a phase diagram such as in Fig. 13.3 clearly reveals the coexistence of the vapor and liquid phases. The relative amounts of each phase can be quantitatively determined as follows. The total number of moles of all the components, n, is given by the following sum of the moles in solution and vapor phases, n_s and n_v, respectively. Therefore,

$$n = n_1^{(1)} + n_2^{(1)} + n_1^{(2)} + n_2^{(2)}$$

$$= n_s + n_v. \tag{13.48}$$

Let the total mole fraction of component 2 at arbitrary point Z between points x_2 and y_2 in Fig. 13.3 be denoted by z_2

$$z_2 = \frac{n_2^{(1)} + n_2^{(2)}}{n}. \tag{13.49}$$

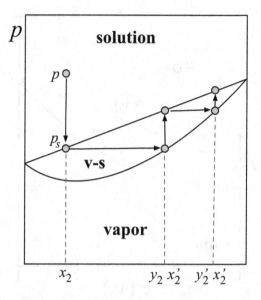

Fig. 13.2. Fractional distillation is illustrated beginning from the solution of concentration x_2 of component 2 to a desired concentration of 2. At vapor composition y_2, the solution composition is x'_2 and the equilibrium vapor composition at x'_2 is y'_2, which is higher than y_2. This process is repeated until the desired enriched composition is reached.

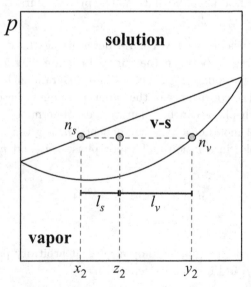

Fig. 13.3. The lever rule determining the composition of arbitrary point z in the two-phase region.

The total number of moles of component 2 can then be written as $n_2 = nz_2$ and can further be written as

$$n_2 = nz_2 = n_s z_2 + n_v z_2. \tag{13.50}$$

On the other hand, we have the following relation for n_2:

$$nz_2 = n_s x_2 + n_v y_2. \tag{13.51}$$

By equating Eqs. (13.50) and (13.51), we obtain

$$n_s l_s = n_v l_v, \tag{13.52}$$

where

$$l_s = (x_2 - z_2), \quad l_v = (y_2 - z_2) \tag{13.53}$$

with l_s and l_v denoting the distances from z_2 as shown in Fig. 13.3. Equation (13.52) is known as the lever rule.

13.4. Two-Component, Two-Phase Equilibrium Reconsidered

In Chapter 8, we have considered phase equilibria between two phases composed of a binary mixture and derived a set of differential equations comparable to the Clapeyron equation. In particular, we have derived a pair of differential equations for concentration dependences of vapor pressure. Integration of the differential equations for real solutions is rather complex and cannot be generally performed analytically even if the chemical potentials are completely known as a function of concentrations. However, it is possible to integrate them analytically for ideal solutions. In this section, we consider a binary ideal solution which is in equilibrium with its vapor. Since the governing equations can be solved analytically, it makes a good illustrative example for the application of the theory of two-phase equilibria of a binary mixture developed in Chapter 8. Therefore, we take up the same subject discussed in the previous section and derive the vapor pressure equations (13.45) and (13.47) by integrating Eq. (8.72). The object under consideration here is two phases of a binary solution and vapor which are in equilibrium at a given T. The vapor and the solution are assumed to be ideal.

Let us designate the vapor phase as phase 2 and the solution phase as phase 1. In order to simplify the notation, we will write

$$x_1 = x_1^{(1)}, \quad x_2 = x_2^{(1)},$$
$$y_1 = x_1^{(2)}, \quad y_2 = x_2^{(2)}.$$

Equation (8.72) now reads

$$\left(\frac{\partial p}{\partial y_2}\right)_T = -\frac{\sum_{i=1}^{2} x_i \phi_i^{(2)}}{\sum_{i=1}^{2} x_i(\bar{v}_i^{(2)} - \bar{v}_i^{(1)})},$$

$$\left(\frac{\partial p}{\partial x_2}\right)_T = -\frac{\sum_{i=1}^{2} y_i \phi_i^{(1)}}{\sum_{i=1}^{2} y_i(\bar{v}_i^{(1)} - \bar{v}_i^{(2)})}. \tag{13.54}$$

Since the vapor is assumed to be ideal, the partial molar volumes of the vapor are equal to the molar volumes, which are given by

$$\bar{v}_1^{(2)} = \frac{RT}{p} \quad \text{and} \quad \bar{v}_2^{(2)} = \frac{RT}{p}.$$

Moreover, it is reasonable to assume

$$\bar{v}_i^{(2)} \gg \bar{v}_i^{(1)} \quad (i = 1, 2).$$

Under these approximations, the denominators in Eq. (13.54) are very much simplified:

$$\sum_{i=1}^{2} x_i(\bar{v}_i^{(2)} - \bar{v}_i^{(1)}) \simeq \sum_{i=1}^{2} x_i \bar{v}_i^{(2)} = \frac{RT}{p} \sum_{i=1}^{2} x_i = \frac{RT}{p},$$

$$\sum_{i=1}^{2} y_i(\bar{v}_i^{(1)} - \bar{v}_i^{(2)}) \simeq -\sum_{i=1}^{2} y_i \bar{v}_i^{(2)} = -\frac{RT}{p} \sum_{i=1}^{2} y_i = -\frac{RT}{p}. \tag{13.55}$$

Since both the solution and vapor are ideal, the chemical potentials are those of ideal solutions and gases. By using such chemical potentials and the definition (8.59) for $\phi_i^{(\alpha)}$, we obtain

$$\phi_1^{(1)} = -\frac{RT}{x_1}, \quad \phi_2^{(1)} = \frac{RT}{x_2},$$

$$\phi_1^{(2)} = -\frac{RT}{y_1}, \quad \phi_2^{(2)} = \frac{RT}{y_2}. \tag{13.56}$$

When these results are used, the numerators in Eq. (13.54) take the following simple forms:

$$\sum_{i=1}^{2} x_i \phi_i^{(2)} = \frac{RT(x_2 - y_2)}{y_1 y_2},$$

$$\sum_{i=1}^{2} y_i \phi_i^{(1)} = -\frac{RT(x_2 - y_2)}{x_1 x_2}. \tag{13.57}$$

Upon substitution of Eqs. (13.55) and (13.56) into Eq. (13.57), there follows a pair of relatively simple pressure–concentration equations

$$\left(\frac{\partial p}{\partial x_2}\right)_T = -p\frac{x_2 - y_2}{x_1 x_2},$$

$$\left(\frac{\partial p}{\partial y_2}\right)_T = -p\frac{x_2 - y_2}{y_1 y_2}. \tag{13.58}$$

When these two equations are added side by side, there follows a partial differential equation for pressure

$$x_1 x_2 \left(\frac{\partial p}{\partial x_2}\right)_T - y_1 y_2 \left(\frac{\partial p}{\partial y_2}\right)_T = 0. \tag{13.59}$$

This partial differential equation is solved as follows. The characteristic equations[3] for the partial differential equation (13.59) are

$$\frac{dx_2}{dt} = x_1 x_2,$$

$$\frac{dy_2}{dt} = -y_1 y_2, \tag{13.60}$$

$$\frac{dp}{dt} = 0,$$

where t is a parameter. The first two equations imply the ordinary differential equations, which may be combined to the equation

$$\frac{dx_2}{x_2(1 - x_2)} = -\frac{dy_2}{y_2(1 - y_2)}. \tag{13.61}$$

This equation is easily integrated to yield the solution

$$C\frac{x_2}{1 - x_2} = \left(\frac{y_2}{1 - y_2}\right)^{-1}, \tag{13.62}$$

where C is the integration constant. Solving this equation for y_2, we find

$$y_2 = \frac{Cx_2}{1 + (C - 1)x_2}. \tag{13.63}$$

By integrating the third characteristic equation, we trivially find

$$p = A, \tag{13.64}$$

---

[3] See, for example, F. John, *Partial Differential Equations* (Springer, Berlin, 1978).

where A is an integration constant. It may be regarded as a function of x_2 alone because of the relation (13.63) between x_2 and y_2. Differentiating it with x_2, we find

$$\frac{dA}{dx_2} = \frac{dp}{dx_2} = -\frac{x_2 - y_2}{x_1 x_2} p, \tag{13.65}$$

for which we have made use of the first equation in Eq. (13.58). Eliminating y_2 on the right-hand side of this equation by using Eq. (13.63), we obtain

$$\frac{dp}{dx_2} = \frac{(C-1)p}{1 + (C-1)x_2}.$$

Integrating this equation, we finally obtain

$$p = B[1 + (C-1)x_2], \tag{13.66}$$

where B is the integration constant. To fix the integration constants, we consider the case of $x_2 = 0$. Then the pressure will be equal to p_1^0, since the system consists of component 1 alone in that case. We find

$$B = p_1^0. \tag{13.67}$$

If $x_2 = 1$, then $p = p_2^0$, which implies that C is given by the expression

$$C = \frac{p_2^0}{p_1^0}. \tag{13.68}$$

Eliminating B and C from Eq. (13.66) with Eqs. (13.67) and (13.68), we finally obtain the following formula for p:

$$p = p_1^0 \left[1 + \frac{(p_2^0 - p_1^0)x_2}{p_1^0} \right]. \tag{13.69}$$

Replacing C in Eq. (13.63) with Eq. (13.68) and inverting the relationship of y_2 and x_2, we obtain

$$x_2 = \frac{y_2 p_1^0}{p_2^0 + (p_1^0 - p_2^0)y_2}.$$

When this is substituted into Eq. (13.69), there follows the formula for $p(y_2)$:

$$p(y_2) = \frac{p_1^0 p_2^0}{p_2^0 + (p_1^0 - p_2^0)y_2}. \tag{13.70}$$

These are the results we are looking for. We see that Eqs. (13.69) and (13.70) are precisely what we have obtained in the previous section, namely, Eqs. (13.45) and (13.47), by using a more elementary method. To be sure, the present method is more complicated and circumspect, but its aim is to illustrate a method of solution for Eq. (8.72) or (8.71) in the case of real solutions, which must be solved by one means or another for the pressure–concentration relations of real solutions. In the case of real solutions, the principle and method are the same, but mathematically more complicated because of the more complicated concentration dependence of chemical potentials. Such equations may be solved numerically on a computer, as is often done, for example, in material sciences or engineering.

13.5. Margules Expansions

The thermodynamics of real solutions requires activity coefficients expressed as functions of concentrations. Activity coefficients for binary solutions may be expanded in power series in mole fractions:

$$RT \ln f_1 = A_1 x_2 + B_1 x_2^2 + C_1 x_2^3 + \cdots,$$
$$RT \ln f_2 = A_2 x_1 + B_2 x_1^2 + C_2 x_1^3 + \cdots. \tag{13.71}$$

These expansions are equivalent to the virial expansion for a gas and are called the Margules expansions. The expansion coefficients, which may be temperature- and pressure-dependent, are not all independent since the chemical potentials must satisfy the Gibbs–Duhem equation. Therefore, it is possible to eliminate dependent coefficients by making use of the Gibbs–Duhem equation. If T and p are kept constant, the Gibbs–Duhem equation for a binary solution may be written as

$$RT x_1 d \ln f_1 + RT x_2 d \ln f_2 = 0, \tag{13.72}$$

when Eq. (13.4) is made use of for the chemical potentials. By using the identity $x_1 + x_2 = 1$ and substituting the Margules expansions into the aforementioned equation, we obtain the equation

$$A_2 + (2B_2 - A_2)x_1 + (3C_2 - 2B_2)x_1^2 - 3C_2 x_1^3 + \cdots \tag{13.73}$$
$$= (A_1 + 2B_1 + 3C_1)x_1 - (2B_1 + 6C_1)x_1^2 + 3C_1 x_1^3 + \cdots.$$

Comparison of the terms of like power on both sides of the equation yields the following set of algebraic equations from the first order through the third in x_1

$$A_2 = 0,$$

$$2B_2 - A_2 = A_1 + 2B_1 + 3C_1,$$

$$3C_2 - 2B_2 = -(2B_1 + 6C_1),$$

$$C_2 = -C_1,$$

$$\vdots$$

When these equations are solved, there follows the following relations between coefficients:

$$A_1 = A_2 = 0,$$

$$B_2 = B_1 + \frac{3}{2}C_1,$$

$$C_2 = -C_1, \tag{13.74}$$

$$\vdots$$

Therefore, the Margules expansions now take the forms

$$RT \ln f_1 = B_1 x_2^2 + C_1 x_2^3 + \cdots ,$$

$$RT \ln f_2 = \left(B_1 + \tfrac{3}{2}C_1\right) x_1^2 - C_1 x_1^3 + \cdots . \tag{13.75}$$

Note that the expansions are consistent with Convention I which we have chosen for the reference states of chemical potentials (see Section 11.4). The coefficients can be determined by measuring various properties of the solution. They are generally dependent on temperature and pressure.

13.6. Regular Solutions

Statistical mechanical considerations show that if a mixture of two liquids is to behave ideally, the two liquids must consist of similar molecules. The forces acting on any molecule are then quite similar to the forces in the pure liquids. Under these conditions, the partial vapor pressure or the fugacity of each component, which is a measure of molecular tendency to escape from the solution, is expected to obey Raoult's law. Such is actually the case for some solutions, since liquid solutions known to behave ideally consist of similar molecules, for example, ethylene bromide and propylene bromide, n-hexane and n-heptane, n-butyl chloride and bromide, ethyl bromide and iodide, benzene and toluene, and so on. If the constituents of a mixture differ appreciably in their nature, a deviation from ideal behavior is generally the rule. For example, the vapor pressures show either positive or negative deviations from Raoult's law for such solutions. Here, we discuss a special class of nonideal solutions before we take up the subject of real solutions.

In 1929, J. H. Hildebrand observed that there is a great similarity in the thermodynamic behavior of a class of nonideal solutions which are characterized by the absence of hydrogen bonding and acid–base association. He called them regular solutions. Regular solutions differ from ideal solutions in that the intermolecular forces between different components in the solution are no longer similar and molecules may be unequal unlike in the case of ideal solutions. Nevertheless, the differences are sufficiently small so that the entropy of mixing is nearly that of ideal solutions. We shall therefore define regular solutions as follows: *a solution is called regular if the partial molar entropies of components in the solution are those of ideal solutions*

$$\bar{s}_i = s_i^l - R \ln x_i. \tag{13.76}$$

This implies that for regular solutions, the entropy of mixing is that of ideal solutions since

$$\Delta s = \sum_{i=1}^{c} x_i (\bar{s}_i - s_i^l) = -R \sum_{i=1}^{c} x_i \ln x_i. \tag{13.77}$$

This definition of regular solutions gives rise to a number of characteristics. We list them here.

1. *The partial molar specific heat of component i is equal to the molar specific heat of pure liquid i:*

$$\bar{c}_{pi} = c_{pi}^l.$$

This follows from Eq. (13.76) since

$$\bar{c}_{pi} = T \left(\frac{\partial \bar{s}_i}{\partial T} \right)_{p, x_i} = T \left(\frac{\partial s_i^l}{\partial T} \right)_p = c_{pi}^l.$$

2. *For regular solutions, the activity coefficients are given in terms of enthalpies.*

This is in contrast to real solutions for which the chemical potentials are given in terms of activity coefficients. Since the chemical potentials for pure liquids may be written as

$$\mu_i^l = h_i^l - T s_i^l,$$

we find

$$\mu_i = \bar{h}_i - T\bar{s}_i$$
$$= \mu_i^l + RT \ln x_i + \bar{h}_i - h_i^l. \tag{13.78}$$

When this is compared with Eq. (13.4) in an appropriate convention for the reference state, we identify

$$f_i = \exp\left(\frac{\bar{h}_i - h_i^l}{RT}\right). \tag{13.79}$$

This equation suggests that it is possible to calculate the activity coefficients from the calorimetric data on the solution if the solution is regular. We will elaborate on this point a little later.

3. *The partial molar volumes of regular solutions are given by the pressure derivatives of* $\bar{h}_i - h_i^l$.

From Eq. (13.78), we find

$$\bar{v}_i - v_i^l = \left(\frac{\partial \mu_i}{\partial p}\right)_{T, x_i} - \left(\frac{\partial \mu_i^l}{\partial p}\right)_T$$
$$= \left[\frac{\partial(\bar{h}_i - h_i^l)}{\partial p}\right]_{T, x_i}. \tag{13.80}$$

Therefore, the volume change of mixing is given by the formula

$$\Delta v = \sum_{i=1}^{c} x_i(\bar{v}_i - v_i)$$
$$= \left(\frac{\partial \Delta h}{\partial p}\right)_T$$
$$= RT \sum_{i=1}^{c} x_i \left(\frac{\partial}{\partial p} \ln f_i\right)_{T, x_i}, \tag{13.81}$$

where Δh is the heat of mixing.

Differentiating Eq. (13.80) with respect to T, we find

$$\left(\frac{\partial \bar{v}_i}{\partial T}\right)_{p,x_i} - \left(\frac{\partial v_i^l}{\partial T}\right)_p = \frac{\partial}{\partial T}\left[\frac{\partial}{\partial p}(\bar{h}_i - h_i^l)\right]_{T,x_i}$$

$$= \frac{\partial}{\partial p}(\bar{c}_{pi} - c_{pi}^l)$$

$$= 0, \tag{13.82}$$

the last equality being due to property 1 listed above. Therefore, we obtain

$$\left(\frac{\partial \Delta v}{\partial T}\right)_{p,x} = 0 \tag{13.83}$$

for regular solutions.

For regular solutions, the Margules expansions become expansions for partial molar enthalpies. Because of the relation (13.79) between the activity coefficients and partial molar enthalpies, we obtain

$$\bar{h}_1 - h_1^l = B_1 x_2^2 + C_1 x_2^3 + \cdots,$$

$$\bar{h}_2 - h_2^l = \left(B_1 + \frac{3}{2}C_1\right)x_1^2 - C_1 x_1^3 + \cdots. \tag{13.84}$$

Therefore, it is possible to determine the coefficients from calorimetric data of the solution in the case of a regular solution. If the solution is dilute, then it is possible to neglect the cubic or higher terms in Eq. (13.84) in which case we may write

$$\bar{h}_1 - h_1^l = B_1 x_2^2, \quad \bar{h}_2 - h_2^l = B_1 x_1^2. \tag{13.85}$$

If these approximations are used for calculating the heat of mixing for a binary regular solution, there follows a rather simple formula for the heat of mixing

$$\Delta h = \sum_{i=1}^{2} x_i(\bar{h}_i - h_i^l) = B_1 x_1 x_2. \tag{13.86}$$

To this approximation, B_1 therefore is seen as four times the equimolar heat of mixing:

$$B_1 = 4\Delta h\left(x_1 = x_2 = \frac{1}{2}\right). \tag{13.87}$$

This clearly provides a way to measure B_1 if C_1 is negligible.

13.7. Real Solutions

Most solutions are neither ideal nor regular, and thus possess none of the simplifying features of ideal and regular solutions. Here, we study real solutions. We shall pay particular attention to dilute real solutions because of its relative simplicity which, nevertheless, does not lose the essential features of real solutions.

If two or more constituents markedly different in their nature are mixed together, the solution shows noticeable deviations from the ideal behavior if the concentrations of the solutes are finite. For example, the vapor pressures show either positive or negative deviations. If the deviation is positive, the partial pressure or fugacity of each constituent is greater than it should be if Raoult's law were obeyed, and if the deviation is negative, it is just the opposite. It has been experimentally observed that as the mole fraction of a given constituent of a solution tends to unity, the fugacity of the constituent approaches the value for an ideal solution. That is, in a dilute solution, the behavior of the solvent approaches that required by Raoult's law, although it may depart markedly from the ideal behavior at higher concentrations. Since the approach to the ideal behavior is asymptotic, it is not expected that Raoult's law would hold for the solute as well unless the system as a whole exhibits no deviation from the ideal behavior over the whole range of composition. Although the solute in the dilute solution does not necessarily obey Raoult's law, as $x_1 \to 1$, it does conform to the simple relation

$$p_{f2} = kx_2, \qquad (13.88)$$

where k is a constant. This constant becomes the fugacity p_{f2}^0 of the pure solute if the solution is ideal. The aforementioned expression embodies Henry's law for the solutes. We may thus define *a dilute solution as one in which the solvent obeys Raoult's law whereas the solutes obey Henry's law*. It is possible to see that Eq. (13.88) must hold for the solute. We consider it for a binary solution.

At constant T and p, the Gibbs–Duhem equation for the solution is

$$x_1 d\mu_1 + x_2 d\mu_2 = 0, \qquad (13.89)$$

where x_1 and x_2 are the mole fractions of the solvent and the solute. Since the chemical potentials of the species in the vapor phase are given by

$$\mu_1^v = \mu_1^*(T) + RT \ln p_{f1},$$
$$\mu_2^v = \mu_2^*(T) + RT \ln p_{f2}, \qquad (13.90)$$

and by the equilibrium conditions

$$\mu_1 = \mu_1^v \quad \text{and} \quad \mu_2 = \mu_2^v,$$

the Gibbs–Duhem equation (13.89) may be written as

$$\frac{\partial \ln p_{f1}}{\partial \ln x_1} = \frac{\partial \ln p_{f2}}{\partial \ln x_2}. \tag{13.91}$$

If Raoult's law holds for the solvent, namely, component 1, then

$$\frac{\partial \ln p_{f1}}{\partial \ln x_1} = 1,$$

and there follows from Eq. (13.90) the equation

$$d \ln p_{f2} = d \ln x_2. \tag{13.92}$$

Integration of this differential equation yields

$$p_{f2} = k x_2, \tag{13.93}$$

where k is a constant. This is Henry's law for the solute. Therefore, we see that Henry's law is the necessary condition for the solvent to show the ideality (Figs. 13.4 and 13.5).

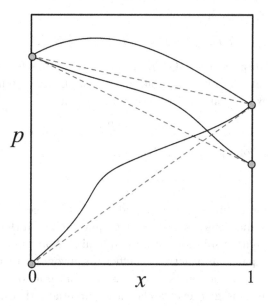

Fig. 13.4. Schematic display of positive deviation from the ideal behavior.

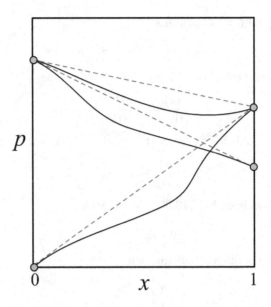

Fig. 13.5. Schematic display of a negative deviation from the ideal behavior.

For binary solutions, the Margules expansions can still be applied. If the solution is dilute, the expansions may be truncated, for example, they may be limited to the first term:

$$RT \ln f_1 = B x_2^2, \quad RT \ln f_2 = B x_1^2, \tag{13.94}$$

where we have dropped the subscript from the coefficient B. As an improvement of the treatment of binary solutions by van Laar, Scatchard proposed the following expressions for the activity coefficients:

$$RT \ln f_1 = A v_1 \phi_2^2, \quad RT \ln f_2 = A v_2 \phi_1^2, \tag{13.95}$$

where ϕ_i is the volume fraction of species i

$$\phi_i = \frac{n_i v_i}{n_1 v_1 + n_2 v_2} \quad (i = 1, 2), \tag{13.96}$$

v_i are the molar volumes of pure liquids, and A is a constant parameter. The Scatchard equations (13.95) correlate quite well with the experimental data for many binary solutions of normal liquids. Note that the Scatchard equations are expansions of activity coefficients in volume fractions, whereas the Margules expansions are expansions in mole fractions. In any case, since the expansions give excess chemical potentials, thermodynamics of real solutions ultimately amounts to measuring activity coefficients as a function of temperature, pressure, and concentrations in appropriate units. Scatchard equations turn out to be appropriate to use for macromolecular solutions because mole fractions are a misleading measure of concentration effects for thermodynamic properties of such solutions. The reason is that macromolecules have a large number of monomer units which contribute to thermodynamic properties as if they are independent molecules. Therefore, if there are molecules with large molecular weights, it is more appropriate to use volume fractions instead of mole fractions to measure deviations from the ideal behavior.

If one activity coefficient of a binary solution is known, the other activity coefficient can be obtained from it by making use of the Gibbs–Duhem equation. If T and p are kept constant, there follows from Eqs. (13.90) and (13.91) the equation

$$x_1 d \ln f_1 = -x_2 d \ln f_2. \qquad (13.97)$$

Integration of this equation gives either f_1 or f_2 in terms of the other. We first find that one of the Margules expansions implies the other. Since $f_2 \to 1$ as $x_2 \to 1$ according to the convention adopted for activity coefficients (Convention I), integrating the aforementioned equation yields

$$\ln f_2 = -\int_1^{x_2} dx_2 \, \frac{x_1}{x_2} \frac{\partial \ln f_1}{\partial x_2} = \int_0^{1-x_2} dx_1 \, \frac{x_1}{x_2} \frac{\partial \ln f_1}{\partial x_2}. \qquad (13.98)$$

Substitution of the Margules expansion for $\ln f_1$ into the right-hand side and integration of the resulting series yield the series

$$RT \ln f_2 = \left(B_1 + \frac{3}{2} C_1 \right) x_1^2 - C_1 x_1^3 + \cdots. \qquad (13.99)$$

This is the other expansion of Eq. (13.75). The reverse is also true. If the information on the solute is more readily available, the procedure used here may be reversed and we find

$$\ln f_1 = -\int_0^{x_2} \frac{x_1}{x_2} d \ln f_2. \qquad (13.100)$$

If an analytic formula for $\ln f_2$ is not available, the right-hand side can be graphically integrated by plotting x_2/x_1 against $\ln f_2$ as in Fig. 13.6 and then measuring the area under the curve.

We now examine the entropy of mixing for real solutions with the example of a binary real solution. We will use the Margules expansions for the

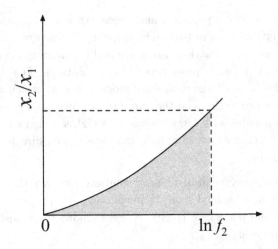

Fig. 13.6.   Graphical integration for $\ln f_2$.

activity coefficients. The chemical potentials are then given by the formulas

$$\mu_1 = \mu_1^l(T, p) + RT \ln x_1 + B_1 x_2^2 + C_1 x_2^3 + \cdots,$$
$$\mu_2 = \mu_2^l(T, p) + RT \ln x_2 + B_2 x_1^2 + C_2 x_1^3 + \cdots, \qquad (13.101)$$

where B_2 and C_2 are related to B_1 and C_1 as in Eq. (13.74). The coefficients are functions of T and p in general. Here, Convention I is used for the chemical potentials. The partial molar entropies are then given by

$$\bar{s}_1 = s_1^l - R \ln x_1 - \left(\frac{\partial B_1}{\partial T}\right)_p x_2^2 - \left(\frac{\partial C_1}{\partial T}\right)_p x_2^3 - \cdots,$$

$$\bar{s}_2 = s_2^l - R \ln x_2 - \left(\frac{\partial B_2}{\partial T}\right)_p x_1^2 - \left(\frac{\partial C_2}{\partial T}\right)_p x_1^3 - \cdots. \qquad (13.102)$$

Therefore, the entropy of mixing for the binary solution is

$$\Delta s = -R \sum_{i=1}^{2} x_i \ln x_i - \left(\frac{\partial B_1}{\partial T}\right)_p x_1 x_2$$

$$- \frac{1}{2}\left(\frac{\partial C_1}{\partial T}\right)_p (1 + x_2)\, x_1 x_2 + \cdots. \qquad (13.103)$$

The first term on the right-hand side is positive semidefinite, but the rest of the terms is not positive unless the derivatives of B_1 and C_1 are all negative. Therefore, unless the aforementioned derivatives are all negative or the first term on the right-hand side is larger than the rest of the terms, the entropy of mixing for an isolated system cannot be concluded to be positive semidefinite, as is for the case of ideal solutions formed in the same condition. The terms involving the derivatives simply give a measure of the temperature effect. This analysis amplifies the statement made in Chapter 11 and Section 13.2 in connection with the entropy of mixing and the second law of thermodynamics which is often illustrated with the positive entropy of mixing.

13.8. Osmotic Coefficient of Bjerrum

Since Eq. (13.100) is not in a convenient form to integrate numerically because of the singular behavior of x_2^{-1}, it is useful to cast it into a more suitable form. This can be achieved by using the osmotic coefficient introduced by Bjerrum. It is defined by

$$\phi = -r^{-1}\ln a_1, \tag{13.104}$$

where

$$r = \frac{x_2}{x_1}. \tag{13.105}$$

Rearrangement of Eq. (13.104) and taking a differential yields

$$d\ln a_1 = -rd\phi - \phi dr = -rd\phi - r\phi d\ln r. \tag{13.106}$$

The Gibbs–Duhem equation for a two-component system at constant T and p is

$$x_2 d\ln a_2 = -x_1 d\ln a_1. \tag{13.107}$$

By combining (13.106) and (13.107), we find a differential form for a_2:

$$d\ln a_2 = d\phi + \phi d\ln r. \tag{13.108}$$

Since

$$a_2 = x_2 f_2,$$

and, as $x_2 \to 0$,

$$f_2 \to 1 \quad \text{and} \quad x_1 \to 1,$$

we find

$$\frac{a_2}{r} = \frac{x_2 f_2}{x_2} x_1 = x_1 f_2 \to 1. \tag{13.109}$$

Therefore, it is useful to recast Eq. (13.108) in the form

$$d\ln\left(\frac{a_2}{r}\right) = d\phi + (\phi - 1)d\ln r. \tag{13.110}$$

When this equation is integrated from $r = 0$ to an arbitrary value of r, we find

$$\ln\left(\frac{a_2}{r}\right) = \phi(r) - \phi(0) + \int_0^r dr\frac{\phi - 1}{r}, \tag{13.111}$$

for which we have made use of the equality

$$\ln\left(\frac{a_2}{r}\right)\bigg|_{r=0} = 0.$$

This holds since $a_2 \to x_2$ as $r \to 0$, namely, as $x_2 \to 0$. We now examine the behavior of ϕ as $r \to 0$. Since we have

$$RT\ln f_1 = Bx_2^2 + \cdots,$$

and

$$\ln x_1 = \ln\left(\frac{x_1}{x_1 + x_2}\right) = -\ln(1 + r),$$

the osmotic coefficient ϕ may be put in the form

$$\phi = -r^{-1}\ln a_1$$
$$= r^{-1}\ln(1 + r) + k'x_1x_2 + \cdots, \tag{13.112}$$

where

$$k' = -\frac{B}{RT}.$$

It is also possible to write

$$x_1x_2 = rx_1^2$$
$$= r(1 + r)^{-2}$$
$$= r(1 - 2r + \cdots).$$

Upon substituting this result and expanding $\ln(1+r)$ into the Taylor series, we obtain the following limiting form for ϕ:

$$\phi = 1 + \left(k' - \frac{1}{2}\right)r + O(r^2).$$

This implies that

$$\lim_{r \to 0}\frac{\phi - 1}{r} = k' - \frac{1}{2}, \tag{13.113}$$

and

$$\phi(0) = 1. \tag{13.114}$$

Therefore, the integral of Eq. (13.106) is convergent at $r = 0$.

The aforementioned result may be easily adapted for data in the units of molality. For a solvent of molecular weight M_1,

$$r = \frac{mM_1}{1000},$$

and

$$\phi = -\frac{1000}{mM_1} \ln a_1.$$

With these results, it is then easy to find

$$\ln \gamma_2 = \ln \left(\frac{a_2}{m} \right)$$

$$= \phi - 1 + \int_0^m dm \, \frac{\phi - 1}{m}. \tag{13.115}$$

This equation forms a basis of measurements for activity coefficients which will be discussed in later sections.

As we have seen with regular solutions, studies of changes in thermo-dynamic functions can provide information on parameters appearing in the expansions such as the Margules expansions or the Scatchard equations for activity coefficients. Studies generally involve the free energy, enthalpy, and entropy of mixing which show marked deviations from ideal behavior. Here, we calculate them in the case of binary solutions.

We define the Gibbs free energy of mixing by the equation

$$\Delta g = \sum_{i=1}^2 x_i(\mu_i - \mu_i^l). \tag{13.116}$$

Substitution of chemical potentials (13.90) produces it in the form

$$\Delta g = RT \sum_{i=1}^2 x_i \ln x_i + RT \sum_{i=1}^2 x_i \ln f_i. \tag{13.117}$$

The first term on the right-hand side of Eq. (13.117) is the Gibbs free energy of mixing for the ideal solution and the second term represents the deviation from the ideal behavior. We call it the excess Gibbs free energy of mixing:

$$\Delta g_{\text{ex}} = RT \sum_{i=1}^{2} x_i \ln f_i. \tag{13.118}$$

The excess free energy and other excess quantities are the quantities to which we will pay attention in the following. If the Margules expansions (13.94) are substituted into Eq. (13.118), we find the excess free energy in a relatively simple form:

$$\Delta g_{\text{ex}} = B x_1 x_2. \tag{13.119}$$

We observe that this is in the same form as the heat of mixing (13.86) for regular solutions. The coefficient B is generally dependent on temperature, although the dependence is weak. The excess entropy of mixing is then derived by taking the temperature derivative of Δg_{ex}:

$$\Delta s_{\text{ex}} = -x_1 x_2 \frac{dB}{dT}. \tag{13.120}$$

The entropy of mixing is given by the expression

$$\Delta s = -R \sum_{i=1}^{2} x_i \ln x_i - x_1 x_2 \frac{dB}{dT}. \tag{13.121}$$

This is an approximation of the more general formula in Eq. (13.103). If $B(T)$ is independent of T, the entropy of mixing is then that of an ideal solution.

The heat of mixing can be derived by using Eqs. (13.117), (13.119), and (13.121) in the relation

$$\Delta h = \Delta g + T \Delta s.$$

It gives rise to the equation

$$\Delta h = x_1 x_2 \left(B - T \frac{dB}{dT} \right). \tag{13.122}$$

Since the heat of mixing is equal to zero for ideal solutions, the entire Δh is an excess quantity unlike other excess quantities. If Δh is differentiated with T, we obtain the change of heat capacity upon mixing:

$$\Delta C_{pm} = -x_1 x_2 T \frac{d^2 B}{dT^2}. \tag{13.123}$$

Similar quantities can be derived by using the Scatchard equations instead of the Margules expansions. We find, by using Eq. (13.95), the excess free energy in the form

$$\Delta g_{ex} = Av\phi_1\phi_2, \tag{13.124}$$

where v is the mean volume

$$v = x_1v_1 + x_2v_2.$$

Since the coefficient A is also a function of temperature, the entropy of mixing can be obtained from Eq. (13.124) and the procedure for other quantities is the same as before.

Henry's law constant may be determined by measuring the solute fugacity in the limit of infinite dilution. This coefficient can be related to the coefficient B. In order to see this connection, let us return to Eq. (13.90), which may be cast in the form

$$p_{f1} = x_1 f_1 p_{f1}^0, \quad p_{f2} = x_2 f_2 p_{f2}^0, \tag{13.125}$$

where p_{fi}^0 $(i = 1, 2)$ are the fugacities of pure liquids. (see Eqs. (13.40) and (13.41)). Substituting the Margules expansions (13.90) into Eq. (13.125), we obtain

$$p_{f1} = x_1 p_{f1}^0 \exp\left(\frac{Bx_2^2}{RT}\right),$$
$$p_{f2} = x_2 p_{f2}^0 \exp\left(\frac{Bx_1^2}{RT}\right). \tag{13.126}$$

Therefore, in the limit of $x_1 \to 1$ Raoult's law

$$p_{f1} = x_1 p_{f1}^0 \tag{13.127}$$

is recovered for the solvent, whereas Henry's law

$$p_{f2} = kx_2 \tag{13.128}$$

is recovered for the solute, where

$$k = p_{f2}^0 \exp\left(\frac{B}{RT}\right). \tag{13.129}$$

Henry's law constant k can thus be determined from the coefficient B and vice versa.

13.9. Determination of Activity Coefficients

We have indicated some ways to determine activity coefficients by measuring excess thermodynamic quantities of mixing. There are other ways to determine them. These methods usually exploit phase equilibria and in this sense, they are basically different from those based on excess quantities of mixing. We discuss them in the following.

13.9.1. *Vapor Fugacity*

Relation (13.125) between the vapor fugacity and the activity coefficient points to a way to determine the activity coefficient. To make the discussion general, we cast Eq. (13.125) in a general form

$$p_{fi} = f_i x_i p_{fi}^0 \quad (i = 1, 2, \ldots, r). \tag{13.130}$$

This formula indicates that it is possible to determine the activity coefficient by measuring the fugacity at a given composition of the solution.

Let us assume that the vapor pressure is such that the Lewis–Randall rule is applicable to the fugacity. If the mole fraction of the component i in the vapor phase is y_i, then

$$p_{fi} \simeq y_i p_{fi}^0, \tag{13.131}$$

and, therefore, Eq. (13.130) reduces to the form

$$f_i \simeq \frac{y_i}{x_i}. \tag{13.132}$$

In this approximation, the activity coefficient is simply the ratio of the concentrations of the substance in the vapor phase and in the solution. It is then sufficient to measure the compositions of the solution and the vapor in order to determine the activity coefficients.

13.9.2. *Freezing Point Depression*

Two-phase equilibria between a binary solution and its solid solvent component can be used for measuring activity coefficients. For example, consider ice in equilibrium with an aqueous solution of glycerol. We designate the solvent as component 1 and the solute as component 2. If T and p are kept constant, the equilibrium condition for the phase equilibrium is

$$\mu_1^s = \mu_1,$$

where μ_1^s is the chemical potential of pure solid of component 1. Since

$$\mu_1 = \mu_1^l(T,p) + RT\ln(x_1 f_1),$$

the equilibrium condition may be written as

$$T^{-1}(\mu_1^s - \mu_1^l) = R\ln(x_1 f_1). \tag{13.133}$$

This expression does not clearly indicate how the activity coefficient depends on temperature. In order to make it a little more obvious, it is necessary to obtain an expression for the rate of change in the activity coefficient with regard to temperature. To obtain such an expression, let us differentiate Eq. (13.133) with respect to T at constant p. We then find

$$\frac{\partial}{\partial T}\ln(x_1 f_1) = \frac{\Delta h_m}{RT^2}, \tag{13.134}$$

where Δh_m is the molar heat of melting:

$$\Delta h_m = h_1^l - h_1^s.$$

We have used the relation

$$\left(\frac{\partial}{\partial T}\frac{\mu}{T}\right)_p = -\frac{h}{T^2}$$

to obtain Eq. (13.134) from Eq. (13.132). Let us denote by T_m the melting temperature at which the solid and liquid phases of pure substance 1 are in equilibrium in the absence of the solute. We now integrate Eq. (13.134) from the melting point T_m to an arbitrary temperature T to obtain

$$\ln(x_1 f_1) = \ln(x_1 f_1)|_{T=T_m} + \int_{T_m}^{T} dT' \frac{\Delta h_m}{RT'^2}. \tag{13.135}$$

From Eq. (13.133),

$$\ln(x_1 f_1)|_{T=T_m} = (RT_m)^{-1}[\mu_1^s(T_m,p) - \mu_1^l(T_m,p)], \tag{13.136}$$

but since at $T = T_m$ the solid and liquid phases of pure substance 1 are in equilibrium, the following equilibrium condition holds:

$$\mu_1^s(T_m,p) = \mu_1^l(T_m,p).$$

Consequently, by the definition of T_m, $\ln(x_1 f_1)|_{T=T_m} = 0$ identically, and Eq. (13.135) reduces to the equation

$$\ln(x_1 f_1) = \int_{T_m}^{T} dT \frac{\Delta h_m}{RT^2}. \tag{13.137}$$

Since $x_1 f_1 \leq 1$, the integral on the right is negative. Since Δh_m is positive by definition, the integrand is positive. Therefore, the only way the integral is negative is that the upper end T of the integral be less than the lower end T_m which is the melting point of pure solid of component 1, namely, the solvent:

$$T \leq T_m.$$

In other words, since T is the temperature at which the pure solid of component 1 (solvent) is in equilibrium with the solution, the presence of the solute lowers the melting point of the solid consisting of the solvent species. This phenomenon is called the freezing point depression.

Equation (13.137) provides a procedure to determine the activity coefficient by measuring the lowering of the melting point from the value of the melting point that the pure solid would have if it were in equilibrium with the pure solvent. To see this, we perform the integration by using the following expansion for the heat of melting:

$$\Delta h_m = \Delta h_m^0 + \Delta c_{pm}(T - T_m) + \cdots, \tag{13.138}$$

where

$$\Delta c_{pm} = \left(\frac{\partial \Delta h_m}{\partial T} \right)_p.$$

We also expand T^{-2}:

$$T^{-2} = T_m^{-2} \left(1 + 2\frac{\Delta T}{T_m} + \cdots \right), \tag{13.139}$$

where

$$\Delta T = T_m - T.$$

By substituting these expansions into Eq. (13.137) and performing the integration, we obtain to second order in ΔT the equation

$$\ln(x_1 f_1) = -\frac{\Delta h_m^0}{RT_m^2}\Delta T - \left(\frac{\Delta h_m^0}{RT_m} - \frac{\Delta c_{pm}}{2R} \right)\left(\frac{\Delta T}{T_m} \right)^2. \tag{13.140}$$

Here, ΔT denotes the degree of deviation in the melting point from that of the pure solvent–solid equilibrium and is called the freezing point depression. The value of ΔT depends on the concentration of the solution in equilibrium with the solid and measuring ΔT can provide the concentration dependence of f_1. If $\Delta T / T_m \ll 1$, then it is possible to neglect the second-order term in Eq. (13.140) and hence the equation is simplified:

$$\ln(x_1 f_1) = -\frac{\Delta h_m^0}{R T_m^2} \Delta T. \tag{13.141}$$

This may be used to a good approximation if the solution is dilute. Since at infinite dilution $f_1 \to 1$, it is possible to put $f_1 = 1$ to an approximation in Eq. (13.141) if the solution is sufficiently dilute:

$$\ln x_1 = -\frac{\Delta h_m^0}{R T_m^2} \Delta T. \tag{13.142}$$

Since the logarithmic function can be expanded in the Taylor series of x_2

$$\ln x_1 = \ln(1 - x_2) = -x_2 - \frac{1}{2} x_2^2 + \cdots ,$$

Eq. (13.142) may be further approximated to the form

$$x_2 \simeq \frac{\Delta h_m^0}{R T_m^2} \Delta T. \tag{13.143}$$

In this approximation, the freezing point depression is directly proportional to the concentration of the solute if the solution is sufficiently dilute. Of course, Eqs. (13.142) and (13.143) are not useful for determining the activity coefficient, but Eq. (13.143) may be employed for determining the molecular weight of the solute.

If we denote by W_1 and W_2 the weights of the solute and solvent component in the solution, because the solution is dilute, the mole fraction x_2 may then be approximated by the formula

$$x_2 \simeq \frac{n_2}{n_1} = \frac{W_2 M_1}{W_1 M_2},$$

where M_1 and M_2 are the molecular weights of the solvent and the solute, respectively. We may then write Eq. (13.143) in the form

$$M_2 = \frac{R T_m^2 M_1 W_2}{\Delta h_m^0 W_1} \Delta T^{-1}, \tag{13.144}$$

when the solution is sufficiently dilute. Strictly speaking, this formula is applicable if the solution is very dilute so that the activity coefficient is equal to unity and $\ln x_2$ may be replaced with $-x_2$, if Δh_m is constant with respect to T, and if ΔT is not too large compared with T_m.

To derive the formulas given above for freezing point depression, we have made an important assumption that the solid phase consists of the pure solvent component only. However, this may not be generally the case. It is sometimes possible that the solid phase consists of a solid solution of a composition different from that of the liquid solution. Even in this case, the equilibrium condition is still

$$\mu_1(s) = \mu_1(l),$$

where $\mu_1(s)$ and $\mu_1(l)$ are the chemical potential of component 1 in the solid and liquid phases, respectively. The former may be given in the form

$$\mu_1(s) = \mu_1^s(T,p) + RT \ln(x_1^s f_1^s), \tag{13.145}$$

where x_1^s is the mole fraction of component 1 in the solid solution, and f_1^s the corresponding activity coefficient. When Eq. (13.145) is used in the aforementioned equilibrium condition, Eq. (13.133) is now modified to the following form:

$$T^{-1}(\mu_1^s - \mu_1^l) = R \ln \left(\frac{x_1 f_1}{x_1^s f_1^s} \right). \tag{13.146}$$

By proceeding in the same manner as for Eq. (13.137), we obtain

$$\ln \left(\frac{x_1 f_1}{x_1^s f_1^s} \right) = \int_{T_m}^{T} dT \, \frac{\Delta h_m}{RT^2}. \tag{13.147}$$

If Eq. (13.138) is employed for Δh_m, we obtain from Eq. (13.147)

$$\ln \left(\frac{x_1 f_1}{x_1^s f_1^s} \right) = -\frac{\Delta h_m^0}{RT_m^2} \Delta T - \left(\frac{\Delta h_m^0}{RT_m} - \frac{\Delta c_{pm}}{2R} \right) \left(\frac{\Delta T}{T_m} \right)^2 \tag{13.148}$$

to second order in ΔT. The argument of the logarithmic function on the left-hand side represents a distribution of the solvent component between the two phases. We define the partition coefficient K_{part} for the two solutions by the ratio

$$K_{\text{part}} = \frac{x_1^s f_1^s}{x_1 f_1}. \tag{13.149}$$

Since in the case of very dilute solutions

$$f_1 \to 1 \quad \text{and} \quad f_1^s \to 1,$$

we find

$$\lim_{x_1 \to 1} K_{\text{part}} = \frac{x_1^s}{x_1} = K_{\text{part}}^0, \tag{13.150}$$

which shows that K_{part} indeed indicates the degree of partitioning of component 1 between the solid and liquid solutions. Therefore, measurement of ΔT can supply the partitioning coefficient. If the solution is sufficiently dilute and $\Delta T / T_m \ll 1$, we may approximate Eq. (13.148) as follows:

$$x_2 - x_2^s = \frac{\Delta h_m^0}{RT_m^2} \Delta T, \tag{13.151}$$

or

$$x_2 \left(1 - \frac{x_2^s}{x_2} \right) = \frac{\Delta h_m^0}{RT_m^2} \Delta T. \tag{13.152}$$

If the liquid phase is richer in component 2 so that $x_2^s / x_2 < 1$, then the left-hand side is positive and hence $\Delta T > 0$; that is, the freezing point is lowered by the presence of a solute, but if the solid phase is richer in component 2, then $x_2^s / x_2 > 1$ and consequently $\Delta T < 0$, which means that the freezing point is elevated by the presence of the solute in both the solutions. These phenomena are consequences of the second law of thermodynamics, inasmuch as the equilibrium conditions are consequences thereof.

13.9.3. *Boiling Point Elevation*

Activity coefficients may be determined from boiling point measurements. The treatment for the method of determination is completely analogous to that of the method of freezing point depression. We consider a solution in equilibrium with the vapor of the solvent at a given pressure and temperature. It is assumed that the solute is nonvolatile so that the vapor consists of the solvent species only. The solvent will be designated as component 1 and the solute as component 2. By the material equilibrium condition, the chemical potential of the solvent components in the vapor phase must be equal to its chemical potential in the solution. Thus, we have the equation

$$\ln(x_1 f_1) = \frac{\mu_1(v) - \mu_1^l(T, p)}{RT}, \tag{13.153}$$

where $\mu_1(v)$ is the chemical potential of the solvent (component 1) of the vapor in equilibrium with the solution phase. Differentiating this equation with T at constant p and x_1 yields the equation

$$\frac{\partial}{\partial T} \ln(x_1 f_1) = -\frac{\Delta \bar{h}_v}{RT^2}, \tag{13.154}$$

where

$$\Delta \bar{h}_v = \bar{h}_1^v - h_1^l, \tag{13.155}$$

with $\bar{h}_1^v$ and h_1^l denoting the partial molar and molar enthalpy of the solvent in the vapor and the solution, respectively. Since the vapor consists of component 1 (i.e., the solvent) only, $\bar{h}_1^v = h_1^v$. Therefore,

$$\Delta \bar{h}_v = \Delta h_v$$
$$= h_1^v - h_1^l. \tag{13.156}$$

Replacing $\Delta \bar{h}_v$ with Δh_v in Eq. (13.154) and integrating the equation, we obtain

$$\ln(x_1 f_1) = \ln(x_1 f_1)|_{T=T_b} - \int_{T_b}^{T} dT \frac{\Delta h_v}{RT^2}, \tag{13.157}$$

where T_b is the boiling point of the pure solvent liquid. Then, by the choice of standard state we have made for the solvent chemical potential μ_1, we find

$$\ln(x_1 f_1)|_{T=T_b} = 0. \tag{13.158}$$

Furthermore, the heat of vaporization may be expanded in temperature around T_b:

$$\Delta h_v = \Delta h_v^0(T_b) + \Delta c_{pv} \Delta T + \cdots,$$

where

$$\Delta T = T - T_b,$$

$$\Delta c_{pv} = \left(\frac{\partial \Delta h_v}{\partial T} \right)_{p, T=T_b}.$$

Substituting this expansion into Eq. (13.157) and carrying out the integration finally yields the equation

$$\ln(x_1 f_1) = -\frac{\Delta h_v^0}{RT_b^2} \Delta T + \left(\frac{\Delta h_v^0}{RT_b} - \frac{\Delta c_{pv}}{2R} \right) \left(\frac{\Delta T}{T_b} \right)^2 + \cdots, \tag{13.159}$$

for which we have also made use of the expansion

$$T^{-1} = T_b^{-1} \left(1 + \frac{\Delta T}{T_b} \right)^{-1} = T_b^{-1} \left[1 - \frac{\Delta T}{T_b} + \left(\frac{\Delta T}{T_b} \right)^2 - \cdots \right].$$

Equation (13.159) plays the same role for the boiling point elevation as Eq. (13.133) does for the freezing point depression. It can be used for determining not only the activity of the solvent but also the molecular weight of the solute in the case of sufficiently dilute binary solutions. The approximation equivalent to Eq. (13.137) is

$$x_2 = \frac{\Delta h_v^0}{RT_b^2} \Delta T, \tag{13.160}$$

which implies that $\Delta T > 0$, that is, the boiling point must be elevated owing to the presence of the solute.

Let us remove the assumption that the solute is nonvolatile. Then the vapor consists of a mixture of the same species as those comprising the solution, and the chemical potential of the solvent species in the vapor phase is given by the form

$$\mu_1(v) = \mu_1^*(T) + RT \ln p_{f1}$$
$$= \mu_1^0(T, p) + RT \ln(y_1^v f_1^v), \tag{13.161}$$

where y_1^v is the mole fraction and f_1^v is the fugacity coefficient of component 1 in the vapor. Since at equilibrium

$$\mu_1(v) = \mu_1(l)$$
$$= \mu_1^l(T, p) + RT \ln(x_1 f_1),$$

there follows the equation

$$\ln\left(\frac{x_1 f_1}{x_1^v f_1^v}\right) = \frac{\mu_1^0(T, p) - \mu_1^l(T, p)}{RT}.$$

Note that $\mu_1^0(T, p)$ is the chemical potential of the vapor consisting of component 1 only. The right-hand side may be written as before:

$$\frac{\mu_1^0(T, p) - \mu_1^l(T, p)}{RT} = -\int_{T_b}^{T} dT \frac{\Delta h_v}{RT^2}.$$

Therefore, we finally obtain

$$\ln\left(\frac{x_1 f_1}{y_1^v f_1^v}\right) = -\int_{T_b}^{T} dT \frac{\Delta h_v}{RT^2}. \tag{13.162}$$

The left-hand side is not necessarily negative since it is possible that

$$x_1 f_1 > y_1^v f_1^v.$$

In this case, the boiling point can be lowered, namely,

$$T < T_b.$$

This situation is the counterpart of the freezing point elevation which occurs when the solid phase consists of a solid solution richer than the liquid solution phase with regard to the solute species, as we have discussed in the previous section.

The effect of removing the nonvolatility assumption for the solutes may be examined in another way if Eq. (13.154) is used. Since the vapor is a mixture, the partial molar enthalpy $\bar{h}_1^v$ is no longer equal to the molar enthalpy of pure component 1, h_1^v. In this case, the $\Delta \bar{h}_v$ may be decomposed into the heat of vaporization and the heat of mixing:

$$\Delta \bar{h}_v = \Delta h_v + \Delta h_{\text{mix}},$$

$$\Delta h_v = h_1^v - h_1^l, \tag{13.163}$$

$$\Delta h_{\text{mix}} = \bar{h}_1^v - h_1^v,$$

where h_1^v denotes the molar enthalpy of the pure gaseous solvent component. Therefore, Δh_v is obviously the heat change accompanying the vaporization of 1 mole of the pure liquid solvent and Δh_{mix} represents the change in enthalpy of the solvent component when a gaseous mixture is formed with the constituents of the solution. According to our previous calculations (see Eq. (9.36))

$$\Delta h_{\text{mix}} = \int_0^p dp \left[\Delta \bar{v}_1 - T \left(\frac{\partial \Delta \bar{v}_1}{\partial T} \right)_{T, x_j} \right].$$

Upon integration of Eq. (13.154) where $\Delta \bar{h}_1$ is given by Eq. (13.163), there follows the equation

$$\ln(x_1 f_1) = - \int_{T_b}^{T} dT' \frac{\Delta h_v + \Delta h_{\text{mix}}}{RT'^2},$$

for which Eq. (13.158) is made use of. The integral on the right-hand side may be evaluated in an approximation. For example, the Lewis–Randall rule is applicable at pressures less than about 100 atm. Under this rule,

$$\Delta \bar{v}_1 = \bar{v}_1 - v_1 \simeq 0,$$

which means that

$$\Delta h_{\text{mix}} \simeq 0.$$

If the vapor pressure is not too high, the aforementioned approximation allows us to ignore Δh_{mix} and recover the result for the case of the vapor consisting of the solvent only:

$$\ln(x_1 f_1) = -\int_{T_b}^{T} dT' \, \frac{\Delta h_v}{RT'^2},$$

which is Eq. (13.154).

13.9.4. Osmotic Pressure

Let us imagine a liquid solution separated from the pure solvent by nondeformable, diathermal membrane which is permeable to the solvent molecules only (see Fig. 13.7). We assume that the two phases are in thermal equilibrium. Since one phase is a solution, the chemical potential of the solvent in the solution is generally different from that of the pure solvent if the pressures and temperatures are the same in both phases. Since we are assuming thermal equilibrium between both phases, material equilibrium of the solvent component between the two phases can be possible only if there is a difference in pressure in the two phases. This pressure difference is called the *osmotic pressure*.

We define it more precisely as follows: the osmotic pressure is the excess pressure that must be applied to a solution to prevent the passage of the solvent into the solution through a semipermeable membrane which separates the two liquids and is permeable to the solvent molecules only. Thus, if we denote the pressure on the pure solvent phase by p^0 and the osmotic pressure by π, then the pressure on the solution necessary to maintain material equilibrium is

$$p = p^0 + \pi. \tag{13.164}$$

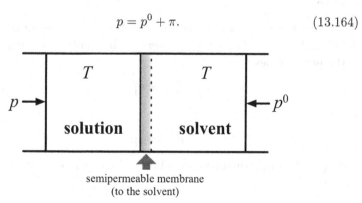

Fig. 13.7.   Schematic figure for osmosis.

This is the equilibrium condition for pressure. Since there is thermal equilibrium, the only other equilibrium condition to consider is

$$\mu_1(T, p, x_1) = \mu_1^l(T, p^0). \tag{13.165}$$

Since

$$\left(\frac{\partial \mu_1}{\partial p}\right)_{T, x_1} = \bar{v},$$

integrating it back from p^0 to p yields the equation

$$\mu_1(T, p, x_1) - \mu_1(T, p^0, x_1) = \int_{p^0}^{p} dp\, \bar{v}_1(p). \tag{13.166}$$

Because of the equilibrium condition (13.165), the left-hand side may be replaced with

$$\mu_1(T, p^0) - \mu_1(T, p^0, x_1).$$

Since the solvent chemical potential of the solution at p^0 is

$$\mu_1(T, p^0, x_1) = \mu_1^l(T, p^0) + RT \ln\left[x_1 f_1(p^0)\right] \tag{13.167}$$

and $\mu_1(T, p^0) = \mu_1^l(T, p^0)$ by the convention on the chemical potential of the pure liquid (the solvent), Eq. (13.165) is given by the form

$$RT \ln\left[x_1 f_1(p^0)\right] = -\int_{p^0}^{p} dp \bar{v}_1(p). \tag{13.168}$$

Since $x_1 f_1 \leq 1$, the left-hand side of (13.168) is negative and hence it is possible to deduce $p - p^0 = \pi \geq 0$. That is, the osmotic pressure must be positive.

To proceed further, it is necessary to have the pressure dependence of the partial molar volume of the solvent in the solution. We may expand $\bar{v}_1$ in the power series of $p - p^0 = \pi$:

$$\bar{v}_1 = \bar{v}_1^0 - \kappa_1^0 \bar{v}_1^0 \pi + \cdots, \tag{13.169}$$

where

$$\bar{v}_1^0 = \bar{v}_1(p^0),$$

and κ_1^0 is the isothermal compressibility of the pure solvent defined by the formula

$$\kappa_1^0 = -\frac{1}{\bar{v}_1^0}\left(\frac{\partial \bar{v}_1^0}{\partial p^0}\right)_T.$$

Use of Eq. (13.169) for the integral in Eq. (13.168) yields

$$RT \ln[x_1 f_1(p^0)] = -\bar{v}_1^0 \pi + \frac{1}{2}\kappa_1^0 \bar{v}_1^0 \pi^2 + \cdots, \tag{13.170}$$

for which κ_1^0 is assumed to be constant. This equation provides a way to determine the activity coefficient by measuring the osmotic pressure. Note that the activity coefficient is determined at the value of pressure p^0.

If the liquid is incompressible, then $\kappa_1^0 = 0$, and to first order in osmotic pressure Eq. (13.170) gives rise to the equation

$$RT \ln x_1 f_1 = -\bar{v}_1^0 \pi. \tag{13.171}$$

If the solution is very dilute so that $f_1 \simeq 1$, then it is possible to approximate Eq. (13.171) as follows:

$$\ln x_1 = -\bar{v}_1^0 \pi. \tag{13.172}$$

If the solution is binary and dilute, then

$$\ln x_1 \simeq -x_2$$

and, consequently, Eq. (13.172) reduces to the form

$$\pi \bar{v}_1^0 = RTx_2, \tag{13.173}$$

which is the van't Hoff equation for osmotic pressure.

If the Margules expansion is employed for $RT \ln f_1$ in the case of a binary solution, Eq. (13.171) becomes

$$\pi \bar{v}_1^0 = x_2 RT \left[1 + \left(\frac{1}{2} - \frac{B_1}{RT}\right) x_2 + \cdots \right], \tag{13.174}$$

for which we have expanded $\ln x_1$ in the Taylor series of x_2.

Measurement of the osmotic pressure can also provide the molecular weight of the solute molecule. Since $x_2 \simeq n_2/n_1$ if the solution is sufficiently dilute, we obtain from Eq. (13.173)

$$M_2 = \frac{RTW_2 M_1}{\pi \bar{v}_1^0 W_1}, \tag{13.175}$$

where we are using the same notation as in Eq. (13.144). In polymer physical chemistry, measurement of osmotic pressure is routinely made use of to determine polymer molecular weights with the help of Eq. (13.175).

13.9.5.   *Solubility*

The activity coefficients of solutes can be obtained by measuring their solubility in the solvent. In order to make our discussion simple, let us consider a binary solution which is in equilibrium with the solid solute. The solute will be designated as component 2, and the solvent as component 1. The temperature and pressure are kept constant. The remaining equilibrium condition is then

$$\mu_2^s(T,p) = \mu_2(T,p,x_2), \qquad (13.176)$$

where $\mu_2^s(T,p)$ is the chemical potential of the pure solid solute. But the solute chemical potential in the solution is given by

$$\mu_2(T,p,x_2) = \mu_2^l(T,p) + RT \ln(x_2 f_2). \qquad (13.177)$$

Combining this equation with the equilibrium condition yields

$$RT \ln(x_2 f_2) = \mu_2^s(T,p) - \mu_2^l(T,p).$$

Differentiating it with T at constant p and x_2, we obtain the differential equation

$$\left(\frac{\partial \ln x_2 f_2}{\partial T} \right)_{p,x_2} = \frac{\Delta h_m}{RT^2}, \qquad (13.178)$$

where Δh_m is the molar heat of melting of the solute:

$$\Delta h_m = h_2^l - h_2^s$$

with h_2^l and h_2^s denoting the molar enthalpy of pure liquid 2 and pure solid 2, respectively. We may now choose a temperature T_0 such that the solubility of the solute is very low and, consequently, the solution behaves practically as an ideal solution. We will denote the solubility at that temperature by x_2^0. Then at $x_2 = x_2^0$ the activity coefficient f_2 may be put equal to unity. By integrating Eq. (13.178) from T_0 to an arbitrary temperature T, we obtain the following equation for f_2:

$$\ln f_2 = \ln \left(\frac{x_2^0}{x_2} \right) + \int_{T_0}^{T} dT \, \frac{\Delta h_m}{RT^2}. \qquad (13.179)$$

The integral may be evaluated easily by applying the expansion (13.138) for Δh_m. Thus, we see that the heat of melting data of the solute and the solubility data at two temperatures enable us to determine the activity coefficient.

Equation (13.179) may be cast into a more conventional and familiar form if the solubility constant K_{sol} is defined by

$$K_{\text{sol}} = x_2 f_2. \tag{13.180}$$

By the definition of T_0 made earlier, the solubility constant K_{sol}^0 at $T = T_0$ is given by

$$K_{\text{sol}}^0 = x_2^0. \tag{13.181}$$

Then with the help of (13.180), Eq. (13.179) can be expressed in the form

$$\ln\left(\frac{K_{\text{sol}}}{K_{\text{sol}}^0}\right) = \int_{T_0}^{T} dT\,\frac{\Delta h_m}{RT^2}, \tag{13.182}$$

which is reminiscent of the van't Hoff equation for a chemical equilibrium constant.

If the solid phase is a solid solution of, say, two components 1 and 2, then the chemical potential of component 2 of the solid may be written as

$$\mu_2(s) = \mu_2^s(T, p) + RT \ln(y_2 f_2^s), \tag{13.183}$$

where y_2 is the mole fraction of component 2, and f_2^s its activity coefficient. It is then easy to show that

$$\left[\frac{\partial}{\partial T} \ln\left(\frac{x_2 f_2}{y_2 f_2^s}\right)\right]_p = \frac{\Delta h_m}{RT^2}.$$

Upon integration, this yields

$$\ln\left(\frac{x_2 f_2}{y_2 f_2^s}\right) = \ln\left(\frac{x_2 f_2}{y_2 f_2^s}\right)_{T=T_0} + \int_{T_0}^{T} dT\,\frac{\Delta h_m}{RT^2}. \tag{13.184}$$

Choose T_0 as the melting point of the pure solid solute into pure liquid. Since the activity coefficients f_2^s and f_2 in such states are equal to unity, we have

$$\lim_{x_2, y_2 \to 1} \ln\left(\frac{x_2 f_2}{y_2 f_2^s}\right)_{T=T_0} = 0.$$

Therefore, there follows the equation

$$\ln \left(\frac{x_2 f_2}{y_2 f_2^s} \right) = \int_{T_0}^{T} dT \, \frac{\Delta h_m}{RT^2}. \tag{13.185}$$

This equation describes the temperature dependence of the solubility of the binary solid solution in solvent 1. Depending on whether the ratio $(x_2 f_2 / y_2 f_2^s)$ is larger than unity or not, $T > T_0$ or $T < T_0$.

Problems

(1) Give in full the derivation of the statement that heat is evolved upon mixing two liquids which form a system exhibiting negative deviations from ideal behavior. The deviation is said to be negative if the actual partial pressure of each constituent is less than it should be if Raoult's law were obeyed. The deviation is said to be positive in the opposite case.

(2) On the basis of the lowest order Margules expansions, show that for a liquid mixture exhibiting positive deviations from Raoult's law the activity coefficient of each constituent must be greater than unity, whereas for negative deviations, it must be less than unity.

(3) The vapor pressure of liquid ethylene is 40.6 atm at 273.15 K and 24.8 atm at 253.15 K. Estimate the ideal solubility of the gas in a liquid at 298.15 K and 1 atm pressure. How many grams of ethylene should dissolve in 1 kg of benzene under these conditions if the gas and solution behave ideally.

(4) The freezing point of benzene is 278.55 K and its latent heat of melting is 126.36 J g^{-1}. A solution containing 6.054 g of triphenylmethane in 1 kg of benzene has a freezing point which is 273.2763 K below that of the pure solvent. Calculate the molecular weight of the solute.

(5) Mixtures of benzene and toluene behave almost ideally; at 303.15 K, the vapor pressure of pure benzene is 0.1182 mHg and that of pure toluene is 0.0367 mHg. Determine the partial pressures and weight composition of the vapor in equilibrium with a liquid mixture consisting of equal weights of the two constituents.

(6) For a benzene–cyclohexane mixture, the Margules expansion for the activity coefficient f_b of benzene is experimentally found to fit the following expression:

$$RT \ln f_b = (3800 - 8T)x_c^2,$$

where x_c is the mole fraction of cyclohexane. Calculate the entropy and enthalpy of the isothermal and isobaric dilution (mixing) when 1 mole of pure benzene is added to two moles of a solution containing 80 moles % cyclohexane at 300 K and 1 bar.

(7) For a sucrose solution at 273 K the osmotic pressure π is determined experimentally as a function of the concentration of sucrose as shown in Fig. 13.8. Show how such information can be used to calculate the activity coefficient of sucrose in the water–sucrose solution as a function of composition.

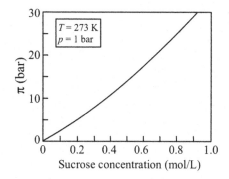

Fig. 13.8.   Osmotic pressure vs. concentration.

Chapter 14

Thermodynamics of Surfaces

In Chapter 13 on thermodynamics of phase equilibria, we have assumed that the densities of matter as well as the energy and entropy densities of phases in equilibrium are uniform up to a mathematical surface that separates the contiguous phases. This presumes, for example, densities of two contiguous phases can be discontinuous across the surface. However, in reality, this is not generally true evidently with respect to density and with respect to energy and entropy, but only an idealized representation. There exist interfaces between contiguous phases which are different from the bulk phases themselves. In this chapter, we shall discuss the thermodynamics of interfaces separating chemically inert phases. The interfacial phenomena manifest themselves more significantly with the smaller size of a phase smaller than the contiguous phases. The phenomena range over a wide ranging subject field, and surface effects cannot be ignored in numerous fields like colloids, soil science, petroleum engineering, botany, etc. Here, in this chapter, we will only present the gist of the thermodynamics of interfaces as an elementary introduction to the subject matter. For more thorough treatments, the readers are referred to numerous monographs[1] on the subjects mentioned.

Microscopically, molecules close to the interface experience an environment different from what the molecules in the bulk phase do.

---

[1]For example, see A. W. Adamson, *Physical Chemistry of Surfaces* (Wiley, New York, 1990).

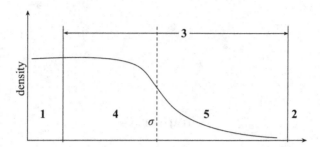

Fig. 14.1.   Interfacial profile of density. The density, for example, changes continuously from one bulk phase to the other across the interface. The numerals refer to the parts of the volume enclosing the surface σ. (see Section 14.1).

The interface generally extends over a thin region of a few or a few tens of molecular diameters in thickness between the contiguous phases, and the physical properties of the system vary continuously across the interface from one bulk phase to the other. See Fig. 14.1 for an example for density profile. Hence, if the phases are internally in equilibrium, the physical properties of phases can be assumed uniform up to the thin interfacial region.

The foundations of thermodynamics of interface and surface were essentially laid by J. W. Gibbs in his monumental work on the thermodynamics of heterogeneous phases some 140 years ago. We follow his treatment in this chapter. For a more complete treatment, we refer to his work.[2]

14.1.   Dividing Surface

We consider a surface of discontinuity in the fluid which is in equilibrium under the assumption that gravity effects are absent. In the neighborhood of the physical surface of discontinuity, choose a point and a geometrical surface of equidensity with respect to a component that passes through the point. Thus, all points on the surface correspond to that of the same local density with respect to the component chosen. This surface is called the dividing surface. It will be denoted by σ. Its position is arbitrary to some extent, but the normal directions are determined owing to the fact that all surfaces drawn in the manner prescribed have the parallel normal directions.

---

[2]J. W. Gibbs, *The Scientific Papers of J. Willard Gibbs, Vol. 1: Thermodynamics,* pp. 219–328 (Dover, New York, 1961).

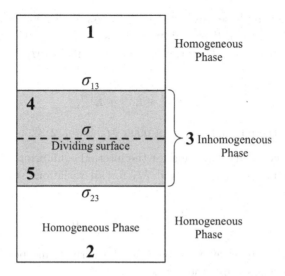

Fig. 14.2. Cross-section of the interface. The lines represented by σ_{12} and σ_{23} on both sides of σ marks the thin film. The surfaces σ_{12} and σ_{23} are placed sufficiently far into 1 and 2.

Imagine a closed surface that cuts through the dividing surface perpendicularly on each side and extends well into the homogeneous phases on each side of the dividing surface. This surface is parallel to the normal direction of σ. The space enclosed by the closed surface is then divided into three parts by placing two surfaces σ_{13} and σ_{23} placed sufficiently far from σ so that the surfaces are outside the influence of the inhomogeneous film (bounded by σ_{13} and σ_{23} in the figure) surrounding σ (see Fig. 14.2). The homogeneous parts beyond σ_{13} and σ_{23} outside the film are designated 1 and 2, and that part containing the film is 3. The corresponding masses are denoted by M_1 and M_2, and that part enclosing the dividing surface σ by M_3. The area of σ is a. The part of 3 enclosed by σ and σ_{13} is designated 4, and that part enclosed by σ and σ_{23} is designated 5. Their masses are, respectively, denoted by M_4 and M_5 which obviously make up the mass of the inhomogeneous phase M_3: $M_3 = M_4 + M_5$.

14.2. Gibbs Relation for Interface

To carry on the study of thermodynamics of interfaces, it is required to have the fundamental thermodynamic equation for interface. To this end, we now consider the interface to be diathermal and perfectly permeable to all the species in the two phases involved. *We also assume energy is assignable to each mass element and thus continuous as is the density.* It is then possible to assume that the total internal energy E of mass $(M_1 + M_2 + M_3)$ is given by

$$E = E^{(1)} + E^{(2)} + E^{(3)}, \qquad (14.1)$$

where $E^{(i)}$ is the internal energy of mass M_i $(i = 1, 2, 3)$.

Now, the necessary condition for the internal equilibrium of the system composed of masses M_1, M_2, and M_3 for all variations subject to fixed boundaries is

$$\delta E^{(1)} + \delta E^{(2)} + \delta E^{(3)} \geq 0. \qquad (14.2)$$

The boundaries are assumed to be rigid, diathermal, and perfectly permeable. This inequality may then be written as

$$T^{(1)}\delta S^{(1)} + \sum_{j=1}^{c} \mu_j^{(1)} \delta n_j^{(1)} + T^{(2)}\delta S^{(2)} + \sum_{j=1}^{c} \mu_j^{(2)} \delta n_j^{(2)}$$

$$+ T^{(3)}\delta S^{(3)} + \sum_{j=1}^{c} \mu_j^{(3)} \delta n_j^{(3)} \geq 0. \qquad (14.3)$$

This variation is subject to the constraints

$$\delta S^{(1)} + \delta S^{(2)} + \delta S^{(3)} = 0,$$

$$\sum_{j=1}^{c} \delta n_j^{(\alpha)} = 0 \quad (\alpha = 1, 2, 3). \qquad (14.4)$$

The necessary and sufficient conditions for internal equilibrium of the system are then

$$T^{(1)} = T^{(2)} = T^{(3)} := T, \qquad (14.5)$$

$$\mu_j^{(1)} = \mu_j^{(2)} = \mu_j^{(3)} := \mu_j \quad (1 \leq j \leq c). \qquad (14.6)$$

This conclusion follows upon applying the method used for the equilibrium conditions of heterogeneous phases in Chapter 8.

On account of the boundaries including σ dividing region 3 into regions 4 and 5, which are assumed rigid (in other word, fixed), the volumes of

M_4 and M_5 remain constant. Therefore, by virtue of the conditions (14.5) and (14.6), while there holds for M_3 the equation

$$\delta E^{(3)} = T\delta S^{(3)} + \sum_{j=1}^{c} \mu_j \delta n_j^{(3)}, \tag{14.7}$$

for M_4 and M_5, there hold the equations

$$\delta E^{(4)} = T\delta S^{(4)} + \sum_{j=1}^{c} \mu_j \delta n_j^{(4)}, \tag{14.8}$$

$$\delta E^{(5)} = T\delta S^{(5)} + \sum_{j=1}^{c} \mu_j \delta n_j^{(5)}. \tag{14.9}$$

Upon subtracting Eqs. (14.8) and (14.9) from Eq. (14.7), we obtain the equation (differential form) for the interface

$$\delta E^s = T\delta S^s + \sum_{j=1}^{c} \mu_j \delta n_j^s, \tag{14.10}$$

where

$$E^s = E^{(3)} - E^{(1)} - E^{(2)},$$
$$S^s = S^{(3)} - S^{(1)} - S^{(2)}, \tag{14.11}$$
$$n_j^s = n_j^{(3)} - n_j^{(1)} - n_j^{(2)} \quad (1 \leq j \leq c).$$

It is emphasized that Eq. (14.10) holds, provided that the surfaces bounding M, including the surface σ, are fixed (rigid). The E^s S^s, and n_j^s therefore define the excess energy, entropy, and densities of mass M_3 over the values of energy, entropy, and densities that mass M_3 would have if the quantities had the same uniform values up to the dividing surface σ as those of M_1 and M_2.

It is convenient to define excess quantities per unit area a of σ:

$$E_\sigma = E^s/a, \quad S_\sigma = E^s/a, \quad \Gamma_j^\sigma = n_j^s/a, \tag{14.12}$$

which are, respectively, the surface energy, surface entropy, and surface concentration. Thus, we obtain the thermodynamic relation for surface

$$\delta E_\sigma = T\delta S_\sigma + \sum_{j=1}^{c} \mu_j \delta \Gamma_j^\sigma, \tag{14.13}$$

subject to the conditions that the bounding surfaces of M_4 and M_5 are fixed. The presence and appearance of E_σ, S_σ, and Γ_j^σ is the difference between the previous theory of heterogeneous equilibria (Chapter 8) and the present. Now, we remove the condition that σ is fixed and examine the effects of variations in the position and form of the surface σ.

The quantities E^s, S^s, $n_1^s, \ldots, n_c^s$ and E_σ, S_σ, $\Gamma_1^\sigma, \ldots, \Gamma_c^\sigma$ are determined partly by the state of the system under consideration and partly by the form and position of the dividing surface σ. However, the positions of the surfaces drawn in the homogeneous regions cannot affect the values of E^s, etc. or E_σ, etc. The reason is: if the plane dividing surface is moved, a distance ϵ toward M_1, $n_i^{(1)}$ is unchanged while $n_i^{(4)}$ is increased by $\epsilon a \Gamma_i^{(4)}$ and $n_i^{(5)}$ is decreased by $-\epsilon a \Gamma_i^{(5)}$ where $\Gamma_i^{(4)}$ and $\Gamma_i^{(5)}$ are densities of the species i per a in M_4 and M_5, respectively. Therefore,

$$\Delta \Gamma_i^\sigma = \epsilon(\Gamma_i^{(4)} - \Gamma_i^{(5)}) = \epsilon(\Gamma_i^{(1)} - \Gamma_i^{(2)}) = 0, \tag{14.14}$$

which means that the location of the dividing surface does not affect the values of Γ_i^σ. One can argue similarly for E^s and S^s.

However, Eq. (14.13) is modified if the position and shape of σ are varied because it is expected that E^s should be a function of S^s and n_j^s as well as the characteristics of surface σ, namely, the area a and the principal curvatures c_1 and c_2 of σ. Therefore, we look for E^s such that

$$E^s = E^s \left(S^s, n_1^s, \ldots, n_c^s, a, c_1, c_2\right). \tag{14.15}$$

With the definitions of derivatives

$$T = \left(\frac{\partial E^s}{\partial S^s}\right)_{\Gamma^\sigma a, c_1, c_2}, \quad \mu_j^s = \left(\frac{\partial E^s}{\partial n_j^s}\right)_{S^\sigma, \Gamma^{\sigma\prime}, a, c_1, c_2}, \tag{14.16}$$

$$\gamma = \left(\frac{\partial E^s}{\partial a}\right)_{S^\sigma, \Gamma^\sigma, a, c_1, c_2}, \tag{14.17}$$

$$C_1 = \left(\frac{\partial E^s}{\partial c_1}\right)_{S^\sigma, \Gamma^\sigma, a, c_2}, \tag{14.18}$$

$$C_2 = \left(\frac{\partial E^s}{\partial c_2}\right)_{S^\sigma, \Gamma^\sigma, a, c_1}, \tag{14.19}$$

we obtain the variational form for E^s

$$\delta E^s = T\delta S^s + \sum_{j=1}^{c} \mu_j \delta n_j^s + \gamma \delta a + C_1 \delta c_1 + C_2 \delta c_2. \tag{14.20}$$

Here, C_1 and C_2 are the constitutive properties of the interface, as are T, γ, and μ_j. This variational form may be rearranged to the form

$$\delta E^s = T\delta S^s + \sum_{j=1}^{c} \mu_j \delta n_j^s + \gamma \delta a$$

$$+ \frac{1}{2}(C_1 + C_2)\,\delta\,(c_1 + c_2) + \frac{1}{2}(C_1 - C_2)\,\delta\,(c_1 - c_2). \tag{14.21}$$

For a planar surface, since the curvatures vanish, i.e., $c_1 = c_2 = 0$, the fourth and fifth terms on the right of Eq. (14.21) vanish. For nearly planar surfaces, it can be shown that

$$C_1 + C_2 = 0, \quad \delta\,(c_1 - c_2) = 0, \tag{14.22}$$

and hence the second line on the right of Eq. (14.21) can be omitted. Equation (14.22) is shown to be true in the following.

We imagine a planar surface σ of area $(x+\delta x)(y+\delta y)$ (see Fig. 14.3). In the neighborhood of this surface on the positive side of curvature, imagine a surface σ' of area $(x + \delta x')(y + \delta y')$ and another surface σ'' of area $(x + \delta x'')(y + \delta y'')$ at distance λ from σ'. These surfaces are constructed such that their normal vectors of σ' and σ'' are parallel. The principal curvature vectors vary from zero to $1/R_1$ and $1/R_2$, respectively. That is,

$$\delta c_1'' = \frac{1}{R_1}, \quad \delta c_2'' = \frac{1}{R_2}. \tag{14.23}$$

Then it follows

$$\delta c_1' = \frac{1}{R_1 + \epsilon}, \quad \delta c_2' = \frac{1}{R_2 + \epsilon}. \tag{14.24}$$

Upon eliminating R_1 and R_2 by using Eq. (14.23), we obtain

$$\delta c_1' = \frac{\delta c_1''}{1 + \epsilon \delta c_1''}, \quad \delta c_2' = \frac{\delta c_2''}{1 + \epsilon \delta c_2''}. \tag{14.25}$$

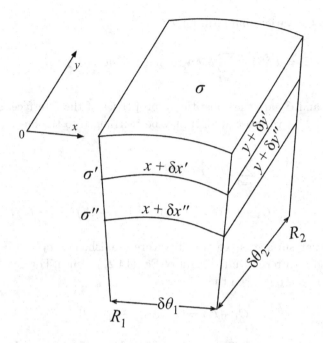

Fig. 14.3. Surface σ is planar and parallel to the dividing surface, and surfaces σ' and σ'' are spherical with the curvature vectors downward. R_1 and R_2 are the principal radii of curvature of surface σ'' and the principal radii of curvature of σ' are $R_1 + \epsilon$ and $R_2 + \epsilon$. $\delta\theta_1$ is the angle between R_1 and its displacement in the x-direction whereas $\delta\theta_2$ is the corresponding angle for R_2 and its displacement in the y-direction.

Therefore, we obtain to the first order in variations

$$\delta\left(c_1' + c_2'\right) = \delta\left(c_1'' + c_2''\right), \tag{14.26}$$

where the term of $O(\epsilon\delta c_1''^2 + \epsilon\delta c_2''^2)$ is neglected. Since the varied areas are $(x + \delta x')(y + \delta y')$ and $(x + \delta x'')(y + \delta y'')$, respectively, the angles between the principal curvature vectors before and after the variation $\delta\theta_1$ and $\delta\theta_2$ are, respectively,

$$\delta\theta_1 = \frac{x + \delta x'}{R_1 + \epsilon} = \frac{x + \delta x''}{R_1}, \tag{14.27}$$

$$\delta\theta_2 = \frac{y + \delta y'}{R_2 + \epsilon} = \frac{y + \delta y''}{R_2}. \tag{14.28}$$

From these relations follow, to the first order in variations, the relations

$$\delta x' = \delta x'' + \epsilon x \delta c_1'', \tag{14.29}$$

$$\delta y' = \delta y'' + \epsilon y \delta c_2''. \tag{14.30}$$

With these relations, we find the variation in area a' of σ':

$$\delta a' = (x + \delta x')(y + \delta y') - xy \tag{14.31}$$

$$= x\delta y' + y\delta x' + \text{ higher order},$$

which, upon use of relations (14.29) and (14.30), gives rise to the equation

$$\delta a' = \delta a'' + a\epsilon \delta (c_1'' + c_2''). \tag{14.32}$$

For imaginary surfaces σ' and σ'',

$$\delta E^\sigma - T\delta S^s - \sum_{j=1}^{c} \mu_j \delta n_j^\sigma$$

remains unchanged since M_3 remains unchanged. Therefore, we obtain for planar surfaces, for which $\delta(c_1 - c_2) = 0$, the equation

$$\gamma'\delta a' + \tfrac{1}{2}(C_1' + C_2')\delta(c_1' + c_2') = \gamma''\delta a'' + \tfrac{1}{2}(C_1'' + C_2'')\delta(c_1'' + c_2''). \tag{14.33}$$

Upon use of Eq. (14.26) and the fact that $\gamma'' = \gamma' := \gamma$ for a plane surface (this is shown in Section 14.5 below), we obtain

$$C_1' + C_2' + 2\gamma a\epsilon = C_1'' + C_2''. \tag{14.34}$$

If σ' is held fixed in position, the quantities C_1', C_2', and γ have fixed values. Therefore, ϵ can be chosen such that

$$C_1'' + C_2'' = 0, \tag{14.35}$$

as promised earlier (see Eq. (14.22)).

For a nearly planar surface, evidently

$$\delta(c_1 - c_2) = O(\epsilon), \quad C_1 + C_2 = O(\epsilon), \tag{14.36}$$

according to Eqs. (14.34) and (14.35).

Therefore, if the inhomogeneous film surrounding the surface σ is sufficiently thin compared with the principal radii of curvature, then $\frac{1}{2}(C_1 + C_2)\delta(c_1 + c_2)$ is negligible. Therefore, the dividing surface can be chosen such that this condition is met for a planar or nearly planar surface and it is possible to omit the term

$$\frac{1}{2}(C_1 + C_2)\,\delta\,(c_1 + c_2) + \frac{1}{2}(C_1 - C_2)\,\delta\,(c_1 - c_2)$$

in Eq. (14.21). For such surfaces Eq. (14.21) reduces to the differential form

$$\delta E^s = T\delta S^s + \sum_{j=1}^{c} \mu_j \delta n_j^s + \gamma \delta a. \tag{14.37}$$

This is the fundamental thermodynamic equation for planar surfaces σ. Equation (14.37) is called *the Gibbs relation for interface*. The meaning of γ is now clearly seen as the *surface tension*, and $\gamma\delta a$ as the *surface work*.

14.3.   Nearly Planar Surface and Surface Tension

We now consider the effects caused by curvature, especially, on pressure. We shall assume that the surface tension is not affected by surface curvature. This assumption is justifiable as long as the radius of curvature is large compared with the thickness of the film.

For this purpose, we now investigate the conditions for heterogeneous equilibrium on the pressures of the homogeneous phases on either side of surface σ. The equilibrium criterion for M_3 (region 3) is

$$(\delta E^{(3)})_{S^{(3)}, n^{(3)}} \geq 0 \tag{14.38}$$

when the bounding surface is rigid. Since

$$\delta E^{(3)} = a\delta E_\sigma + \delta E^{(4)} + \delta E^{(5)}, \tag{14.39}$$

where

$$\delta E^{(\alpha)} = T\delta S^{(\alpha)} - p^{(\alpha)}\delta V^{(\alpha)} + \sum_{j=1}^{c} \mu_j \delta n_j^{(\alpha)} \quad (\alpha = 4, 5), \tag{14.40}$$

substitution of Eqs. (14.39) and (14.40) into inequality (14.38) gives rise to the equilibrium conditions for pressure

$$p^{(4)} = p^{(1)}, \quad p^{(5)} = p^{(2)}. \tag{14.41}$$

Since the properties of M_1 and M_2 are the same as those of the bulk properties of phases on either side of σ by construction, if we denote the phases by α and β for 1 and 2, respectively, then we obtain the inequality

$$\gamma \delta a - p^{(\alpha)} \delta V^{(4)} - p^{(\beta)} \delta V^{(5)} \geq 0. \tag{14.42}$$

Since all variations are either positive or negative, this gives rise to the conclusion

$$\gamma \delta a - p^{(\alpha)} \delta V^{(4)} - p^{(\beta)} \delta V^{(5)} = 0. \tag{14.43}$$

If $\delta a = 0$, then since

$$\delta V^{(4)} + \delta V^{(5)} = 0,$$

we conclude

$$p^{(\alpha)} = p^{(\beta)}. \tag{14.44}$$

Thus, the pressure is found to have the same value on both sides of σ in the case of an undeformed plane surface.

In the case of a curved surface, if the dividing surface is uniformly moved by a distance $\delta \epsilon$ in the direction of its normals, in Eq. (14.43) the change in area is given by

$$\delta a = a \left(c_1 + c_2 \right) \delta \epsilon, \tag{14.45}$$

while

$$\delta V^{(4)} = a \delta \epsilon, \quad \delta V^{(5)} = -a \delta \epsilon. \tag{14.46}$$

By Eq. (14.43), we obtain

$$[\gamma(c_1 + c_2) - (p^{(\alpha)} - p^{(\beta)})]a \delta \epsilon = 0.$$

Since $\delta \epsilon$ is arbitrary there follows the equation

$$p^{(\alpha)} - p^{(\beta)} = \gamma \left(c_1 + c_2 \right) \tag{14.47}$$

with the curvatures positive when their centers lie in the phase α. If the surface is part of a spherical surface of curvature $c = 1/R$, then

$$p^{(\alpha)} - p^{(\beta)} = 2\gamma c$$
$$= \frac{2\gamma}{R}. \tag{14.48}$$

According to this formula, the pressure inside a spherical drop of liquid of radius R is greater than the external pressure by $2\gamma/R$. Therefore, the equilibrium pressure of a smaller sphere is greater than that of a larger sphere of the liquid at the same temperature and external pressure, and the difference in pressure is determined by the surface tension and the radius of the drop.

Relation (14.48) can be applied to the following situation of a thin spherical shell having two concentric surfaces, a typical example being a bubble of detergent. Let the outer surface have a radius of R_1 and the inner surface a radius of R_2. The pressure $p^{(\mathrm{film})}$ in the film is greater than the pressure $p^{(\beta)}$ in the exterior phase by $2\gamma/R_1$, namely,

$$p^{(\mathrm{film})} - p^{(\beta)} = \frac{2\gamma}{R_1},$$

whereas the pressure within the bubble $p^{(\beta)}$ is greater than $p^{(\mathrm{film})}$, namely,

$$p^{(\alpha)} - p^{(\mathrm{film})} = \frac{2\gamma}{R_2}.$$

Therefore,

$$p^{(\alpha)} - p^{(\beta)} = 2\gamma \left(\frac{1}{R_1} + \frac{1}{R_2} \right). \tag{14.49}$$

Let

$$R = \tfrac{1}{2}(R_1 + R_2), \quad \varepsilon = R_1 - R_2.$$

Then there holds the Young–Laplace equation

$$p^{(\alpha)} - p^{(\beta)} = \frac{4\gamma}{R} \left[1 + O\left(\frac{\varepsilon^2}{R^2} \right) \right]. \tag{14.50}$$

The pressure inside the bubble is greater than the exterior pressure and the bubble is stabilized by the surface tension γ.

We consider another related example: the vapor pressure of a spherical drop of a liquid. When the vapor phase is in equilibrium with the liquid phase at T, there holds the Poynting equation

$$\frac{dp^{(v)}}{dp^{(l)}} = \frac{v_l}{v_v}, \tag{14.51}$$

where $p^{(v)}$ and $p^{(l)}$ are the vapor pressure and the pressure within the liquid, respectively, and v_l and v_v are molar volumes of the vapor and the liquid, respectively. For the sake of simplicity, let us assume that the vapor is ideal. Then

$$\frac{dp^{(v)}}{dp^{(l)}} = v_l \frac{p^{(v)}}{RT}.$$

(14.52)

Integrating this equation, under the assumption of incompressibility of the liquid to an approximation, from the state of a planar liquid to a spherical liquid of radius r, we obtain

$$\ln \frac{p^{(v)}}{p_0} = \frac{v_l}{RT}(p_l - p_0).$$

(14.53)

Note that initially $p_0 = p_l^{(\text{initial})}$ in the planar state of the liquid at equilibrium with its vapor. Using Eq. (14.48), this can be written in the form

$$\ln \frac{p^{(v)}}{p_0} = \frac{v_l}{RT}\left[\frac{2\gamma}{r} + (p^{(v)} - p_0)\right],$$

(14.54)

where r is the radius of the liquid drop. Since r is assumed small, this may be approximated as follows:

$$\ln \frac{p^{(v)}}{p_0} = \frac{2\gamma v_l}{RTr}.$$

(14.55)

This expression shows that the smaller the drop, the larger the vapor pressure becomes. This formula enables us to estimate the vapor pressure above a spherical drop from its properties such as the surface tension, temperature, molar volume of the liquid, and the size.

14.4.   Gibbs–Duhem Relation for Interface

The excess Helmholtz free energy A^s and the excess Gibbs free energy G^s of interface are, respectively, defined by

$$A^s = E^s - TS^s,$$

(14.56)

and

$$G^s = E^s - TS^s - \gamma a.$$

(14.57)

Upon using Eq. (14.37) in these expressions, we obtain

$$\delta A^s = -S^s \delta T + \sum_{j=1}^{c} \mu_j \delta n_j^s + \gamma \delta a,$$

(14.58)

and

$$\delta G^s = -S^s \delta T + \sum_{j=1}^{c} \mu_j \delta n_j^s - a\delta\gamma.$$ (14.59)

Since it is known experimentally that, as is the Gibbs free energy of the bulk phases, G^s is a first-order homogeneous function of densities $\{n_j^s : j = 1,\ldots,c\}$, it is possible to express it as

$$G^s = \sum_{j=1}^{c} \mu_j n_j^s.$$ (14.60)

In view of Eq. (14.59), it follows

$$\sum_{j=1}^{c} n_j^s \delta\mu_j = -S^s \delta T - a\delta\gamma.$$ (14.61)

Alternatively, this may be written as

$$\sum_{j=1}^{c} \Gamma_j^\sigma \delta\mu_j = -S_\sigma \delta T - \delta\gamma.$$ (14.62)

See Eq. (14.12) for the definition of Γ_j^σ. Equation (14.62) is the Gibbs–Duhem equation for interface, which is the integrability condition for the differential form (14.59). Just as the Gibbs–Duhem equation for bulk phases, Eq. (14.61) plays an important role in the thermodynamics of interfaces.

14.5.   Location of the Dividing Surface and Surface Tension

For the derivation of the interfacial equations and, in particular, Eq. (14.37), we have introduced a dividing surface at an arbitrary location in the vicinity of the physical surface of discontinuity. Therefore, the surface tension and other interfacial quantities involved are expected to depend on the location of the dividing surface. Contrary to the expectation, this is not true, that is, *they are independent of the location of the dividing surface*. We have made use of this fact in the previous sections.

In this section, we show the statement is indeed true before proceeding to further discussions of interfacial thermodynamics. We will specialize our treatment to a plane interface.

Let us now consider the difference between the values of various extensive properties of the surface as the plane dividing surface is moved a distance ϵ into, say, phase β from location 1 to location 2, keeping the area and state of the system fixed. The changes in volume of the phases α and β are therefore

$$\Delta V^{(\alpha)} = \epsilon a, \quad \Delta V^{(\beta)} = -\epsilon a. \tag{14.63}$$

The difference in E^s arising from the change in the location of the dividing surfaces is

$$\begin{aligned}
\Delta E^s &= E_2^s - E_1^s \\
&= (E^{(3)} - E_2^{(\alpha)} - E_2^{(\beta)}) - (E^{(3)} - E_1^{(\alpha)} - E_1^{(\beta)}) \\
&= -(E_2^{(\alpha)} - E_1^{(\alpha)}) - (E_2^{(\beta)} - E_1^{(\beta)}),
\end{aligned} \tag{14.64}$$

where $E_i^{(\alpha)}$ and $E_i^{(\beta)}$ $(i = 1, 2)$ are the internal energies of phases α and β with the dividing surface located at position i, respectively. Similarly,

$$\Delta S^s = -(S_2^{(\alpha)} - S_1^{(\alpha)}) - (S_2^{(\beta)} - S_1^{(\beta)}), \tag{14.65}$$

$$\Delta n_j^s = -(n_{j2}^{(\alpha)} - n_{j1}^{(\alpha)}) - (n_{j2}^{(\beta)} - n_{j1}^{(\beta)}). \tag{14.66}$$

By the integrability condition of the Gibbs equation, namely, the Gibbs–Duhem equation, there holds Eq. (14.57), and we find

$$\Delta G^s = \Delta E^s - T\Delta S^s + \sum_{j=1}^{c} \mu_j \Delta n_j^s - a\Delta\gamma, \tag{14.67}$$

but since

$$\begin{aligned}
\Delta G^s &= G_2^s - G_1^s \\
&= -(G_2^{(\alpha)} - G_1^{(\alpha)}) - (G_2^{(\beta)} - G_1^{(\beta)}) \\
&= -(E_2^{(\alpha)} - E_1^{(\alpha)}) + T(S_2^{(\alpha)} - S_1^{(\alpha)}) - \sum_{j=1}^{c} \mu_j (n_{j2}^{(\alpha)} - n_{j1}^{(\alpha)}) \\
&\quad -(p_2^{(\alpha)} - p_1^{(\alpha)})a\epsilon \\
&\quad -(E_2^{(\beta)} - E_1^{(\beta)}) + T(S_2^{(\beta)} - S_1^{(\beta)}) - \sum_{j=1}^{c} \mu_j (n_{j2}^{(\beta)} - n_{j1}^{(\beta)})
\end{aligned}$$

$$-(p_2^{(\beta)} - p_1^{(\beta)})a\epsilon$$

$$= \Delta E^s - T\Delta S^s + \sum_{j=1}^{c} \mu_j \Delta n_j^s - (p_2^{(\alpha)} - p_1^{(\alpha)})a\epsilon - (p_2^{(\beta)} - p_1^{(\beta)})a\epsilon,$$

$$(14.68)$$

balancing Eqs. (14.67) and (14.68) yields

$$(p_1^{(\alpha)} - p_2^{(\alpha)})\epsilon + (p_1^{(\beta)} - p_2^{(\beta)})\epsilon = \Delta\gamma. \qquad (14.69)$$

Since $p_2^{(\alpha)} - p_1^{(\alpha)} = 0$ and $p_2^{(\beta)} - p_1^{(\beta)} = 0$ on account of mechanical equilibrium condition, we conclude that

$$\Delta\gamma = \gamma_2 - \gamma_1 = 0.$$

This proves that *the surface tension is independent of the position of the dividing surface.* Recall that this result was used for Eq. (14.34) in Section 14.2.

14.6.   Gibbs Phase Rule Including Interface

The two-phase system of c-components with an interface obeys a phase rule, which is modified from the Gibbs phase rule in the absence of interfaces. The intensive variables of the bulk phases α and β are $T^{(\alpha)}, p^{(\alpha)}, x_1^{(\alpha)}, \ldots, x_{c-1}^{(\alpha)}$ and $T^{(\beta)}, p^{(\beta)}, x_1^{(\beta)}, \ldots, x_{c-1}^{(\beta)}$ whereas the intensive variables for the interface are $T^\sigma, \gamma, \Gamma_1^\sigma, \ldots, \Gamma_{c-1}^\sigma$. Therefore, there are $3(c+1)$ variables in total. The number of constraining equations — the equilibrium conditions — are $2(c+1)$ according to Eqs. (14.5) and (14.6). Therefore, the number of degree of freedom is

$$f = 3(c+1) - 2(c+1) = c+1. \qquad (14.70)$$

This is the Gibbs phase rule for two-phase equilibrium for a c-component system with an interface. It is in contrast to the two-phase equilibrium of the same system without an interface for which $f = c$.

14.7.   Thermodynamics of Interface

If variations are limited to those in which the varied state is one of equilibrium, then the variation sign δ can be replaced with the differential sign d.

In this case, in place of Eq. (14.37) we have the differential form

$$dE^s = TdS^s + \sum_{j=1}^{c} \mu_j dn_j^s + \gamma da, \qquad (14.71)$$

and the Gibbs–Duhem equation is written as

$$S^s dT + \sum_{j=1}^{c} n_j^s d\mu_j = -ad\gamma, \qquad (14.72)$$

or

$$S_\sigma dT + \sum_{j=1}^{c} \Gamma_j^\sigma d\mu_j = -d\gamma. \qquad (14.73)$$

This equation is in fact called the Gibbs adsorption equation.

14.7.1. *Invariance of Derivatives to the Position of the Dividing Surface*

It was shown that the surface tension is invariant to the position of the dividing surface. The derivatives of the surface tension is also invariant to the position of the dividing surface. It is shown below: It is possible to choose the dividing surface such that one of Γ_j^σ is equal to zero except when the corresponding quantities in the two phases are identical. For example, if the dividing surface is chosen such that Γ_1^σ vanishes, the Gibbs adsorption equation becomes

$$d\gamma = -S_\sigma dT - \sum_{j=2}^{c} \Gamma_{j(1)}^\sigma d\mu_j, \qquad (14.74)$$

where the subscript 1 in $\Gamma_{j(1)}^\sigma$ refers to the densities when the dividing surface is chosen such that $\Gamma_1^\sigma = 0$. It follows from Eq. (14.74)

$$\left(\frac{\partial \gamma}{\partial \mu_j} \right)_{T,\mu'} = -\Gamma_{j(1)}^\sigma, \qquad (14.75)$$

where $\mu' = \{\mu_k; \mu_k \neq \mu_j\}$. We would like to show this derivative is invariant to the position of the dividing surface. We consider a two-component system for simplicity. If $T = $ constant, it follows from Eq. (14.74)

$$dγ = -Γ_1^σ dμ_1 - Γ_2^σ dμ_2,$$ (14.76)

but

$$dp^{(ω)} = ρ_1^{(ω)} dμ_1 + ρ_2^{(ω)} dμ_2 \ (ω = α, β),$$ (14.77)

where $ρ_i^{(ω)} = n_i^{(ω)}/V^{(ω)}$. Since $p^{(α)} = p^{(β)}$ at two-phase equilibrium, it follows

$$dμ_1 = \frac{ρ_2^{(β)} - ρ_2^{(α)}}{ρ_1^{(α)} - ρ_1^{(β)}} dμ_2.$$ (14.78)

Substitution of this into Eq. (14.76) yields

$$\left(\frac{∂γ}{∂μ_2}\right)_T = -\left[Γ_2^σ - \frac{ρ_2^{(α)} - ρ_2^{(β)}}{ρ_1^{(α)} - ρ_1^{(β)}} Γ_1^σ\right].$$ (14.79)

We note that $Γ_2^σ$ and $Γ_1^σ$ are the excess quantities of the components per unit area of the surface when the dividing surface is located at the surface of tension. Since $ε(ρ_1^{(α)} - ρ_1^{(β)})$ is the amount by which $Γ_1^σ$ is increased when the dividing surface is moved normally a distance $ε$ toward the inside $α$ and $Γ_1^σ$ is the excess quantity of component 1 when the dividing surface is given the position of the surface of tension, the distance that the dividing surface must be moved toward $α$ to make the excess quantity of 1 at the surface equal to zero is

$$ε = -\frac{Γ_1^σ}{ρ_1^{(α)} - ρ_1^{(β)}}.$$ (14.80)

The amount that must be added to $Γ_2^σ$ is then

$$-\frac{(ρ_2^{(α)} - ρ_2^{(β)})Γ_1^σ}{ρ_1^{(α)} - ρ_1^{(β)}}.$$ (14.81)

Therefore, we conclude that

$$Γ_{2(1)}^σ = Γ_2^σ - \frac{(ρ_2^{(α)} - ρ_2^{(β)})}{ρ_1^{(α)} - ρ_1^{(β)}} Γ_1^σ.$$ (14.82)

and Eq. (14.79) is identical with Eq. (14.75) in the case of a two-component system. This shows the invariance of the derivatives to the position of the dividing surface.

14.7.2. *Various Thermodynamic Relations for Interface*

Earlier, we have introduced the Helmholtz and Gibbs free energies for interface. It is also convenient to define the enthalpy for the interface. Formally and in analogy to the enthalpy of a bulk phase, it may be defined by

$$H^s = E^s + p^s V^s,$$

but since the excess volume V^s of a surface of discontinuity is equal to zero by definitions of the excess quantities, we find

$$H^s = E^s. \tag{14.83}$$

Since

$$dE^s = TdS^s + \sum_{j=1}^{c} \mu_j dn_j^s + \gamma da, \tag{14.84}$$

it follows

$$dH^s = TdS^s + \sum_{j=1}^{c} \mu_j dn_j^s + \gamma da. \tag{14.85}$$

From the definitions of A^s and G^s follow the differential forms — the fundamental equations

$$dA^s = -S^s dT + \sum_{j=1}^{c} \mu dn_j^s + \gamma da, \tag{14.86}$$

and

$$dG^s = -S^s dT + \sum_{j=1}^{c} \mu_j dn_j^s - ad\gamma. \tag{14.87}$$

Various thermodynamic relations for interface can be derived from these fundamental equations — the Gibbs relations for interface. We will examine some of the thermodynamic relations in the case of a binary mixture for simplicity.

14.7.3. *Liquid–Vapor Equilibrium*

14.7.3.1. *Density Dependence of Surface Tension γ*

In Section 14.7.1, we have calculated the relation of surface tension to the interfacial density (see Eq. (14.75)). Here, we consider the case of a binary solution at equilibrium with its vapor phase. At constant temperature, the following equations holds:

$$d\gamma = -\Gamma^\sigma_{2(1)}d\mu_2, \tag{14.88}$$

$$d\gamma = -\Gamma^\sigma_{1(2)}d\mu_1, \tag{14.89}$$

$$-d\gamma = \Gamma^\sigma_1 d\mu_1 + \Gamma^\sigma_2 d\mu_2. \tag{14.90}$$

We designate component 2 for the solute and component 1 for the solvent. To make further progress in thermodynamics, it is necessary to know the chemical potentials — the constitutive relations. We shall assume the chemical potentials are given by the form

$$\mu_i\left(p, T, x\right) = \mu_i^0\left(p, T\right) + RT \ln a_i \quad (i = 1, 2), \tag{14.91}$$

$$a_i = x_i f_i, \tag{14.92}$$

where x_i is the mole fraction and f_i the activity coefficient of component i. Note that this is the same as the chemical potential of component i in the bulk owing to the equilibrium conditions. An appropriate convention should be used for the activity coefficients (see Chapter 11). Then the excess densities $\Gamma^\sigma_{2(1)}$ and $\Gamma^\sigma_{1(2)}$ can be calculated from the information on the surface tension and the bulk properties:

$$\Gamma^\sigma_{1(2)} = -\frac{1}{RT}\left(\frac{\partial \gamma}{\partial \ln x_1 f_1}\right)_T, \tag{14.93}$$

$$\Gamma^\sigma_{2(1)} = -\frac{1}{RT}\left(\frac{\partial \gamma}{\partial \ln x_2 f_2}\right)_T. \tag{14.94}$$

Equation (14.94) shows how the excess interfacial density of the solute — that is, the adsorption of component 2 — can be determined by measuring the variation of the surface tension with respect to the solute concentration in the bulk phase. Note that for dilute solutions, the activity coefficients may be taken as equal to unity. These equations show that if the solute is positively adsorbed in the surface, namely, $\Gamma^\sigma_{2(1)}$ is positive, the surface tension decreases with increasing concentration of solute in the dilute solution limit. However, in the case of $\Gamma^\sigma_{2(1)}$ negative, since Γ^σ_2 cannot be negative, there holds the bound in solute density

$$\frac{(\rho_2^{(\alpha)} - \rho_2^{(\beta)})\Gamma^\sigma_1}{\rho_1^{(\alpha)} - \rho_1^{(\beta)}} \geq 0, \tag{14.95}$$

which puts a bound on the increase in surface tension.

It is possible to make use of vapor pressure to express Eqs. (14.93) and (14.94). If the vapor may be regarded as ideal, then since at constant T,

$$d\mu_i = RTd\ln p_i, \tag{14.96}$$

where p_i is the vapor pressure, it follows that

$$\Gamma^\sigma_{1(2)} = -\frac{1}{RT}\left(\frac{\partial\gamma}{\partial\ln p_1}\right)_T = -\frac{p_1}{RT}\left(\frac{\partial\gamma}{\partial p_1}\right)_T, \tag{14.97}$$

$$\Gamma^\sigma_{2(1)} = -\frac{1}{RT}\left(\frac{\partial\gamma}{\partial\ln p_2}\right)_T = -\frac{p_2}{RT}\left(\frac{\partial\gamma}{\partial p_2}\right)_T. \tag{14.98}$$

From the Gibbs–Duhem equation for the vapor phase at constant temperature and pressure,

$$x_1 d\ln p_1 + x_2 d\ln p_2 = 0, \tag{14.99}$$

where x_1 and x_2 are the mole fractions of vapor species, and from Eqs. (14.97) and (14.98) follows the relation

$$x_2\Gamma^\sigma_{1(2)} = -x_1\Gamma^\sigma_{2(1)}. \tag{14.100}$$

Since for the situation under consideration Eq. (14.90) can be written

$$-d\gamma = \Gamma^\sigma_1 RTd\ln p_1 + \Gamma^\sigma_2 RTd\ln p_2, \tag{14.101}$$

with the help of Eq. (14.99) Eq. (14.101) may be rearranged to the form

$$-d\gamma = \left(\Gamma^\sigma_2 - \frac{x_2}{x_1}\Gamma^\sigma_1\right)RTd\ln p_2 = \left(\Gamma^\sigma_1 - \frac{x_1}{x_2}\Gamma^\sigma_2\right)RTd\ln p_1. \tag{14.102}$$

It then implies

$$\Gamma^\sigma_{1(2)} = \Gamma^\sigma_1 - \frac{x_1}{x_2}\Gamma^\sigma_2, \tag{14.103}$$

$$\Gamma^\sigma_{2(1)} = \Gamma^\sigma_2 - \frac{x_2}{x_1}\Gamma^\sigma_1. \tag{14.104}$$

This shows $\Gamma^\sigma_{1(2)}$ and $\Gamma^\sigma_{2(1)}$ can be measured from the variation in γ with respect to the vapor pressure. These results obtained for binary mixtures can be easily generalized to multicomponent mixtures.

14.7.3.2. *Temperature Dependence of Surface Tension γ*

Since by the equilibrium conditions,

$$\mu_j = \mu_j^{(\alpha)} = \mu_j^{(\beta)}, \tag{14.105}$$

and the temperature derivative of μ_j may be expressed by the equation

$$\left(\frac{\partial \mu_j}{\partial T}\right)_{x^{(\alpha)}} = -s_j^{(\alpha)} + \left(\frac{\partial p}{\partial T}\right)_{x^{(\alpha)}} \overline{v}_j^{(\alpha)}, \tag{14.106}$$

differentiating the Gibbs–Duhem equation for interface at constant $x^{(\alpha)}$ yields the rate of temperature variation for surface tension γ:

$$\left(\frac{\partial \gamma}{\partial T}\right)_{x^{(\alpha)}} = -S_\sigma + \sum_{j=1}^{c} \Gamma_j^\sigma s_j^{(\alpha)} - \sum_{j=1}^{c} \Gamma_j^\sigma \overline{v}_j^{(\alpha)} \left(\frac{\partial p}{\partial T}\right)_{x^{(\alpha)}}. \tag{14.107}$$

Furthermore, since

$$d\mu_j^{(\alpha)} = -s_j^{(\alpha)} dT + \overline{v}_j^{(\alpha)} dp + \sum_{i=1}^{c-1} \left(\frac{\partial \mu_j^{(\alpha)}}{\partial x_i^{(\beta)}}\right)_{T,p,x_{j\neq i}^{(\alpha)}} dx_i^{(\beta)}, \tag{14.108}$$

from the second equality of Eq. (14.105) follows

$$\sum_{j=1}^{c} x_j^{(\beta)} \Delta \overline{s}_j dT - \sum_{j=1}^{c} x_j^{(\beta)} \Delta \overline{v}_j dp + \sum_{i=1}^{c} \mu_i^{(\alpha\beta)} dx_i^{(\beta)} = 0, \tag{14.109}$$

where

$$\Delta \overline{s}_j = \overline{s}_j^{(\beta)} - \overline{s}_j^{(\alpha)}, \tag{14.110}$$

$$\Delta \overline{v}_j = \overline{v}_j^{(\beta)} - \overline{v}_j^{(\alpha)}, \tag{14.111}$$

$$\mu_i^{(\alpha\beta)} = \sum_{j=1}^{c} x_j^{(\beta)} \left(\frac{\partial \mu_j^{(\alpha)}}{\partial x_i^{(\beta)}}\right)_{T,p,x_{j\neq i}^{(\alpha)}}. \tag{14.112}$$

Therefore, we obtain from Eq. (14.109) the Clapeyron equation for the two-phase mixture

$$\left(\frac{\partial p}{\partial T}\right)_{x^{(\alpha)}} = \frac{\sum_{j=1}^{c} x_j^{(\beta)} \Delta \bar{s}_j}{\sum_{j=1}^{c} x_j^{(\beta)} \Delta \bar{v}_j} = \frac{\Delta s}{\Delta v}, \tag{14.113}$$

where Δs and Δv are the entropy and volume change per mole of the mixture:

$$\Delta s = \sum_{j=1}^{c} x_j^{(\beta)} \Delta \bar{s}_j, \quad \Delta v = \sum_{j=1}^{c} x_j^{(\beta)} \Delta \bar{v}_j.$$

Upon substitution of this derivative into Eq. (14.107), we obtain the rate of temperature variation of γ

$$\left(\frac{\partial \gamma}{\partial T}\right)_{x^{(\alpha)}} = \sum_{j=1}^{c} \Gamma_j^{\sigma} s_j^{(\alpha)} - S_{\sigma} - \sum_{i=1}^{c} \Gamma_i^{\sigma} \bar{v}_i^{(\alpha)} \frac{\Delta s}{\Delta v}. \tag{14.114}$$

In the case of single-component system, this equation reduces to

$$\frac{d\gamma}{dT} = \frac{\left(s^{(\alpha)} v^{(\beta)} - v^{(\alpha)} s^{(\beta)}\right)}{v^{(\beta)} - v^{(\alpha)}} \Gamma^{\sigma} - S_{\sigma}. \tag{14.115}$$

If the surface is chosen such that $\Gamma^{\sigma} = 0$, then

$$\frac{d\gamma}{dT} = -S_{\sigma}. \tag{14.116}$$

This relation makes it possible to evaluate the surface entropy.

Experimentally, it is known that in the vicinity of the critical temperature T_c, the surface tension varies with T as[3]

$$\gamma = \gamma_0 \left(1 - T/T_c\right)^{11/9}. \tag{14.117}$$

That is, the surface tension vanishes at $T = T_c$. According to Guggenheim, this form was originally suggested by J. van der Waals who took 1.234 for the exponent. Thus, in the neighborhood of T_c, the surface entropy is

---

[3]See A. Ferguson, *Trans. Faraday Soc.* **19**, 408 (1923); E. A. Guggenheim, *J. Chem. Phys.* **13**, 253 (1945); L. Riedel, *Chem.-Ing. Tech.* **27**, 209 (1955). M. Yu. Gorbachev, *Phys. Chem. Liq.* **39**, 315 (2001).

deduced to behave as

$$S_\sigma = \frac{11\gamma_0}{9T_c}\left(1 - T/T_c\right)^{2/9}.$$

(14.118)

Since $H_\sigma = E_\sigma$ as shown earlier, we find

$$E_\sigma = G_\sigma + TS_\sigma,$$

(14.119)

or it may be written as

$$E_\sigma = \gamma - T\frac{d\gamma}{dT}.$$

(14.120)

Therefore, in the neighborhood of $T_c > T$,

$$E_\sigma = \gamma_0\left(1 - T/T_c\right)^{2/9}\left(1 + \frac{2T}{9T_c}\right),$$

(14.121)

according to the empirical formula (14.117). Consequently, the surface excess energy decreases with increasing T and eventually vanishes at T_c.

Since the surface specific heat C_σ may be defined by

$$C_\sigma = \frac{dE_\sigma}{dT},$$

(14.122)

we note that it is possible to deduce from formula (14.122) that in the neighborhood of T_c the surface specific heat is singular and negative:

$$C_\sigma = -\frac{22\gamma_0}{81T_c^2}\frac{T}{\left(1 - T/T_c\right)^{7/9}} \quad (T < T_c).$$

(14.123)

This negative surface excess heat capacity near the critical temperature $(T_c > T)$ is in contrast to the bulk heat capacities. Since the surface excess energy in essence is a relative energy of the surface over the bulk energy, the surface specific heat C_σ is the rate of change in the relative energy with respect to temperature. Therefore, C_σ given in Eq. (14.123) is not in violation of the second law of thermodynamics, which demands a positive heat capacity as we have shown in Chapter 5.

Chapter 15

Electrolyte Solutions

The general idea taken in the treatment of activities and activity coefficients of nonelectrolytes given in Chapter 13 can be applied to electrolyte solutions. An electrolyte is called strong if it completely dissociates into constituent ions in solution. Otherwise, it is called weak, weak acids or bases being examples for weak electrolytes. We will discuss solutions of strong electrolytes in this chapter.

Nonideal behavior of nonelectrolyte solutions arises from generally finite-ranged molecular interactions of solute molecules, which are, for simplicity, assumed to be dissolved in an inert neutral solvent. Therefore, if the solution is dilute, then the interactions between solute molecules are rather weak because of large spatial separations of solute molecules in dilute solutions.

However, the situation changes drastically in the case of ionic solutions, since the interionic interactions are no longer negligible even at a large separation of ions in an infinitely dilute solution owing to the long range nature of Coulomb interactions between ions in the solution. Therefore, for example, ionic solutions would not obey Henry's law even at a concentration at which nonelectrolyte solutions would exhibit almost ideal behavior. We have seen that according to the Margules expansions for the activities of nonionic binary solutions the logarithm of an activity coefficient starts with a quadratic term in concentration, but in the case of ionic solutions it is experimentally known that the density expansion for the same molar quantity starts with the square root of the ionic strength of the solution or, roughly speaking, with $m^{1/2}$ where m is the molality of the solution. This behavior descending from the long-rangedness of the Coulomb interactions can be understood by means of the Debye–Hückel theory, which will be discussed in the next chapter.

Our aim here is to discuss methods of phenomenologically determining activities and activity coefficients of ions. Therefore, we shall first develop some notions and definitions necessary for the purpose in this chapter.

15.1. Mean Activity and Mean Activity Coefficient

For electrolyte solutions the standard state of each ionic species is chosen such that the activity of an ion becomes equal to the concentration at infinite dilution at 1 atm pressure and the temperature of the solution. For electrolyte solutions it is conventional to express the concentrations in terms of molalities.

Let us consider an electrolyte represented by the formula $C_{\nu_+}^{z_+} A_{\nu_-}^{z_-}$ which upon dissociation produces ν_+ cations C^{z+} of charge number z_+ (positive), and ν_- anions A^{z-} of charge number z_- (negative), where the subscripts $+$ and $-$ refer to the cation and anion, respectively. As an example, take $BaSO_4$ for which $\nu_+ = 1$, $\nu_- = 1$, $z_+ = 2$, and $z_- = -2$. The dissociation of an electrolyte may be looked upon as a chemical reaction, thus

$$C_{\nu_+}^{z_+} A_{\nu_-}^{z_-} \rightleftharpoons \nu_+ C^{z+} + \nu_- A^{z-}. \tag{15.1}$$

The electroneutrality demands

$$\nu_+ z_+ + \nu_- z_- = 0. \tag{15.2}$$

The chemical potential of each ionic species is looked for in the form

$$\mu_+ = \mu_+^0(T, p) + RT \ln a_+, \tag{15.3}$$

and

$$\mu_- = \mu_-^0(T, p) + RT \ln a_-, \tag{15.4}$$

where a_+ and a_- are the activities of cation and anion, respectively. These activities may be expressed in terms of the activity coefficient γ_+ or γ_- and the molality m_+ or m_- :

$$a_+ = \gamma_+ m_+; \quad a_- = \gamma_- m_-. \tag{15.5}$$

The standard chemical potentials are then defined by the limits

$$\mu_+^0 = \lim_{m_+ \to 0}(\mu_+ - RT\ln m_+),$$
$$\mu_-^0 = \lim_{m_- \to 0}(\mu_- - RT\ln m_-), \tag{15.6}$$

which imply that

$$\lim_{m_+ \to 0}\gamma_+ = 1,$$
$$\lim_{m_- \to 0}\gamma_- = 1. \tag{15.7}$$

Since we are assuming complete dissociation of the electrolyte, the chemical potential $\mu_\pm$ of $C_{\nu_+}^{z+}A_{\nu_-}^{z-}$ must be the sum of ionic chemical potentials:

$$\mu_\pm = \nu_+\mu_+ + \nu_-\mu_-. \tag{15.8}$$

Let us denote the molality of the electrolyte solution by m. Then, obviously

$$m_+ = \nu_+m \quad \text{and} \quad m_- = \nu_-m. \tag{15.9}$$

Substitution of Eqs. (15.3) and (15.4) into Eq. (15.8) yields the chemical potential $\mu_\pm$ in the form

$$\mu_\pm = \mu_\pm^0 + RT\ \ln[(\gamma_+m_+)^{\nu_+}(\gamma_-m_-)^{\nu_-}], \tag{15.10}$$

where

$$\mu_\pm^0 = \nu_+\mu_+^0 + \nu_-\mu_-^0. \tag{15.11}$$

The ionic chemical potentials μ_+ and μ_- or the ionic activity coefficients γ_+ and γ_- are not separately measurable in the laboratory. They are in fact only collectively measured. It is therefore convenient to define the mean activity coefficient by the geometric mean of γ_+ and γ_-

$$\gamma_\pm = (\gamma_+^{\nu_+}\gamma_-^{\nu_-})^{1/\nu}, \tag{15.12}$$

where

$$\nu = \nu_+ + \nu_-, \tag{15.13}$$

and the mean molality by the geometric mean of m_+ and m_-

$$m_\pm = (m_+^{\nu_+}m_-^{\nu_-})^{1/\nu}. \tag{15.14}$$

With these definitions, the chemical potential $\mu_\pm$ may be given in the form

$$\mu_\pm = \mu_\pm^0 + \nu RT\ln(\gamma_\pm m_\pm). \tag{15.15}$$

This form of chemical potential is operational and can be measured in the laboratory by using the methods discussed in Chapter 13. It can also be measured by electrochemical methods which will be discussed later. When Eq. (15.9) is made use of, the mean molality may be written in terms of the molality m of the electrolyte:

$$m_{\pm} = (\nu_+^{\nu_+} \nu_-^{\nu_-})^{1/\nu} m. \tag{15.16}$$

The mean quantities defined above suggest that the mean activity may be defined similarly:

$$a_{\pm} = (a_+^{\nu_+} a_-^{\nu_-})^{1/\nu} = \gamma_{\pm} m_{\pm}. \tag{15.17}$$

In order to better understand the meaning of the mean activity let us return to Eq. (15.8) and interpret it somewhat differently. This equation may be regarded as the chemical equilibrium condition for the chemical reaction (15.1) if we write the chemical potential for the electrolyte in the form

$$\mu_{\pm} = \mu_2^0 + RT \ln a_2, \tag{15.18}$$

where a_2 is the activity of the electrolyte and μ_2^0 is the standard chemical potential. Lewis and Randall proposed to choose this such that

$$\mu_2^0 = \mu_{\pm}^0. \tag{15.19}$$

This choice is tantamount to choosing the standard state such that the equilibrium constant of chemical reaction (15.1) becomes unity. This is easily seen as follows. Since the equilibrium constant is

$$K = \frac{a_+^{\nu_+} a_-^{\nu_-}}{a_2}, \tag{15.20}$$

and it is also given in terms of standard chemical potentials by the formula

$$K = \exp\left(\frac{\mu_{\pm}^0 - \mu_2^0}{RT}\right), \tag{15.21}$$

Eq. (15.19) implies that $K = 1$ and thus

$$a_2 = a_+^{\nu_+} a_-^{\nu_-} = a_{\pm}^{\nu}. \tag{15.22}$$

The second equality follows from Eq. (15.17). It clearly shows that $a_{\pm}$ is the geometric mean of a_+ and a_-.

Mean activities and mean activity coefficients can be determined by applying the methods discussed in Chapter 13 in addition to the electrochemical methods we are going to discuss in Chapter 17. Since electrolytes are not generally volatile, vapor pressure measurements cannot be employed for direct measurement of activity coefficients. However, indirect measurements may be carried out for the solvent by using various methods already mentioned for nonelectrolytes and then employing the Gibbs–Duhem equation to calculate the activity coefficient of the solute.

15.2. Isopiestic Method

This is a classical method for activity of an electrolyte solution, but we discuss it for the historical interest and for the principle made use of activity measurements for electrolytes. Here, we shall discuss a method of vapor pressure measurement which is called the isopiestic method. This method enables us to determine the activity of an electrolyte solution whose solvent vapor is in equilibrium with that of a solution for which the activity is accurately known over a concentration range. To make the discussion simple, we consider two binary electrolytic solutions consisting of different electrolytes. As usual, the solvent will be designated as 1 and two electrolytes as 2 and 3, respectively. Thus, one solution consists of components 1 and 2, and the other consists of components 1 and 3. We shall assume that the (1–3) solution is the reference solution for which the activities are already known. We will denote this solution by solution 2. The other, the (1–2) solution, will be denoted by solution 1. The two solutions are maintained at a given temperature and are enclosed in a single cover such that they acquire a common vapor pressure, that is, an equilibrium vapor pressure (see Fig. 15.1). When an equilibrium is reached, the solutions are analyzed for their compositions. The equilibrium conditions are then

$$\mu_1(v) = \mu_1(sol.\,1), \quad \mu_1(v) = \mu_1(sol.\,2). \tag{15.23}$$

Let us denote the activities of the solvent in solutions 1 and 2 by $a_1^{(1)}$ and $a_1^{(2)}$, respectively. Then the chemical potentials may be written as

$$\mu_1(sol.\,1) = \mu_1^l(T,p) + RT \ln a_1^{(1)},$$
$$\mu_1(sol.\,2) = \mu_1^l(T,p) + RT \ln a_1^{(2)}, \tag{15.24}$$

with obvious meaning for other symbols. The equilibrium conditions therefore imply the equation

$$a_1^{(1)} = a_1^{(2)}, \tag{15.25}$$

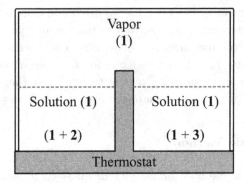

Fig. 15.1. Isopiestic method of determining activity coefficients.

which is another form of the equilibrium condition for the system. This may be rewritten in the form

$$x_1^{(1)} f_1^{(1)} = x_1^{(2)} f_1^{(2)}, \tag{15.26}$$

where the superscripts refer to the solutions. Therefore, the activity coefficient $f_1^{(1)}$ at the concentration $x_1^{(1)}$ can be determined from the known value of $f_1^{(2)}$ at $x_1^{(2)}$. To determine the solute activity coefficient, we now define osmotic coefficients ϕ_1 and ϕ_2 for the two solutions:

$$\phi_1 = -\left(\frac{1000}{\nu_2 m_2 M_1}\right) \ln a_1^{(1)},$$

$$\phi_2 = -\left(\frac{1000}{\nu_3 m_3 M_1}\right) \ln a_1^{(2)}, \tag{15.27}$$

where ν_2 is the number of ions produced by 1 mole of electrolyte 2, ν_3 the number of ions produced by 1 mole of electrolyte 3, and m_2 and m_3 are the molalities of electrolytes 2 and 3. Then the Gibbs–Duhem equations for the two solutions are given by

$$\left(\frac{1000}{M_1}\right) d\ln a_1^{(1)} = -m_2 d\ln a_2^{(1)}$$

$$= -\nu_2 m_2 d\ln(\gamma_{\pm 2} m_{\pm 2}),$$

$$\left(\frac{1000}{M_1}\right) d\ln a_1^{(2)} = -m_3 d\ln a_2^{(2)} \tag{15.28}$$

$$= -\nu_3 m_3 d\ln(\gamma_{\pm 3} m_{\pm 3}).$$

When the equilibrium condition (15.25) is applied to the two equations in Eq. (15.29), there follows the equation

$$\nu_2 m_2 d \ln(\gamma_{\pm 2} m_{\pm 2}) = \nu_3 m_3 d \ln(\gamma_{\pm 3} m_{\pm 3}). \tag{15.29}$$

Since

$$d \ln m_{\pm 2} = d \ln m_2, \quad d \ln m_{\pm 3} = d \ln m_3,$$

Eq. (15.29) may be put into a more convenient form

$$\begin{aligned} d \ln \gamma_{\pm 2} &= \mathcal{R} \, d \ln(\gamma_{\pm 3} m_{\pm 3}) - d \ln m_2 \\ &= d \ln \gamma_{\pm 3} + (\mathcal{R} - 1) d \ln(\gamma_{\pm 3} m_{\pm 3}) + d \ln \mathcal{R}, \end{aligned} \tag{15.30}$$

where $\mathcal{R}$ is the isopiestic ratio defined by

$$\mathcal{R} = \frac{\nu_3 m_3}{\nu_2 m_2}. \tag{15.31}$$

By integrating Eq. (15.30), we find the activity coefficient by the isopiestic method:

$$\ln \gamma_{\pm 2} = \ln \gamma_{\pm 3} + \ln \mathcal{R} + \int_0^{m_2} d(\gamma_{\pm 3} m_{\pm 3}) \frac{(\mathcal{R} - 1)}{\gamma_{\pm 3} m_{\pm 3}}. \tag{15.32}$$

Therefore, a plot of $[(\mathcal{R} - 1)/\gamma_{\pm 3} m_{\pm 3}]$ vs. $\gamma_{\pm 3} m_{\pm 3}$ and a graphical integration of the curve will yield the integral and thus the mean activity coefficient of electrolyte 2. It is sometimes necessary to combine the isopiestic method with other methods such as osmotic pressure and freezing point measurements to cover the required range of concentration, since it is not possible to obtain accurate values for $\mathcal{R}$ below 0.1 m in concentration.

15.3. Activity Coefficient from Freezing Point Measurement

Since freezing point measurements of electrolytes give highly accurate activity coefficients, they are an important source of information. If the expansions (13.138) and (13.139) are used in Eq. (13.137), it can be recast into the form more suitable for discussion:

$$d \ln a_1 = -(RT_m^2)^{-1}[\Delta h_m^0 + b\theta + O(\theta^2)]d\theta, \qquad (15.33)$$

where a_1 is the activity of the solvent,

$$\theta = T_m - T,$$

$$b = \Delta h_m^0 - \tfrac{1}{2}T_m^2 \Delta C_{pm},$$

and we have retained only two terms in the expansion. Since the Gibbs–Duhem equation may be written at constant T and p as

$$d \ln a_2 = -\frac{1000}{mM_1}d \ln a_1,$$

the equation for a_2 is

$$d \ln a_2 = \frac{1000}{RT_m^2 M_1}\left(\Delta h_m^0 + b\theta\right)\frac{d\theta}{m}. \qquad (15.34)$$

With the definitions

$$\lambda = \frac{RT_m^2 M_1}{1000\,\Delta h_m^0},$$

$$\eta = \frac{1000b}{RT_m^2 M_1},$$

and, with the aid of Eqs. (15.22) and (15.17), we obtain the following equation for the activity coefficient of the electrolyte

$$d \ln \gamma_\pm = (\nu\lambda m)^{-1}d\theta + \eta(\nu m)^{-1}\theta d\theta - d \ln m. \qquad (15.35)$$

Integration of this equation will supply the mean activity coefficient. Since the freezing point depression θ is a function of m, the integration requires measurements of θ over a range of m.

Since $\theta = \nu\lambda m$ at very low concentrations as can easily be inferred from Eq. (13.142), Eq. (15.35) is not in a suitable form for integration because of the logarithmic behavior of the first and last terms. The following transformation proposed by Lewis is found to be useful for such a purpose:

$$\Omega = 1 - \frac{\theta}{\nu\lambda m}.$$

Then

$$d\Omega = \theta(\nu\lambda m^2)^{-1}dm - (\nu\lambda m)^{-1}d\theta,$$

so that

$$d \ln \gamma_\pm = -\Omega d \ln m - d\Omega + \eta\theta(\nu m)^{-1}d\theta. \qquad (15.36)$$

By integrating this equation, we obtain

$$\ln \gamma_{\pm} = -\Omega - \int_0^m \Omega \, d\ln m + \int_0^m \nu^{-1} \eta \theta \left(\frac{d\theta}{dm} \right) d\ln m. \qquad (15.37)$$

This form of equation has no logarithmic singularity and a graphical integration becomes quite straightforward and not prone to inaccuracy.

15.4. Activity Coefficient from Osmotic Pressure Measurement

The osmotic pressure expression (13.167) for nonelectrolytes can also be directly made use of for determining the activity coefficients of electrolytes. We consider a binary solution. Changing the variable from p to the osmotic pressure π, we write Eq. (13.167) in the form

$$RT \ln a_1 = - \int_0^{\pi} dz \, \bar{v}_1, \qquad (15.38)$$

where $\bar{v}_1$ is a function of z: $\bar{v}_1 = \bar{v}_1 \left(p^0 + z \right)$. We introduce the osmotic coefficient of the electrolyte defined by the expression

$$\phi = - \left(\frac{1000}{\nu m M_1} \right) \ln a_1. \qquad (15.39)$$

Note that there appears the total number of ions produced by the electrolyte in the definition of osmotic coefficient, since the molality is increased by the factor ν in the case of a strong electrolyte. Replacing with ϕ the activity term on the left-hand side of Eq. (15.38), we obtain

$$\left(\frac{\nu M_1 RT}{1000} \right) m\phi = \int_0^{\pi} dz \, \bar{v}_1. \qquad (15.40)$$

Since the solvent is generally negligibly compressible, the partial molar volume on the right in Eq. (15.40) may be regarded as a constant and the integration may be easily performed. We then obtain the osmotic pressure

$$\pi = \left(\frac{\nu M_1 RT}{1000 \bar{v}_1^0} \right) m\phi, \qquad (15.41)$$

where $\bar{v}_1^0 = \bar{v}_1(p^0)$. Compare this expression for the equivalent formula for nonelectrolytes (see Section 13.9.4). The terminology osmotic coefficient originated from this equation. This equation suggests that measurements

of the osmotic pressure yield the values of the osmotic coefficient which can
be used to find the activity coefficient of the electrolyte by applying the
method leading to Eq. (13.111):

$$\ln \gamma_{\pm} = \phi - 1 + \int_0^m dm \frac{\phi - 1}{m}, \qquad (15.42)$$

where ϕ is given by Eq. (15.39).

The osmotic pressure effect can be significantly noticeable even for
dilute solutions, but its measurement requires a reliable semipermeable
membrane. Recent developments in polymeric membranes and zeolites
could make this effect achieve its potential for many systems of interest
and usefulness in electrochemistry.

15.5.   Additional Remarks on Activity Coefficients of Electrolyte Solutions

The three examples of methods measuring activity coefficients of electrolyte
solutions discussed in the last three sections do not exhaust the methods
to determine the electrolytic solution activity coefficients. One may use any
colligative properties for which thermodynamic theories may be developed
in analogy to those for neutral (nonionic) solutions described in the earlier
chapters. For this purpose, we may also make use of the thermodynamics
of electrochemical cells described in Chapter 17.

As alluded to earlier in the introductory remark of this chapter, the
concentration dependence of activity coefficients of electrolyte solutions is
known to be markedly different from the concentration dependence of neutral solutes, which will be discussed in Chapter 16 where the Debye–Hückel
theory is discussed. As a matter of fact, the Debye–Hückel theory predicts
that unlike the neutral solutes, whose logarithm of activity coefficient vanishes with vanishing concentration, it depends on concentration like

$$\log \gamma_{\pm} = -B\sqrt{2m} \qquad (15.43)$$

in the limit of $m \to 0$, where m is the molality of the electrolyte and B is
a constant depending on temperature and other parameters characteristic
to the electrolyte. This result of the Debye–Hückel theory has been amply
verified experimentally to be valid.[1]

---

[1]See, for example, H. S. Harned and B. B. Owen, *The Physical Chemistry of Electrolyte
Solutions* (Reinhold, New York, 1958) and numerous references cited therein.

Chapter 16

Debye–Hückel Theory of Strong Electrolyte Solutions

We have indicated in Chapter 15 that solutions of strong electrolytes exhibit nonideal behavior even at very low concentrations at which nonelectrolyte solutions would normally obey Henry's law. We have seen that in the case of binary solutions, the Margules expansions for nonelectrolytic solutions begin with the square of a mole fraction. That is, for example, $\ln f_1$ tends to zero quadratically with respect to the concentration of the solute. This indicates that the activity coefficients tend to unity faster for neutral molecules than electrolyte solutions. The situation is drastically different for electrolytes. While studying thermodynamics of strong electrolyte solutions, G. N. Lewis in 1913 observed that in contrast to nonelectrolyte solutions, the mean excess free energies of electrolytes in solutions tend to zero as $m^{1/2}$. This means that the nonideality of the solutions persists down to much lower concentrations than for nonelectrolyte solutions.

In order to make this statement more precise, let us introduce the concept of ionic strength which was defined by Lewis as

$$\Gamma = \frac{1}{2} \sum_i m_i z_i^2. \tag{16.1}$$

Here, m_i is the molality of ionic species i and z_i is its charge number. Lewis showed experimentally that

$$\ln \gamma_{\pm} = -B\sqrt{\Gamma}, \tag{16.2}$$

where B is a constant dependent on the temperature and other properties such as the dielectric constant of the solution.

This was a vexing phenomenon at that time until Debye and Hückel came up with a theory[1] in 1923 on Eq. (16.2) as a limiting law for the activity coefficient. Their theory is a statistical mechanical theory. Such a statistical mechanical theory would not normally have a place in a textbook that treats thermodynamics phenomenologically. However, their theory has far-reaching implications for many physical phenomena related to ionic solutions and plasmas. In fact, it was the first theory which showed how one might treat interactions in many particle systems in an average way (mean field theory), particularly, for charged particle systems and consequently has had a considerable influence on the thinking in physical chemistry and physics of electrolytes and plasmas. For this reason, we make a break from our basic philosophy of describing phenomenology in this textbook and discuss their theory in an elementary manner by using a method of statistical mechanics.

One of the most important basic assumptions in the Debye–Hückel theory is that all deviations from ideality in ionic solutions are caused by the Coulomb interactions between ions produced by the electrolyte(s) in the solution. In the second assumption, it is postulated that the solvent is a structureless dielectric continuum of dielectric constant D. In the third assumption, ions are assumed to be hard spheres of radius r_i with the charge distributed uniformly on the surface. Therefore, we imagine that an ionic solution is a continuous medium in which charged ions are immersed, while interacting with each other through long-ranged Coulomb potentials.

16.1. Ionic Atmosphere

Ions in the solution do not remain in one position, but continuously and randomly move around — i.e., execute Brownian motions — while mutually interacting, owing to the thermal agitation exerted on them by the medium. Since ions of opposite charges tend to attract each other, if we take an ion and count the distribution of ions around it, there will be a tendency to gather more ions of the opposite charge, but as we move farther out from the ion, the clustering of oppositely charged ions will gradually diminish

---

[1]P. Debye and E. Hückel, *Phys. Z.* **185**, 305 (1923).

and then the distribution of positively and negatively charged ions will be evened out and become that of the bulk. This region of space of an uneven charge distribution around the ion in question is called the ionic atmosphere.

Since the ionic atmosphere is, on the average, charged oppositely to the ion in question, there is a Coulomb interaction between the ion and its ionic atmosphere, and the first task is then to calculate this interaction potential Φ. Since the charges in the ionic atmosphere are distributed and these distributions depend on the concentration of ions, the potential is expected to depend on the concentration of the solution. The potential on an ion is determined by the charge distribution around it through the Poisson equation

$$-\nabla^2 \Phi(\mathbf{R}) = 4\pi \rho / D, \qquad (16.3)$$

where ρ is the charge distribution and D is the dielectric constant of the solvent.

To solve Eq. (16.3) for Φ, it is necessary to know the charge distribution ρ in the ionic atmosphere. Debye and Hückel assumed that the distribution is given by the Boltzmann distribution function. If the potential on a charge e_i at position $\mathbf{R}$ is $\Phi(\mathbf{R})$, then the potential energy is

$$V(\mathbf{R}) = e_i \Phi(\mathbf{R}). \qquad (16.4)$$

According to the assumption just made, the number n_i' of ion i of charge e_i at position $\mathbf{R}$ is distributed according to the formula

$$n_i' = n_i \exp\left[-\frac{V(\mathbf{R})}{k_B T}\right] = n_i \exp\left[-\frac{e_i \Phi(\mathbf{R})}{k_B T}\right], \qquad (16.5)$$

where n_i is the bulk number density of ion i defined by

$$n_i = \frac{\text{total number of ionic species } i \text{ in the solution}}{\text{volume of the solution}}. \qquad (16.6)$$

The charge distribution ρ is then given by the expression

$$\rho = \sum_i n_i e_i \exp\left[-\frac{e_i \Phi(\mathbf{R})}{k_B T}\right], \qquad (16.7)$$

where the summation is over all ionic species. For example, if the solution is made up by a univalent electrolyte only, then

$$\rho = ne \left\{ \exp \left[\frac{e\Phi(\mathbf{R})}{k_B T} \right] - \exp \left[-\frac{e\Phi(\mathbf{R})}{k_B T} \right] \right\}, \tag{16.8}$$

where n is the number density of the electrolyte in the solution and e is the charge of an electron (the absolute value).

16.2. Mean Potential and the Excess Free Energy

When Eqs. (16.3) and (16.7) are combined, there follows the Poisson–Boltzmann equation:

$$\nabla^2 \Phi(\mathbf{R}) = -\frac{4\pi}{D} \sum_i n_i e_i \exp \left[-\frac{e_i \Phi(\mathbf{R})}{k_B T} \right]. \tag{16.9}$$

This equation unfortunately is nonlinear[2] with respect to Φ and not solvable in an analytic form. However, at low concentrations, the Boltzmann factor may be expanded and the nonlinear terms in the expansion may be neglected to a good approximation. Then there follows a linearized equation

$$\nabla^2 \Phi(\mathbf{R}) = -\frac{4\pi}{D} \sum_i n_i e_i \left[1 - \frac{e_i \Phi(\mathbf{R})}{k_B T} \right]. \tag{16.10}$$

Since by the electroneutrality

$$\sum_i n_i e_i = 0,$$

we finally obtain

$$\nabla^2 \Phi(\mathbf{R}) = \frac{4\pi}{D k_B T} \sum_i n_i e_i^2 \Phi(\mathbf{R}). \tag{16.11}$$

---

[2]If the Poisson–Boltzmann (PB) equation is regarded as a differential equation that by itself defines $\Phi(\mathbf{R})$, then it is indeed nonlinear. However, from the more complete statistical mechanical viewpoint of electrolytes Φ or, more generally, the charge density in Eq. (16.7) obeys, for example, Fokker–Planck equations for pair correlation functions to which the PB equation is coupled. In this extended viewpoint, the PB equation is no longer nonlinear. L. Onsager solved such coupled differential equations in the low density limit in his work on conductance of electrolytes. See L. Onsager and R. M. Fuoss, *J. Phys. Chem.* **34**, 2689 (1932).

It is convenient to define a parameter

$$\kappa = \left(\frac{4\pi}{D k_B T} \sum_i n_i e_i^2 \right)^{1/2}, \tag{16.12}$$

which is called the Debye screening constant or the inverse Debye length. It has the dimension of reciprocal length and gives a measure of the size of the ionic atmosphere in the solution as we will see later.

Since ions are assumed to be hard spheres, the ionic atmospheres around them must be spherically symmetric and consequently the potential $\Phi(\mathbf{R})$ must be spherically symmetric. This means that Φ does not depend on the orientation of position vector $\mathbf{R}$. In this case, it is convenient to use the spherical coordinate system, and we cast the linearized Poisson–Boltzmann equation in the form

$$\frac{1}{R^2} \frac{d}{dR} R^2 \frac{d}{dR} \Phi(R) = \kappa^2 \Phi(R). \tag{16.13}$$

This ordinary differential equation is easily solved. Its general solution given by

$$\Phi(R) = A \frac{e^{-\kappa R}}{R} + A' \frac{e^{\kappa R}}{R}, \tag{16.14}$$

where A and A' are the integration constants. They are determined as follows. First, let us observe that the potential must vanish as $R \to \infty$, since the ionic distribution becomes that of the bulk solution, namely, uniform with respect to the positive and negative charges, as R is increased from the position of the ion in question. This boundary condition can be satisfied by the solution (16.14) only if $A' = 0$. Therefore, the acceptable solution is

$$\Phi(R) = \frac{A}{R} \exp(-\kappa R). \tag{16.15}$$

The remaining constant A will be determined by making use of the electroneutrality of the ionic atmosphere. The potential obtained here has an exponential factor which decreases with increasing R and its rate of decrease is determined by the Debye screening factor κ. If κ is equal to zero, the potential is like a Coulomb potential. This potential is called the Debye potential.

The charge e_i on ion i must be exactly balanced by other ions in the ionic atmosphere. Since the charge density in the volume element $4\pi R^2 dR$

around the central ion i in the ionic atmosphere is

$$4\pi\rho R^2 dR,$$

the electroneutrality condition is

$$e_i = -\int_{r_i}^{\infty} dR\ R^2 4\pi\rho. \tag{16.16}$$

Since to the linear approximation in Φ the charge density in Eq. (16.17) is given by the expression

$$\rho = -\sum_i \frac{n_i e_i^2}{k_B T} \Phi(R), \tag{16.17}$$

substitution of Eq. (16.17) into Eq. (16.16) yields the electroneutrality condition in the form

$$e_i = AD\kappa^2 \int_{r_i}^{\infty} dR\ R\ \exp(-\kappa R). \tag{16.18}$$

Upon integration, it gives rise to the equation for A:

$$A = \frac{e_i \exp(\kappa r_i)}{D(1 + \kappa r_i)}.$$

Upon substitution of this result into Eq. (16.18), we finally obtain the potential in the form

$$\Phi(R) = \frac{e_i \exp[-\kappa(R - r_i)]}{D(1 + \kappa r_i)R}. \tag{16.19}$$

At $R = r_i$ the potential has the form

$$\Phi(r_i) = \frac{e_i}{Dr_i} - \frac{e_i \kappa}{D(1 + \kappa r_i)}$$

$$:= \Phi_s + \Phi_c. \tag{16.20}$$

The first term on the right Φ_s is the potential created by the charge of ion i itself which is uniformly distributed on the surface of the sphere, and this part is obviously independent of other ions. The second part Φ_c is the contribution from the ionic atmosphere and depends on the concentrations of other ions present in the solution. This is that part of the potential contributing to the excess chemical potential, which depends on the concentrations of ions in the surrounding the ion i in the solution.

Since by the first assumption, the excess chemical potential arises from the Coulomb interaction, this excess free energy must be equal to the work done to bring an ion against the potential Φ_c. This work may be calculated by the following device. The work is divided up into two parts: one part is that of bringing a neutralized ion to the point $R = r_i$, and the other is that of charging the neutralized ion brought to $R = r_i$ to the full ionic charge e_i. The first part is small compared to the second in magnitude and therefore can be neglected. Then the total work is that of charging the ion on the sphere of radius r_i. It is given by

$$W_{\text{electrical}} = \int_0^{e_i} de\, \Phi_c(r_i)$$

$$= -\frac{e_i^2 \kappa}{2D(1 + \kappa r_i)}. \tag{16.21}$$

This electrical work must be equal to the excess free energy of the ion, which may be identified by the excess chemical potential $k_B T \ln \gamma_i$:

$$k_B T \ln \gamma_i = -\frac{z_i^2 e^2 \kappa}{2D(1 + \kappa r_i)}, \tag{16.22}$$

where γ_i is the activity coefficient of ion i and z_i is its charge number: $e_i = z_i e$ where z_i is positive for cations and negative for anions.

The mean activity coefficient of the electrolyte $C_{\nu_+}^{z_+} A_{\nu_-}^{z_-}$ is now easily calculated with Eq. (16.22):

$$\nu k_B T \ln \gamma_\pm = \nu_+ k_B T \ln \gamma_+ + \nu_- k_B T \ln \gamma_-$$

$$= -\frac{(\nu_+ z_+^2 + \nu_- z_-^2) e^2 \kappa}{2D(1 + \kappa r)},$$

for which we have replaced the ionic radii r_i with their mean value $r = (r_+ + r_-)/2$ in the denominator. Since $\nu_+ z_+ + \nu_- z_- = 0$ by the electroneutrality of the electrolyte, we find

$$\nu_+ z_+^2 + \nu_- z_-^2 = \nu_+ z_+ (z_+ - z_-),$$

and

$$\frac{\nu_+ z_+^2 + \nu_- z_-^2}{\nu_+ + \nu_-} = \frac{\nu_+ z_+ (z_+ - z_-)}{\frac{\nu_+ z_+}{z_+} + \frac{\nu_- z_-}{z_-}} = -z_+ z_- = |z_+ z_-|.$$

Using this relation, we rewrite the mean activity coefficient in the form

$$\ln \gamma_\pm = -\frac{|z_+ z_-| e^2 \kappa}{2D k_B T (1 + \kappa r)}. \tag{16.23}$$

This is the Debye–Hückel expression for the mean excess chemical potential of an electrolyte. To put Eq. (16.23) into a more useful form, we express the Debye screening factor κ in molality and, more specifically, the ionic strength Γ:

$$\kappa = \left(\frac{8\pi e^2 N^2 \rho_0}{1000 DRT} \Gamma \right)^{1/2}, \tag{16.24}$$

where ρ_0 is the density of the solvent and N is the Avogadro number. It must be stressed that κ is proportional to the square root of ionic strength Γ. The radius of the ionic atmosphere therefore decreases with increasing ionic strength, and the excess chemical potential is approximately proportional to $\Gamma^{1/2}$. If the concentration is so small that $\kappa r \ll 1$, then the κr term in the denominator can be neglected, and we obtain the excess chemical potential in a form similar to the formula found empirically by Lewis: at 298.15 K,

$$\ln \gamma_\pm = -B' |z_+ z_-| \sqrt{\Gamma}$$

$$= -0.509 |z_+ z_-| \sqrt{\Gamma}, \tag{16.25}$$

where B' is given by

$$B' = \frac{e^2}{2Dk_B T} \left(\frac{8\pi e^2 \rho_0 N^2}{1000 DRT} \right)^{1/2}. \tag{16.26}$$

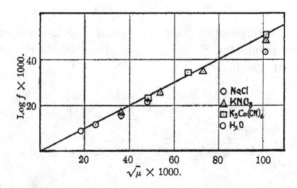

Fig. 16.1. Mean activity vs. ionic strength (denoted by μ). The solid line is the Debye–Hückel theory and the symbols are experiment. Reproduced with permission from N. J. Brönsted and V. K. La Mer, J. Amer. Chem. Soc. **46**, 557 (1924) © 1924 American Chemical Society.

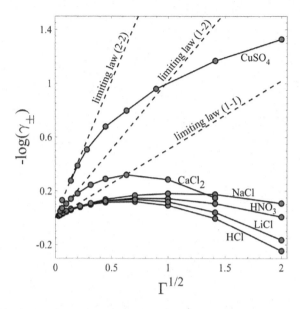

Fig. 16.2.   The broken lines are the Debye–Hückel theory limiting law that is valid for very dilute electrolyte solutions. The concentration range of validity is generally of the order of 0.01 per mole or less. The figure is prepared with experimental data available in various literature sources.

This gives the molecular theoretic meaning of the empirical constant B in the Lewis formula for the excess free energy of an electrolyte.

The Debye–Hückel theory correctly provides the limiting law for the concentration dependence of mean activity coefficients (see Figs. 16.1 and 16.2). As can be expected on the basis of the assumptions and approximations made and evident from Fig. 16.2, the theory is limited to the low concentration regime, more specifically, below about 0.01 m in concentration. Above this value of molality, the theory begins to show considerable deviations. This limitation should be expected from the linearization approximation made to Poisson–Boltzmann equation, giving rise to the linear differential equation (16.11). Improving the theory beyond the limitation is not trivial and poses theoretical challenge, which stems from the fact that since Coulomb potentials are infinite-ranged and the integrals involved in the statistical mechanical treatment are divergent and thus require unconventional theoretical stratagems. This subject is beyond the scope of this chapter. The readers are referred to numerous modern theories of electrolyte solutions in the literature.

Chapter 17

Galvanic Cells and Other Classes of Cells

Chemical reactions involving an electrolytic compound with another element or a compound thereof involve a chemical energy change before and after the chemical reactions. Electric cells — commonly known as electric batteries — utilize the chemical energy difference associated with the reaction to convert it into an electrical current. This process of energy conversion can be described by means of thermodynamics. There are a number of energy conversion devices known at present that can be classified into electric cells in a broad stroke of classification.

In this chapter, galvanic cells will be principally discussed as a classical prototype of electrochemical cells used as an energy storage device. They are historically the oldest kind of energy storage devices. In recent years, many energy storage or energy conversion devices have been invented, which are often designed to circumvent the limitations of galvanic cells or as alternatives to the galvanic cells. However, since for many cases of alternative cells their underlying theories also require dynamical theories of transport processes beyond the scope of equilibrium thermodynamics, we will limit the discussion on other kinds of energy conversion devices to photovoltaic and betavoltaic cells, which permit us to discuss about them without invoking theories of transport processes involved.

17.1. Reversible Galvanic Cells and Reversible Electrodes

A galvanic cell is an electrochemical device consisting of more than two phases of electrolytes and two electrodes in which a chemical reaction occurs and the free energy of the chemical reaction is transformed into a current which is conducted through the two electrodes of the cell. In reversible cells, chemical reactions are made to occur infinitesimally slowly by imposing an opposing potential. See Fig. 17.1 for an example of galvanic cells.

There are several types of electrodes used in constructing galvanic cells. Regardless of the types of electrodes employed, they are based on the same fundamental principle that an oxidized and reduced state of an element are always involved. Here, oxidation refers to the loss of electrons and reduction to the gain of electrons by the element or the ion. The first type of electrode consists of an element in contact with a solution of its own ion as in the case of a metallic electrode which is immersed in a solution of its soluble salt. Another example is the hydrogen electrode which may be in contact with a solution of, say, HCl. Since hydrogen is nonconducting, an inert metallic conductor, for example, Pt or Au, is used to maintain electrical contact. The second type of electrode consists of a metal and its salt of low solubility which is in contact with a solution of the anion of the salt, for example, Ag,

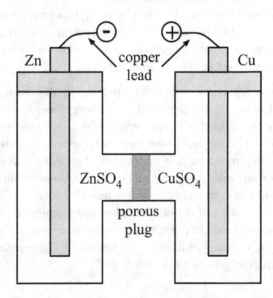

Fig. 17.1.   A galvanic cell without a liquid junction.

AgCl(s), MCl(solution) where M is an element which forms a halide. The third type of electrode consists of an inert metal such as gold and platinum immersed in a solution which contains both oxidized and reduced states of an oxidation–reduction system, for example, Fe^{2+} and Fe^{3+}.

Galvanic cells are assemblies of two electrodes of the types just described and in each electrode, the following chemical reaction occurs:

$$M \rightleftharpoons M^{z\pm} \pm ze.$$

17.2.   Electrochemical Potentials

The thermodynamic formalism for reversible galvanic cells and electrochemical systems in general is the same as for nonelectrolytes except for the fact that the electroneutrality condition

$$\sum_i n_i e_i = 0 \tag{17.1}$$

must be imposed, where n_i is the mole number and e_i is the charge on ionic species i. Therefore, $z_i = e_i/e$ is the charge number of the ion, e being the unit of charge. In order to appreciate the significance and magnitude of the effect in the event that it is locally violated because of local fluctuations in concentrations, let us consider a sphere of radius 1 cm which contains an excess charge[1] of 10^{-12} moles. The excess charge will be distributed on the surface of the sphere and the potential of this charge distribution on the surface, denoted ψ, is given by the expression

$$\psi = \frac{Q}{4\pi\epsilon_0 R},$$

where Q is the charge on the sphere, R is the radius, and ϵ_0 is the vacuum permittivity. Since $4\pi\epsilon_0 = 1.11 \times 10^{-10}\,CV^{-1}m^{-1}$, for $Q = 10^{-12} \times 0.96 \times 10^5\,C$ and $R = 10^{-2}\,m$ we find

$$\psi = \frac{0.96 \times 10^{-7}}{1.11 \times 10^{-10}} \times 10^2\,V = 0.86 \times 10^5\,V.$$

This simple calculation indicates that a fluctuation in ionic concentration on the order of a picomole results in a potential of an order of $10^5\,V$, but

---

[1]The absolute value of the charge on an electron is $e = 1.6021 \times 10^{-19}$ Coulombs and the charge of ionic species of charge number 1 is 1 Faraday defined as $F = 0.9687 \times 10^5$ Coulombs mol^{-1} for a mole of ions.

the concentration change of 10^{-12} moles is not accessible to a laboratory detection. Therefore, two phases of practically the same chemical composition may have different electrical potentials because of undetectable charge fluctuations. For this reason, the chemical potential may be written in two parts: one depending on the composition and the other depending on the electrical work and thus on the electrical potential:

$$\mu_{ei} = \mu_i + z_i F \psi. \tag{17.2}$$

Here, μ_i is the composition part of the chemical potential and ψ is the electrical potential and F denotes Faraday for a mole of charge ($1F = 9.64931 \times 10^4$ C). The μ_{ei} is called the electrochemical potential of ionic species i. This division of μ_{ei} into two parts, however, has no operational meaning, since it is not possible to measure the two parts separately. The difference in the electrochemical potentials of two phases α and β of an identical chemical composition is then

$$\mu_{ei}^\alpha - \mu_{ei}^\beta = z_i F(\psi^\alpha - \psi^\beta), \tag{17.3}$$

where the quantities superscripted with α and β refer to those in phases α and β, respectively. Phases α and β are the conducting wires attached to the two electrodes of the cell. The potential difference $(\psi^\alpha - \psi^\beta)$ is called the electromotive force (EMF) of the cell. It must be stressed that *the potential differences $(\psi^\alpha - \psi^\beta)$ and thus the electromotive forces are definable only if the two phases are identical in chemical composition.* The two phases are in equilibrium with respect to ionic species i only if

$$\mu_{ei}^\alpha = \mu_{ei}^\beta, \tag{17.4}$$

and, therefore, the potential difference vanishes at equilibrium.

If the potential is variable, then the Gibbs relation for a phase is

$$dE = TdS - pdV + qd\psi + \sum_i \mu_{ei} dn_i, \tag{17.5}$$

where

$$q = \sum_i n_i z_i F. \tag{17.6}$$

This is a generalized form of the Gibbs relation used for nonelectrolytes for electrochemical cells. If the potentials ψ^α and ψ^β are constant regardless of phases, then it takes the form

$$dE = TdS - pdV + \sum_i \mu_{ei} dn_i. \tag{17.7}$$

The internal energy is generally dependent on the electric energy, and an electrochemical energy may be defined so as to include the electrical contribution. This is, of course, suggested by the fact that the electrochemical potential may be given as a derivative of the internal energy,

$$\mu_{ei} = \left(\frac{\partial E}{\partial n_i} \right)_{S,V,n_i,\psi}. \tag{17.8}$$

It is customary to define the electrochemical energy by the relation

$$E_e = E + \sum_i n_i z_i F \psi, \tag{17.9}$$

where E denotes the internal energy in the absence of potential ψ and the second term on the right arises from the electrical work on the ionized species. This division is as arbitrary as is for the electrochemical potential partitioned as in Eq. (17.2). With the electrochemical energy so defined, the fundamental equation may be written as

$$dE_e = TdS - pdV + \psi dq + \sum_i \mu_{ei} dn_i. \tag{17.10}$$

It is possible to develop a thermodynamic theory of electrochemical cells by using Eq. (17.5) or Eq. (17.10). However, we will not have an occasion to use them in this chapter.

17.3. Galvanic Cells without Liquid Junction

Galvanic cells are classified into two major categories of cells with and without liquid junctions. Liquid junctions occur when cells consist of electrodes in contact with a liquid solution of the salt of the element constituting the electrode; they are a source of irreversibility. This kind of irreversibility can be minimized if a salt bridge is employed. They generally give rise to a noticeable junction potential which complicates the matter. Here, we consider cells without liquid junctions.

17.3.1. *Cell Diagrams and the Sign Convention*

We have defined the electromotive force of a cell by the potential difference between two wires of the same metal connected to the two electrodes of the cell. However, there is still a sign ambiguity in the definition and it is necessary to fix the sign by adopting a convention. To illustrate it, we will take an example as follows.

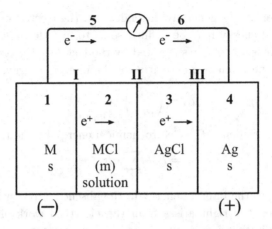

Fig. 17.2.   Silver(Ag)-metal galvanic cell.

We denote a galvanic cell with a diagram which indicates various components comprising the cell. We call it the cell diagram (see Fig. 17.2). Let us consider a cell denoted by the following cell diagram:

$$M(s)|MCl(m)|AgCl(s)|Ag(s),\qquad\qquad (17.11)$$

where M is either a metal or an element which can form a compound with chlorine, for example, H_2. Different phases are separated by a vertical bar in between and the letter m or s in the parentheses indicates the state or concentration of the substance in question. In constructing a cell diagram, the cell is arranged such that the following convention applies.

Convention. *The oxidizing reaction occurs in the electrode on the left and the reducing reaction occurs in the electrode on the right.*

In this convention, the chemical reaction occurring in the cell is written such that the direction of the reaction corresponds to that of positive electricity flowing through the cell from left to right (or that of negative electricity flowing from right to left). Therefore, the chemical reaction for the cell diagram written is

$$M(s) + AgCl(s) = MCl(s) + Ag(s),\qquad\qquad (17.12)$$

where M is oxidized whereas Ag^+ is reduced. When put in terms of the direction of electron movement through the wire connecting the two electrodes, the convention is equivalent to saying that electrons are transferred from the left electrode to the right *through the wire connecting the electrodes.*

To facilitate discussions on electromotive force, we schematically represent the cell (17.11) as a system of heterogeneous phases as in Fig. 17.2. Boundary I is assumed to be permeable to ion M^+ only, Boundary II to ion Cl^- only, and Boundary III to ion Ag^+ only. That is, *the boundaries are semipermeable.* If the temperature and pressure are the same for all phases, the equilibrium conditions for the heterogeneous systems are

$$\mu_{eM^+}(1) = \mu_{eM^+}(2),$$
$$\mu_{eCl^-}(2) = \mu_{eCl^-}(3), \tag{17.13}$$
$$\mu_{eAg^+}(3) = \mu_{eAg^+}(4),$$

where the number in the parentheses indicates the phase in which the substance in question is found. Therefore, $\mu_{eM^+}(1)$ denotes the electrochemical potential of ion M^+ in phase 1 and so on. In phases 1 and 4,

$$\mu_{eAg}(4) = \mu_{Ag}(4), \quad \mu_{eM}(1) = \mu_M(1), \tag{17.14}$$

owing to the fact that M and Ag are neutral, but since M and Ag are decomposable into ions and an electron, the following relations hold:

$$\mu_{eM}(1) = \mu_{eM^+}(1) + \mu_{ee^-}(1),$$
$$\mu_{eAg}(4) = \mu_{eAg^+}(4) + \mu_{ee^-}(4). \tag{17.15}$$

By combining these relations with the equilibrium conditions (17.13) and making use of Eq. (17.2) for ions, we find

$$\mu_M(1) - \mu_{ee^-}(1) = \mu_{M^+}(2) + F\psi(2),$$
$$\mu_{Cl^-}(2) - F\psi(2) = \mu_{Cl^-}(3) - F\psi(3),$$
$$\mu_{Ag}(4) - \mu_{ee^-}(4) = \mu_{Ag^+}(3) + F\psi(3).$$

Subtraction of the last equations from the first and rearranging the terms yield

$$\mu_{ee^-}(4) - \mu_{ee^-}(1) = [\mu_{Ag}(4) + \mu_{M^+}(2) + \mu_{Cl^-}(2)]$$
$$- [\mu_M(1) + \mu_{Ag^+}(3) + \mu_{Cl^-}(3)]$$
$$= [\mu_{Ag}(4) + \mu_{MCl}(2)] - [\mu_M(1) + \mu_{AgCl}(3)]. \tag{17.16}$$

Note that the right-hand side of the second equality of Eq. (17.16) is simply the affinity of the chemical reaction (17.12); see Eqs. (12.10) and (12.11) for the definition of affinity. We also observe that if the wire connecting the two electrodes is of a composition different from those of the electrodes, then there hold the following equilibrium conditions for electrons in phase 1 and phase 5 or phase 4 and phase 6:

$$\mu_{ee^-}(1) = \mu_{ee^-}(5), \quad \mu_{ee^-}(4) = \mu_{ee^-}(6),$$

where $\mu_{ee^-}(5)$ and $\mu_{ee^-}(6)$ are the electrochemical potentials of the wire attached to the electrodes in phases 1 and 4, respectively. Of course, the assumption here is that the boundaries between the electrodes and wire are permeable to the electrons only. Then the left-hand side of Eq. (17.16) may be written as

$$\mu_{ee^-}(4) - \mu_{ee^-}(1) = [\mu_{ee^-}(4) - \mu_{ee^-}(6)] + \mu_{ee^-}(6)$$
$$- [\mu_{ee^-}(1) - \mu_{ee^-}(5)] - \mu_{ee^-}(5)$$
$$= \mu_{ee^-}(6) - \mu_{ee^-}(5)$$
$$= - F[\psi(6) - \psi(5)].$$

The electromotive force of the cell is defined by

$$\epsilon = \psi(6) - \psi(5) = \frac{\mu_{ee^-}(4) - \mu_{ee^-}(1)}{- F}. \tag{17.17}$$

This is *the American convention* for EMF. It is opposite to *the European convention*. To put it simply as a summary, we adopt the sign of the EMF of a cell as the electrochemical potential $\mu_{ee^-}(4)$ of the right electrode minus the electrochemical potential $\mu_{ee^-}(1)$ of the left electrode divided by $-F$:

$$\epsilon = \frac{\mu_{ee^-}(\text{right}) - \mu_{ee^-}(\text{left})}{-F}. \tag{17.18}$$

When this equation is combined with Eq. (17.16) and the chemical potentials are written as

$$\mu_i = \mu_i^0 + RT \ln a_i, \tag{17.19}$$

where a_i is the activity of species i and μ_i^0 is its appropriately chosen standard chemical potential, there follows the equation for EMF

$$\epsilon = \epsilon^0 - \frac{RT}{F} \ln \left(\frac{a_{MCl} a_{Ag}}{a_M a_{AgCl}} \right). \tag{17.20}$$

Here, ϵ^0 is the standard electromotive force defined by

$$\epsilon^0 = -\frac{1}{F}\left[\mu_{Ag}^0 + \mu_{MCl}^0 - (\mu_M^0 + \mu_{AgCl}^0)\right]. \tag{17.21}$$

Since we may put

$$a_{Ag}(s) = a_M(s) = a_{AgCl}(s) = 1$$

according to the standard states chosen for activity (see Eq. (15.17) for the definition of mean activity and the related equations.), Eq. (17.20) may be written in the form

$$\epsilon = \epsilon^0 - \frac{RT}{F}\ln a_{MCl}. \tag{17.22}$$

It is called Nernst's equation. Since we may write the activity a_{MCl} in terms of mean activity coefficient and mean molarity

$$a_{MCl} = (\gamma_\pm m_\pm)^2,$$

the Nernst equation may be given in terms of the mean activity coefficient $\gamma_\pm$ and molality $m_\pm$ of electrolyte $MCl(m)$:

$$\epsilon = \epsilon^0 - \frac{2RT}{F}\ln(\gamma_\pm m_\pm). \tag{17.23}$$

In the case of the galvanic cell, we have just considered that there is only 1 mole of electrons transferred in the course of the chemical reaction, since the metals are univalent.

In order to generalize the result obtained above to a case where more than 1 mole of electrons are transferred by the reaction, we consider a galvanic cell of the following type:

$$L(s)|LX_n(m)|RX_n(s)|R(s), \tag{17.24}$$

where L and R are elements constituting the electrodes and LX_n and RX_n are the electrolytic compounds of L and R with univalent ion X. The cell reaction is then

$$L(s) + RX_n(s) = LX_n(m) + R(s), \tag{17.25}$$

and there are n moles of electrons transferred from L to R by the reaction. That is, L is oxidized by giving up n moles of electrons, whereas R is reduced by gaining n moles of electrons. The cell may be represented by the same figure as for Eq. (17.11) (see Fig. 17.2). Now, by carrying out an analysis similar to that for the galvanic cell (17.11), we find

$$n[\mu_{ee^-}(4) - \mu_{ee^-}(1)] = [\mu_{LX_n}(2) + \mu_R(4)] - [\mu_L(1) + \mu_{RX_n}(3)]. \tag{17.26}$$

The EMF ϵ is still defined by Eq. (17.17). Therefore, substitution of Eqs. (17.17) and (17.19) into Eq. (17.26) gives rise to Nernst's equation in the form

$$\epsilon = \epsilon^0 - \frac{RT}{nF} \ln a_{\mathrm{LX}_n}, \qquad (17.27)$$

for which we have put the activities of pure solid substances equal to 1 in accordance with the standard states chosen and the standard EMRF is given by the equation

$$-nF\epsilon^0 = \left(\mu_{\mathrm{R}}^0 + \mu_{\mathrm{LX}_n}^0\right) - \left(\mu_{\mathrm{L}}^0 + \mu_{\mathrm{RX}_n}^0\right).$$

In terms of the mean activity coefficient and molality of LX_n, the activity a_{LX_n} is given by the expression

$$a_{\mathrm{LX}_n} = (\gamma_{\pm} m_{\pm})^{\nu}, \qquad (17.28)$$

where

$$\nu = \nu_+ + \nu_- = n + 1. \qquad (17.29)$$

Therefore, the Nernst equation for the cell (17.24) is

$$\epsilon = \epsilon^0 - \frac{\nu RT}{nF} \ln(\gamma_{\pm} m_{\pm}), \qquad (17.30)$$

where ν is given by Eq. (17.29). This is the desired generalization for the case of cell (17.25).

17.3.2. Fuel Cells

In this section, we discuss the basic thermodynamic aspect of fuel cells. Fuel cells are open systems and operate differently from the galvanic cells, which are closed systems as is evident from the discussion given earlier. Nevertheless, its thermodynamic principles are the same as for the galvanic cells. For this reason, the thermodynamic basis of fuel cells can be discussed in the same manner as for the conventional galvanic cells, although at a quick glance, the topic might give an impression that it is somewhat being misplaced in this chapter for galvanic cells. In any event, the topic is of current technological relevance and importance at present time, but we do not need more than what we have developed for thermodynamics of galvanic cells to study it. Therefore, we discuss the gist of it in this section — especially, the thermodynamic efficiency aspect despite the appearance of this topic.

A fuel cell is an electrochemical device generating electrical current from fuel supplied on the anode side and an oxidant supplied on the cathode side, which reacts in the presence of an electrolyte and metal serving also as a catalyst. The reaction products are continuously drawn out of the system while the electrolytes remain within the cell. For this reason, the fuel cell is an open system, running as long as the fuel is supplied, whereas the galvanic cells are closed and stop functioning requiring recharging. We take the example of a hydrogen–oxygen fuel cell for the purpose of illustration.

The hydrogen–oxygen fuel cell consists of two electrodes between which a polymer electrolyte (semipermeable) membrane is enclosed that physically separates the anode and cathode. The reaction product, water, is formed on the cathode side and continuously drawn out of it. The electrons generated by the reaction at the anode side is impermeable to the membrane, but drawn out of the cell on the anode side through a metallic wire (circuit) leading to the cathode. Thus, the circuit is completed and electric current is generated. Therefore, the mode of operation of a fuel cell is seen to be evidently different from those of galvanic cells discussed earlier in this chapter. Nevertheless, the same thermodynamic principles apply as for galvanic cells.

In this section, we employ a simplified model representing the essential point of the hydrogen–oxygen fuel cell to illustrate the thermodynamics involved. Expressed schematically according to the mode of cell diagram adopted in this chapter, the hydrogen–oxygen fuel cell may be written as

$$|(\text{pt})\text{H}_2 - \text{electrolyte}|\text{M}|\text{O}_2 - \text{electrolyte}(\text{pt})|, \qquad (17.31)$$

where M is a semipermeable polymer electrolyte membrane that passes charged masses (protons) but not electrons which get conducted through the metallic wires connected to the platinum (electrodes) completing the electrical circuit. See Fig. 17.3 for the schematic diagram of the system. At the cathode-electrolyte interface, catalyzed by electrolytes, the half reaction

$$\tfrac{1}{2}\text{O}_2 + 2\text{e}^- \rightleftharpoons \text{O}^{2-} \qquad (17.32)$$

occurs, whereas at the anode-electrolyte interface, also catalyzed by the electrolytes, the half reaction

$$\text{H}_2 + \text{O}^{2-} \rightleftharpoons \text{H}_2\text{O} + 2\text{e}^- \qquad (17.33)$$

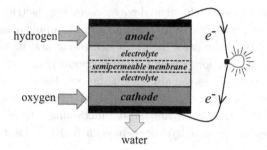

Fig. 17.3. Schematic diagram for a hydrogen–oxygen fuel cell. The electrolyte is impermeable to electrons which are conducted through the wires attached to the electrodes. The broken lines represent the boundaries of the semipermeable membrane permeable to hydrogen ions (protons). This completes the circuit with the external conducting wires.

occurs. In this particular case, there are two equivalents of electrons involved. Therefore, the overall chemical reaction is

$$H_2 + \tfrac{1}{2}O_2 + \text{electrolyte} \rightleftharpoons H_2O + \text{electrolyte}. \qquad (17.34)$$

This shows that, in principle, any exothermic chemical reaction can be made use of to construct a fuel cell, although there may be a number of practical aspects involved to make it practically feasible.

Expressed generally, a fuel cell involves a chemical reaction

$$\sum_i \nu_i M_i = 0, \qquad (17.35)$$

and electrochemical potentials of charged species

$$\mu_{ei} = \mu_i + z_i F \psi. \qquad (17.36)$$

If the charge transfer involved is denoted by q and the potential difference by ϵ, then the work done when charge dq is transported under ε is

$$dW = \epsilon dq, \qquad (17.37)$$

and the internal energy is given by

$$dE = dQ - dW. \qquad (17.38)$$

If the process is reversible, we have already established that dQ may be given in terms of entropy change dS:

$$dQ = TdS, \qquad (17.39)$$

and the fundamental thermodynamic relation is given by

$$TdS = dE + \epsilon dq, \tag{17.40}$$

if there is no pressure–volume work. It should be noted that fuel cells operate at constant pressure and volume. Since dE is a total differential in manifold (T, q), Eq. (17.40) can be written as

$$TdS = \left(\frac{\partial E}{\partial T}\right)_q dT + \left[\epsilon + \left(\frac{\partial E}{\partial q}\right)_T\right] dq. \tag{17.41}$$

Since S is also an exact differential, there follows the condition

$$\left[\frac{\partial}{\partial q}\frac{1}{T}\left(\frac{\partial E}{\partial T}\right)_q\right]_T = \left\{\frac{\partial}{\partial T}\frac{1}{T}\left[\epsilon + \left(\frac{\partial E}{\partial q}\right)_T\right]\right\}_q, \tag{17.42}$$

which yields

$$\left(\frac{\partial \epsilon}{\partial T}\right)_q = \frac{1}{T}\left[\epsilon + \left(\frac{\partial E}{\partial q}\right)_T\right]. \tag{17.43}$$

This means

$$TdS = C_q dT + T\left(\frac{\partial \epsilon}{\partial T}\right)_q dq. \tag{17.44}$$

Note that the specific heat C_q is given by

$$C_q = \left(\frac{\partial E}{\partial T}\right)_q = T\left(\frac{\partial S}{\partial T}\right)_q. \tag{17.45}$$

With the definition of enthalpy as usual

$$H = E + pV, \tag{17.46}$$

we find at $d(pV) = 0$,

$$dH = dQ - \epsilon dq = C_q dT + \left[T\left(\frac{\partial \epsilon}{\partial T}\right)_q - \epsilon\right] dq, \tag{17.47}$$

and, for an isothermal process, the enthalpy change is given by

$$\Delta H = \left[T\left(\frac{\partial \epsilon}{\partial T}\right)_q - \epsilon\right]\Delta q. \tag{17.48}$$

If n equivalents per mole of chemical are consumed by the fuel cell, then the charge transfer involved is

$$\Delta q = nF. \tag{17.49}$$

Hence, we find the enthalpy change may be expressible as follows:

$$\Delta H = nF \left[T \left(\frac{\partial \epsilon}{\partial T} \right)_q - \epsilon \right] = nFT^2 \left[\frac{\partial}{\partial T} \left(\frac{\epsilon}{T} \right) \right]_q. \tag{17.50}$$

Since $\Delta H < 0$ for the process under consideration, it follows

$$T \left(\frac{\partial \epsilon}{\partial T} \right)_q < \epsilon, \tag{17.51}$$

or, upon integration for a range of T,

$$\frac{\epsilon}{T} - \left(\frac{\epsilon}{T} \right)_{T_0} = \int_{T_0}^{T} dT' \frac{\Delta H}{nFT'^2} < 0. \tag{17.52}$$

The free energy change is given by

$$\Delta G = \Delta (H - TS), \tag{17.53}$$

which is the available work. The efficiency of "engine" — energy conversion device — is of interest to us. It is defined by the work obtained per input of heat. Therefore, the efficiency of the fuel cell can be defined by

$$\eta_f = \frac{\Delta G}{(\Delta H)_{\max}}, \tag{17.54}$$

where $(\Delta H)_{\max}$ is the maximum available heat input into the fuel cell. For an isothermal process, we obtain for ΔG accompanying the fuel cell:

$$\Delta G = \Delta H - T\Delta S$$
$$= nF \left[T \left(\frac{\partial \epsilon}{\partial T} \right)_q - \epsilon \right] - T \left(\frac{\partial \epsilon}{\partial T} \right)_q nF$$
$$= -nF\epsilon. \tag{17.55}$$

Hence, the efficiency is expressible by the formula

$$\eta_f = \frac{\epsilon}{\left[\epsilon - T \left(\frac{\partial \epsilon}{\partial T} \right)_q \right]_{\max}} \tag{17.56}$$

in terms of measurables. In this regard, it is useful to recognize that fuel cells do not involve a Carnot cycle like internal combustion engines. Nevertheless, it is subject to the Carnot theorem and the thermodynamic laws.

The Nernst equation can also be calculated by using the electrochemical potentials in the same form as for a galvanic cell by following a similar procedure as for the latter. For example, for the hydrogen–oxygen fuel cell considered above, we find

$$\epsilon = \epsilon^0 - \frac{RT}{2F} \ln \left(\frac{a_{H_2O}}{a_{H_2} a_{O_2}^{1/2}} \right), \tag{17.57}$$

where a_i is the activity of species i and ϵ^0 is the standard potential difference (emf). It is seen to be in the usual form for the Nernst equation, indicating that the same thermodynamic principles are in operation for the fuel cell considered. The same principles apply to other types of fuel cells.

17.4. Lithium-Ion Cells

Unlike conventional electrochemical cells which, when fully discharged, become useless, lithium-ion cells are rechargeable, thereby can be made operational again, and thus are useful for a long time for practical devices. For this reason, it has been made use of in numerous commercial electronic devices. We first would like to discuss how a lithium-ion cell works and point out the point of difference in its mode operation from the conventional electrochemical cells (i.e., galvanic cells).

Lithium-ion cells (batteries), like the galvanic cells discussed earlier, can be thermodynamically described by the same kind of thermodynamic principles. However, they are constructed such that, after discharge, they can be recharged. Like the galvanic cells, they consist of positive and negative electrodes and an electrolyte facilitating diffusion of lithium ions. The positive electrode is a lithium metal oxide and the electrolyte is a lithium salt dissolved in an organic solvent. The negative electrode generally consists of carbon (e.g., graphite). The electrolyte is typically a mixture of a high ionic conductivity and low viscosity organic carbonate or diethyl carbonate solvent containing complexes of lithium ions. It provides a conducting medium for lithium ions to migrate between the two electrodes. Both anode and cathode (electrodes) permit lithium ions to move in and out of their interiors. This process of moving in and out of the electrodes is called *intercalation* when the ions are inserted, and *deintercalation* when the ions are extracted from the electrodes. During insertion (i.e., *intercalation*), ions migrate into the electrodes, whereas during extraction (i.e., *deintercalation*) — the reverse process of the insertion — ions migrate back out of the electrode. When electrons flow through a closed circuit, a useful work is gained. Acquisition of the useful work is thermodynamically understood based on the same thermodynamic principles as the ordinary galvanic cells.

We will take the LiCoO$_2$–carbon (C$_6$) cell as an illustrative example for lithium-ion cells. Here, C$_6$ represents graphite. The graphite constitutes the anode and LiCoO$_2$ the cathode, and the electrolyte employed is a lithium salt that separates the cathode and anode. See Fig. 17.4 for a schematic drawing of LiCoO$_2$–carbon (C$_6$) cell. If a lithium-ion cell is discharging, the positive Li$^+$ ion migrates from the negative electrode (e.g., graphite C$_6$, more generally C$_x$ denoted below) and enters the positive electrode made of a lithium salt (LiCoO$_2$ in the present example considered here). When the cell is charging, the reverse process occurs.

The half cell chemical reaction on the cathode is

$$Li_{1-x}CoO_2 + xLi^+ + xe^- \leftrightarrows LiCoO_2, \qquad (17.58)$$

whereas the half cell chemical reaction on the anode is

$$xLiC_6 \leftrightarrows xLi^+ + xe^- + xC_6. \qquad (17.59)$$

The reactions are expressed in units of moles as we have in the case of galvanic cells. Thus, oxidation occurs on the anode (−), whereas reduction

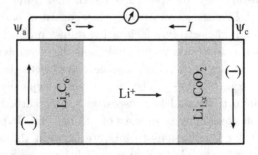

Fig. 17.4.   Schematics of lithium-ion battery.

occurs on the cathode. The overall net reaction is therefore

$$Li_{1-x}CoO_2 + xLiC_6 \rightleftarrows LiCoO_2 + xC_x. \qquad (17.60)$$

In terms of the cell diagram introduced earlier in this chapter, the lithium-ion cell is denoted by

$$LiC_6(s)|Li_{1-x}CoO_2(x)(salt)|LiCoO_2(s), \qquad (17.61)$$

where the left-hand side represents the anode ($-$) and the right-hand side the cathode($+$). At the anode the lithium is oxidized whereas at the cathode the lithium ion is reduced. In the lithium-ion cell, the lithium ions are transported to and from the positive or negative electrodes on oxidation of cobalt Co (transition metal) in $Li_{1-x}CoO_2$ from Co^{3+} to Co^{4+} during charging, and on reduction from Co^{4+} to Co^{3+} during discharge. The voltage difference on discharge is typically 3.6 V for the lithium-ion batteries whereas it is 1.5 V for the zinc battery, for example, offering a favorable advantage over a zinc battery. It is known that reaction (17.60) is reversible if $x < 0.5$.

Because of the technological importance of lithium batteries at present, research activity on the subject is intense and there is considerable room for study to improve the ionic transport process in the lithium batteries and research in the subject matter will continue in the future.

The electrochemical potentials are denoted by μ_{ei}^c and μ_{ei}^a, respectively, for the cathode and anode species, and the cathode and anode potentials are denoted by ψ^c and ψ^a. The chemical potential difference in species i at the cathode and anode is expressible in terms of potentials by the formula

$$\Delta\mu_i = \mu_{ei}^c - \mu_{ei}^a = z_i F(\psi^c - \psi^a),$$

where $z_i = ne^-$ is the charge density. The Nernst equation for reaction (17.60) is then given by

$$\epsilon = \epsilon^0 - \frac{RT}{xF} \ln\left(\frac{a_{Li^+}}{a_{LiCoO}}\right), \qquad (17.62)$$

where a_{Li^+} represents the activity of $Li_{1-x}CoO_2$ formed by oxidation of LiC_6 in the electrolyte phase. Since we may put

$$a_{LiCoO_2} = 1, \qquad (17.63)$$

the Nernst equation for the electromotive force of the cell is given by

$$\epsilon = \epsilon^0 - \frac{RT}{xF} \ln a_{Li^+}. \tag{17.64}$$

By means of the Nernst equation, it is now possible to investigate the thermodynamics of rechargeable lithium-ion cells. The procedure of investigation is completely parallel to the ordinary galvanic cells. It should be kept in mind that this result is under the assumption of vanishing dissipative processes in the system. However, charge transport processes through a medium are unavoidably accompanied by energy dissipation, and it will be advantageous to reduce the energy dissipation and, consequently, the charge transport made more facile and as less dissipative in energy as possible. In this regard, it should be noted that charge transport and associated energy dissipation belong to the realm of irreversible thermodynamics.

17.5. Photovoltaic Cells

The photovoltaic cell is an energy conversion device (cell), which has a wide practical application, but fundamentally different from the galvanic cell in the underlying physical process. Nevertheless, the thermodynamics for the electromotive force of the cell may be put in basically the same form as the galvanic cell despite the fundamentally different molecular theoretic cause.

In a photovoltaic solar cell, the solar photon energy is captured and made use of with the help of a semiconductor. If the solar photon energy is equal to or higher than the semiconductor band gap E_g, the solar photon excites and electron is promoted into the conduction band, thereby forming an electron–hole pair. The excess of the solar photon energy is lost as low-frequency phonons or lattice vibrations. A single photon, however, almost never generates two electron–hole pairs even if there is an enough energy to do so. The solar spectrum mainly consists of photons with energy less than $3.5\,eV$, which is of the same order of magnitude as a typical semiconductor bandgap. The basic idea of photovoltaic cell is schematically illustrated in Fig. 17.5 in analogy to the cell diagram for the galvanic cell employed in the earlier section of this chapter.

If it is assumed that every photon with energy greater than E_g creates one electron–hole pair with energy E_g' and if the irreversibility and hence the accompanying energy dissipation is neglected, it is possible to derive the

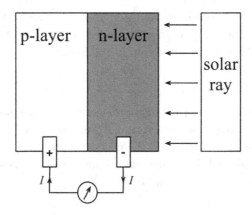

Fig. 17.5.   Schematic diagram for the photovoltaic cell.

Shockley–Queisser limit[2] for the efficiency of a single-band gap solar cell

$$\eta_{SQ} = I[V(\max)]V(\max)P_{inc}$$

$$= \frac{15E_g}{\pi^4 k_B T_s} \int_{x_g}^{\infty} dx \frac{x^2}{e^x - 1} \left(x_g = \frac{E_g}{k_B T}\right), \qquad (17.65)$$

where I is the current, V the voltage, P_{inc} the incident power of radiation, and $E_g := h\nu_g$ the band gap of the semiconductor. The limit may be optimized[3] to 33% with the bandgap of Si, which is 1.1 eV.

If we may figuratively express the elementary process in a photovoltaic cell by using the notation employed in thermodynamics of electric cells in this work, "the overall chemical reaction" in the photovoltaic cell may be written as follows:

$$VB(p\text{-}n) + h\nu \to e^- + CB(p\text{-}n)^+, \qquad (17.66)$$

where VB(p-n) denotes the valence bond of the semiconductor — more precisely, it should be the valence bond energy level — and CB(p-n) denotes the conduction band of the semiconductor. The electron gets promoted (i.e., excited) into the conduction band on absorption of excitation energy E_g, and the so-generated electron does a useful work when the circuit is closed. In the terminology of galvanic cells, it may be said that "VB(p-n)

---

[2]See W. Shocklet and H. J. Queisser, *J. Appl. Phys.* **32**, 510 (1961).
[3]See G. Crabtree and N. Lewis, *Phys. Today*, **March**, 37 (2007).

is oxidized by $h\nu$ liberating e^- and a hole of an electron–hole pair in the semiconductor." By using this model, it is readily calculated that the Nernst equation for EMF ϵ is easy to obtain in the form

$$\epsilon = -\frac{RT}{F}\ln a_-, \tag{17.67}$$

where a_- is the activity of conduction electron. It may be approximated by the density of conduction electron n_-. This Nernst equation provides a mathematical tool to calculate all the necessary thermodynamic properties of the photovoltaic cell, including the thermodynamic efficiency, which we have already described in the earlier sections.

17.6.   Betavoltaic Cells

A betavoltaic cell works similar to the photovoltaic cell. Both produce electric current from a semiconductor p-n junction's built-in electric field acting on the electrons and holes. In the case of a betavoltaic cell, electron–hole pairs are produced by beta particles emitted by a radioactive isotope. Beta particles typically have energies of 10–100 keV, depending on the radioisotope employed.

One energetic beta particle can create many electron–hole pairs, but much of its energy is lost as high-frequency optical and low-frequency acoustic phonons. A schematic diagram for these processes are illustrated in Fig. 17.6.

The elementary physical process in a betavoltaic cell is similar to (17.66). Hence, the thermodynamic description of the process is basically the same as for the photovoltaic cell. Consequently, the electromotive force for the cell is given by Eq. (17.67).

17.7.   Donnan Membrane Equilibrium

Membrane equilibria in nonelectrolyte solutions have been considered in connection with equilibrium conditions in the previous chapters. The same theory can be applied to membrane equilibria in electrolyte solutions, especially when there are species present in the solution, which are not capable of passing through the membrane.

An example would be a solution of charged polymers and NaCl. Imagine two solutions separated by a semipermeable membrane. One solution consists of species R^-, C^+, and A^- in aqueous solution whereas the other

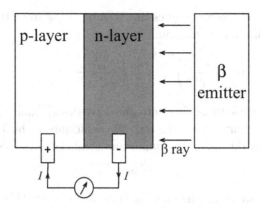

Fig. 17.6. Schematic diagram for betavoltaic cell. The betavoltaic cell works similarly to the photovoltaic cell except that the beta ray source (radioactive isotope, e.g., tritium tritide of Ti or Sc or promethium ($^{147}Pm_2O_3$) replaces the solar ray). The p-layer and n-layer represent the junctions of the semiconductor diaode. When they are closed by an external circuit, a useful work is gained.

solution consists of C^+, and A^- in the same solvent. The membrane is impermeable to the species R^- whereas C^+, A^-, and the solvent are permeable through the membrane. The solution containing the species R^- will be denoted by α, and the other solution by β. We will assume that the temperature is uniform throughout the system and that all the components are incompressible. Because of the impermeability of the species R^-, there will be a pressure difference — an osmotic pressure — and the equilibrium conditions are

$$p_\alpha v_1 + \mu_1^0 + RT \ln a_1^{(\alpha)} = p_\beta v_1 + \mu_1^0 + RT \ln a_1^{(\beta)}, \qquad (17.68)$$

$$p_\alpha(v_c + v_a) + \mu_\pm^0 + RT \ln(a_+^{(\alpha)} a_-^{(\alpha)}) = p_\beta(v_c + v_a) + \mu_\pm^0 + RT \ln(a_+^{(\beta)} a_-^{(\beta)}), \qquad (17.69)$$

where the subscript 1 denotes the solvent, p_α and p_β are the pressures of solutions α and β, v_1, v_c, and v_a are the molar volumes of the solvent, cation, and anion, respectively, and the rest of notation is standard.

Rearranging the equilibrium conditions, we obtain

$$p_\alpha - p_\beta = \frac{RT}{v_1} \ln \left(\frac{a_1^{(\beta)}}{a_1^{(\alpha)}} \right), \qquad (17.70)$$

$$p_\alpha - p_\beta = \frac{RT}{v_c + v_a} \ln \left(\frac{a_+^{(\beta)} a_-^{(\beta)}}{a_+^{(\alpha)} a_-^{(\alpha)}} \right). \qquad (17.71)$$

Elimination for the osmotic pressure $\Pi = p_\alpha - p_\beta$ from these equations yields the condition of equilibrium

$$\left(\frac{a_1^{(\beta)}}{a_1^{(\alpha)}}\right)^{v_c+v_a} = \left(\frac{a_+^{(\beta)}a_-^{(\beta)}}{a_+^{(\alpha)}a_-^{(\alpha)}}\right)^{v_1}. \tag{17.72}$$

If the solutions are sufficiently dilute, the activities $a_1^{(\alpha)}$ and $a_1^{(\beta)}$ are approximately equal to unity. Since the activity coefficients of the ions may be set equal to unity in this case, the equilibrium condition is reduced to the form

$$n_+^{(\alpha)}n_-^{(\alpha)} = n_+^{(\beta)}n_-^{(\beta)}, \tag{17.73}$$

where $n_+^{(\alpha)}$ and so on are the mole numbers of the ions. This is the Donnan membrane equilibrium condition.

The difference in concentrations of ions in solutions α and β gives rise to a potential difference, which is given by the formula

$$\epsilon = \frac{RT}{z_+F}\ln\left(\frac{a_+^{(\beta)}}{a_+^{(\alpha)}}\right), \tag{17.74}$$

where we have put z_+ for generality, but $z_+ = 1$ for the case considered here. Equation (17.74) provides a method of determining the concentration imbalance by measuring the potential difference produced by the presence of an impermeable species in a solution. For example, one can study the thermodynamic properties of the species R^- by measuring the potential difference.

Chapter 18

Thermodynamics of Electric and Magnetic Fields

In the chapters on electrolytes and galvanic cells, electric fields entered in the thermodynamic formalism in the discussion of motions of strong electrolytes and their thermodynamic properties. Polarizable materials or dielectrics, although electrically neutral and nonconducting, can be altered in their properties by the presence of an electric field, static or time-dependent (i.e., dynamic). Similarly, a magnetic field can affect the thermodynamic properties of matter. In this chapter, we consider the thermodynamics of dielectrics subjected to a static electric field and also the thermodynamics of magnetically polarizable materials in a static magnetic field. We will use the Gaussian units for electromagnetic fields and related quantities in the discussion. Since in the case of static fields, the electric and magnetic fields are independent of each other, their effects can be discussed separately. We exclude conductors from the discussion.

18.1. Dielectrics in Electrostatic Field

We consider a capacitor of parallel plates α and β of area a separated by distance L. The volume of the capacitor is then $V = aL$. Assume that the surface charge density of the negative plate α is $-\sigma$, whereas the surface charge density of the positive plate β is σ. See Fig. 18.1.

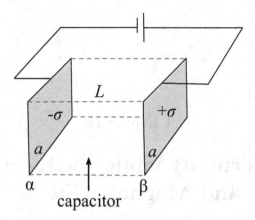

Fig. 18.1. A dielectric in a capacitor. Parallel plates α and β of area a have charge densities $-\sigma$ and $+\sigma$, respectively. They are separated at distance L and connected to a battery. The electric field strength is $\mathbf{E} = \Delta\phi/L$ where $\Delta\phi$ is the potential difference $\Delta\phi = \phi^{(\beta)} - \phi^{(\alpha)}$.

The potential difference between the plates is $\Delta\phi = \phi^{(\beta)} - \phi^{(\alpha)}$. The electric field strength $\mathbf{E}$ is then given by[1]

$$\mathbf{E} = \frac{\Delta\phi}{L}. \tag{18.1}$$

Recall that in the case of continuous potential, this may be written as

$$\mathbf{E} = -\nabla\phi.$$

It is well known in electrostatics that if $\mathbf{P}$ is the total polarization of the dielectric and $\mathbf{p}$ is the polarization density defined by

$$\mathbf{p} = \frac{\mathbf{P}}{V}, \tag{18.2}$$

then the dielectric displacement $\mathbf{D}$ is given by

$$\mathbf{D} = \mathbf{E} + 4\pi\mathbf{p}. \tag{18.3}$$

For the capacitor under consideration $\mathbf{D}$ is given by

$$\mathbf{D} = 4\pi\sigma, \tag{18.4}$$

---

[1]It should be noted that the boldface letters $\mathbf{E}$, $\mathbf{P}$, and $\mathbf{p}$ used below are not vectors which they generally are. The boldface letters are used to distinguish them from lightface letters used for other quantities such as internal energy, pressure, etc.

and if we write

$$\mathbf{D} = \varepsilon\mathbf{E}, \tag{18.5}$$

where ε is the dielectric constant of the nonconducting matter in the capacitor, then it follows from Eqs. (18.3)–(18.5) the polarization density $\mathbf{p}$ may be written

$$\mathbf{p} = \chi\mathbf{E} \tag{18.6}$$

with χ defined by

$$\chi = \frac{\varepsilon - 1}{4\pi}. \tag{18.7}$$

Here, χ is called the electric susceptibility per unit volume of the dielectric. It should be noted that relation (18.6) holds for substances that are not ferroelectric. For ferroelectric materials, there appears a term nonvanishing at $\mathbf{E} = 0$, that is,

$$\mathbf{p} = \mathbf{p}_0 + \chi\mathbf{E}, \tag{18.8}$$

where $\mathbf{p}_0$ represents the polarization density at $\mathbf{E} = 0$. In this chapter, we will not consider ferroelectrics.

If the charge density is uniform over the plates, the work to bring an infinitesimal charge $d\sigma$ from infinity onto the plate is then given by

$$dW_e = a(\phi^{(\beta)} - \phi^{(\alpha)})d\sigma = V\mathbf{E}d\sigma, \tag{18.9}$$

where $V = aL$. Therefore, upon adding this work to the internal energy change of the dielectric in the absence of the field, the overall internal energy change in the dielectric is given by

$$dE = dE_0 + dW_e = dE_0 + V\mathbf{E}d\sigma, \tag{18.10}$$

where dE_0 is the internal energy change in the absence of the electric field. For systems in equilibrium or for a reversible process, this change may be written as

$$dE_0 = TdS - pdV + \sum_{i=1}^{c} \mu_i dn_i. \tag{18.11}$$

If the field is constant, from Eqs. (18.3)–(18.5) follows

$$Vd\sigma = d\mathbf{P}. \tag{18.12}$$

Combining Eq. (18.11) and Eq. (18.12) into Eq. (18.10), we obtain the thermodynamic fundamental equation—the Gibbs relation

$$dE = TdS - pdV + \sum_{i=1}^{c} \mu_i dn_i + \mathbf{E}d\mathbf{P}. \tag{18.13}$$

From this differential form follows the following constitutive derivatives for dielectrics in the presence of an electric field:

$$T^{-1} = \left(\frac{\partial S}{\partial E}\right)_{V,n,\mathbf{P}}, \tag{18.14}$$

$$p = -\left(\frac{\partial E}{\partial V}\right)_{S,n,\mathbf{P}}, \tag{18.15}$$

$$\mu_i = \left(\frac{\partial E}{\partial n_i}\right)_{S,V,n',\mathbf{P}}, \tag{18.16}$$

$$\mathbf{E} = \left(\frac{\partial E}{\partial \mathbf{P}}\right)_{S,V,n}. \tag{18.17}$$

Here, the prime on n in the subscript denotes excluding n_i. The enthalpy, Helmholtz free energy, and Gibbs free energy are defined as usual:

$$H = E + pV, \tag{18.18}$$

$$A = E - TS, \tag{18.19}$$

$$G = H - TS. \tag{18.20}$$

Using these definitions, we obtain from Eq. (18.13) the differential forms— the fundamental equations, equivalent to Eq. (18.13):

$$dH = TdS + Vdp + \sum_{i=1}^{c} \mu_i dn_i + \mathbf{E}d\mathbf{P}, \tag{18.21}$$

$$dA = -SdT - pdV + \sum_{i=1}^{c} \mu_i dn_i + \mathbf{E}d\mathbf{P}, \tag{18.22}$$

$$dG = -SdT + Vdp + \sum_{i=1}^{c} \mu_i dn_i + \mathbf{E}d\mathbf{P}. \tag{18.23}$$

If variables T, p, $n_1, \ldots, n_c$, and $\mathbf{E}$ are treated as independent variables, it is convenient to define a function

$$\mathcal{F} = G - \mathbf{E} \cdot \mathbf{P} \tag{18.24}$$

to change the characteristic variables. Consequently, $d\mathcal{F}$ is given by the differential form

$$d\mathcal{F} = -S\,dT + V\,dp + \sum_{i=1}^{c} \mu_i\,dn_i - \mathbf{P}\,d\mathbf{E}, \tag{18.25}$$

in which T, p, $n_1, \ldots, n_c$, and $\mathbf{E}$ are characteristic variables instead of T, p, $n_1, \ldots, n_c$, and $\mathbf{P}$. Thermodynamic observables M are then regarded as functions of the new characteristic variables, and the partial molar quantities of M are defined by

$$\overline{M}_i = \left(\frac{\partial M}{\partial n_i}\right)_{T,p,n',\mathbf{E}}. \tag{18.26}$$

According to this definition, the chemical potentials are the partial molar free energy $\mathcal{F}$:

$$\mu_i = \left(\frac{\partial \mathcal{F}}{\partial n_i}\right)_{T,p,n',\mathbf{E}}. \tag{18.27}$$

The Maxwell relations between derivatives can be generalized to the case of electric fields. In particular, by using the differential form (18.25), we obtain

$$\left(\frac{\partial S}{\partial p}\right)_{T,n,\mathbf{E}} = -\left(\frac{\partial V}{\partial T}\right)_{p,n,\mathbf{E}}, \tag{18.28}$$

$$\left(\frac{\partial V}{\partial \mathbf{E}}\right)_{T,p,n} = -\left(\frac{\partial \mathbf{P}}{\partial p}\right)_{T,V,\mathbf{E}}, \tag{18.29}$$

$$\left(\frac{\partial \mu_i}{\partial \mathbf{E}}\right)_{T,p,n} = -\left(\frac{\partial \mathbf{P}}{\partial n_i}\right)_{T,p,n'\mathbf{E},} = -\overline{\mathbf{P}}_i, \tag{18.30}$$

where $\overline{\mathbf{P}}_i$ is the partial molar polarization. Since μ_i depends on T, p, $n_1, \ldots, n_{c-1}$, and $\mathbf{E}$, there follows from Eq. (18.25) and Eq. (18.27) the differential form for chemical potential μ_i:

$$d\mu_i = -\bar{s}_i dT + \bar{v}_i dp + \sum_{j=1}^{c-1} \mu_{ij} dn_j - \overline{\mathbf{P}}_i d\mathbf{E}, \tag{18.31}$$

where $\bar{s}_i$ and $\bar{v}_i$ are, respectively, the partial molar entropy and volume of component i

$$\bar{s}_i = -\left(\frac{\partial \mu_i}{\partial T}\right)_{p,n,\mathbf{E}}, \tag{18.32}$$

$$\bar{v}_i = \left(\frac{\partial \mu_i}{\partial p}\right)_{T,n,\mathbf{E}}, \tag{18.33}$$

and

$$\mu_{ij} = \left(\frac{\partial \mu_i}{\partial n_j}\right)_{T,p,n',\mathbf{E}}. \tag{18.34}$$

Of particular interest is the field dependence of $\bar{s}_i$ and $\bar{v}_i$. They can be examined by using the derivative relations

$$\left(\frac{\partial \bar{s}_i}{\partial \mathbf{E}}\right)_{T,p,n} = \left(\frac{\partial \overline{\mathbf{P}}_i}{\partial T}\right)_{p,n,\mathbf{E}}, \tag{18.35}$$

$$\left(\frac{\partial \bar{v}_i}{\partial \mathbf{E}}\right)_{T,p,n} = -\left(\frac{\partial \overline{\mathbf{P}}_i}{\partial p}\right)_{T,n,\mathbf{E}}. \tag{18.36}$$

These derivatives may be employed to examine the field dependence of thermodynamic functions of dielectrics.

18.2. Field Dependence of Thermodynamic Quantities

The molecular theory of dielectrics suggests that the electric susceptibility is the statistical average of dipole moments of the material. Consequently, it is generally a function of $\mathbf{E}$ and may be expressed in a power series of electric field $\mathbf{E}$. Since our discussion will be limited to a small field strength regime, we may consider χ constant with respect to $\mathbf{E}$.

In the case of weak fields, thermodynamic functions are generally quadratic with respect to $\mathbf{E}$. We can verify this is indeed true. We begin with the chemical potential. Integrating Eq. (18.30) over $\mathbf{E}$ while keeping other variables fixed, we obtain

$$\mu_i\left(T, p, n, \mathbf{E}\right) = \mu_i\left(T, p, n, 0\right) - \int_0^{\mathbf{E}} \overline{\mathbf{P}}_i d\mathbf{E}$$

$$= \mu_i\left(T, p, n, 0\right) - \frac{1}{2}\overline{\chi}_i \mathbf{E}^2, \tag{18.37}$$

for which we have assumed χ is independent of $\mathbf{E}$ and

$$\overline{\chi}_i = \left(\frac{\partial \chi}{\partial n_i}\right)_{T, p, n'}, \tag{18.38}$$

the partial molar electric susceptibility. The $\mu_i\left(T, p, n, 0\right)$ is the chemical potential of component i in the absence of $\mathbf{E}$. Therefore, the total electric susceptibility is given by

$$\chi = \sum_{i=1}^{c} \overline{\chi}_i n_i. \tag{18.39}$$

This shows that the free energy $\mathcal{F}$ is quadratic in $\mathbf{E}$ in the low field regime considered, as are the chemical potentials.

18.2.1. *Electrostriction*

In the same manner as for the chemical potentials, we find from Eq. (18.35)

$$\overline{s}_i\left(\mathbf{E}\right) = \overline{s}_i^0 + \frac{1}{2}\left(\frac{\partial \overline{\chi}_i}{\partial T}\right)_{p, n} \mathbf{E}^2, \tag{18.40}$$

where $\overline{s}_i^0 = \overline{s}_i(T, p, n, \mathbf{E} = 0)$. Therefore, the total entropy is given by

$$S\left(\mathbf{E}\right) = S^0 + \frac{\mathbf{E}^2}{2}\sum_{i=1}^{c} n_i \left(\frac{\partial \overline{\chi}_i}{\partial T}\right)_{p, n}. \tag{18.41}$$

Similarly, from Eq. (18.36) follows the partial molar volume

$$\overline{v}_i\left(\mathbf{E}\right) = \overline{v}_i^0 - \frac{1}{2}\left(\frac{\partial \overline{\chi}_i}{\partial p}\right)_{T, n} \mathbf{E}^2, \tag{18.42}$$

and

$$V = V^0 - \frac{\mathbf{E}^2}{2}\sum_{i=1}^{c} n_i \left(\frac{\partial \overline{\chi}_i}{\partial p}\right)_{T, n} = V^0 - \frac{\mathbf{E}^2}{2}\left(\frac{\partial \chi}{\partial p}\right)_{T, n}. \tag{18.43}$$

Equation (18.43) is the formula for electrostriction given as a function of T, p, n, and $\mathbf{E}$. The sign of $(\partial\chi/\partial p)_{T,n}$ can be positive or negative, but the volume change is not reversible because it is proportional to $\mathbf{E}^2$, and hence the volume change is not affected by reversing the direction of the field.

18.2.2. *Electrocaloric Effects*

It is interesting to calculate the effect of $\mathbf{E}$ on temperature. To this end, we consider the partial molar enthalpy. Since from the definition of H, the partial molar enthalpy is given by

$$\overline{h}_i = \mu_i + \overline{s}_i T + \mathbf{E}\overline{\mathbf{P}}_i,$$

upon using Eqs. (18.37), (18.40), and (18.42) we find the partial molar enthalpy is given by the expression

$$\overline{h}_i = \overline{h}_i^0 + \frac{1}{2}\left[\overline{\chi}_i + T\left(\frac{\partial\overline{\chi}_i}{\partial T}\right)_{p,n}\right]\mathbf{E}^2, \tag{18.44}$$

and hence the enthalpy change arising from turning on $\mathbf{E}$ is found to be quadratic with respect to $\mathbf{E}$:

$$\Delta H = \frac{1}{2}\left[\chi + T\left(\frac{\partial\chi}{\partial T}\right)_{p,n}\right]\mathbf{E}^2. \tag{18.45}$$

Let us now examine the effect on temperature by $\mathbf{E}$. Since it is convenient to take S, p, $n_1,\ldots, n_c$, and $\mathbf{E}$ for variables, if we wish to study the field dependence of temperature, Eq. (18.21) is rearranged to the form

$$d(H - \mathbf{E}\mathbf{P}) = TdS + Vdp + \sum_{i=1}^{c}\mu_i dn_i - \mathbf{P}d\mathbf{E}, \tag{18.46}$$

which gives the Maxwell relation

$$\left(\frac{\partial T}{\partial \mathbf{E}}\right)_{S,p,n} = -\left(\frac{\partial \mathbf{P}}{\partial S}\right)_{p,n,\mathbf{E}}. \tag{18.47}$$

Since

$$\left(\frac{\partial \mathbf{P}}{\partial S}\right)_{p,n,\mathbf{E}} = \left(\frac{\partial \chi}{\partial S}\right)_{p,n,\mathbf{E}} \mathbf{E} \tag{18.48}$$

$$= \left(\frac{\partial v\overline{\chi}}{\partial T}\right)_{p,n,\mathbf{E}} \left(\frac{\partial T}{\partial s}\right)_{p,n,\mathbf{E}} \mathbf{E}$$

$$= \frac{T}{c_{p\mathbf{E}}} \left(\frac{\partial v\overline{\chi}}{\partial T}\right)_{p,n,\mathbf{E}} \mathbf{E},$$

where $\overline{\chi} = \chi/V$ is the electric susceptibility per volume, it follows that

$$\left(\frac{\partial T}{\partial \mathbf{E}}\right)_{S,p,n} = -\frac{T}{c_{p\mathbf{E}}} \left(\frac{\partial v\overline{\chi}}{\partial T}\right)_{p,n,\mathbf{E}} \mathbf{E}, \tag{18.49}$$

where $c_{p\mathbf{E}}$ is the molar isobaric specific heat at constant $\mathbf{E}$

$$c_{p\mathbf{E}} = T\left(\frac{\partial s}{\partial T}\right)_{p,n,\mathbf{E}} \tag{18.50}$$

and v is the molar volume. We note that from Eq. (18.41), it follows that $c_{p\mathbf{E}}$ takes the quadratic form with respect to $\mathbf{E}$

$$c_{p\mathbf{E}} = c_p^0 + c_p^{(1)}\mathbf{E}^2, \tag{18.51}$$

where c_p^0 is the specific heat at $\mathbf{E} = 0$ and

$$c_p^{(1)} = \frac{1}{2}T\frac{\partial}{\partial T}\left(\frac{\partial v\overline{\chi}}{\partial T}\right)_{p,n,\mathbf{E}}. \tag{18.52}$$

Integrating Eq. (18.49), we find the $\mathbf{E}$-dependence of temperature:

$$T = T_0 - \int_0^{\mathbf{E}} \frac{\mathbf{E}}{(c_p^0 + c_p^{(1)}\mathbf{E}^2)}\left(\frac{\partial v\overline{\chi}}{\partial T}\right)_{p,n,\mathbf{E}} d\mathbf{E}. \tag{18.53}$$

Here, the derivative $(\partial v\overline{\chi}/\partial T)_{p,n,\mathbf{E}}$ is generally negative. Therefore, the effect of the field is to raise the temperature of the dielectric above the temperature at $\mathbf{E} = 0$. Reversely, if the magnitude of $\mathbf{E}$ is reduced reversibly and adiabatically from $\mathbf{E}$, the dielectric will be cooled by decreasing the field strength.

Before closing the discussion on electric fields, we remark that the equilibrium conditions for heterogeneous phases consisting of dielectrics are the same as for the case of external electric fields absent. The differences in various thermodynamic quantities from those of field-free systems originate from the field dependence of constitutive relations, such as the chemical potentials, which we have found depend on $\mathbf{E}$ quadratically in the leading order. Working out such $\mathbf{E}$-dependences for thermodynamic quantities is left to the readers as an exercise.

18.3. Static Magnetic Fields

If a paramagnetic substance is placed in a static magnetic field, the magnetic work is done. It is given by

$$dW_m = \mathbf{H}d\mathbf{M}, \tag{18.54}$$

where $\mathbf{H}$ is the magnetic field and $\mathbf{M}$ is the magnetization of the substance. Therefore, the fundamental equation for the internal energy is given by the differential form

$$dE = TdS - pdV + \sum_{i=1}^{c} \mu_i dn_i + \mathbf{H}d\mathbf{M}. \tag{18.55}$$

We may define an analog of the enthalpy in the presence of a magnetic field

$$\mathcal{H} = E + pV - \mathbf{H}\mathbf{M}. \tag{18.56}$$

Similarly, an analog of the Helmholtz free energy may be defined by

$$\mathcal{A} = E - TS - \mathbf{H}\mathbf{M}, \tag{18.57}$$

and the magnetic field equivalent of $\mathcal{F}$ defined earlier in the case of the electric field by the expression

$$\mathcal{F} = E + pV - TS - \mathbf{H}\mathbf{M} = \mathcal{H} - TS. \tag{18.58}$$

Combined with the fundamental equation for the internal energy E Eq. (18.55), these definitions give rise to the fundamental equations for the respective quantities:

$$d\mathcal{H} = TdS + Vdp + \sum_{i=1}^{c} \mu_i dn_i - \mathbf{M}d\mathbf{H}, \tag{18.59}$$

$$d\mathcal{A} = -SdT - pdV + \sum_{i=1}^{c} \mu_i dn_i - \mathbf{M}d\mathbf{H}, \tag{18.60}$$

$$d\mathcal{F} = -SdT + Vdp + \sum_{i=1}^{c} \mu_i dn_i - \mathbf{M}d\mathbf{H}. \tag{18.61}$$

Thus, $\mathcal{H}$, $\mathcal{A}$, and $\mathcal{F}$ are characteristic functions of $(S, p, n_i, \mathbf{H})$, $(T, V, n_i, \mathbf{H})$, and $(T, p, n_i, \mathbf{H})$, respectively. From Eq. (18.61), the chemical potential is seen to be the partial molar free energy of species i:

$$\mu_i = \left(\frac{\partial \mathcal{F}}{\partial n_i}\right)_{T,p,n',\mathbf{H}}. \tag{18.62}$$

Thus, $\mathcal{F}$ is a first degree homogeneous function of n_i:

$$\mathcal{F} = \sum_{i=1}^{c} n_i \mu_i. \tag{18.63}$$

Furthermore, we find the Gibbs–Duhem equation for paramagnetic substances is given by

$$\sum_{i=1}^{c} n_i d\mu_i = -SdT + Vdp - \mathbf{M}d\mathbf{H}. \tag{18.64}$$

Since the chemical potentials may be regarded as functions of T, p, $n_1, \ldots, n_{c-1}$, $\mathbf{H}$, we obtain

$$d\mu_i = -\bar{s}_i dT + \bar{v}_i dp + \sum_{j=1}^{c-1} \mu_{ij} dn_j - \overline{\mathbf{M}}_i d\mathbf{H}, \tag{18.65}$$

where

$$\mu_{ij} = \left(\frac{\partial \mu_i}{\partial n_j}\right)_{T,p,n',\mathbf{H}}. \tag{18.66}$$

Various Maxwell relations can be derived from the differential forms (18.55) and (18.59)–(18.61). For example, we obtain from Eq. (18.59) one of the Maxwell relations

$$\left(\frac{\partial T}{\partial \mathbf{H}}\right)_{S,p,n} = -\left(\frac{\partial \mathbf{M}}{\partial S}\right)_{p,n,\mathbf{H}}, \tag{18.67}$$

which will be made use of for studying the magnetocaloric effect. From Eq. (18.61) follow the relations

$$\left(\frac{\partial V}{\partial \mathbf{H}}\right)_{T,p,n} = -\left(\frac{\partial \mathbf{M}}{\partial p}\right)_{T,n,\mathbf{H}}, \tag{18.68}$$

$$\left(\frac{\partial S}{\partial \mathbf{H}}\right)_{T,p,n} = -\left(\frac{\partial \mathbf{M}}{\partial T}\right)_{p,n,\mathbf{H}}, \tag{18.69}$$

which may be used to examine the effects on volume and entropy of the magnetic field. From electrodynamics, the magnetic induction $\mathbf{B}$ may be written as

$$\mathbf{B} = \mu_{\mathrm{mp}}\mathbf{H}, \tag{18.70}$$

where μ_{mp} is called the magnetic permeability. Some authors call $\mathbf{B}$ the magnetic field strength. Since $\mathbf{M}$ is related to $\mathbf{B}$ and $\mathbf{H}$ by the relation

$$\mathbf{B} = \mathbf{H} + 4\pi\mathbf{M}, \tag{18.71}$$

we may write

$$\mathbf{M} = \chi_m \mathbf{H}, \tag{18.72}$$

where

$$\chi_m = \frac{\mu_{\mathrm{mp}} - 1}{4\pi}. \tag{18.73}$$

It is called the magnetic susceptibility. Recall that analogous relations and quantities were defined for the case of an electric field. As in the case of electric field, the thermodynamic functions are quadratic with respect to $\mathbf{H}$ if the field strength is weak. In the present case, μ_{mp} and the magnetic susceptibility χ_m are independent of the magnetic field strength. We will henceforth limit our discussion to such cases only.

18.3.1. *Magnetostriction*

Similar to the case of the electric effect on the volume of a dielectric, the volume of a paramagnetic substance can be affected by a magnetic field.

Integrating Eq. (18.68) from $\mathbf{H} = 0$ to $\mathbf{H}$, we find

$$V = V_0 - \frac{1}{2}\left(\frac{\partial \chi_m}{\partial p}\right)_{T,n} \mathbf{H}^2, \tag{18.74}$$

where V_0 is the volume in the absence of the magnetic field. The derivative $(\partial \chi_m/\partial p)_{T,n}$ is not large, but the volume can in principle change when the paramagnetic substance is subject to a magnetic field. This is the magnetic field analog of electrostriction.

18.3.2. *Magnetocaloric Effects*

In Chapter 6 on the third law of thermodynamics, we briefly discussed how low temperature could be achieved by applying the diamagnetic cooling method—the Debye–Giauque method, but its thermodynamic basis was not elucidated. Here, we will discuss the thermodynamic foundation of the method. We begin the discussion by calculating the field effect on entropy.

The field effect on entropy can be calculated from Eq. (18.69). By integrating Eq. (18.69) over $\mathbf{H}$, we obtain

$$S(T, p, \mathbf{H}) = S(T, p, 0) - \int_0^{\mathbf{H}} \left(\frac{\partial V \overline{\chi}_m}{\partial T}\right)_{p,n,\mathbf{H}} \mathbf{H} d\mathbf{H}. \tag{18.75}$$

This can be further evaluated. Since V and $\overline{\chi}_m$ are constant in the linear order in $\mathbf{H}$, we obtain

$$s(T, p, \mathbf{H}) = s(T, p, 0) - \frac{1}{2}v\left[\alpha \overline{\chi}_m + \left(\frac{\partial \overline{\chi}_m}{\partial T}\right)_{p,n}\right] \mathbf{H}^2, \tag{18.76}$$

where s is the molar entropy and α is the isobaric expansion coefficient

$$\alpha = V^{-1}\left(\frac{\partial V}{\partial T}\right)_{p,n,\mathbf{H}}. \tag{18.77}$$

Therefore, the isobaric specific heat at constant $\mathbf{H}$ is given by

$$c_{p\mathbf{B}} = T\left(\frac{\partial s}{\partial T}\right)_{p,n,\mathbf{H}} = c_p + c_p^{(1)}\mathbf{H}^2, \tag{18.78}$$

where c_p is the isobaric specific heat in the absence of **H** and

$$c_p^{(1)} = \frac{1}{2} T \frac{\partial}{\partial T} \left\{ v \left[\alpha \overline{\chi}_m + \left(\frac{\partial \overline{\chi}_m}{\partial T} \right)_{p,n} \right] \right\}. \qquad (18.79)$$

We may make use of Eq. (18.67). Upon integrating this equation over **H**, we obtain

$$T = T_0 - \int_0^{\mathbf{H}} \left(\frac{\partial \chi_m}{\partial S} \right)_{p,n,\mathbf{B}} \mathbf{H} d\mathbf{H}. \qquad (18.80)$$

Since

$$\left(\frac{\partial \chi_m}{\partial S} \right)_{p,n,\mathbf{H}} = \left(\frac{\partial \overline{\chi}_m}{\partial T} \right)_{p,n} \left(\frac{\partial T}{\partial s} \right)_{p,n,\mathbf{H}} = \frac{1}{c_p \mathbf{H}} \left(\frac{\partial \overline{\chi}_m}{\partial T} \right)_{p,n}, \qquad (18.81)$$

where $\overline{\chi}_m$ is the molar magnetic susceptibility

$$\overline{\chi}_m = \frac{\chi_m}{n}, \qquad (18.82)$$

we find

$$T = T_0 - \left(\frac{\partial \overline{\chi}_m}{\partial T} \right)_{p,n} \int_0^{\mathbf{B}} \frac{\mathbf{H}}{c_p \mathbf{H}} d\mathbf{H} \qquad (18.83)$$

$$\simeq T_0 - \frac{1}{2c_p} \left(\frac{\partial \overline{\chi}_m}{\partial T} \right)_{p,n} \mathbf{H}^2 \qquad (18.84)$$

if $(\partial \overline{\chi}_m / \partial T)_{p,n}$ is assumed to be constant with respect to **H**. If not, it should be kept under the integral sign. Since $(\partial \overline{\chi}_m / \partial T)_{p,n}$ is generally negative, $\Delta T = T(\mathbf{H}') - T(\mathbf{H})$ is negative if $\mathbf{H}' < \mathbf{H}$. That is, by lowering the magnetic field strength the temperature of the system is decreased. This is the thermodynamic basis of the diamagnetic cooling method by Giauque and Debye discussed in Chapter 6.

Part II
Thermodynamics of Irreversible Processes — Irreversible Thermodynamics

Although our principal interest in Part I (consisting of Chapters 2–18) lies in equilibrium (reversible) processes in matter, we have treated the first and second laws of thermodynamics from the general viewpoint that encompasses reversible as well as irreversible processes occurring in the system. This manner of treatment represents a departure from the conventional treatment in classical thermodynamics textbooks, in which the processes are limited to reversible processes only. The main reason for the departure is that the thermodynamic laws, at least, the first two laws, comprehensively apply to all natural macroscopic phenomena including both reversible and irreversible processes. As a matter of fact, reversible processes are idealized limits of irreversible phenomena occurring in real natural processes. Therefore, it is preferable to formulate the thermodynamic laws more comprehensively than usually done in the classical equilibrium thermodynamics practiced in the literature.

Having formulated the thermodynamic laws in such a general context in Part I, it is incumbent upon us to show how one might use the general form of the thermodynamic laws to properly treat irreversible phenomena. This subject area treating irreversible processes as wide as possible is by no means exhausted, but still developing, and we believe it will continue to do so as horizon of our scientific curiosity expands and as we explore new phenomena in macroscopic matter. For a fuller treatment of the thermodynamics of irreversible processes discussed here, the readers are referred to monographs available in the literature.[1] A more comprehensive treatise should be prepared in the future as the subject matter develops and the theory is applied to more practical problems.

In Part II, we discuss a general treatment of the thermodynamics of nonequilibrium processes in the leading chapter and, in succeeding chapters, we provide some examples of non-Newtonian flow in a Lennard-Jones fluid and chemical oscillations and pattern formation in a reacting fluid, and other topics. Most of them have appeared in one form or another in the literature.

---

[1] For example, see B. C. Eu, *Generalized Thermodynamics: The Thermodynamics of Irreversible Processes and Generalized Hydrodynamics* (Kluwer, Dordrecht, 2002); D. Jou and J. Casas-Vazquez, *Extended Irreversible Thermodynamics* (Springer, Berlin, 2012).

Chapter 19

Thermodynamics
of Nonequilibrium Processes

19.1. Extended Gibbs Relation for Calortropy

Recall that the extended notion of entropy, which was enunciated for only reversible processes by R. Clausius, was given the terminology *calortropy* (*heat evolution*) in Chapter 4. In irreversible thermodynamics, calortropy is regarded as dependent not only on conserved variables (e.g., density, momentum,[1] internal energy, and angular momentum) as is for entropy, but also on nonconserved variables, which are representative of irreversible processes occurring in the system in nonequilibrium. The most commonly appearing nonconserved variables are stress or strain, heat flow, mass flow, which are generally objects of study in fluid mechanics familiar to us. Since irreversible processes are generally time-dependent and spatially nonhomogeneous, they are described in space–time. Here, our discussion will be limited to nonrelativistic phenomena.[2] For relativistic treatment of irreversible

---

[1]Momentum, however, does not explicitly appear in the nonrelativistic classical thermodynamic formalism.

[2]See, for example, the recent monograph by B. C. Eu, *Kinetic Theory of Nonequilibrium Ensembles, Irreversible Thermodynamics, and Generalized Hydrodynamics, Volume 1 Nonrelativistic Theories* (Springer, Switzerland, 2016).

phenomena in spacetime, the reader is referred to a recent monograph[3] by one of us on the subject.

In the thermodynamic theory of irreversible phenomena we describe below, the calortropy Ψ of a system of volume V is expressed in terms of its local density depending on position and time. We therefore define the calortropy density $\widehat{\Psi}$ in relation to the global calortropy Ψ as follows:

$$\Psi(t) = \int_V d\mathbf{r} \rho \widehat{\Psi}(\mathbf{r}, t), \qquad (19.1)$$

where ρ denotes the mass density at position $\mathbf{r}$ and time t. In irreversible thermodynamics, this local calortropy density is subject to the local form of the second law of thermodynamics. This aspect was already discussed in terms of Ψ in Chapter 4, where the second law of thermodynamics was presented.

The union of manifold spanned by the collection $\mathfrak{P}$ of conserved and nonconserved variables and the manifold $\mathfrak{T}$ of variables tangent to the surface of Ψ spanned by the former set is called the thermodynamic space (manifold): $\mathfrak{M} := \mathfrak{P} \cup \mathfrak{T}$. We denote the aforementioned conserved variables by ρ, $\rho\mathbf{u}$ (momentum density), $\rho\mathcal{E}$ (internal energy density or just energy density), c_a (mass fraction; ratio of mass density ρ_a of species a to ρ) in the case of a mixture, and the nonconserved variables[4] are collectively denoted by $\Phi_a^{(q)} = \rho\widehat{\Phi}_a^{(q)}$ $(q \geq 1)$ for $a = 1, \ldots, c$ for a c-component mixture.

We will denote the pressure (stress) tensor by $\mathbf{P}$, the heat flux by $\mathbf{Q}$, mass fluxes by $\mathbf{J}_a$, etc. Both $\mathbf{P}$ and $\mathbf{Q}$ may be decomposed into their species components:

$$\mathbf{P} = \sum_{a=1}^c \mathbf{P}_a, \quad \mathbf{Q} = \sum_{a=1}^c \mathbf{Q}_a. \qquad (19.2)$$

Other as-yet-unspecified nonconserved variables may be similarly decomposed into their species components. These nonconserved variables are suitably ordered in a set. More specifically, in this work, they are ordered as in

---

[3]B. C. Eu, *Kinetic Theory of Nonequilibrium Ensembles, Irreversible Thermodynamics, and Generalized Hydrodynamics, Volume 2 Relativistic Theories* (Springer, Switzerland, 2016).

[4]In the present discussion, we will not take volume transport phenomena into account. Therefore, the molar volume and its flux will not be included in $\mathfrak{M}$ for simplicity of formalism. Since their effects are generally of the second order with regard to the density gradient, their neglect is tolerable. For a theory of volume transport phenomena, see B. C. Eu, *J. Chem. Phys.* **129**, 094502, 134509 (2008).

Table 19.1.   Nonconserved variables.

Name	$\Phi_a^{(q)}$	Hydrodynamic Symbols
Shear stress	$\Phi_a^{(1)}$	$[\mathbf{P}_a]^{(2)} = \frac{1}{2}(\mathbf{P}_a + \mathbf{P}_a^t) - \frac{1}{3}\delta \mathrm{Tr}\mathbf{P}_a$
Excess normal stress	$\Phi_a^{(2)}$	$\Delta_a = \frac{1}{3}\mathrm{Tr}\mathbf{P}_a - p_a$
Heat flux	$\Phi_a^{(3)}$	$\mathbf{Q}_a' = \mathbf{Q}_a - \widehat{h}_a \mathbf{J}_a$
Mass flux	$\Phi_a^{(4)}$	$\mathbf{J}_a = \rho_a(\mathbf{u}_a - \mathbf{u}) := \rho_a \widehat{\mathbf{J}}_a$
$\vdots$	$\vdots$	$\vdots$

Table 19.1. In this table, p_a denotes the local equilibrium pressure, $\widehat{h}_a$ the local enthalpy density, and δ the unit second-rank Cartesian tensor, which may be represented by the unit matrix[5]

$$\delta = \begin{pmatrix} 1 & 0 & 0 \\ 0 & 1 & 0 \\ 0 & 0 & 1 \end{pmatrix}. \tag{19.3}$$

Therefore, $[\mathbf{P}_a]^{(2)}$ is the traceless symmetric part of the stress (pressure) tensor, and $\mathbf{Q}_a'$ is the apparent heat flux in excess of heat carried by mass diffusion.

19.1.1.   Differential Form for Calortropy

The differential form for the calortropy density $\widehat{\Psi}$ is exact[6] (total) in the thermodynamic manifold conjugate to the tangent manifold $\mathfrak{T}$ spanned by (intensive) variables $(T, p, \widehat{\mu}_a, X_a^{(q)} : 1 \leq a \leq c; q \geq 1)$. It is expressed by the exact differential form

$$d\widehat{\Psi} = T^{-1}\left(d\mathcal{E} + pdv - \sum_{a=1}^{c}\widehat{\mu}_a dc_a + \sum_{q\geq 1}\sum_{a=1}^{c}X_a^{(q)}d\widehat{\Phi}_a^{(q)}\right). \tag{19.4}$$

Here, T, p, $\widehat{\mu}_a$, $X_a^{(q)}$ are the (local) temperature, pressure, chemical potentials, generalized potentials, and the last terms are scalar products of either vector or tensors; we have omitted the symbols for scalar products or tensor contraction for brevity of notation. The elements of tangent manifold $\mathfrak{T}$ generally depend on position and time. The terminology tangent manifold

---

[5] Here, by tensors, we mean Cartesian tensors.
[6] This is a mathematical term implying that the differential form is integrable in the manifold of the variables in terms of which the differential is expressed.

originates from the fact that the variables of the manifold are related to the tangent to the surface $\widehat{\Psi}$ in manifold $\mathfrak{P}$. The first three elements of the tangent manifold $\mathfrak{T}$ have the classical thermodynamic counterparts, whereas the last set of elements $X_a^{(q)}$ are new to the present generalized formalism. The generalized potentials $X_a^{(q)}$ vanish as the system approaches equilibrium. Therefore, the generalized potentials are a measure of displacement from equilibrium. It should be emphasized that the quantities involved in the differential one-form (19.4) are local and time-dependent owing to the fact that the variables making up differentials on the right obeys the local field equations — generally partial differential equations in space and time shown below.

In the present formalism, the evolution of $\widehat{\Psi}$ is described by the evolution of variables in manifold $\mathfrak{P}$, which obey their own evolution equations — often called the constitutive equations in irreversible thermodynamics — characteristic of nonequilibrium matter under consideration, which are going through a nonequilibrium process. It can be shown that the conserved variables $\mathcal{E}$, $v := 1/\rho$, c_a, and $\mathbf{u}$ obey the conservation laws

$$\rho d_t v = \nabla \cdot \mathbf{u}, \tag{19.5}$$

$$\rho d_t c_a = -\nabla \cdot \mathbf{J}_a \quad (1 \le a \le c), \tag{19.6}$$

$$\rho d_t \mathbf{u} = -\nabla \cdot \mathbf{P} + \rho \mathbf{F}^{(ex)}, \tag{19.7}$$

$$\rho d_t \mathcal{E} = -\nabla \cdot \mathbf{Q} - \mathbf{P} : \nabla \mathbf{u} + \sum_{a=1}^{c} \mathbf{J}_a \cdot \mathbf{F}_a^{(ex)}, \tag{19.8}$$

where d_t defined by the formula

$$d_t = \partial_t + \mathbf{u} \cdot \nabla \tag{19.9}$$

denotes the substantial time derivative in the coordinate system moving with the fluid velocity $\mathbf{u}$; $\mathbf{F}_a^{(ex)}$ is the external force[7] (i.e., body force) density on species a; and

$$\mathbf{F}^{(ex)} = \sum_{a=1}^{c} c_a \mathbf{F}_a^{(ex)} \tag{19.10}$$

is the total external force density. Equation (19.5) is the mass conservation law (mass balance equation); Eq. (19.6) is the mass fraction conservation law for which we have assumed that there is no chemical reaction present in

---

[7]If the external forces are electromagnetic, then Maxwell's electrodynamic equations must be added to the set (19.5)–(19.8).

the system; Eq. (19.7) is the momentum conservation law; and Eq. (19.8) is the internal energy conservation law for which we have excluded the effect of radiation. In nonrelativistic irreversible thermodynamics, the momentum conservation law does not directly enter the differential form for $\widehat{\Psi}$, although the fluid velocity **u** is one of the conserved variables obeying Eq. (19.7). For this reason, we have not included **u** in manifold $\mathfrak{P}$ explicitly. If there exist chemical reactions in the system, Eq. (19.6) acquires a source term representative of the chemical reactions involved. In that event, concentration balance equations, strictly speaking, are no longer evolution equations for conserved variables. Instead, they should be included in the set for the nonconserved variables.

The evolution equations of the nonconserved variables in $\mathfrak{P}$ may be compactly expressed in the form

$$\rho d_t \widehat{\boldsymbol{\Phi}}_a^{(q)} = -\nabla \cdot \boldsymbol{\psi}_a^{(q)} + \mathcal{Z}_a^{(q)} + \Lambda_a^{(q)} \quad (1 \leq a \leq c; \; q \geq 1), \qquad (19.11)$$

where $\boldsymbol{\psi}_a^{(q)}$ is the flux of $\boldsymbol{\Phi}_a^{(q)}$, $\mathcal{Z}_a^{(q)}$ is the kinematic term which is generally nonlinear with respect to the variables in $\mathfrak{P}$ and **u** or variables spanning $\mathfrak{T}$, and $\Lambda_a^{(q)}$ is the dissipation term which is the seat of energy dissipation arising from process $\boldsymbol{\Phi}_a^{(q)}$. The dissipation terms are closely related to the Rayleigh dissipation function,[8] as will be shown later. The kinematic terms $\mathcal{Z}_a^{(q)}$ are also dependent on the external forces and contain thermodynamic forces driving the irreversible processes. The set of differential equations (19.5)–19.8) and (19.11), which describe macroscopic variables in manifold $\mathfrak{P}$, is called *generalized hydrodynamic equations*.

They not only reduce to the classical hydrodynamic equations of Navier and Stokes, Fourier, and Fick as the system tends to equilibrium from a state far removed from equilibrium, but they are also consistent with the laws of thermodynamics, i.e., thermodynamically consistent, which is a terminology used in this work to mean that *the theory strictly conforms to the laws of thermodynamics,* because the calortropy depends on the variables

---

[8] Lord Rayleigh, *Theory of Sound* (Dover, New York, 1949). In his book, Rayleigh introduces the dissipation function in terms of a scalar quadratic form of velocities. Here, we are generalizing the Rayleigh dissipation functions in the form of scalar products (or, more precisely, contraction to scalars of tensors) of nonconserved fluxes and their "conjugate" dissipation terms making up the source terms of the evolution equations of the nonconserved variables. Therefore, the Rayleigh dissipation function in the present theory reduces to the original Rayleigh dissipation function in a rather special case of velocities. Here, we borrow the term for want of better terminology.

of the thermodynamic manifold and it strictly conforms to the constraints of the thermodynamic laws. This aspect was already discussed in Chapter 4 where the concept of calortropy is discussed in detail. We will have more occasions to elaborate on this statement.

Phenomenologically, the precise forms for the constitutive equations represented by Eq. (19.11) depend on the nature and level of description for the process of interest, as all constitutive equations are in macroscopic thermal physics. For example, if we are interested in phenomena occurring near equilibrium, Eq. (19.11) may be approximated by linear equations[9] with respect to thermodynamic gradients such as ∇T, $\nabla \mathbf{u}$, ∇p, and ∇c_a. And, furthermore, if the processes are steady and thus time-independent, then the time derivative term on the left of Eq. (19.11) can be omitted to a good approximation, or, more precisely, in the *adiabatic approximation*. In this approximation the nonconserved variables evolve on faster time scales than the conserved variables and hence they reach steady state well before the conserved variables reach their steady state. Therefore, on the time scale of variation of nonconserved variables, the conserved variables remain constant, hence the term *adiabatic*. Such an approximation results in linear irreversible thermodynamics if constitutive equations for conserved variables are linearized with respect to the thermodynamic forces and nonconserved variables (i.e., fluxes). Typical examples for such steady-state linear constitutive equations (i.e., evolution equations) for fluxes are the Fourier's law of heat conduction, Newton's law of viscosity, Fick's law of diffusion, and so on. Some specific forms for the constitutive equations will be given when we discuss examples for the application of the formalism presented here, especially, in connection with linear irreversible processes.

Furthermore, if the external forces are electromagnetic, then Maxwell's electrodynamic equations must be appended[10] to the constitutive equations (19.11). In this regard, it is helpful to recall that the Poisson–Boltzmann equation enters the theory of transport processes in electrolytes in the

---

[9]For example, see B. C. Eu, *Kinetic Theory and Irreversible Thermodynamics* (Wiley, New York, 1992) for discussions on constitutive equations in irreversible thermodynamics.

[10]There is a question regarding the form for the electromagnetic momentum balance equation. See B. C. Eu, *Phys. Rev. A* **33**, 4121 (1986) and also see footnote 5 of this chapter; B. C. Eu, *Kinetic Theory and Irreversible Thermodynamics* (Wiley, New York, 1992) cited earlier.

aforementioned context in which we discuss the thermodynamics of strong electrolytes; see Chapter 16.

It is important to point out that the evolution equations for nonconserved variable set $(\boldsymbol{\Phi}_a^{(1)}, \boldsymbol{\Phi}_a^{(2)}, \boldsymbol{\Phi}_a^{(3)}, \boldsymbol{\Phi}_a^{(4)})$ discussed in the present chapter are limited to the conventional nonconserved variable set, namely, shear stresses, excess normal stresses, heat fluxes, and diffusion fluxes. Therefore, the set excludes the nonconserved variables representing the molar volume (i.e., mean Voronoi volume) and the volume flux — flux of molar volume — and other possible nonconserved variables if the irreversible processes under consideration require them. Both molar volume and volume flux obey their own evolution equations, which should be added to the evolution equations of the set $(\boldsymbol{\Phi}_a^{(1)}, \boldsymbol{\Phi}_a^{(2)}, \boldsymbol{\Phi}_a^{(3)}, \boldsymbol{\Phi}_a^{(4)})$, Eq. (19.11), if irreversible processes in liquids are considered. However, in the conventional treatment the molar volume and its flux are not included.

Inclusion of such additional nonconserved variables significantly alters the fluid mechanics and irreversible thermodynamics of dense fluids and liquids. This new aspect of fluid mechanics and irreversible thermodynamics of dense fluids and liquids is discussed in detail in the monograph by Eu cited in footnote 2 of this chapter.

In the case of dense fluids and liquids, it is not so simple to guess the kinematic terms $\mathcal{Z}_a^{(q)}$ if the system is far removed from equilibrium because they usually involve mean values of derivatives of intermolecular forces and the virial tensors multiplied by velocity vectors, etc. Derivations of $\mathcal{Z}_a^{(q)}$ in dense fluids and liquids from the kinetic equation of nonequilibrium ensembles is given in the reference cited in footnote 2 of this chapter.

19.1.2. *Variables of the Tangent Manifold*

The extended Gibbs relation (19.4) implies that the tangents (in manifold $\mathfrak{T}$) to the calortropy surface $\widehat{\Psi}$ in $\mathfrak{P}$ are given by the derivatives of $\widehat{\Psi}$ with respect to the variables belonging to $\mathfrak{P}$:

$$T^{-1} = \left(\frac{\partial \widehat{\Psi}}{\partial \mathcal{E}} \right)_{v,c,\widehat{\Phi}}, \tag{19.12}$$

$$p = T \left(\frac{\partial \widehat{\Psi}}{\partial v} \right)_{T,c,\widehat{\Phi}}, \tag{19.13}$$

$$\widehat{\mu}_a = -T \left(\frac{\partial \widehat{\Psi}}{\partial c_a} \right)_{T,v,c',\widehat{\Phi}} , \tag{19.14}$$

$$X_a^{(q)} = T \left(\frac{\partial \widehat{\Psi}}{\partial \widehat{\Phi}_a^{(q)}} \right)_{T,v,c,\widehat{\Phi}'} . \tag{19.15}$$

The prime on c and $\widehat{\Phi}$ in the subscript to the derivatives in (19.14) and (19.15) means the exclusion of c_a, and similarly for $\widehat{\Phi}'$. Phenomenologically, the left-hand sides of these derivatives are related to constitutive relations, such as the caloric equation of state, the equation of state, the chemical potentials, and the generalized potentials which may be given in terms of the variables spanning manifolds $\mathfrak{P}$ and $\mathfrak{T}$. Development of such constitutive relations is one of the principal tasks in the phenomenological theory of irreversible thermodynamics in accordance with experiments.

Integrating the aforementioned constitutive relations in $\mathfrak{P}$, the calortropy density surface $\widehat{\Psi}$ is obtained as *a storage of constitutive information of the substance in nonequilibrium*. In irreversible thermodynamics, such a task necessarily involves solutions of the coupled set of partial differential equations (19.5)–(19.11), which constitute the generalized hydrodynamic equations mentioned earlier. It should be remarked that the generalized hydrodynamic equations reduce to the classical hydrodynamic equations—the Navier–Stokes–Fourier–Fick equations appearing in classical hydrodynamics,[11] if the irreversible processes occur near equilibrium, so that evolution equations for nonconserved variables are linearized with respect to the thermodynamic forces, such as velocity gradients, temperature gradients, and concentration gradients. We thereby recover the thermodynamic theory of linear irreversible processes. This will be discussed in a subsequent chapter.

19.2. Nonequilibrium Maxwell Relations

As the second derivatives of entropy do in equilibrium thermodynamics, the second derivatives of calortropy with respect to its dependent variables give rise to symmetrical relations, producing what we call nonequilibrium Maxwell relations. For example, by differentiating Eq. (19.15) with respect

---

[11]For example, see L. D. Landau and E. M. Lifshitz, *Fluid Mechanics* (Pergamon, Oxford, 1958).

to $\widehat{\Phi}_b^{(r)}$ at fixed T, v, and c and interchanging the order of differentiations with respect to $\widehat{\Phi}_a^{(q)}$ and $\widehat{\Phi}_b^{(r)}$, we obtain the symmetry relation between the derivatives

$$\left(\frac{\partial X_a^{(q)}}{\partial \widehat{\Phi}_b^{(r)}}\right)_{T,v,c,\widehat{\Phi}''} = \left(\frac{\partial X_b^{(r)}}{\partial \widehat{\Phi}_a^{(q)}}\right)_{T,v,c,\widehat{\Phi}''}, \tag{19.16}$$

and similarly for other derivatives with respect to conserved variables, which would give nonequilibrium extensions of the equilibrium Maxwell relations. We do not bother to write them out explicitly here. Some of them are, in fact, made use of in subsequent discussions of this chapter. Note that the relation (19.16) does not have an equilibrium counterpart because it is made up of nonequilibrium attributes only.

Since the generalized potentials $X_a^{(q)}$ may be expressible in terms of a power series in nonconserved variables with the scalar coefficients consisting of conserved variables as well as nonconserved variables, the relations such as Eq. (19.16) can provide some symmetrical relations between the coefficients. Such symmetry relations can be rather useful for determining, or extracting from them, the experimental information on the nonconserved variables not as yet measured in the laboratory or difficult to access by current experimental methods. For example, it is generally possible to express on the basis of symmetry reason the generalized potentials $X_a^{(q)}$ in the form

$$X_a^{(q)} = -g_{qa}(T,v,c,\widehat{\Phi}_b^{(s)})\widehat{\Phi}_a^{(q)}, \tag{19.17}$$

where $g_{qa}(T,v,c,\widehat{\Phi}_b^{(s)})$ is a scalar-valued function of variables indicated. Then the nonequilibrium Maxwell's relation (19.16) provides the symmetry relation of the partial derivatives of functions g_{qa} and g_{rb}, with respect to $\widehat{\Phi}_a^{(q)}$ and $\widehat{\Phi}_b^{(r)}$, respectively:

$$\left(\frac{\partial g_{qa}}{\partial \widehat{\Phi}_b^{(r)}}\right)_{T,v,c,\widehat{\Phi}'} = \left(\frac{\partial g_{rb}}{\partial \widehat{\Phi}_a^{(q)}}\right)_{T,v,c,\widehat{\Phi}'}. \tag{19.18}$$

This may be regarded as an irreversible thermodynamic example of reciprocal relation for the coefficient functions g_{qa} and g_{rb}. For processes near equilibrium, g_{qa} may be approximated by a function independent of $\widehat{\Phi}_a^{(q)}$, but dependent on the conserved variables only or temperature. Equation (19.18) provides a way to determine g_{qa} from the information on g_{rb} and vice versa.

19.3.   Legendre Transforms of Thermodynamic Functions

To develop and investigate the constitutive relations for the variables spanning the tangent manifold $\mathfrak{T}$, it is convenient to transform the internal energy
to another kind as we usually have done in equilibrium
thermodynamics. Thus, we define the extended enthalpy by the Legendre
transformation

$$\mathcal{H} = \mathcal{E} + pv, \tag{19.19}$$

and the extended Gibbs free energy by the transformation

$$\mathcal{G} = \mathcal{H} - T\widehat{\Psi}. \tag{19.20}$$

It is remarkable that $\mathcal{H}$ is isomorphic to the equilibrium enthalpy and so is
$\mathcal{G}$ except that the calortropy density $\widehat{\Psi}$ is replacing the entropy density $\widehat{S}$.
The extended Gibbs relations for $\mathcal{H}$ and $\mathcal{G}$ are then given respectively by
the differential forms

$$d\mathcal{H} = Td\widehat{\Psi} + vdp + \sum_{a=1}^{c} \widehat{\mu}_a dc_a - \sum_{q\geq 1}\sum_{a=1}^{c} X_a^{(q)} d\widehat{\Phi}_a^{(q)}, \tag{19.21}$$

and

$$d\mathcal{G} = -\widehat{\Psi}dT + vdp + \sum_{a=1}^{c} \widehat{\mu}_a dc_a - \sum_{q\geq 1}\sum_{a=1}^{c} X_a^{(q)} d\widehat{\Phi}_a^{(q)}. \tag{19.22}$$

The differential form (19.21) implies that the following derivatives hold:

$$T = \left(\frac{\partial \mathcal{H}}{\partial \widehat{\Psi}}\right)_{p,c,\widehat{\Phi}}, \tag{19.23}$$

$$v = \left(\frac{\partial \mathcal{H}}{\partial p}\right)_{T,c,\widehat{\Phi}}, \tag{19.24}$$

$$\widehat{\mu}_a = \left(\frac{\partial \mathcal{H}}{\partial c_a}\right)_{T,p,c',\widehat{\Phi}}, \tag{19.25}$$

$$X_a^{(q)} = -\left(\frac{\partial \mathcal{H}}{\partial \widehat{\Phi}_a^{(q)}}\right)_{T,p,c,\widehat{\Phi}'}, \tag{19.26}$$

which give rise to the extended Maxwell relations:

$$\left(\frac{\partial T}{\partial \widehat{\Phi}_a^{(q)}}\right)_{\widehat{\Psi},p,c,\widehat{\Phi}'} = -\left(\frac{\partial X_a^{(q)}}{\partial \widehat{\Psi}}\right)_{p,c,\widehat{\Phi}} = -\frac{T}{c_{p\Phi}}\left(\frac{\partial X_a^{(q)}}{\partial T}\right)_{p,c,\widehat{\Phi}}, \quad (19.27)$$

$$\left(\frac{\partial v}{\partial \widehat{\Phi}_a^{(q)}}\right)_{\widehat{\Psi},p,c,\widehat{\Phi}'} = \left(\frac{\partial X_a^{(q)}}{\partial p}\right)_{T,c,\widehat{\Phi}}, \quad (19.28)$$

$$\left(\frac{\partial \widehat{\mu}_a}{\partial \widehat{\Phi}_a^{(q)}}\right)_{\widehat{\Psi},p,c,\widehat{\Phi}'} = -\left(\frac{\partial X_a^{(q)}}{\partial c_a}\right)_{T,p,c,\widehat{\Phi}}, \quad (19.29)$$

$$\left(\frac{\partial X_a^{(q)}}{\partial \widehat{\Phi}_b^{(r)}}\right)_{\widehat{\Psi},p,c,\widehat{\Phi}'} = \left(\frac{\partial X_b^{(r)}}{\partial \widehat{\Phi}_a^{(q)}}\right)_{T,p,c,\widehat{\Phi}'}. \quad (19.30)$$

In Eq. (19.27) $c_{p\Phi}$ is the extended isobaric heat capacity at constant $\widehat{\Phi} = \left\{\widehat{\Phi}_a^{(q)}\right\}$:

$$c_{p\Phi} = T\left(\frac{\partial \widehat{\Psi}}{\partial T}\right)_{p,c,\widehat{\Phi}}. \quad (19.31)$$

It is therefore the isobaric heat capacity of the fluid under fixed values of $\widehat{\Phi}_a^{(q)}$, namely, a nonequilibrium system which is displaced from equilibrium.

The differential form (19.22) gives rise to derivatives similar to Eqs. (19.23)–(19.26) and nonequilibrium Maxwell relations similar to Eqs. (19.27)–(19.30). These relations may be employed to calculate the $\widehat{\Phi}$-dependence of the constitutive relations for the variables in $\mathfrak{T}$, including the generalized potentials $X_a^{(q)}$.

19.4. Nonequilibrium Effects

19.4.1. *Nonequilibrium Effect on Temperature*

We calculate the dependence of temperature on the nonconserved variables $\widehat{\Phi}_a^{(q)}$ in nonequilibrium fluids. Upon integrating Eq. (19.27) over $\widehat{\Phi}_a^{(q)}$ along the path of constant p and $\{c_a\}$, the $\widehat{\Phi}_a^{(q)}$-dependence of temperature is obtained:

$$T = T_0 + \sum_{q \geq 1} \sum_{a=1}^{c} \int_0^{\widehat{\Phi}_a^{(q)}} d\widehat{\Phi}_a^{(q)} \frac{T}{c_{p\Phi}} \left(\frac{\partial g_{qa}}{\partial T} \right)_{p,c,\widehat{\Phi}} \widehat{\Phi}_a^{(q)}, \tag{19.32}$$

where T_0 is the temperature of the system at $\widehat{\Phi}_a^{(q)} = 0$, namely, at equilibrium at which the generalized potentials have vanished. Thus, in the linear regime, where g_{qa} is independent of $\widehat{\Phi}_a^{(q)}$ and also the constant-Φ heat capacity $c_{p\Phi}$ is approximately constant, namely, $c_{p\Phi} \simeq c_p$, temperature T is quadratic with respect to $\widehat{\Phi}_a^{(q)}$:

$$\Delta T = T - T_0 \approx \sum_{q \geq 1} \sum_{a=1}^{c} \frac{1}{2c_p} \left(\frac{\partial g_{qa}}{\partial T} \right)_{p,c} \widehat{\Phi}_a^{(q)2}. \tag{19.33}$$

If $(\partial g_{qa}/\partial T)_{p,c}$ is positive, the nonequilibrium processes generally tend to increase temperature with increasing $\widehat{\Phi}_a^{(q)}$, namely, $\Delta T \geq 0$.

19.4.2. *Nonequilibrium Effect on Pressure*

The effect on pressure can be calculated similarly to the effect of irreversible processes on temperature. For this purpose it is convenient to define the extended Helmholtz free energy:

$$\mathcal{A} = \mathcal{E} - T\widehat{\Psi}, \tag{19.34}$$

which is a Legendre transformation and gives rise to the differential one-form in manifold $\mathfrak{P} \cup \mathfrak{T}$

$$d\mathcal{A} = -\widehat{\Psi}dT - pdv + \sum_{a=1}^{c} \widehat{\mu}_a dc_a - \sum_{q \geq 1} \sum_{a=1}^{c} X_a^{(q)} d\widehat{\Phi}_a^{(q)}. \tag{19.35}$$

From this follows the extended Maxwell relation

$$\left(\frac{\partial p}{\partial \widehat{\Phi}_a^{(q)}} \right)_{T,v,c,\widehat{\Phi}'} = \left(\frac{\partial X_a^{(q)}}{\partial v} \right)_{T,c,\widehat{\Phi}}. \tag{19.36}$$

Integration of this equation over $\widehat{\Phi}_a^{(q)}$ along a path of constant $T, v, c := \{c_a\}$ yields the pressure in the form

$$p = p_0 - \sum_{q \geq 1} \sum_{a=1}^{c} \int_0^{\widehat{\Phi}_a^{(q)}} \left(\frac{\partial g_{ab}}{\partial v} \right)_{T,c,\widehat{\Phi}} \widehat{\Phi}_a^{(q)} d\widehat{\Phi}_a^{(q)}, \tag{19.37}$$

where $p_0 = p(\widehat{\Phi} = 0)$, the pressure at equilibrium. In the linear regime of irreversible processes, since g_{ab} is independent of $\widehat{\Phi}_a^{(q)}$, it follows that the nonequilibrium correction to pressure is quadratic with respect to nonconserved variables $\widehat{\Phi}_a^{(q)}$:

$$p = p_0 - \frac{1}{2}\sum_{q\geq 1}\sum_{a=1}^{c}\left(\frac{\partial g_{ab}}{\partial v}\right)_{T,c,\widehat{\Phi}}\widehat{\Phi}_a^{(q)2}. \tag{19.38}$$

Other derivative relations (i.e., Maxwell relations) can be similarly integrated to obtain the constitutive relations in the nonequilibrium regime.

19.4.3. *Nonequilibrium Effect on Chemical Potentials*

The differential one-form for $\mathcal{G}$ (19.22) implies that chemical potentials are the extended partial molar Gibbs free energy

$$\widehat{\mu}_a = \left(\frac{\partial \mathcal{G}}{\partial c_a}\right)_{T,p,c',\widehat{\Phi}}, \tag{19.39}$$

which in turn means that $\mathcal{G}$ is a first-degree homogeneous function of concentrations $\{c_a : 1 \leq a \leq c\}$ because $\widehat{\mu}_a$ is a partial molar $\mathcal{G}$, the nonequilibrium Gibbs free energy. That is,

$$\mathcal{G} = \sum_{a=1}^{c}\widehat{\mu}_a c_a. \tag{19.40}$$

Upon substitution into Eq. (19.22), it gives rise to the extended Gibbs–Duhem equation

$$\sum_{a=1}^{c}c_a d\widehat{\mu}_a = -\widehat{\Psi}dT + vdp - \sum_{q\geq 1}\sum_{a=1}^{c}X_a^{(q)}d\widehat{\Phi}_a^{(q)}, \tag{19.41}$$

which turns out to be the integrability condition[12] of the differential one-form (19.22). Note that this verifies that the extended Gibbs–Duhem equation is indeed the integrability condition for the extended Gibbs relation for nonequilibrium Gibbs free energy $\mathcal{G}$.

---

[12]See footnote 1 and also M. Chen and B. C. Eu, *J. Math. Phys.* **34**, 3012 (1993).

Equation (19.41) also suggests that the chemical potentials may be regarded as functions of $(T, p, c_a, \widehat{\Phi}_a^{(q)})$. Therefore, we find for the differential form for $\widehat{\mu}_a$:

$$d\widehat{\mu}_a = -\overline{\Psi}_a dT + \overline{v}_a dp + \sum_{a=1}^{c} \sum_{b=1}^{c-1} \widehat{\mu}_{ab} dc_b - \sum_{q \geq 1} \sum_{b=1}^{c} \overline{X}_{b;a}^{(q)} d\widehat{\Phi}_b^{(q)}, \qquad (19.42)$$

where the coefficients to the differentials are given by the derivatives of $\widehat{\mu}_a$:

$$\overline{\Psi}_a = -\left(\frac{\partial \widehat{\mu}_a}{\partial T}\right)_{p,c,\widehat{\Phi}} = -\left(\frac{\partial \widehat{\Psi}}{\partial c_a}\right)_{T,p,c',\widehat{\Phi}}, \qquad (19.43)$$

$$\overline{v}_a = \left(\frac{\partial \widehat{\mu}_a}{\partial p}\right)_{T,c,\widehat{\Phi}} = \left(\frac{\partial v}{\partial c_a}\right)_{T,p,c',\widehat{\Phi}}, \qquad (19.44)$$

$$\overline{X}_{b;a}^{(q)} = -\left(\frac{\partial \widehat{\mu}_a}{\partial \widehat{\Phi}_b^{(q)}}\right)_{T,p,c,\widehat{\Phi}'} = \left(\frac{\partial X_b^{(q)}}{\partial c_a}\right)_{T,p,c',\widehat{\Phi}}, \qquad (19.45)$$

$$\widehat{\mu}_{ab} = \left(\frac{\partial \widehat{\mu}_a}{\partial c_b}\right)_{T,p,c',\widehat{\Phi}}. \qquad (19.46)$$

Therefore, $\overline{\Psi}_a$, $\overline{v}_a$, and $\overline{X}_a^{(q)}$ are found to be the partial molar calortropy, volume, and generalized potential.

Integrating Eq. (19.42) over $\widehat{\Phi}_b^{(q)}$ along a path of constant T, p, and $\{c_a\}$, we obtain nonequilibrium chemical potentials

$$\widehat{\mu}_a = \widehat{\mu}_a^{eq} - \sum_{q \geq 1} \sum_{b=1}^{c} \int_0^{\widehat{\Phi}_b^{(q)}} \overline{X}_{b;a}^{(q)} d\widehat{\Phi}_b^{(q)}, \qquad (19.47)$$

where $\widehat{\mu}_a^{eq} := \widehat{\mu}_a(\widehat{\Phi}_a^{(q)} = 0)$. Since it is possible to write

$$\overline{X}_{b;a}^{(q)} = -\overline{g}_{qb;a} \widehat{\Phi}_b^{(q)}, \qquad (19.48)$$

where $\overline{g}_{qb;a}$ is the partial molar g_{qb}, in the linear regime where g_{qa} is independent of $\widehat{\Phi}_a^{(q)}$, the chemical potentials are quadratic with respect to $\widehat{\Phi}_a^{(q)}$:

$$\widehat{\mu}_a = \widehat{\mu}_a^{eq} + \frac{1}{2} \sum_{q \geq 1} \sum_{b=1}^{c} \overline{g}_{qb;a} \widehat{\Phi}_b^{(q)2}. \qquad (19.49)$$

It should be noted that Eq. (19.48) is a phenomenological constitutive equation for the partial molal generalized potential $X_a^{(q)}$, and (19.49) the phenomenological constitutive equation for nonequilibrium chemical potential $\widehat{\mu}_a$.

By using the formalism presented here in this section, it is possible to develop and phenomenologically investigate the thermodynamics of irreversible processes in nonequilibrium fluids in a manner parallel to equilibrium thermodynamics discussed in the previous chapters in Part I of this work. Thus, one might state that this section summarizes the formalism for the operational side of irreversible thermodynamics developed in this work.

19.4.4. *Nonequilibrium Effect on Equilibrium Constants*

Since we have excluded the possibility of chemical reactions, the present formalism, strictly speaking, does not apply to the case of chemical reactions. However, even if chemical reactions are not excluded, they generally appear only in the mass fraction balance equations as a source term. Moreover, they generally occur on a faster time scale than the hydrodynamic processes involving the nonconserved variables owing to the fact that chemical bond exchange between molecules over the order of bond length, which is much shorter than intermolecular interaction ranges (at least, one order). Therefore, even though there are irreversible processes with regard to the nonconserved variables, the chemical equilibrium condition is still given by

$$\sum_{a=1}^{r} \omega_a \widehat{\mu}_a = 0, \qquad (19.50)$$

where ω_a are the stoichiometric coefficients of the chemical reaction. Here, we are assuming there is a single chemical reaction.

Assuming that irreversible processes are occurring near equilibrium, we insert Eq. (19.49) into Eq. (19.50) to obtain

$$\sum_{a=1}^{r} \omega_a \left[\widehat{\mu}_a^0 + (RT/\rho) \ln (x_a f_a) \right] + \frac{1}{2} \sum_{q \geq 1} \sum_{b=1}^{c} \Delta g_{qb} \widehat{\Phi}_b^{(q)2} = 0, \qquad (19.51)$$

where we have used the phenomenological formula for chemical potentials

$$\widehat{\mu}_a^{eq} = \widehat{\mu}_a^0 + (RT/\rho) \ln(x_a f_a) \quad (\rho = \text{ density}; \ f_a = \text{activity coefficient}) \qquad (19.52)$$

and Δg_{qb} is the change in g_{qb} arising from the chemical reaction:

$$\Delta g_{qb} = \sum_{a=1}^{r} \bar{g}_{qb;a} \omega_a. \tag{19.53}$$

Therefore, the equilibrium constant is given by

$$K(T) = K^0(T) \exp\left[-\frac{1}{2} \sum_{q\geq 1} \sum_{b=1}^{c} \Delta g_{qb} \widehat{\Phi}_b^{(q)2}\right], \tag{19.54}$$

where $K^0(T)$ is the equilibrium constant in the absence of nonequilibrium processes $\widehat{\Phi}_b^{(q)}$:

$$K^0(T) = \exp\left[-\Delta G^0/RT\right], \tag{19.55}$$

$$\Delta G^0 = \sum_{a=1}^{r} \omega_a \rho \widehat{\mu}_a^0 = \sum_{a=1}^{r} \omega_a \mu_a^0, \tag{19.56}$$

where ΔG^0 is the standard Gibbs free energy change for the reaction.

This result indicates the equilibrium constant is modified by the presence of nonconserved variables undergoing irreversible processes in the fluid, for example, when the fluid is sheared, or compressed by a shock wave, or subjected to a temperature gradient or concentration gradients giving rise to viscose flow, thermal conduction, diffusion, etc.

Chapter 20

Linear Irreversible Thermodynamics

Having formulated a general thermodynamic theory of irreversible processes, we now would like to see in what manner the theory of linear irreversible thermodynamics in the existing literature emerges from the general theory. We would like to show the general theory of irreversible thermodynamics and generalized hydrodynamics presented in Chapter 19 can give rise to the conventionally known theory of linear irreversible thermodynamics.[1] Let us recall the basic premises of the linear theory of irreversible thermodynamics originally formulated by L. Onsager, J. Meixner,[2] I. Prigogine,[3] and their followers[4]: the existence of Clausius entropy and its perfect differential form. The assumption required for such a differential form is the local equilibrium hypothesis, which allows one to interpret differentials of conserved variables in the differential form for the entropy as substantial derivatives obeying macroscopic (hydrodynamic) conservation laws. Secondly, it is assumed that the thermodynamic force–flux relations for fluxes (i.e., stresses, heat fluxes, and diffusion fluxes) are simply linear

---

[1]S. R. de Groot and P. Mazur, *Nonequilibrium Thermodynamics* (North-Holland, Amsterdam, 1962); I. Prigogine, *Thermodynamics of Irreversible Processes*, third edition (Interscience, New York, 1967); R. Haase, *Irreversible Thermodynamics* (Dover, New York, 1969).

[2]For example, see J. Meixner and H. G. Reik, *Thermodynamic der irreversiblen Processes*, Handbuch der Physik, Vol. III, S. Flügge, ed. (Springer Berlin, 1959).

[3]I. Prigogine, *Thermodynamics of Irreversible Processes*, third edition (Interscience, New York, 1967)

[4]S. R. de Groot and P. Mazur, *Nonequilibrium Thermodynamics* (North-Holland, Amsterdam, 1962); R. Haase, *Irreversible Thermodynamics* (Dover, New York, 1969).

with respect to spatial linear gradients of conserved variables (i.e., velocity, density, temperature or energy). With this pair of assumptions the aforementioned linear theory of irreversible processes is formulated in the linear irreversible thermodynamics. In the following, it is shown under what circumstances the aforementioned assumptions and approximations can be made use of to obtain the linear theory of irreversible processes — linear irreversible thermodynamics—from the general theory presented.

20.1. Approximations and Assumptions for Linear Irreversible Thermodynamics

20.1.1. *Linearized Generalized Potentials*

First of all, linear approximation must be taken for the generalized potentials $X_a^{(q)}$, which then become linear with respect to Φ_a. It should be noted that the Curie principle applies to the linear approximation. In the case of a single-component fluid, the generalized potentials $X^{(q)}$ are linearized with respect to the nonconserved variables $\Phi^{(q)} := (\Pi, \Delta, Q)$ — a finite membered set — in the forms

$$X^{(1)} = -\frac{\Pi}{2p}, \quad X^{(2)} = -\frac{3\Delta}{2p}, \quad X^{(3)} = -\frac{Q}{\widehat{C}_p T p}. \tag{20.1}$$

There is no diffusion flux present in the case of a single-component fluid. Note that in these approximate forms the coefficients to the fluxes (nonconserved variables) in Eq. (19.17) are

$$g^{(1)} = \frac{1}{2p}, \quad g^{(2)} = \frac{3}{2p}, \quad g^{(3)} = \frac{1}{\widehat{C}_p T p}. \tag{20.2}$$

As a matter of fact, if statistical mechanics is made use of, the generalized potentials in Eq. (20.1) can be deduced from Eqs. (19.12)–(19.15) or (19.16)–(19.18) in Chapter 19 as follows: Since the thermodynamic potential pv can be identified with the local nonequilibrium grand canonical partition function in the nonequilibrium statistical mechanics approach (see footnote 2 of Chapter 19), if the nonequilibrium grand canonical partition function is expanded in power series of nonconserved variables $\widehat{\Phi}^{(q)}$, then the formulas in Eq. (20.1) follow to the lowest order approximation in $\widehat{\Phi}^{(q)}$ for the generalized potentials presented. Note also similar approximations for $X^{(q)}$ must be applied to the generalized potentials appearing in the generalized hydrodynamic equations (19.5)–(19.8) and, in particular, (19.11) of Chapter 19.

If the fluid is a mixture, then a similar set of approximations can be taken for Eqs. (19.12)–(19.15). First, observe that in the case of a multicomponent fluid mixture, the nonconserved variables taken into account are extended to include diffusion fluxes: $(\mathbf{\Pi}_a, \mathbf{\Delta}_a, \mathbf{Q}'_a, \mathbf{J}_a : c \geq a \geq 1)$, where

$$\mathbf{Q}'_a := \mathbf{Q}_a - \widehat{h}_a \mathbf{J}_a \tag{20.3}$$

with $\widehat{h}_a$ denoting the enthalpy density per unit mass of species a. The vector $\mathbf{Q}'_a$ then defines the net heat flux of species a over and above $\mathbf{Q}_a$ relative to the heat flow $\widehat{h}_a \mathbf{J}_a$ attributed to the mass diffusion flow. The approximations for the generalized potentials of moments chosen are then given as follows:

$$X_a^{(1)} = -\frac{\mathbf{\Pi}_a}{2p_a}, \; X_a^{(2)} = -\frac{3\mathbf{\Delta}_a}{2p_a}, \; X_a^{(3)} = -\frac{\mathbf{Q}'_a}{\widehat{C}_{pa}Tp_a}; \; X_a^{(4)} = -\frac{\mathbf{J}_a}{\rho_a}. \tag{20.4}$$

These expressions turn out to be applicable to quite a wide range of irreversible processes and will be used for generalized potentials in the applications of the generalized hydrodynamics in subsequent discussions as well.

20.1.2. *Adiabatic Approximation*

In addition to the linearization of generalized potentials $X_a^{(q)}$, the quasi-steady-state approximation, which was alternatively called *the adiabatic approximation* discussed earlier, is taken to make the evolution equations for the nonconserved variables steady in the time scale of evolution of more slowly changing conserved variables. In the quasi-steady-state approximation, the generalized hydrodynamic equations, which are hyperbolic partial differential equations, become parabolic partial differential equations for conserved variables owing to the fact that in the time scale of conserved variable evolution the nonconserved variables remain constant in time. Nevertheless, they can still be highly nonlinear partial differential equations in the position coordinate space. Upon linearizing thus obtained steady-state nonlinear constitutive equations for fluxes with respect to the thermodynamic forces and fluxes (i.e., nonconserved variables), the classical hydrodynamic equations of Navier, Stokes, Fourier, and Fick, well-known in the hydrodynamics literature, are obtained from the nonlinear set of steady-state partial differential equations obtained for the nonconserved variables. From the mathematical standpoint, this procedure is based on a two-time scale concept that the conserved variables evolve more slowly than the nonconserved variables. It is in fact the first step of the approach taken when nonlinear set of differential equations is analyzed for their linear stability of the solutions under the assumption that two different time scales of temporal evolutions of the dependent variables in the differential equation set of conserved and nonconserved variables are involved. Thus, in the adiabatic approximation we obtain the steady-state constitutive equations for nonconserved variables

$$-\nabla \cdot \boldsymbol{\psi}_a^{(q)} + \mathcal{Z}_a^{(q)} + \Lambda_a^{(q)} = 0. \tag{20.5}$$

This approximation is called the adiabatic approximation, a terminology arising from the fact that during the evolutions of nonconserved variables the conserved variables remain fixed in time. This equation, upon linearization of the kinematic term and the dissipation term as well as $\boldsymbol{\psi}_a^{(q)}$ with respect to the generalized potentials, thermodynamic forces, and nonconserved variables, result in the desired constitutive equations for nonconserved variables (fluxes) in the linear irreversible thermodynamics. The detailed forms for Eq. (20.5) are shown below for nonconserved variables.

20.2.   A Single-component Fluid

For single-component (pure) fluids, we take the following set of approximations in the steady-state nonconserved variable evolution equations (20.4): First, ignoring the higher-order nonconserved variables,[5] we limit the number of nonconserved variables to $\boldsymbol{\Pi}$, $\boldsymbol{\Delta}$, $\mathbf{Q}$ in them — a step which is tantamount to a closure of the macroscopic variable set. This closure implies that "the fluxes $\psi^{(q)}$ of $\boldsymbol{\Pi}$, $\boldsymbol{\Delta}$, $\mathbf{Q}$" vanish for all q values

$$\psi^{(q)} = 0 \tag{20.6}$$

except for $q = 2$ for the bulk viscous flow — namely, the excess normal stress. It means that, by the closure, the divergence term $\nabla \cdot \psi^{(q)}$ vanishes for $q = 1, 3$ in the nonconserved variable evolution equations, namely, for shear stress and heat flux: the rationale behind this procedure $\psi^{(q)}$ for $q = 1, 3$ is that they are of higher order moments in the molecular theory of the processes. Second, in Eq. (19.11), the kinematic terms $\mathcal{Z}^{(q)}$ are linearized with respect to the thermodynamic forces, namely, the gradients of ρ, $\mathbf{u}$, and $\ln T$, and finally, the dissipation terms $\Lambda^{(q)}$ are also linearized with

---

[5]In other words, the discussion is limited to experiments relevant to nonconserved variables $\boldsymbol{\Pi}$, $\boldsymbol{\Delta}$, $\mathbf{Q}$.

respect to $\mathbf{\Pi}$, $\mathbf{\Delta}$, and $\mathbf{Q}$. Then, it can be readily shown that the steady-state nonconserved variable evolution equations in the case of a single-component fluid are approximated by the linearized constitutive equations for $\mathbf{\Pi}$, $\mathbf{\Delta}$, and $\mathbf{Q}$:

$$0 = -2p\,[\nabla\mathbf{u}]^{(2)} - \frac{p}{\eta_0}\mathbf{\Pi}, \tag{20.7}$$

$$0 = -\frac{2}{3}p\nabla\cdot\mathbf{u} - \frac{2p}{3\eta_b^0}\mathbf{\Delta}, \tag{20.8}$$

$$0 = -Tp\nabla\ln T - \frac{\widehat{C}_p Tp}{\lambda_0}\mathbf{Q}, \tag{20.9}$$

which can then be trivially rearranged to the well-known constitutive relations in the linear theory of transport processes[6]

$$\mathbf{\Pi} = -2\eta_0\,[\nabla\mathbf{u}]^{(2)}, \tag{20.10}$$

$$\mathbf{\Delta} = -\eta_b^0\nabla\cdot\mathbf{u}, \tag{20.11}$$

$$\mathbf{Q} = -\lambda_0\nabla\ln T. \tag{20.12}$$

Here, the parameters η_0, η_b^0, and λ_0 are phenomenological constants that will turn out to be the phenomenological transport coefficients: viscosity, bulk viscosity, and thermal conductivity, respectively. The parameter $\widehat{C}_p$ is the constant pressure heat capacity per unit mass of the fluid. If we resort to the kinetic theory[7] of fluids, the aforementioned phenomenological coefficients can be related to molecular theory formulas for transport coefficients in the Chapman–Enskog approach to the kinetic theory, but in the present phenomenological theory approach they are simply taken for phenomenologically (i.e., experimentally) determinable transport coefficients. It must be noted that in this work our interest does not lie in the molecular theory of irreversible thermodynamics and transport processes in fluids, but the thermodynamic theory of irreversible transport processes.

---

[6]See footnotes 2 and 3.

[7]For a dilute gas kinetic theory, see S. Chapman and T. G. Cowling, *The Mathematical Theory of Nonuniform Gases* (Cambridge University Press., London, 1972); J. H. Ferziger and H. G. Kaper, *Mathematical Theory of Transport Processes in Gases* (North-Holland, Amsterdam, 1972). For a dense fluid kinetic theory, see B. C. Eu, *Kinetic Theory and Irreversible Thermodynamics* (Wiley, New York, 1992) and B. C. Eu, *Kinetic Theory of Nonequilibrium Ensembles, Irreversible Thermodynamics, and Generalized Hydrodynamics, Volume 1 Nonrelativistic Theories and Volume 2 Relativistic Theories* (Springer, Switzerland, 2016).

The three constitutive equations listed above are, respectively, the well-known Newtonian law of viscosity, law of bulk viscosity, and Fourier's law of heat conduction. Therefore, the phenomenological coefficients therein are recognized as the Newtonian shear viscosity, bulk viscosity, and Fourier thermal conductivity of the fluid. They are examples for linear transport coefficients.

Upon substituting the constitutive equations (20.7)–(20.9) into the momentum balance equation and energy balance equation, we obtain the classical hydrodynamic equations[8] of Navier, Stokes, and Fourier:

$$\rho d_t \mathbf{u} = -\nabla p + 2\eta_0 \nabla \cdot [\nabla \mathbf{u}]^{(2)} + \eta_b \nabla^2 \mathbf{u} + \mathbf{F}, \qquad (20.13)$$

$$\rho \widehat{C}_p d_t T = \lambda_0' \nabla^2 T + \eta_0 [\nabla \mathbf{u}]^{(2)} : [\nabla \mathbf{u}]^{(2)} + \eta_b (\nabla \cdot \mathbf{u})^2. \qquad (20.14)$$

Here, it should be noted that $\lambda_0' = \lambda_0/T$ and the transport coefficients are assumed to be constant in space, although they are generally variable with respect to position $\mathbf{r}$ because they depend on T and fluid density, which depend on $\mathbf{r}$ if the nonequilibrium fluid is spatially inhomogeneous. These equations affirm that from the viewpoint of irreversible thermodynamics *the classical hydrodynamics of Navier, Stokes, and Fourier is at the level of linear irreversible thermodynamics* that assumes the linear thermodynamic force–flux relations (20.7)–(20.9) for the linear steady-state constitutive relations for nonconserved variables $\mathbf{\Pi}$, $\mathbf{\Delta}$, and $\mathbf{Q}$, (20.7)–(20.9). It should be noted that the linearized dissipation terms in Eqs. (20.7)–(20.9) do not contain cross-terms between variables $\mathbf{\Pi}$, $\mathbf{\Delta}$, and $\mathbf{Q}$ because the three variables are of different spatial symmetries, being a second-rank tensor, scalar, and vector, respectively, and therefore their linear constitutive equations are not coupled. This aspect (Curie principle) will be elaborated at the end of the next section.

20.3. A Multi-component Fluid

If the closure for the nonconserved variable set is made for a mixture such that only variables $(\mathbf{\Pi}_a, \mathbf{\Delta}_a, \mathbf{Q}_a', \mathbf{J}_a : a \geq 1; c \geq a \geq 1)$ are retained for the macroscopic variables, all the terms $\psi_a^{(q)}$ making up the divergence terms in the nonconserved variable evolution equations must be set equal

---

[8] For example, see L. D. Landau and E. M. Lifshitz, *Fluid Mechanics* (Pergamon, Oxford, 1958).

to zero except for the cases of $q = 2$ (excess normal stress) and $q = 4$ (diffusion flux), since $\psi_a^{(2)}$ and $\psi_a^{(4)}$ are, respectively, a vector and a tensor of rank 2, which can be expressed in terms of $\mathbf{Q}_a'$ and $\mathbf{J}_a$, and hence are within the closed macroscopic variable set. They are in fact expressible in linear combination of $\mathbf{\Pi}_a$, $\mathbf{\Delta}_a$, $\mathbf{Q}_a'$, $\mathbf{J}_a$ and conserved variables subjected to symmetry requirements.

The kinematic terms $\mathcal{Z}_a^{(q)}$ are linearized with respect to the thermodynamic forces — gradients of conserved variables — and external forces as well as $\mathbf{\Pi}_a$, $\mathbf{\Delta}_a$, $\mathbf{Q}_a'$, $\mathbf{J}_a$. Thus, for linear processes we approximate them as follows:

$$\mathcal{Z}_a^{(1)} = -2p_a \left[\nabla \mathbf{u}\right]^{(2)}, \quad \mathcal{Z}_a^{(2)} = -\tfrac{2}{3}p_a \nabla \cdot \mathbf{u}, \tag{20.15}$$

$$\mathcal{Z}_a^{(3)} = -\widehat{C}_{pa} T p_a \nabla \ln T, \quad \mathcal{Z}_a^{(4)} = -\rho_a \chi_a, \tag{20.16}$$

where the thermodynamic force χ_a for diffusion of species a is now defined by the formula

$$\chi_a = -\left(\nabla \widehat{\mu}_a\right)_T + \widehat{\mathbf{F}}_a - \widehat{\mathbf{F}} + v \nabla p \tag{20.17}$$

with the subscript T on $(\nabla \widehat{\mu}_a)$ denoting the spatial gradient of chemical potential $\widehat{\mu}_a$ at fixed T.

In the case of a mixture, it should be recalled that, according to the definition of mass diffusion flux made earlier, all diffusion fluxes $\mathbf{J}_a$ ($c \geq a \geq 1$) are not linearly independent since diffusion fluxes of species are defined by

$$\mathbf{J}_a = \rho_a \left(\mathbf{u}_a - \mathbf{u}\right), \tag{20.18}$$

and fluid velocity $\mathbf{u}$ defined by the formula

$$\rho \mathbf{u} = \sum_{a=1}^{c} \rho_a \mathbf{u}_a, \tag{20.19}$$

$\mathbf{u}_a$ being the velocity of species a. In other words, $\mathbf{u}$ is the mean velocity of the mixture weighted by $c_a := \rho_a/\rho$. Therefore, it follows that

$$\sum_{a=1}^{c} \mathbf{J}_a = 0. \tag{20.20}$$

Consequently, there is one diffusion flux, say, $\mathbf{J}_c$ depending on the rest of the diffusion fluxes, $\mathbf{J}_a$ ($c - 1 \geq a \geq 1$). For this reason, the dependent diffusion flux appearing in the nonconserved variable evolution equations (constitutive equations) of the generalized hydrodynamic equations must be eliminated. Hence, the evolution equation for the dependent diffusion flux $\mathbf{J}_c$ is removed from the aforementioned set as well as the dependent flux $\mathbf{J}_c$ from the evolution equations for $\Phi_a^{(3)} = \mathbf{Q}'_a$ and $\Phi_a^{(4)} = \mathbf{J}_a$. This elimination procedure results in modifications of the dissipation terms in the evolution equations for $\Phi_a^{(3)}$ and $\Phi_a^{(4)}$. For more details of the dissipation functions, see the monograph cited in footnote 2 of this chapter. The new phenomenological coefficients are linear combinations of old phenomenological coefficients. In any case, the so-modified phenomenological equations may be written as given below by Eqs. (20.30)–(20.31) or Eqs. (20.23)–(20.24) if the adiabatic approximations are applied.

It should be recognized that the evolution equations for macroscopic variables cannot be coupled if the macroscopic variables are of different spatial symmetries and if the evolution equations are linear, namely, for example, the equations for vectorial macroscopic variables cannot couple with evolution equations for scalar or tensorial macroscopic variables in the linear regime of processes. This mathematical (symmetry) principle under spatial rotation is known in irreversible thermodynamics literature as *the Curie principle*,[9] a term which we will simply accept as it stands without further examination of its origin. When linearized with respect to the nonconserved variables, the dissipation terms in the nonconserved evolution equations can therefore couple with each other only if they belong to the same symmetry group. Thus, we see that, in the approximation limiting the nonconserved variables to $(\mathbf{\Pi}_a, \mathbf{\Delta}_a, \mathbf{Q}'_a, \mathbf{J}_a : c \geq a \geq 1)$, which is a closed set, only two classes of equations belonging to the heat fluxes and mass diffusion fluxes are coupled in the case of linear processes. This principle is behind the three decoupled constitutive equations (20.7)–(20.9) or Eqs. (20.30)–(20.31) or Eqs. (20.23)–(20.24).

By taking into account the approximations enumerated above for the nonconserved evolutions and the Curie principle, we may take the following steady-state linearized evolution equations for the nonconserved variables

---

[9]S. R. de Groot and P. Mazur in their monograph entitled *Non-equilibrium Thermodynamics* (North-Holland, Amsterdam, 1962) attribute the origin of the terminology to P. Curie, Oeuvres (Paris, 1908), p. 129. We simply follow their attribution of the term without verifying the source. It is a term for a symmetry principle that must be obeyed, under spatial rotation, by a set of linear differential or algebraic equations appearing in irreversible thermodynamics.

as follows:

$$0 = -2p_a \left[\nabla \mathbf{u}\right]^{(2)} - 2p_a \sum_{b=1}^{c} \mathfrak{B}_{ab} \Pi_b, \tag{20.21}$$

$$0 = -\frac{2}{3} p_a \nabla \cdot \mathbf{u} - \frac{2p_a}{3} \sum_{b=1}^{c} \mathfrak{V}_{ab} \Delta_b, \tag{20.22}$$

$$0 = -\widehat{C}_{pa} T p_a \nabla \ln T - \widehat{C}_{pa} p_a T \left[\sum_{b=1}^{c} \mathfrak{T}_{ab} \mathbf{Q}'_a + \sum_{b=1}^{c-1} \mathfrak{H}_{ab} \mathbf{J}_b \right], \tag{20.23}$$

$$0 = -\rho_a \chi'_a - \rho_a \left[\sum_{b=1}^{c} \mathfrak{K}_{ab} \mathbf{Q}'_b + \sum_{b=1}^{c-1} \mathfrak{D}_{ab} \mathbf{J}_b \right], \tag{20.24}$$

where

$$\chi'_a = \chi_a + (d_t \mathbf{u} - \widehat{\mathbf{F}}_a), \tag{20.25}$$

and constant coefficients $\mathfrak{B}_{ab}$, $\mathfrak{V}_{ab}$, $\mathfrak{T}_{ab}$, $\mathfrak{H}_{ab}$, $\mathfrak{K}_{ab}$, and $\mathfrak{D}_{ab}$ are phenomenological parameters. In the kinetic theory (i.e., molecular theory) approach, the coefficients may be given in terms of collision bracket integrals $\mathfrak{R}_{ab}^{(qs)}$, which appear in the dissipation functions $\Lambda_a^{(q)}$ in Eq. (19.11). Explicitly put in terms of the collision bracket integrals $\mathfrak{R}_{ab}^{(qs)}$, they may be expressed as follows:

$$\mathfrak{B}_{ab} = \frac{\mathfrak{R}_{ab}^{(11)}}{\beta g}, \quad \mathfrak{V}_{ab} = \frac{\mathfrak{R}_{ab}^{(22)}}{\beta g}, \quad \mathfrak{T}_{ab} = \frac{\mathfrak{R}_{ab}^{(33)}}{\beta g},$$

$$\mathfrak{H}_{ab} = \frac{\mathfrak{R}_{ab}^{(34)} - \mathfrak{R}_{ac}^{(34)}}{\beta g}, \quad \mathfrak{K}_{ab} = \frac{\mathfrak{R}_{ab}^{(43)} - \mathfrak{R}_{ac}^{(43)}}{\beta g}, \tag{20.26}$$

$$\mathfrak{D}_{ab} = \frac{\mathfrak{R}_{ab}^{(44)} + \mathfrak{R}_{cc}^{(44)} - \mathfrak{R}_{ac}^{(44)} - \mathfrak{R}_{bc}^{(44)}}{\beta g},$$

where g is defined by parameters characterizing the length of mean free path

$$g = \frac{1}{n^2 d^2} \sqrt{\frac{m}{2 k_B T}} \tag{20.27}$$

with m and d are the mean reduced mass and the mean size parameter of molecules of the fluid. The relations in Eq. (20.26) are obtained by means of a kinetic theory, and the collision bracket integrals $\mathfrak{R}_{ab}^{(qs)}$ therein are given in terms of collision operators characterizing intermolecular interactions within the fluid.[10]. However, the coefficients $\mathfrak{B}_{ab}$, etc., are treated here as phenomenological parameters, which may be determined from experiments in the phenomenological irreversible thermodynamics discussed here. The inverses of the coefficient matrices $\mathfrak{B} := (\mathfrak{B}_{ab})$ and $\mathfrak{V} := (\mathfrak{V}_{ab})$ will turn out to be related to shear viscosity and bulk viscosity when Eqs. (20.21) and (20.22) are solved for $\mathbf{\Pi}_b$ and Δ_b; and when Eqs. (20.23) and (20.24) are algebraically solved for $\mathbf{Q}'_b$ and $\mathbf{J}_b$, the coefficient matrix inverses are related to thermal conductivity, diffusive thermal conductivity, thermal diffusion coefficient, and diffusion coefficient of component a — more precisely put, elements ab of the matrix of respective phenomenological parameters $\mathfrak{R}_{ab}^{(1)}$, etc. Note that Eqs. (20.23) and (20.24) are coupled, whereas Eqs. (20.21) and (20.22) are not only decoupled from each other, but also from Eqs. (20.23) and (20.24) according to the Curie principle. It should be noted that the dependent diffusion flux has been already eliminated in Eqs. (20.23) and (20.24).

Together with the conservation laws for mass, concentrations, momentum, and internal energy — a subset of the generalized hydrodynamic equations — the steady-state constitutive equations (20.21)–(20.24) form the classical hydrodynamic equations for the fluid mixture. The first two equations (20.21) and (20.22) are Newton's laws for viscous and dilatation or compression phenomena (bulk viscous phenomena) of a fluid mixture and the last two equations (20.23) and (20.24) describe the thermodiffusion phenomena of the fluid mixture. The constitutive equations, (20.21)–(20.23) but for (20.24), do not contain the external forces because the external forces are usually contained in the higher order terms compared to the nonconserved variable in question in the equations. However, in the case of Eq. (20.24), the external force is a part of the driving force for diffusion, which is a vector as the diffusion flux is, and hence must be kept in the equation.

20.4. Nonsteady Linear Constitutive Equations

If the adiabatic approximation is not applied to the linearized nonconserved variable evolution equations for the mixture, the time-dependent

---

[10]See, for example, B. C. Eu, *Kinetic Theory and Irreversible Thermodynamics* (Wiley, New York, 1992) and also B. C. Eu, *Kinetic Theory of Nonequilibrium Ensembles, Irreversible Thermodynamics, and Generalized Hydrodynamics*, Vol. 1 (Springer, Switzerland, 2016).

linear constitutive equations for nonconserved variables are obtained as follows:

$$\rho\frac{d\widehat{\Pi}_a}{dt} = -2\left[\nabla\mathbf{u}\right]^{(2)} - 2p_a\sum_{b=1}^{c}\mathfrak{B}_{ab}\Pi_b,$$ (20.28)

$$\rho\frac{d\widehat{\Delta}_a}{dt} = -\frac{2}{3}p_a\nabla\cdot\mathbf{u} - \frac{2p_a}{3}\sum_{b=1}^{c}\mathfrak{V}_{ab}\Delta_b,$$ (20.29)

$$\rho\frac{d\widehat{\mathbf{Q}}_a}{dt} = -\widehat{C}_{pa}Tp_a\nabla\ln T - \widehat{C}_{pa}p_aT\left(\sum_{b=1}^{c}\mathfrak{T}_{ab}\mathbf{Q}'_a + \sum_{b=1}^{c-1}\mathfrak{H}_{ab}\mathbf{J}_b\right),$$ (20.30)

$$\rho\frac{d\widehat{\mathbf{J}}_a}{dt} = -\rho\chi'_a - \rho_a\left(\sum_{b=1}^{c}\mathfrak{K}_{ab}\mathbf{Q}'_b + \sum_{b=1}^{c-1}\mathfrak{D}_{ab}\mathbf{J}_b\right).$$ (20.31)

Therefore, Eqs. (20.21)–(20.24) would define the steady state of the set of linear partial differential equations (20.28)–(20.31) for the nonconserved variables. These linear differential equations would give the terminology adiabatic approximation a more precise physical meaning. As the steady state of Eqs. (20.28)–(20.31) in the moving coordinate system defines the steady-state fluxes defined by Eqs. (20.21)–(20.24) and, if we define deviations of variables $(\mathbf{\Pi}_a, \mathbf{\Delta}_a, \mathbf{Q}'_a, \mathbf{J}_a : c \geq a \geq 1)$ from their steady-state values by

$$\delta\mathbf{\Pi}_a = \mathbf{\Pi}_a - \mathbf{\Pi}^{\mathrm{s}}_a, \ \delta\mathbf{\Delta}_a = \mathbf{\Delta}_a - \mathbf{\Delta}^{\mathrm{s}}_a, \ \delta\mathbf{Q}'_a = \mathbf{Q}'_a - \mathbf{Q}'^{\mathrm{s}}_a,$$

$$\delta\mathbf{J}_a = \mathbf{\Pi}_a - \mathbf{J}^{\mathrm{s}}_a,$$ (20.32)

where variables superscripted with letter s are the steady-state variables obeying Eqs. (20.21)–(20.24), the fluctuations from the steady state $\delta\mathbf{\Pi}_a$, etc., then obey the relaxation equations of fluctuations resulting from Eqs. (20.28)–(20.31):

$$\rho\frac{d\delta\widehat{\mathbf{\Pi}}_a}{dt} = -2p_a\sum_{b=1}^{c}\mathfrak{B}_{ab}\delta\mathbf{\Pi}_b,$$ (20.33)

$$\rho\frac{d\delta\widehat{\Delta}_a}{dt} = -\frac{2p_a}{3}\sum_{b=1}^{c}\mathfrak{V}_{ab}\delta\Delta_b,$$ (20.34)

$$\rho \frac{d\delta\widehat{\mathbf{Q}}_a}{dt} = -\widehat{C}_{pa}p_aT \left(\sum_{b=1}^{c} \mathfrak{T}_{ab}\delta\mathbf{Q}_a' + \sum_{b=1}^{c-1} \mathfrak{H}_{ab}\delta\mathbf{J}_b \right), \qquad (20.35)$$

$$\rho \frac{d\delta\widehat{\mathbf{J}}_a}{dt} = -\rho_a \left(\sum_{b=1}^{c} \mathfrak{K}_{ab}\delta\mathbf{Q}_b' + \sum_{b=1}^{c-1} \mathfrak{D}_{ab}\delta\mathbf{J}_b \right). \qquad (20.36)$$

Since the phenomenological coefficients $\mathfrak{B}_{ab}$, etc., give rise to positive linear transport coefficients in accordance with the second law of thermodynamics (see Eq. (4.48) or Eq. (4.69) in Chapter 4) and since various relaxation times consequently are all positive, it can be easily concluded that the steady-state defined by Eqs. (20.21)–(20.24), therefore, is stable and the nonequilibrium state converges to the steady state as $t \to \infty$. Hence, we conclude

$$\lim_{t\to\infty} (\mathbf{\Pi}_a, \mathbf{\Delta}_a, \mathbf{Q}_a', \mathbf{J}_a : c \geq a \geq 1) = (\mathbf{\Pi}_a^{\mathrm{s}}, \mathbf{\Delta}_a^{\mathrm{s}}, \mathbf{Q}_a'^{\mathrm{s}}, \mathbf{J}_a^{\mathrm{s}} : c \geq a \geq 1).$$

$$(20.37)$$

In other words, fluctuations $\delta\widehat{\mathbf{\Pi}}_a$, etc., vanish as $t \to \infty$ and the steady-state values are recovered for the nonconserved variables if the nonconserved variables obey Eqs. (20.28)–(20.31). Hence, the steady states of Eqs. (20.28)–(20.31) are stable. However, this does not mean that the dynamics of the differential equation system (20.28)–(20.31) together with conservation laws is equally so simple. As a matter of fact, they allow the emergence of plethora of dynamics of the combined system of differential equations — linearized generalized hydrodynamic equations — that can be rather rich and complex, ranging from simple laminar flow to chaos. For various aspects of such hydrodynamic phenomena, the reader is referred to the literature[11] on the classical hydrodynamics and also the literature[12] on the generalized hydrodynamics for more details. In the following, we would like to show, first, a couple of less commonly studied examples of applications of linear nonconserved variable constitutive equations and, then, some examples of experimental and theoretical results describable by generalized hydrodynamic equations.

---

[11] See, for example, numerous monographs on classical hydrodynamics such as L. D. Landau and E. M. Lifshitz, *Fluid Mechanics* (Pergamon, London, 1959); G. K. Batchelor, *Fluid Dynamics*(Cambridge University Press, London, 1967) H. Schlichting, *Boundary-Layer Theory*, seventh edn. (McGraw-Hill, New York, 1979).

[12] See, for example, B. C. Eu, *Kinetic Theory of Nonequilibrium Ensembles, Irreversible Thermodynamics, and Generalized Hydrodynamics, Volume 1 Nonrelativistic Theories and Volume 2 Relativistic Theories* (Springer, Heidelberg, 2016) and references cited therein.

20.5. Viscoelasticity of Fluids

When disturbed from their equilibrium state, fluids usually tend to relax back to their equilibrium state. Such tendencies sometimes are not easily observed by naked eyes, especially, in the case of gases and require careful and appropriate experimental methods to observe them. However, such relaxation behaviors can be easily observed if the fluid is complex. One typical example of such relaxation behaviors is viscoelasticity of some polymeric liquids. Simple fluids do not exhibit easily observable viscoelasticity, but by using generalized hydrodynamic equations or a special linearized model based on them, we can illustrate how we may theoretically describe the viscoelasticity of a fluid. It is the main aim of this section as an example of applications of the linearized nonconserved variable evolution equations presented in the previous section.

It should be noted that a proper understanding of viscoelastic behavior of fluids in a more general context would require nonlinear nonconserved variable evolution equations if the phenomena are to be appropriately understood and accounted for with the complexity of molecules (e.g., polymeric liquids) comprising the fluids properly taken into consideration.[13] We will make some comments on this question in the next chapter in which nonlinear transport processes will be discussed.

In addition to the assumptions and approximations made to obtain the linearized nonconserved variables in the previous section, it will be assumed that the temperature is maintained uniform over the system made up of a single-component fluid. It is also assumed that the fluid is incompressible. This assumption makes it possible to ignore excess normal stress and thereby simplifies analysis. Under the assumptions, the density ρ and the pressure p are independent of position and time. It will be further assumed that the viscosity η_0 is also a constant.[14]

In the case considered, there is only a pair of evolution equations to consider: the momentum balance equation and the shear stress evolution

---

[13] For generalized hydrodynamic treatments of viscoelasticity and, more generally, rheology, see B. C. Eu, *Kinetic Theory and Irreversible Thermodynamics* (Wiley, New York, 1992), Chapter 14 and B. C. Eu and R. E. Khayat, *Rheologica Acta* **30**, 204 (1991).

[14] Since the viscosity η_0 is generally a function of temperature and density, it can vary with position in the fluid. Here, we are assuming η_0 is a constant.

equation, which take the forms

$$\rho\frac{\partial \mathbf{u}}{\partial t} = -\nabla \cdot \mathbf{u}, \tag{20.38}$$

$$\rho\frac{\partial \widehat{\mathbf{\Pi}}}{\partial t} = -2p\left[\nabla \mathbf{u}\right]^{(2)} - \frac{\rho p}{\eta_0}\widehat{\mathbf{\Pi}}. \tag{20.39}$$

In passing, we note that these partial differential equations are hyperbolic. Since the equations are linearized, the inertia terms are neglected. In the momentum balance equation, the pressure and the excess normal stress terms are absent owing to the incompressibility assumption made. Therefore, there exists only the shear component in the shear stress tensor $\mathbf{\Pi}$. As a matter of fact, Eq. (20.39) is known as the Maxwell equation for the shear stress, which J. C. Maxwell derived in his kinetic theory of gases.[15]

To make the analysis[16] more specific, we will consider an oscillating Couette flow configuration: The fluid is confined between two infinite parallel plates separated by a distance D in the y-coordinate (vertical to the x-coordinate in the figure) and parallel to the x-axis, each of which oscillates back and forth in the $\pm x$-direction at an amplitude $\pm u_D/2$ in opposite directions and at frequency ω_0 (see Fig. 20.1). The temperature of the system is maintained at constant T. The direction of flow is neutral in the z-direction. Upon analytical continuation, we may regard the velocity and stress tensor as complex variables. Therefore, in the flow configuration just described,

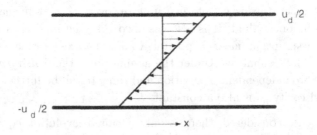

Fig. 20.1. Oscillating Plane Couette flow configuration. Two plates aligned in the x-direction and located at $y = \pm D/2$ on the y-axis perpendicular to the x-axis oscillate back and forth at velocity amplitude $\pm u_c/2$ at frequency ω.

---

[15] J. C. Maxwell, *Philos. Trans. Roy. Soc. London A* **157**, 49 (1867).
[16] For more details, see B. C. Eu, *J. Chem. Phys.* **82**, 4683 (1985).

the traceless symmetric part of the velocity gradient takes the form

$$[\nabla \mathbf{u}]^{(2)} = \gamma e^{i\omega_0 t} \begin{pmatrix} 0 & 1 & 0 \\ 1 & 0 & 0 \\ 0 & 0 & 0 \end{pmatrix}, \qquad (20.40)$$

where

$$\gamma = \frac{1}{2}\frac{\partial u_x}{\partial y}, \quad \mathbf{u} = (u_x, 0, 0). \qquad (20.41)$$

The shear rate γ is constant under the flow considered here. In the flow configuration under the discussion, there are only two differential equations, namely, the x-component of the momentum balance equation and the xy-component of the shear stress evolution equation.

Equations (20.38) and (20.39) under discussion are solved by Fourier transforms:

$$u_x(x,t) = \frac{1}{4\pi^2} \int_{-\infty}^{\infty} d\omega \int_{-\infty}^{\infty} dk \exp(ikx - i\omega t)\hat{u}(k,\omega), \quad (20.42)$$

$$\hat{\Pi}_{xy}(x,t) = \frac{1}{4\pi^2} \int_{-\infty}^{\infty} d\omega \int_{-\infty}^{\infty} dk \exp(ikx - i\omega t)\hat{\Pi}(k,\omega), \quad (20.43)$$

which, upon insertion into the differential equations, give rise to a pair of algebraic equations for the Fourier components

$$\omega\hat{u}(k,\omega) = k\hat{\Pi}(k,\omega), \qquad (20.44)$$

$$i\omega\hat{\Pi}(k,\omega) = \frac{2p\gamma}{\rho}\delta(\omega - \omega_0)\delta(k) + \frac{p}{\eta_0}\hat{\Pi}(k,\omega). \qquad (20.45)$$

These equations yield the solutions

$$\rho\hat{\Pi}(k,\omega) = -\frac{2\eta_0\gamma}{1 - i\omega\tau}\delta(\omega - \omega_0)\delta(k), \qquad (20.46)$$

$$\rho\hat{u}(k,\omega) = -\frac{2\eta_0\gamma k}{\omega(1 - i\omega\tau)}\delta(\omega - \omega_0)\delta(k), \qquad (20.47)$$

where τ is the stress relaxation time defined by

$$\tau = \frac{\eta_0}{p}. \qquad (20.48)$$

Substituting these solutions back into (20.42) and (20.43), we obtain

$$u_x(x,t) = 0, \tag{20.49}$$

and we may write the shear stress in terms of dynamic viscosity $\eta(\omega_0)$, which depends on frequency of oscillation:

$$\Pi_{xy}(x,t) = -2\eta\,(\omega_0)\exp\,(-i\omega_0 t)\,, \tag{20.50}$$

where

$$\eta\,(\omega_0) = \frac{\eta_0}{4\pi^2\,(1 - i\omega_0\tau)} = \eta'\,(\omega_0) + i\eta''\,(\omega_0) \tag{20.51}$$

with the real (η') and imaginary (η'') part, respectively, given by the formulas

$$\eta'\,(\omega_0) = \frac{\eta_0}{4\pi^2\,(1 + \omega_0^2\tau^2)}, \tag{20.52}$$

$$\eta''\,(\omega_0) = \frac{\eta_0\tau\omega_0}{4\pi^2\,(1 + \omega_0^2\tau^2)}. \tag{20.53}$$

The real part of dynamic viscosity $\eta(\omega_0)$ gives the loss modulus, whereas the imaginary part provides the storage modulus of the viscoelastic substance. In the linear approximation made for these results presented here, the loss and storage moduli are independent of the shear amplitude γ. However, it is not true in general. More appropriate nonlinear evolution equations[17] would show that they depend on the shear rate γ. Therefore, the evolution equations (20.38) and (20.39) should be suitably extended to properly account for experimental data in the case of nonlinear systems. See the subsequent chapter for nonconserved variable evolution equations for nonlinear transport processes.

20.6. Ultrasonic Wave Propagation and Dispersion in Simple Gases

When sound wave passes through a medium, it periodically compresses and decompresses the medium. If the medium is viscous and diathermal, such a periodic oscillatory process is accompanied by energy dissipation, which in turn attenuates and disperses the sound waves. Such attenuation and dispersion phenomena can be studied experimentally by measuring

---

[17]See B. C. Eu, *J. Chem. Phys.* **82**, 4683 (1985) for a nonlinear theory treatment of viscoelasticity of a simple fluid.

sound wave dispersion and absorption. They can also be made use of to study the bulk viscosity phenomena of the medium. Sometimes, sound wave dispersion and absorption experiments are, as a matter of fact, the most practicable and reliable methods of measuring the bulk viscosity of simple gases, which tends to be much smaller than the shear viscosities.

If the medium consists of molecules with internal structures and is chemically reactive, measurements of ultrasonic wave dispersion and absorption can also be exploited to measure rates of the chemical reactions occurring in the medium. Such phenomena have been studied in the past by using the classical hydrodynamics, which, as shown earlier, arise if the steady-state approximation is made for the constitutive equations of the nonconserved variables, namely, the Newtonian law of viscosity, Fourier's law of heat conduction, and Fick's law of diffusion are taken for shear stress, heat flux, and diffusion flux. If the nonconserved variables do not relax sufficiently fast compared to the relaxation of conserved variables, then the time-dependent linear constitutive equations (20.28)–(20.31) should be employed for studying nonconserved variable evolutions.

In this section, we first consider the case of a nonreactive monatomic medium of a simple gas and then the reactive medium, disregarding the internal state relaxation of the medium in order to make the analysis as simple as possible.

20.6.1. Nonreactive Simple Gas

For the sake of simplicity in analysis, we will assume that the medium in a near-equilibrium state consists of a single-component monatomic gas with a bulk viscosity. This assumption permits the use of linearized constitutive equations for nonconserved variables. In the linearized constitutive equations, we also assume that density ρ and the heat capacity $\widehat{C}_p$ are those of equilibrium fluid values, ρ_0 and $\widehat{C}_{p0}$, respectively, which are independent of time and position variable $\mathbf{r}$. However, pressure has a nonequilibrium value, so that it contains a nonequilibrium component

$$p = p_0 + p', \tag{20.54}$$

where p_0 is the equilibrium pressure and p' the nonequilibrium correction. Therefore, p_0 being spatially uniform, $\nabla p_0 = 0$. Similarly, $\lambda_0' = \lambda_0/T$ is assumed to be constant, so that $\lambda_0 = \lambda_0' T$ with constant λ_0'. The temperature is displaced from equilibrium value T_0. Then the evolution equations for conserved and nonconserved variables to consider are as follows:

$$\rho_0 \partial_t v = \nabla \cdot \mathbf{u}, \tag{20.55}$$

$$\rho_0 \partial_t \mathbf{u} = -\nabla p' - \nabla \cdot (\mathbf{\Pi} + \Delta \mathbf{U}), \tag{20.56}$$

$$\rho_0 \widehat{C}_{p0} \partial_t v = -\nabla \cdot \mathbf{Q}, \tag{20.57}$$

$$\rho_0 \partial_t \widehat{\mathbf{\Pi}} = -2 p_0 \left[\nabla \mathbf{u}\right]^{(2)} - \frac{p_0}{\eta_0} \mathbf{\Pi}, \tag{20.58}$$

$$\rho_0 \partial_t \widehat{\Delta} = -\frac{2}{3} p_0 \nabla \cdot \mathbf{u} - \frac{2 p_0}{3 \eta_0} \Delta, \tag{20.59}$$

$$\rho_0 \partial_t \widehat{\mathbf{Q}} = -\widehat{C}_{p0} p_0 \nabla T - \frac{\widehat{C}_{p0} p_0}{\lambda_0'} \mathbf{Q}. \tag{20.60}$$

If we wish to obtain the Navier–Stokes–Fourier theory of sound wave dispersion and absorption by a simple gas, it is necessary to set the time derivatives equal to zero on the left-hand side of Eqs. (20.58)–(20.60). The present theory, therefore, takes into account the relaxation effects of the nonconserved variables.

To solve the equations presented above, a Fourier transform method is used. The Fourier transforms of various variables are defined by

$$\tilde{q}(\mathbf{k}, t) = \int d\mathbf{r} \int_{-\infty}^{\infty} d\omega \exp(i\omega t + i\mathbf{k} \cdot \mathbf{r}) q(\mathbf{r}, t), \tag{20.61}$$

where the position variable $\mathbf{r}$ is confined to the volume of the fluid and

$$q(\mathbf{r}, t) = (v, \mathbf{u}, T, \mathbf{\Pi}, \Delta, \mathbf{Q})\,(\mathbf{r}, t) \tag{20.62}$$

with $\mathbf{\Pi}$, Δ, and $\mathbf{Q}$ expressed by their densities $\widehat{\mathbf{\Pi}}$, $\widehat{\Delta}$, and $\widehat{\mathbf{Q}}$:

$$(\mathbf{\Pi}, \Delta, \mathbf{Q}) = \rho_0 \left(\widehat{\mathbf{\Pi}}, \widehat{\Delta}, \widehat{\mathbf{Q}}\right). \tag{20.63}$$

Upon taking the Fourier transform of Eqs. (20.55)–(20.60), we obtain the coupled algebraic set of Fourier transformed equations

$$\rho_0 \omega \tilde{v} = -\mathbf{k} \cdot \tilde{\mathbf{u}}, \tag{20.64}$$

$$\omega \tilde{\mathbf{u}} \cdot \mathbf{k} = k^2 v_0 \tilde{p'} + k^2 \tilde{\Delta} + \mathbf{k}\mathbf{k} : \tilde{\mathbf{\Pi}}, \tag{20.65}$$

$$\widehat{C}_{p0} \omega \tilde{T} + \frac{T_0 \alpha_p}{\kappa_T} \omega \tilde{v} = \mathbf{k} \cdot \tilde{\mathbf{Q}}, \tag{20.66}$$

$$(i\omega + \tau_s^{-1})(\mathbf{kk} : \widetilde{\mathbf{\Pi}}) = -i\tfrac{4}{3}p_0 k^2 \omega \widetilde{v}, \tag{20.67}$$

$$(i\omega + \tau_b^{-1})\widetilde{\mathbf{\Delta}} = -i\tfrac{2}{3}p_0 \omega \widetilde{v}, \tag{20.68}$$

$$(i\omega + \tau_h^{-1})\mathbf{k} \cdot \widetilde{\mathbf{Q}} = i p_0 v_0 \widehat{C}_{p0} k^2 \widetilde{T}, \tag{20.69}$$

where

$$v_0 = 1/\rho_0, \tag{20.70}$$

$$\widetilde{p}' = \frac{\alpha_p}{\kappa_T}\widetilde{T} - \frac{1}{v_0 \kappa_T}\widetilde{v}, \tag{20.71}$$

and parameters τ_s, τ_b, and τ_h are the relaxation times for the shear stress, the excess normal stress, and heat flux. They are defined, respectively, by the formulas

$$\tau_s = \frac{\eta_0}{p_0}, \quad \tau_b = \frac{3\eta_b}{2p_0}, \quad \tau_h = \frac{\lambda_0}{\widehat{C}_{p0}T_0 p_0}, \tag{20.72}$$

and α_p and κ_T are, respectively, the isobaric expansion coefficient and the isothermal compressibility of the medium

$$\alpha_p = v_0^{-1}\left(\frac{\partial v}{\partial T}\right)_p, \quad \kappa_T = -v_0^{-1}\left(\frac{\partial v}{\partial p}\right)_T. \tag{20.73}$$

Eliminating $\widetilde{\mathbf{u}}$, $\widetilde{\mathbf{\Pi}}$, $\widetilde{\mathbf{\Delta}}$, $\widetilde{\mathbf{Q}}$ from Eqs. (20.64)–(20.69), we obtain two algebraic equations for $\widetilde{v}$ and $\widetilde{T}$

$$C_{11}\widetilde{v} + C_{12}\widetilde{T} = 0, \tag{20.74}$$

$$C_{21}\widetilde{v} + C_{22}\widetilde{T} = 0, \tag{20.75}$$

where the coefficients C_{11}, ..., C_{22} are defined by the expressions

$$C_{11} = \omega^3 - \frac{k^2}{\kappa_T} - i\frac{\omega\eta_b k^2}{1 + i\omega\tau_b} - i\frac{4\omega\eta_0 k^2}{3(1 + i\omega\tau_s)}, \tag{20.76}$$

$$C_{12} = \frac{\alpha_p v_0 k^2}{\kappa_T}, \tag{20.77}$$

$$C_{21} = \frac{T_0 \alpha_p \omega}{\kappa_T}, \tag{20.78}$$

$$C_{22} = \widehat{C}_{p0}\omega - i\frac{v_0 k^2 \lambda_0}{T_0(1 + i\omega\tau_h)}. \tag{20.79}$$

The solvability condition of this algebraic set, which is the vanishing determinant of the coefficients of the algebraic set (20.74) and (20.75), gives rise to the dispersion relation and the absorption coefficient of the sound wave.

The solvability condition (i.e., the determinant) is given in the form

$$A_4 k^4 + A_2 k^2 + A_0 = 0, \qquad (20.80)$$

where

$$A_4 = i \frac{v_0 \lambda_0}{T_0 \kappa_T (1 + i\omega\tau_h)} - \frac{v_0 \lambda_0 \omega}{T_0 \kappa_T (1 + i\omega\tau_h)} \left[\frac{\eta_b}{1 + i\omega\tau_b} + \frac{4\eta_0}{3(1 + i\omega\tau_s)} \right], \qquad (20.81)$$

$$A_2 = -v_0 \left[T_0 \left(\frac{\alpha_p}{\kappa_T} \right)^2 + \frac{\widehat{C}_{p0}}{v_0 \kappa_T} + i \frac{\omega \lambda_0}{T_0 (1 + i\omega\tau_h)} \right] \omega$$
$$\qquad - i\widehat{C}_{p0} \left[\frac{\eta_b}{1 + i\omega\tau_b} + \frac{4\eta_0}{3(1 + i\omega\tau_s)} \right] \omega^2, \qquad (20.82)$$

$$A_0 = \widehat{C}_{p0} \omega^3. \qquad (20.83)$$

The algebraic equation (20.80) yields four solutions for the wave number k:

$$k = \pm \sqrt{-\frac{A_2}{2A_4} \pm \sqrt{\left(\frac{A_2}{2A_4} \right)^2 - \frac{A_0}{A_4}}}. \qquad (20.84)$$

The real part of k gives the dispersion relation $D(\omega)$ and the imaginary part the absorption coefficient $\Gamma(\omega)$ of the medium. To identify them, the solutions are separated into the real and imaginary parts:

$$D(\omega) = \operatorname{Re} k, \qquad (20.85)$$

$$\Gamma(\omega) = \operatorname{Im} k. \qquad (20.86)$$

There are four modes of wave possible, but since the expression involved are rather unwieldy, in order to obtain the results as simple and comprehensible as possible we will consider simplified results holding in the limit of ω in which ω is such that

$$\tau_s \omega, \tau_b \omega, \tau_h \omega \ll 1. \tag{20.87}$$

To $O(\omega^2)$ in the limits obeying conditions (20.87), we obtain

$$A_4 = -v_0 \lambda_0' \left(\eta_b + \frac{4}{3}\eta_0 \right) \omega + i\frac{v_0 \lambda_0'}{\kappa_T},$$

$$A_2 = -v_0 \left[T_0 \left(\frac{\alpha_p}{\kappa_T} \right)^2 + \frac{\widehat{C}_{v0}}{v_0 \kappa_T} \right] \omega - i \left[v_0 \lambda_0 + \widehat{C}_{v0} \left(\eta_b + \frac{4}{3}\eta_0 \right) \right] \omega^2,$$

$$A_0 = O\left(\omega^3 \right).$$

In these limits, we find the four wave numbers in the forms

$$k_{\pm\pm} = \pm\sqrt{\frac{-A_2 \pm A_2}{2A_4}}. \tag{20.88}$$

More explicitly written, they are as follows:

$$k_{\pm+} = 0, \tag{20.89}$$

$$k_{\pm-} = \pm\sqrt{\frac{-A_2}{A_4}}$$

$$= \pm i\sqrt{\frac{T_0(\frac{\alpha_p}{\kappa_T})^2 + \frac{\widehat{C}_{v0}}{v_0 \kappa_T}}{\lambda_0'\left(\eta_b + \frac{4}{3}\eta_0\right)}} \sqrt{\frac{1 + i\frac{[v_0\lambda_0 + \widehat{C}_{v0}(\eta_b + \frac{4}{3}\eta_0)]\omega}{[v_0 T_0(\frac{\alpha_p}{\kappa_T})^2 + \frac{\widehat{C}_{v0}}{\kappa_T}]}}{1 - i\frac{1}{\kappa_T(\eta_b + \frac{4}{3}\eta_0)\omega}}}. \tag{20.90}$$

Thus, the dispersion relations are given by

$$D_{\pm-}(\omega) = \mp \text{Im}\sqrt{\frac{1 + i\frac{[v_0\lambda_0 + \widehat{C}_{v0}(\eta_b + \frac{4}{3}\eta_0)]\omega}{[v_0 T_0(\frac{\alpha_p}{\kappa_T})^2 + \frac{\widehat{C}_{v0}}{\kappa_T}]}}{1 - i\frac{1}{\kappa_T(\eta_b + \frac{4}{3}\eta_0)\omega}}}, \tag{20.91}$$

and absorption coefficients by

$$\Gamma_{\pm-}(\omega) = \pm\sqrt{\frac{T_0(\frac{\alpha_p}{\kappa_T})^2 + \frac{\widehat{C}_{v0}}{v_0 \kappa_T}}{\lambda_0'\left(\eta_b + \frac{4}{3}\eta_0\right)}} \text{Re}\sqrt{\frac{1 + i\frac{[v_0\lambda_0 + \widehat{C}_{v0}(\eta_b + \frac{4}{3}\eta_0)]\omega}{[v_0 T_0(\frac{\alpha_p}{\kappa_T})^2 + \frac{\widehat{C}_{v0}}{\kappa_T}]}}{1 - i\frac{1}{\kappa_T(\eta_b + \frac{4}{3}\eta_0)\omega}}}. \tag{20.92}$$

These formulas transcend the classical formulas based on the Navier–Stokes–Fourier theory of classical hydrodynamics. More accurate, but algebraically complicated, results can be obtained with complete formulas (20.84).

They can provide the transport coefficients η_0 and η_b through ultrasonic wave dispersion and absorption experiments in the low frequency limit. For more general results, a method of numerical analysis must be made.

20.6.2. *Ultrasonic Waves in a Rigid Diatomic Gas*

The theory of ultrasonic wave absorption and dispersion we have discussed in the previous section can be applied to a rigid molecular gas with rotational degrees of freedom if a slight modification is made to take the internal degrees of freedom into account. In this section, we consider nitrogen and hydrogen treated as rigid diatomic molecules with the vibrational motion frozen. It is assumed that the temperature of the gas is such that the rotational relaxation is much faster than the hydrodynamic relaxation. It is then possible to ignore the rotational energy relaxation equation in the set of evolution equations for the gas. This model was used to show[18] that the linear forms of generalized hydrodynamic equations can be adequately applied to describe the ultrasonic wave absorption and dispersion phenomena, and we briefly describe the theory and results of analysis in the following.

Under the assumptions stated earlier and, additionally, the assumptions made in the previous section, the linearized conservation laws and constitutive equations for the shear stress, excess normal stress, and heat flux of a diatomic fluid are as follows:

$$\rho_0 \frac{\partial v}{\partial t} = \nabla \cdot \mathbf{u}, \tag{20.93}$$

$$\rho_0 \frac{\partial \mathbf{u}}{\partial t} = -\nabla p - \nabla \cdot (\mathbf{\Pi} + \mathbf{U}\Delta), \tag{20.94}$$

$$\rho_0 \widehat{C}_v \frac{\partial T}{\partial t} = -\frac{T_0 \alpha_p}{\kappa_T} \nabla \cdot \mathbf{u} - \nabla \cdot \widehat{\mathbf{Q}}, \tag{20.95}$$

---

[18]See B. C. Eu and Y. G. Ohr, *Phys. Fluids* **13**, 744 (2001).

which are coupled to the linearized constitutive equations for the nonconserved variables of the diatomic fluid

$$\rho_0 \frac{\partial \widehat{\Pi}}{\partial t} = -\nabla \cdot \psi^{(s)} - 2p_0 \left[\nabla \mathbf{u}\right]^{(2)} - \frac{\rho_0 p_0}{\eta_0} \widehat{\Pi}, \qquad (20.96)$$

$$\rho_0 \frac{\partial \widehat{\Delta}}{\partial t} = -\nabla \cdot \psi^{(b)} - \frac{2p_0 \widehat{C}_{vrot}}{3\widehat{C}_v} \nabla \cdot \mathbf{u} - \frac{2\rho_0 p_0 \widehat{C}_{vrot}}{3\widehat{C}_v \eta_b} \widehat{\Delta}, \qquad (20.97)$$

$$\rho_0 \frac{\partial \widehat{\mathbf{Q}}}{\partial t} = -\nabla \cdot \psi^{(q)} - 2p_0 \widehat{C}_p \nabla T - \frac{\rho_0 p_0 \widehat{C}_p}{\lambda'_0} \widehat{\mathbf{Q}}. \qquad (20.98)$$

In the equations given above, $\widehat{C}_{vrot}$ is the rotational part of the specific heat per mass at constant volume and the rest of symbols are the same as in the previous section.

The closure relations for $\psi^{(s)}$, $\psi^{(b)}$, and $\psi^{(q)}$ are taken (empirically) on the basis of their tensorial properties as follows:

$$\nabla \cdot \psi^{(s)} = -\frac{2p_0 \lambda'_0}{3\rho_0 \widehat{C}_p} \left[\nabla \nabla T\right]^{(2)}, \qquad (20.99)$$

$$\psi^{(b)} = -p_0 v_0 (2\eta_0 \left[\nabla \mathbf{u}\right]^{(2)} + \eta_b^0 \mathbf{U} \nabla \cdot \mathbf{u}), \qquad (20.100)$$

$$\psi^{(q)} = \frac{2\rho_0 \widehat{C}_{vrot}}{3\widehat{C}_v} \widehat{\mathbf{Q}}. \qquad (20.101)$$

These are, in fact, closure relations for the nonconserved variable moment set. As such, they are all within the closed set of variables and may be estimated from the kinetic theory expression for $\psi^{(s)}$, etc., by taking the local equilibrium formula for the distribution function, but here they may be taken as empirical relations.

Equations (20.93)–(20.98), as in the case of the evolution equations for the monatomic gas in the previous subsection, can be solved by Fourier transform. The resulting coupled algebraic equations for the Fourier expansion coefficients provide the dispersion relation, which in fact is the solvability condition of the coupled set given as a function of wave number k. The solvability condition, as a matter of fact, is given by a cubic polynomial of k^2 whose coefficients are functions of frequency ω:

$$B_6 k^{*6} + B_4 k^{*4} + B_2 k^{*2} + B_0 = 0, \qquad (20.102)$$

where the coefficients $B_6, \ldots, B_0$ are given in terms of the following reduced variables defined by the formulas:

$$c_0 = \sqrt{\frac{\gamma k_B T_0}{m}}, \quad \omega^* = \frac{\omega \eta_0 v_0}{c_0^2}, \quad k^* = \frac{k c_0}{\omega}, \quad f_b = \frac{\eta_b}{\eta_0}. \qquad (20.103)$$

The meanings of symbols in the formulas in Eq. (20.103) are as follows: f_b is the Eucken ratio, which is a constant, and γ is the polytropic ratio $7/5$ for the rigid diatomic molecule. If the Boltzmann kinetic theory is employed, the Eucken ratio may be calculated from the statistical mechanical expressions (namely, collision bracket integrals) for transport coefficients if a suitable potential model is assumed.[19] We also note that, by comparing the theoretical and experimental sound wave absorption coefficients in the low frequency regime, the Eucken ratio f_b can be obtained and then the bulk viscosity η_b therefrom, knowing the shear viscosity. This method is, as a matter of fact, one of the most reliable methods of measuring bulk viscosity. The coefficients $B_6, \ldots, B_0$ are defined by the formulas given in terms of the parameters in Eq. (20.103):

$$B_0 = -\tfrac{5}{2},$$

$$B_2 = \tfrac{5}{2} + i\omega^*(\tfrac{5}{2} f_b \xi_b + \tfrac{10}{3}\xi_s + \tfrac{19}{4}\xi_q),$$

$$B_4 = -i\tfrac{95}{28}\omega^*\xi_q - \tfrac{19}{3}\omega^{*2}(\tfrac{2}{7}\xi_s + \tfrac{2}{7}\xi_q + \tfrac{3}{4}f_b\xi_b \qquad (20.104)$$

$$\qquad + \tfrac{3}{14}f_b\xi_q - \xi_s\xi_q - \tfrac{3}{4}f_b\xi_b\xi_q),$$

$$B_6 = \tfrac{361}{30}\omega^{*4}\xi_q(\tfrac{8}{21}\xi_s + f_b\xi_b)(1 + \tfrac{3}{4}f_b),$$

where

$$\xi_s = (1 + i\tfrac{7}{5}\omega^*)^{-1}, \quad \xi_q = (1 + i\tfrac{19}{10}\omega^*)^{-1}, \quad \xi_b = (1 + i\tfrac{21}{4}\omega^*)^{-1}.$$
$$\qquad (20.105)$$

Since the solutions of the cubic equation (20.102) are in rather complicated and uninformative forms, it is convenient to examine their low

---

[19]See, for example, S. Chapman and T. G. Cowling, *The Mathematical Theory of Nonuniform Gases 3rd edition* (Cambridge University Press, London, 1970).

frequency limits: As $\omega^* \to 0$, we find

$$B_6 \to 0,$$

$$B_4 \to -i\frac{95}{28}\omega^* - \left(\frac{205}{56} + \frac{19}{14}f_b\right)\omega^{*2}, \tag{20.106}$$

$$B_2 \to \frac{5}{2} + i\left(\frac{97}{12} + \frac{5}{2}f_b\right)\omega^* + \left(\frac{1643}{120} + \frac{105}{8}f_b^2\right)\omega^{*2},$$

which imply that the limiting solution of the wave number for the sound mode is given by the formula

$$k^* = \pm\left[1 - i\frac{1}{2}\left(\frac{197}{105} + f_b\right)\omega^* - \left(\frac{8563}{4200} + \frac{16}{35}f_b + 3f_b^2\right)\omega^2\right], \tag{20.107}$$

and for the thermal mode by the formula

$$k^* = \pm(1 - i)\sqrt{\frac{7}{19\omega^*}}. \tag{20.108}$$

The real part of k^* gives the dispersion relation and the imaginary part the absorption coefficient. They can be obtained in more rigorous forms if the solvability condition (20.102) is solved numerically. The results so obtained are plotted in Figs. 20.2 and Fig. 20.3 for nitrogen, in which

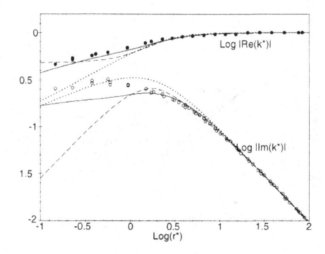

Fig. 20.2. Sound wave absorption coefficient $\log|\operatorname{Im} k^*|$ and dispersion $\log|\operatorname{Re} k^*|$ vs. $\log r^*$ for nitrogen gas. The ordinate is for either $\log|\operatorname{Re} k^*|$ or $\log|\operatorname{Im} k^*|$ in the common scale as indicated. $r^* = 1/\gamma\omega^*$. The prediction by the present theory (solid curve) is compared with the experimental data by Greenspan (symbols), the Navier–Stokes theory (dotted curve), and the result (broken curve) of the Moraal–McCourt moment method. The value of f_b is 0.8. Reproduced with permission from B. C. Eu and Y. G. Ohr, *Phys. Fluids* **13**, 744 (2001) © 2001 American Institute of Physics.

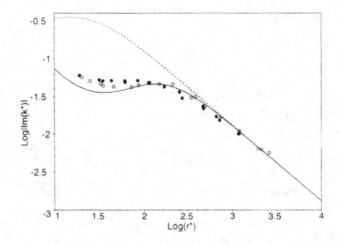

Fig. 20.3. Sound wave absorption coefficient $\log |\operatorname{Im} k^*|$ vs. $\log r^*$ for normal H_2 and para H_2 at $T_0 = 293$ K. The open circles are for normal H_2 and the filled circles are for para H_2. The solid curve denotes the prediction by the present theory, and the broken curve the prediction by the Navier–Stokes theory. $f_b = 35$ and $\eta_0 = 88.2\,\mu P$. Reproduced with permission from B. C. Eu and Y. G. Ohr, *Phys. Fluids* **13**, 744 (2001) © 2001 American Institute of Physics.

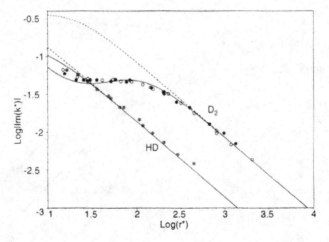

Fig. 20.4. Sound wave absorption coefficient $\log |\operatorname{Im} k^*|$ vs. $\log r^*$ for normal D_2, ortho D_2, and HD at $T_0 = 293$ K. The open circles are for normal D_2 and the filled circles are for ortho D_2, and the dotted circles are for HD. The solid curve denotes the prediction by the present theory, and the broken curve the prediction by the Navier–Stokes theory. $f_b = 22$ for D_2 and $\eta_0 = 123\,\mu P$ for D_2 and $f_b = 2$ for HD and $\eta_0 = 108\,\mu P$. Reproduced with permission from B. C. Eu and Y. G. Ohr, *Phys. Fluids* **13**, 744 (2001) © 2001 American Institute of Physics.

comparison is made with the theoretical results by Moraal and McCourt who solved the Waldmann–Snider equation,[20] which is a quantum mechanical version of the Boltzmann kinetic equation, by a method of moments. In Fig. 20.4, the results for normal D_2, ortho D_2, and HD are also compared with experiments and the theoretical results by Moraal and McCourt[21] who calculated them by the same kinetic theory method as for nitrogen. The comparisons made indicate that the generalized hydrodynamic equations, even at the level of approximation employed, preform better than the macroscopic equations obtained by the moment method employed for solving[22] the Waldmann–Snider equation.

[20]L. Waldmann, *Z. Naturforsch.* **12**a, 661 (1957); **13**a, 609 (1958); *Handbuch der Physik*, Vol. XII, ed. S. Flügge, ed. (Springer, Berlin, 1958), p. 295. R. F. Snider, *J. Chem. Phys.* **32**, 1051 (1960).

[21]Moraal and F. McCourt, *Z. Naturforsch. A* **27**, 583 (1972).

[22]See Moraal and McCourt in footnote cited earlier.

Chapter 21

Nonlinear Irreversible Processes

21.1. Flow of a Non-Newtonian Fluid

The viscosity of a fluid, among other properties of the fluid, is essential for understanding the flow phenomena in macroscopic matter and the processing of materials in engineering. In fact, study of viscosity in fluid flow has a long history in science tracing back to Isaac Newton. The subject field, known as rheology[1] at present, is important in physical chemistry, engineering, and physiology among many other disciplines in science and engineering. We discuss this topic as an example of irreversible phenomena under the purview of the theory of irreversible thermodynamics presented earlier.

Viscous flow phenomena occurring near equilibrium are adequately described by the linear viscosity, namely, the Newtonian viscosity, which is the limiting viscosity at a sufficiently small shear rate. It is well known that description of flow with the Newtonian viscosity is inadequate as the shear rate is increased and the fluid thereby gets increasingly removed from equilibrium. It is generally found that in such a state, the viscosity of the fluid becomes dependent on the shear rate in contrast to the Newtonian viscosity. The viscosity is then said to be non-Newtonian. Non-Newtonian

---

[1] The literature on rheology is rather extensive. To give the reader an introduction to the subject matter, we quote an example, which provides an introduction to the subject: H. A. Barnes, J. F. Hutton, and K. Walters, *An Introduction to Rheology* (Elsevier, Amsterdam, 1989).

viscosities of such fluids are generally nonlinear with respect to the shear rate. In rheology, we mostly deal with non-Newtonian flow phenomena. In this section, we show how these flow phenomena can be treated from the phenomenological theory viewpoint by employing the theory of irreversible processes described earlier.

21.1.1. *Velocity Profile of Flow in a Rectangular Channel*

To make the discussion as simple as possible, we will assume that only a shearing perturbation is present in a single-component simple liquid, which does not support a bulk viscosity; in other words, the excess normal stress Δ is assumed to be equal to zero. Neither is there heat flow present. Then the only relevant nonequilibrium variable is the traceless symmetric part of the stress tensor, and its evolution equation is that of the shear stress $\widehat{\mathbf{\Pi}}$. It takes the form

$$\rho d_t \widehat{\mathbf{\Pi}} = -2p \left[\nabla \mathbf{u} \right]^{(2)} - 2 \left[\mathbf{\Pi} \cdot \nabla \mathbf{u} \right]^{(2)} - \frac{p}{\eta_0} \mathbf{\Pi} q(\kappa), \tag{21.1}$$

for which the pressure tensor $\mathbf{P}$ is decomposed into components

$$\mathbf{\Pi} = \mathbf{P} - p\boldsymbol{\delta} := \rho \widehat{\mathbf{\Pi}} \tag{21.2}$$

with p denoting the hydrostatic pressure, $\boldsymbol{\delta}$ the unit second rank tensor, and $\mathbf{\Pi}$ the traceless symmetric part of $\mathbf{P}$ defined by

$$\mathbf{\Pi} = \tfrac{1}{2}(\mathbf{P} + \mathbf{P}^t) - \tfrac{1}{3}\boldsymbol{\delta} \mathrm{Tr} \mathbf{P} := [\mathbf{P}]^{(2)} \tag{21.3}$$

with the superscript t indicating the transpose. The symbol $[\mathbf{A}]^{(2)}$ stands for the traceless symmetric part of second rank tensor $\mathbf{A}$ as in Eq. (21.3). Other symbols in the equations are as follows: η_0 is the Newtonian shear viscosity independent of the shear rate; ρ is the mass density; $q(\kappa)$ is the nonlinear factor defined by

$$q(\kappa) = \frac{\sinh \kappa}{\kappa}, \tag{21.4}$$

$$\kappa = \frac{\tau}{2\eta_0} \left(\mathbf{\Pi} : \mathbf{\Pi} \right)^{1/2}, \tag{21.5}$$

$$\tau = \frac{(mk_B T)^{1/4} \sqrt{\eta_0}}{p\sigma}. \tag{21.6}$$

In these expressions, m is the reduced mass, σ is the size parameter of the molecule, k_B is the Boltzmann constant, and κ is the Rayleigh dissipation function for shearing. We note that in the kinetic theory treatment of nonlinear transport phenomena, the nonlinear factor $q(\kappa)$ is intimately related to the cumulant expansion of the calortropy production.[2]

For Eq. (21.1), the flux of stress tensor, $\psi^{(p)}$, is set equal to zero because of the assumption that only the shear stress is present. Setting $\psi^{(p)} = 0$, in fact, is equivalent to the closure of the hierarchy of evolution equations for the nonconserved variables. Similarly, $\mathbf{V}^{(1)}$ that originally appears in the full evolution equation for the stress tensor is also ignored because it is a second-rank tensor given in terms of intermolecular forces and peculiar velocities of molecules that belong to the moment other than the shear stress. This type of closure has been found adequate for the description of shock wave phenomena[3] and other flow processes in simple fluids. Therefore, we will also use it in this discussion. Since $\mathbf{V}^{(1)}$ is associated with the velocity times the virial associated with the flow of the fluid, it does not make contribution in the order of approximation we are interested in for the stress tensor in the present discussion. However, it is related to the angular momentum of the fluid.

Flow of a fluid that occurs confined between two plates aligned in parallel to, say, the xz plane in a suitably fixed laboratory coordinate system is called plane Couette flow. On the basis of the evolution equation or the constitutive equation for flow, Eq. (21.1), we consider a plane Couette flow in a flow configuration[4] in a suitably chosen coordinate system; see Fig. 20.1 for flow configuration in which the two parallel plates positioned at $y = D/2$ and $y = -D/2$ on the y-axis and thus being separated by distance D are simply moving in the opposite directions instead of oscillating back and forth. The plates are now assumed separated by distance D, while moving in opposite directions at speed $u_d/2$ and $-u_d/2$, respectively, in the x-direction in the coordinate system. In the flow configuration defined, the flow is in the direction of x, but neutral in the z-direction whereas the velocity gradient of the laminar flow is present in the direction of the y-axis. To simplify the problem, it is assumed that the fluid is incompressible. Since it is also sufficient to consider the case of the normal stress differences being

---

[2]For details of this aspect, see, for example, B. C. Eu, *Nonequilibrium Statistical Mechanics* (Kluwer, Dordrecht, 1998), pp. 175–178 and references cited therein and also B. C. Eu, *Kinetic Theory of Nonequilibrium Ensembles, Irreversible Thermodynamics, and Generalized Hydrodynamics*, Vol. 1 (Springer, Switzerland, 2016) and references cited therein.

[3]It is also discussed in B. C. Eu, *Nonequilibrium Statistical Mechanics* (Kluwer, Dordrecht, 1998), pp. 246–261. See also the reference cited in footnote 1.

[4]L. D. Landau and E. M. Lifshitz, *Fluid Mechanics* (Pergamon, Oxford, 1958).

equal to zero because they are of higher order than the shear stress, we neglect the normal stress differences.

The steady-state constitutive equation for the shear stress tensor is then obtained from Eq. (21.1) in the form

$$\Pi q(\kappa) = -2\eta_0 \gamma, \tag{21.7}$$

where $\Pi := \Pi_{xy} = \Pi_{yx}$ is the shear stress and γ is the shear rate (velocity gradient) defined by

$$\gamma = \frac{1}{2} \frac{\partial u_x}{\partial y} \tag{21.8}$$

with u_x denoting the x-component of the velocity $\mathbf{u}$. If the channel is sufficiently long in the flow (x) direction, translational invariance of the flow holds to a good approximation, making u_x independent of x. But u_x would remain a function of y. It should be noted that if the nonlinear factor $q(\kappa)$ is set equal to unity, then Eq. (21.7) becomes the Newtonian law of viscosity in the classical hydrodynamics of Navier and Stokes: $\Pi = -2\eta_0 \gamma$ for the present flow problem.

The Rayleigh dissipation function κ under the conditions mentioned earlier is given by

$$\kappa = \frac{\tau \sqrt{\mathbf{\Pi} : \mathbf{\Pi}}}{\sqrt{2}\eta_0}. \tag{21.9}$$

The steady-state momentum balance equation consistent with the shear stress equation (21.7) is given by

$$\frac{\partial}{\partial y}\Pi = -p_x, \tag{21.10}$$

where

$$p_x = \frac{\partial p}{\partial x}.$$

This equation is coupled to Eq. (21.7). This pair of equations is solved for the velocity profile $u_x(y)$ for the plane Couette flow. The following two cases are possible for the flow configuration.

21.1.1.1.   *The Case of $p_x = 0$*

For this case, Eqs. (21.7) and (21.10) with $p_x = 0$ are solved subject to the boundary conditions

$$u_x(\pm \tfrac{1}{2}D) = \pm \tfrac{1}{2} u_d, \tag{21.11}$$

where $\pm\frac{1}{2}u_d$ are the speeds of the plates moving in opposite directions along the x-axis. These are called stick boundary conditions because the fluid sticks to the boundary walls. This sticking tendency of the fluid creates a velocity profile in the channel. We are interested in finding the shape of the velocity profile in the flow channel.

The solutions subject to the boundary conditions are easily obtained:

$$u_x = \frac{u_d}{D}y, \tag{21.12}$$

$$\Pi = -\frac{\sqrt{2}\eta_0}{\tau}\ln\left(\frac{\tau u_d}{\sqrt{2}D} + \sqrt{1 + \left(\frac{\tau u_d}{\sqrt{2}D}\right)^2}\right). \tag{21.13}$$

The negative branch is chosen for Π. Note that the shear stress tensor (i.e., xy component) Π is a constant independent of the coordinates; it depends on the shear rate u_d/D only and the velocity profile is linear with respect to y. This conclusion does not remain true if flow configurations are different from the present.

Since in rheology the non-Newtonian viscosity η is defined by the constitutive relation

$$\Pi = -2\eta\gamma, \tag{21.14}$$

it follows from the solution (21.13) obtained that the non-Newtonian viscosity η of the fluid is given by the formula[5]

$$\eta = \eta_0 \frac{\sinh^{-1}\left(\frac{\tau u_d}{\sqrt{2}D}\right)}{\left(\frac{\tau u_d}{\sqrt{2}D}\right)}. \tag{21.15}$$

It should be noted that with u_x given by Eq. (21.12) the shear rate γ is obtained in terms of experimental input as follows:

$$\gamma = \frac{u_d}{2D}. \tag{21.16}$$

Therefore, the non-Newtonian viscosity obtained is dependent on the shear rate according to the formula (21.15). Note that u_d, and hence γ, is an experimental input and a constant independent of density and temperature.

---

[5] This inverse hyperbolic sine function formula for viscosity was derived from Boltzmann kinetic equation for the first time in B. C. Eu, *J. Chem. Phys.* **74**, 3006 (1981) and has been applied to studies of numerous nonlinear transport processes since then.

However, the density and temperature dependences of η are vested in the Newtonian viscosity, which should generally depend on them. The prediction of γ dependence by this formula for η has been found in agreement with molecular dynamics simulation results for Lennard-Jones fluids in the same flow configuration as described. We will return to this question later.

21.1.1.2. The Case of $p_x \neq 0$

It has been possible to obtain simple analytic solutions for the case of flow when the pressure gradient p_x is not equal to zero, but a constant. We consider a flow subject to the boundary conditions

$$u(y)|_{y=\pm D/2} = 0. \tag{21.17}$$

In other words, a non-Newtonian fluid flows in the channel under a constant pressure gradient in the x-direction. Therefore in this case, the boundary walls do not move, but the fluid sticks at the static walls.

Since the shear stress at $y = 0$ (along the axis of the channel) should be equal to zero, integration of Eq. (21.10) yields

$$\Pi(y) = -p_x y. \tag{21.18}$$

Substitution of this result into Eq. (21.7) yields the differential equation for u_x:

$$\frac{\partial u_x}{\partial y} = \frac{2}{\tau} \sinh\left(\frac{\tau p_x}{2\eta_0} y\right). \tag{21.19}$$

The solution of this differential equation is

$$u_x = \alpha u_{\max}\left[\cosh\left(\frac{\delta}{2}\right) - \cosh\left(\frac{\delta y}{D}\right)\right], \tag{21.20}$$

where $u_{\max} = u_x(y = 0)$ and

$$\alpha = -\frac{4\eta_0}{\tau^2 p_x u_{\max}}, \quad \delta = \frac{\tau p_x D}{2\eta_0}. \tag{21.21}$$

On expanding the hyperbolic functions in power series of the arguments in the limit of small δ, this velocity profile reduces to the well-known parabolic Poiseuille profile. This is the limit of either small η_0/τ or small $p_x L$.

It is possible to obtain analytic solutions of the governing equations for the case of flow in which the boundary walls move at $\pm u_d/2$ and hence the boundary conditions are given by Eq. (21.11). We leave this problem as an exercise for the reader.

The flow rate can be calculated with the velocity profile (21.20) obtained. Let the mass flow rate through the channel of cross section $A = D^2$ under the pressure gradient p_x be denoted by Q_v:

$$Q_v = \int_{-D/2}^{D/2} dz \int_{-D/2}^{D/2} dy \rho u_x(y).$$ (21.22)

The integrand is the volume flow per unit volume per unit time. Assuming the fluid is incompressible and using $u_x(y)$ in Eq. (21.20), we obtain

$$Q_v = \frac{\rho D^4 \Delta p}{12 \eta_0 \Delta x} F(\delta),$$ (21.23)

where

$$F(\delta) = \frac{12}{\delta} \left[\cosh\left(\frac{\delta}{2}\right) - \frac{2}{\delta} \sinh\left(\frac{\delta}{2}\right) \right]$$ (21.24)

and Δp is the pressure difference between the head and tail ends of the rectangular tube and Δx is the length of the rectangular tube. As δ tends to zero, $F(\delta) \to 1$ and Q_v reduces to the well-known Hagen–Poiseuille volume flow rate of a Newtonian fluid.

21.1.2. Non-Newtonian Viscosity and Computer Simulations

To assess the quality of the non-Newtonian viscosity formula (21.15) obtained for flow of a dense simple fluid, we compare the present theoretical prediction with the computer simulation results performed on a LJ fluid at varying densities. To this end it is convenient to make use of dimensionless variables reduced with respect to parameters associated with the model fluid. The relevant variables in reduced units are:

$$T^* = k_B T/\epsilon, \quad \rho^* = n\sigma^3,$$

$$\eta^* = \eta \sigma^2 / (m\epsilon)^{1/2}, \quad \eta_0^* = \eta_0 \sigma^2 / (m\epsilon)^{1/2},$$

$$\gamma^* = \gamma \sigma (m/\epsilon)^{1/2} = \frac{u_d}{2D} \sigma \sqrt{\frac{m}{\epsilon}},$$ (21.25)

$$\tau^* = \tau/\sigma \, (m/\epsilon)^{1/2} = \frac{\sqrt{\eta_0^*}}{\rho^* T^{*3/4}},$$

$$\tau_e^* = \sqrt{2}\tau^*.$$

Here ϵ is the well depth of the LJ potential, σ is its contact (size) parameter, m is the mass of the particle, and n is the number density.

Then the non-Newtonian shear viscosity formula is expressible in terms of a universal function of the reduced variable product

$$\tau_e^* \gamma^* = \frac{\sqrt{2\eta_0^*}\gamma^*}{\rho^* T^{*3/4}}, \tag{21.26}$$

where τ_e^* may be regarded as the reduced stress relaxation time. Therefore, if the non-Newtonian viscosity is scaled by the Newtonian viscosity, then the scaled non-Newtonian viscosity becomes a universal function of $\tau_e^* \gamma^*$ (reduced shear rate):

$$\frac{\eta^*}{\eta_0^*} \, (\rho^*, T^*, \gamma^*) = \frac{\sinh^{-1}(\tau_e^* \gamma^*)}{\tau_e^* \gamma^*}, \tag{21.27}$$

which is independent of material parameters and thus indicates that there are rheological corresponding states of ρ^*, T^*, and γ^* since η_0^* is a function of ρ^* and T^* for the LJ liquid. The non-Newtonian viscosity formula (21.15) is known as the Ree–Eyring formula in the rheology literature, which was obtained on the basis of the absolute reaction rate theory[6] of Eyring *et al.* The original Ree–Eyring formula, however, contains a number of adjustable empirical parameters. In contrast to their semiempirical formula, the present non-Newtonian viscosity formula is derived from the kinetic theory of dense fluids[7] and the generalized hydrodynamics derived therefrom and hence, in principle, does not contain empirical parameters at all except for the potential parameters. Therefore, the present formula provides the molecular theory foundation for the Ree–Eyring formula, at least, for simple fluids, if the Newtonian viscosity η_0^* is calculated by means of a molecular theory using an intermolecular force.

---

[6]See S. Glasstone, K. J. Laidler and H. Eyring, *The Theory of Rate Processes* (McGraw-Hill, New York, 1941). Their formula contains adjustable empirical parameters.

[7]See B. C. Eu, *Kinetic Theory and Irreversible Thermodynamics* (Wiley, New York, 1992) and B. C. Eu, *Nonequilibrium Statistical Mechanics* (Kluwer, Dordrecht, 2002) references cited therein, and B. C. Eu, *Kinetic Theory of Nonequilibrium Ensembles, Irreversible Thermodynamics, and Generalized Hydrodynamics* (Springer, Switzerland, 2016), pp. 485–486.

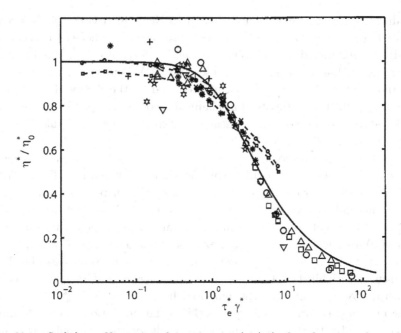

Fig. 21.1. Scaled non-Newtonian shear viscosity (η^*/η_0^*) plotted against the scaled reduced shear rate $(\tau^*\gamma^*)$ for all available molecular simulation data and compared with the universal formula given in Eq. (21.27) at various reduced temperatures and reduced densities. The symbols and filled circles connected by broken lines are MD simulation data reported in the literature whereas the solid curve is the corresponding state (universal) non-Newtonian viscosity predicted by the present theory. Reproduced with permission from R. Laghaei, A. E. Nasrabad, B. C. Eu, *J. Chem. Phys.* **123**, 234507 (2005) © 2005 American Institute of Physics.

The scaled variable $\tau_e^*\gamma^*$ indicates that it will be possible to predict the non-Newtonian viscosity η^* for all reduced shear rates at any density and temperature. We will see that this indeed is the case when all the available molecular dynamics simulation data by various authors are assembled and compared with the prediction by the reduced formula (21.27), as shown in Fig. 21.1, for which η_0^* is calculated by using the density fluctuation theory formula for η_0^*. In the density fluctuation theory,[8] the modified free volume theory for self-diffusion and the generic van der Waals equation of state (i.e., the canonical equation of state described in Chapter 10) are combined to obtain a molecular theory expression for η_0^*. In Fig. 21.1, the

---

[8]See, B. C. Eu, *Transport Coefficients of Fluids* (Springer, Heidelberg, 2006) and references cited therein.

reduced shear rate dependence of the non-Newtonian viscosity thus computed[9] (solid curve) is compared with the results of molecular dynamics (MD) simulation data by Ashurst *et al.*,[10] Heyes *et al.*,[11] and Evans.[12] Various symbols in the figures represent the MD simulation data by various authors, which are reduced with η_0^* predicted by the theory of Newtonian viscosity, which is a statistical mechanical theory developed by Eu and collaborators; see footnote 6. The meanings of the symbols are: $\bigcirc$ at $T^* = 2.5$, $\rho^* = 0.01$; $\square$ at $T^* = 2.5$, $\rho^* = 0.05$; $\triangle$ at $T^* = 2.5$, $\rho^* = 0.1$; $\triangledown$ at $T^* = 1.5$, $\rho^* = 0.05$; open five-pointed star at $T^* = 3.0$, $\rho^* = 0.05$; open six-pointed star at $T^* = 0.85$, $\rho^* = 0.76$. For the triple point data at $T^* = 0.722$, $\rho^* = 0.8449$ the crosses ($+$) are by Heyes *et al.*, whereas asterisks ($*$) are by Evans and the crosses ($\times$) are by Ashurst *et al.*, all of which are scaled by $\eta_0^* = 2.89$ estimated from the simulation data. The left-pointed open triangles ($\triangleleft$) are the data of Heyes and Szczepanski at $T^* = 5.0$, $\rho^* = 0.1$, which are scaled with $\eta_0^* = 0.45$ reported in the NIST reference database. The broken curve with filled circles ($\bullet$) is the simulation results of Evans *et al.* scaled with the Newtonian viscosity computed by Meier *et al.*[13] $\eta_0^* = 3.24$, and the broken curve with filled squares ($\blacksquare$) is the simulation results of Evans *et al.* scaled with their own simulation result for the Newtonian viscosity $\eta_0^* = 3.41$. The value of 3.24 for η_0^* given by Meier *et al.* and Ferrario *et al.*[14] appears to yield better zero shear rate non-Newtonian viscosity, suggesting that the result of Meier *et al.* is more accurate than the older value by Evans *et al.*

In the reduced plot presented in Fig. 21.1, the simulation data are considerably scattered around the theoretical solid curve, suggesting the delicate task of evaluating the non-Newtonian viscosity by molecular dynamics simulation methods. Nevertheless, the simulation data of various authors are still around the universal non-Newtonian viscosity curve (solid curve)

---

[9] See R. Laghaei, A. E. Nasrabad and B. C. Eu, *J. Chem. Phys.* **123**, 234507 (2005).

[10] W. T. Ashurst and W. G. Hoover, *Phys. Rev. A* **11**, 658 (1975); W. G. Hoover, D. J. Evans, R. B. Hickman, A. J. C. Ladd, W. T. Ashurst and B. Moran, *Phys. Rev. A* **22**, 1690 (1980).

[11] D. M. Heyes, J. J. Kim, C. J. Montrose and T. A. Litovitz, *J. Chem. Phys.* **73**, 3987 (1980); D. M. Heyes, *Physica A* **133**, 473 (1985); D. M. Heyes and R. Szczepanski, *J. Chem. Soc. Faraday Trans. II* **83**, 319 (1987).

[12] D. J. Evans, *Phys. Rev. A* **23**, 1988 (1981); D. J. Evans, G. P. Morriss and L. M. Hood, *Mol. Phys.* **68**, 637 (1989).

[13] K. Meier, A. Laesecke and S. Kabelac, *J. Chem. Phys.* **121**, 3671 (2004); *ibid.* **121**, 9526 (2004).

[14] M. Ferrario, G. Ciccotti, L. Holian and J. P. Ryckaert, *Phys. Rev. A* **44**, 6936 (1991).

predicted by the present theory, and it indicates that the non-Newtonian viscosity of the LJ fluid may indeed have a scaling function that is universal for all temperatures and densities.

21.2. Theory of Viscosity and Normal Stress Coefficients

Having studied the simplest case of nonlinear viscous phenomena, in which the normal stresses are absent, we now would like to take the normal stress effects into account. To take the normal stress effects properly taken into consideration, it is necessary to cast the stress tensor evolution equation in the Jaumann[15] derivative form, so that the stress evolution equation properly behaves under rotational transformation. The rules of converting a fixed-frame evolution equation to the corotating-frame version are discussed in the reference cited in the aforementioned footnote. The rules are: (1) *If tensor $\mathfrak{T}$ is of rank 2, the time derivative of $\mathfrak{T}$ in the fixed-frame evolution equation is replaced with its Jaumann derivative $\mathcal{D}\mathfrak{T}/\mathcal{D}t$ as follows*:

$$\frac{d\mathfrak{T}}{dt}\Big|_{\text{fixed}} \Rightarrow \frac{\mathcal{D}\mathfrak{T}}{\mathcal{D}t} = \frac{d\mathfrak{T}}{dt} + [\omega, \mathfrak{T}], \tag{21.28}$$

where

$$[\omega, \mathfrak{T}] := \omega \cdot \mathfrak{T} - \mathfrak{T} \cdot \omega \tag{21.29}$$

with ω defines the vorticity

$$\omega = \tfrac{1}{2}[\nabla \mathbf{u} - (\nabla \mathbf{u})^t] \tag{21.30}$$

and all the vorticity containing terms in the fixed-frame equation are put equal to zero; (2) If $\mathfrak{T}$ is a vector (e.g., heat flux or diffusion flux), the rule of equivalence (21.28) remains the same except that the bracket $[\omega, \mathfrak{T}]$ means a vector product of $\mathbf{W}$ and $\mathfrak{T}$

$$[\omega, \mathfrak{T}] := \mathbf{W} \times \mathfrak{T} \tag{21.31}$$

---

[15] J. Jaumann, *Sitzungsber. Akad. Wiss. Wien (IIa)* **120**, 385 (1911). See W. Prager, *Introduction to Mechanics of Continua* (Ginn, Boston, 1961) for cogent discussions on the necessity of the Jaumann derivative in the continuum theory of matter. See also B. C. Eu, *J. Chem. Phys.* **82**, 4283 (1985) and B. C. Eu, *Kinetic Theory and Irreversible Thermodynamics* (Wiley, New York, 1992).

with **W** *denoting the angular velocity of the fluid*

$$\mathbf{W} = \tfrac{1}{2}\nabla \times \mathbf{u}. \tag{21.32}$$

These rules applied to the evolution equations for nonconserved variable evolution equations in a fixed coordinate-frame of reference assure us the evolution equations satisfying the desired objectivity, and the kinetic theory-based evolution equations for the nonconserved variables thereby become compatible with the objectivity principle in continuum mechanics.[16]

We will consider a single-component nonlinear fluid under the assumption that the nonconserved variables are limited to $(\mathbf{\Pi}, \Delta, \mathbf{Q})$, which may be construed as a closure. We further assume that the fluid is in a uniform temperature field, so that there is no heat flux — this assumption simplifies the set of evolution equations. Therefore, we have only the evolution equations for the shear and normal stresses to take into account. When the rules stated earlier are applied, the shear stress evolution equation and the excess normal stress evolution equation are given by the partial differential equations

$$\rho\frac{d\widehat{\mathbf{\Pi}}}{dt} = -2p[\nabla\mathbf{u}]^{(2)} - 2[\mathbf{\Pi}\cdot[\nabla\mathbf{u}]^{(2)}]^{(2)} - [\omega, \mathbf{\Pi}] - \frac{2}{3}\mathbf{\Pi}\nabla\cdot\mathbf{u} + \Lambda^{(1)}, \tag{21.33}$$

$$\rho\frac{d\widehat{\Delta}}{dt} = -p\frac{d}{dt}\ln(pv^{5/3}) - \frac{2}{3}\mathbf{\Pi} : [\nabla\mathbf{u}]^{(2)} - \frac{2}{3}\rho\widehat{\Delta}\nabla\cdot\mathbf{u} + \Lambda^{(2)}, \tag{21.34}$$

where $\Lambda^{(1)}$ and $\Lambda^{(2)}$ are the dissipation terms, which we will assume the same empirical formulas taken in the previous sections: Following the notation of the monograph on nonequilibrium statistical mechanics,[17] we take the dissipation terms in the forms

$$\Lambda^{(1)} = -2p\mathfrak{R}^{(11)}q(\kappa)\Pi, \tag{21.35}$$

$$\Lambda^{(2)} = -\frac{2p}{3}\mathfrak{R}^{(22)}q(\kappa)\Delta \quad (\Delta = \rho\widehat{\Delta}), \tag{21.36}$$

---

[16]See, for example, C. Eringen, *Continuum Physics*, Vol. 1 (Academic, New York, 1971). Also C. Truesdell and W. Noll, *The Nonlinear Field Theories of Mechanics* (Springer, Berlin, 1965).

[17]B. C. Eu, *Nonequilibrium Statistical Mechanics* (Kluwer, Dordrecht, 1998).

where $q(\kappa)$ denotes the nonlinear factor given in terms of the generalized Rayleigh dissipation function κ — a quadratic form of nonconserved fluxes

$$\kappa = \left(\sum_{a,b=1}^{r} \sum_{q,k\geq 1} \Phi_a^{(q)} \mathfrak{R}_{ab}^{(qk)} \Phi_b^{(k)} \right)^{1/2},$$

$$q(\kappa) = \frac{\sinh \kappa}{\kappa},$$

$$\mathfrak{R}_{ab}^{(qk)} = (\beta g)^{-1} g_a^{(q)} R_{ab}^{(qk)} g_b^{(k)},$$

$$g_a^{(1)} = \frac{1}{2p_a}; \; g_a^{(2)} = \frac{3}{2p_a}; \; g_a^{(3)} = \frac{3}{p_a \widehat{C}_{pa} T}; \; g_a^{(1)} = \frac{1}{\rho_a}$$

and $R_{ab}^{(qk)}$ are the collision bracket integrals (see Eqs. (7.241)–(7.246) in p. 183 in B. C. Eu, *Nonequilibrium Statistical Mechanics* (Kluwer, 1998). Therefore, $\mathfrak{R}_{ab}^{(qk)}$ are parameters representing the basic collision bracket integrals $R_{ab}^{(qk)}$. Hence $\frac{2p}{\eta_0}$ and $\frac{2p}{3\eta_b^0}$ are, respectively, parameters representing the collision bracket integrals $R^{(11)}$ and $R^{(22)}$ as follows:

$$\frac{2p}{\eta_0} = 2p\mathfrak{R}^{(11)} = 2p(\beta g)^{-1} g^{(1)2} R^{(11)} = \frac{1}{2pg\beta} R^{(11)}, \tag{21.37}$$

$$\frac{2p}{3\eta_b^0} = \frac{2p}{3}\mathfrak{R}^{(22)} = \frac{2p}{3}(\beta g)^{-1} g^{(2)2} R^{(22)} = \frac{3p}{2\beta g} R^{(22)}, \tag{21.38}$$

and hence we have the correspondences of η_0 and η_b^0 and the collision bracket integrals:

$$\eta_0 = \frac{4p^2 g\beta}{R^{(11)}}, \tag{21.39}$$

$$\eta_b^0 = \frac{4\beta g}{9R^{(22)}}, \tag{21.40}$$

where $R^{(ii)} (i = 1, 2)$ is a scalar consisting of phenomenological parameters related to the shear viscosity or normal stress coefficients — for example, a collision bracket integral for the fluid consisting of spherical molecules. In the case of a dense fluid, $g\beta$ is replaced by ϵ defined by

$$\epsilon = \frac{1}{l}\sqrt{\frac{2k_B T}{m_r}}; \quad l = \text{mean free path}. \tag{21.41}$$

See Eqs. (6.330)–(6.332), and (6.344) in B. C. Eu, *Kinetic Theory of Nonequilibrium Ensembles, Irreversible Thermodynamics, and Generalized Hydrodynamics*, Vol. 1 (Springer, Switzerland, 2016). Therefore η_0 and η_b^0 may be regarded as the shear viscosity and bulk viscosity in the neighborhood of equilibrium, that is, the Chapman–Enskog viscosities or their dense fluid extensions.

21.2.1. *Unidirectional Flow*

To be specific in a simplest possible manner, we now consider a unidirectional flow in the x-direction with a velocity gradient applied in the direction of y-axis (e.g., a plane Couette flow geometry), but no velocity gradient in the z-direction. In this flow configuration, since the velocity gradient has the component in the y-direction only, it may be represented by the matrix

$$\gamma := -[\nabla \mathbf{u}]^{(2)} = -\gamma \begin{bmatrix} 0 & 1 & 0 \\ 1 & 0 & 0 \\ 0 & 0 & 0 \end{bmatrix}, \tag{21.42}$$

where

$$\gamma := \frac{1}{2} \frac{\partial u_x}{\partial y} \quad (\gamma = \text{shear rate}). \tag{21.43}$$

The pressure tensor may be generally expressed by a matrix

$$\mathbf{P} = \begin{pmatrix} P_{xx} & P_{xy} & P_{xz} \\ P_{yx} & P_{yy} & P_{yz} \\ P_{zx} & P_{zy} & P_{zz} \end{pmatrix}.$$

For the unidirectional flow considered, since there are P_{xz}, P_{zx}, P_{zy}, P_{yz} equal to zero, $\mathbf{P}$ takes the form

$$\mathbf{P} = \begin{pmatrix} P_{xx} & P_{xy} & 0 \\ P_{yx} & P_{yy} & 0 \\ 0 & 0 & P_{zz} \end{pmatrix}. \tag{21.44}$$

The normal stress differences N_1 and N_2 are defined in terms of diagonal components of the stress tensor by the formulas

$$N_1 := P_{xx} - P_{yy} = \rho(\widehat{\Pi}_{xx} - \widehat{\Pi}_{yy}), \tag{21.45}$$

$$N_2 := P_{yy} - P_{zz} = \rho(\widehat{\Pi}_{yy} - \widehat{\Pi}_{zz}). \tag{21.46}$$

The N_1 and N_2 are respectively called primary and secondary normal stress differences in the literature on rheology.

We now note that the pressure tensor is decomposed into various components as follows:

$$\mathbf{P} = \mathbf{\Pi} + [\mathbf{P}, \mathbf{P}^t] + \Delta\mathbf{U} + p\mathbf{U},$$

$$\mathbf{\Pi} = \tfrac{1}{2}(\mathbf{P} + \mathbf{P}^t) - \tfrac{1}{3}\mathbf{U}\mathrm{Tr}\mathbf{P} = [\mathbf{P}]^{(2)},$$

$$\Delta = \tfrac{1}{3}\mathrm{Tr}(\mathbf{P} - \mathbf{P}_0) = \tfrac{1}{3}\mathrm{Tr}\mathbf{P} - p,$$

where $[\mathbf{P}, \mathbf{P}^t]$ represents the asymmetric part of $\mathbf{P}$ denoted by

$$[\mathbf{P}, \mathbf{P}^t] = \tfrac{1}{2}(\mathbf{P} - \mathbf{P}^t),$$

which vanishes in the present case of a simple fluid owing to $\mathbf{P}$ being symmetric. Thus for the present case the pressure tensor is decomposable as follows:

$$\mathbf{P} = \mathbf{\Pi} + [\mathbf{P}, \mathbf{P}^t] + (\Delta + p)\mathbf{U}. \tag{21.47}$$

The first component on the right is traceless symmetric part, the second component the asymmetric part, and the third component the trace part. Let us define the vorticity tensor ω by the formula

$$\omega = \tfrac{1}{2}[\nabla\mathbf{u} - (\nabla\mathbf{u})^t] \tag{21.48}$$

and shear rate tensor γ by the formula

$$\gamma = [\nabla\mathbf{u}]^{(2)} = \tfrac{1}{2}[\nabla\mathbf{u} + (\nabla\mathbf{u})^t] - \tfrac{1}{3}\mathbf{U}\mathrm{Tr}(\nabla\mathbf{u}). \tag{21.49}$$

Since we then find $\mathbf{\Pi}$ in the form

$$\mathbf{\Pi} = \begin{pmatrix} \tfrac{1}{3}(2P_{xx} - P_{yy} - P_{zz}) & P_{xy} & 0 \\ P_{yx} & \tfrac{1}{3}(2P_{yy} - P_{xx} - P_{zz}) & 0 \\ 0 & 0 & \tfrac{1}{3}(2P_{zz} - P_{yy} - P_{xx}) \end{pmatrix}, \tag{21.50}$$

with definitions of the primary (N_1) and secondary (N_2) normal stress differences

$$N_1 = P_{xx} - P_{yy}, \tag{21.51}$$

$$N_2 = P_{yy} - P_{zz}, \tag{21.52}$$

which gives rise to the relation

$$N_1 + N_2 = P_{xx} - P_{zz}, \tag{21.53}$$

we obtain $\mathbf{\Pi}$ in terms of N_1, N_2, and Π_{xy} in the form

$$\mathbf{\Pi} = \begin{pmatrix} \frac{1}{3}(2N_1 + N_2) & \Pi_{xy} & 0 \\ \Pi_{xy} & -\frac{1}{3}(N_1 - N_2) & 0 \\ 0 & 0 & -\frac{1}{3}(N_1 + 2N_2) \end{pmatrix}. \tag{21.54}$$

Now by using the matrix representation of $[\nabla \mathbf{u}]^{(2)}$ for the unidirectional flow assumed, we have the shear rate tensor expressible in the form

$$[\nabla \mathbf{u}]^{(2)} = \gamma \begin{pmatrix} 0 & 1 & 0 \\ 1 & 0 & 0 \\ 0 & 0 & 0 \end{pmatrix} \quad \left(\gamma = \frac{1}{2} \frac{\partial u_x}{\partial y} \right). \tag{21.55}$$

Upon using the results for the tensors enumerated as above, we obtain

$$[\mathbf{\Pi} \cdot [\nabla \mathbf{u}]^{(2)}]^{(2)} = \frac{1}{3}\gamma \begin{pmatrix} \Pi_{xy} & \frac{1}{2}N_1 + N_2 & 0 \\ \frac{1}{2}N_1 + N_2 & \Pi_{xy} & 0 \\ 0 & 0 & -2\Pi_{xy} \end{pmatrix} \tag{21.56}$$

and

$$[\omega, \mathbf{\Pi}] = -\gamma \begin{pmatrix} 2\Pi_{xy} & -N_1 & 0 \\ -N_1 & -2\Pi_{xy} & 0 \\ 0 & 0 & 0 \end{pmatrix}. \tag{21.57}$$

Collecting these results into the evolution equation for $\widehat{\Pi}$, we obtain three independent components $\widehat{\Pi}_{xy}$, $\widehat{N}_1$, and $\widehat{N}_2$ of the stress tensor evolution equation in the following forms:

$$\rho\frac{d}{dt}\widehat{\Pi}_{xy} = -2\gamma p - \frac{p}{\eta_0}\Pi_{xy}q(\kappa) - \frac{2}{3}\gamma(2N_1 + N_2), \qquad (21.58)$$

$$\rho\frac{d}{dt}\widehat{N}_1 = 4\gamma\Pi_{xy} - \frac{p}{\eta_0}N_1 q(\kappa), \qquad (21.59)$$

$$\rho\frac{d}{dt}\widehat{N}_2 = -4\gamma\Pi_{xy} - \frac{p}{\eta_0}N_2 q(\kappa). \qquad (21.60)$$

Here $\widehat{\Pi}_{xy} = \Pi_{xy}/\rho$ and $\widehat{N}_i = N_i/\rho$ $(i = 1, 2)$. In matrix form the evolution equations given above are expressible as

$$\rho\frac{d}{dt}\begin{pmatrix} \widehat{\Pi}_{xy} \\ \widehat{N}_1 \\ \widehat{N}_2 \end{pmatrix} = -2p\gamma\begin{pmatrix} 1 \\ 0 \\ 0 \end{pmatrix} + q(\kappa)\begin{pmatrix} -\dfrac{pq}{\eta_0} & -\dfrac{4}{3}\gamma & -\dfrac{2}{3}\gamma \\ 4\gamma & -\dfrac{pq}{\eta_0} & 0 \\ -4\gamma & 0 & -\dfrac{pq}{\eta_0} \end{pmatrix}\begin{pmatrix} \Pi_{xy} \\ N_1 \\ N_2 \end{pmatrix}.$$
$$\qquad (21.61)$$

These nonlinear evolution equations will be investigated in the following.

21.2.2. Steady-State Solutions

At the steady state — i.e., in the adiabatic approximation — we have the algebraic equations

$$\widehat{\Pi}_{xy} = -\frac{2[\eta_0\gamma/q(\kappa)]}{\left(1 + \dfrac{8\eta_0^2\gamma^2}{3p^2q^2(\kappa)}\right)}, \qquad (21.62)$$

$$\widehat{N}_1 = -\frac{8[\eta_0\gamma/q(\kappa)]^2}{p\left(1 + \dfrac{8\eta_0^2\gamma^2}{3p^2q^2(\kappa)}\right)}, \qquad (21.63)$$

$$\widehat{N}_2 = \frac{8[\eta_0\gamma/q(\kappa)]^2}{p\left(1 + \dfrac{8\eta_0^2\gamma^2}{3p^2q^2(\kappa)}\right)}. \qquad (21.64)$$

From Eqs. (21.63) and (21.64) the relation between the primary and secondary normal stress differences follows:

$$\widehat{N}_1 = -\widehat{N}_2. \tag{21.65}$$

That is, $\widehat{N}_2$ and $\widehat{N}_2$ are opposite in sign and not independent at the approximations and assumptions made for the flow. It, however, should be noted that relation (21.65) holds under the flow condition and assumptions made, but it is not always true. To solve fully the steady constitutive equations, some further approximations must be made, since it is not possible to obtain exact solutions for the given form of Rayleigh dissipation function κ. For this purpose, the nonlinear factor $q(\kappa)$ is examined in more details below.

Since

$$\frac{2p}{\eta_0} = 2p\mathfrak{R}^{(11)}$$

(see Eq. (21.37)), we find that the Rayleigh dissipation function κ is given by the expression — quadratic in $\mathbf{\Pi}$:

$$\kappa = (\Phi^{(1)}\mathfrak{R}^{(11)}\Phi^{(1)})^{1/2} = \frac{1}{\sqrt{\eta_0}}\,(\mathbf{\Pi} : \mathbf{\Pi})^{1/2}\,, \tag{21.66}$$

which in the present case of flow configuration can be written as

$$\kappa = \sqrt{\frac{2}{\eta_0}}\sqrt{\Pi_{xy}^2 + \frac{1}{3}N_1^2}. \tag{21.67}$$

Therefore on substituting $\widehat{\Pi}_{xy}$ and $\widehat{N}_1$ obtained above for formal solutions, we obtain

$$\sinh^{-1}\kappa = \rho\sqrt{\frac{2}{\eta_0}}\,\frac{2\,(\eta_0\gamma)}{\left(1 + \frac{8\eta_0^2\gamma^2}{3p^2q^2}\right)}\left[1 + \frac{8}{p^2}\,(\eta_0\gamma/q)^2\right]^{1/2}. \tag{21.68}$$

This equation does not allow a simple solution for κ in terms of γ unlike the case of the simple non-Newtonian flow in the absence of normal stress differences. The normal stress differences, which are nonlinear with respect to γ, would not lead to simple solutions in terms of γ. So, we must resort to approximate solution methods.

Before we consider approximate solutions, we examine the steady-state equations. Since

$$N_2 = -N_1,$$

we have two independent equations for the components of shear stress tensor $\mathbf{\Pi}$ as we have noted earlier,

$$\frac{p}{\eta_0}\Pi_{xy}q(\kappa) = -2\gamma p - \frac{2}{3}\gamma N_1, \tag{21.69}$$

$$\frac{p}{\eta_0}N_1 q(\kappa) = 4\gamma\Pi_{xy}. \tag{21.70}$$

Eliminating $q(\kappa)$, there follows the equation

$$2\Pi_{xy}^2 = -\left(p + \frac{1}{3}N_1\right)N_1,$$

which is rearranged to a circle in the $[\Pi_{xy}, \frac{1}{\sqrt{6}}(N_1 + \frac{3}{2}p)]$ plane with a radius $\frac{\sqrt{3}}{2\sqrt{2}}p$.

$$\Pi_{xy}^2 + \frac{1}{6}\left(N_1 + \frac{3}{2}p\right)^2 = \frac{3}{8}p^2. \tag{21.71}$$

Approximate solutions of Eqs. (21.69) and (21.70) must satisfy this condition. Therefore, for example, N_1 must be such that

$$N_1 = -\frac{3}{2}p \pm \sqrt{\frac{9}{4}p^2 - 6\Pi_{xy}^2} \tag{21.72}$$

If p is such that the discriminant is negative, then the solution for N_1 becomes complex and hence inadmissible. If the shear stress and pressure are such that the discriminant in Eq. (21.72) is negative, the real solutions to the steady-state equations do not exist. Therefore, the stress circle means that the steady-state shear stress and primary normal stress difference N_1 are confined to a circle in the $[\Pi_{xy}, (N_1 + \frac{3}{2\sqrt{6}}p)]$ plane. Such a pressure p will be called the critical pressure

$$p_c := -\sqrt{\frac{8}{3}}\Pi_{xy}^c. \tag{21.73}$$

21.2.3. *Stress Circle (Ellipse)*

Let us examine the conclusion made with regard to Eqs. (21.71) and (21.72) from another viewpoint. The solutions to the stress circle (or ellipse) may be represented in terms of angle ϑ:

$$\Pi_{xy} = \frac{\sqrt{3}}{2\sqrt{2}} p \cos \vartheta, \tag{21.74}$$

$$N_1 + \frac{3}{2\sqrt{6}} p = \frac{\sqrt{3}}{2\sqrt{2}} p \sin \vartheta, \tag{21.75}$$

where ϑ is therefore defined by

$$\vartheta = \tan^{-1} \left(\frac{N_1 + \frac{3}{2\sqrt{6}} p}{\Pi_{xy}} \right), \tag{21.76}$$

which ranges from 0 to $\pi/2$. This angle characterizes the relative magnitude of the shear stress and the primary normal stress and at the same time their relative sense of orientations:

$$\kappa^2 = \frac{p^2}{4\eta_0} \left(\cos 2\vartheta - 2 \sin \vartheta + 3 \right). \tag{21.77}$$

Thus we see that if the excess primary normal stress over and above $(3p/2\sqrt{6})$ is perpendicular to the shear stress, then the dissipation function is simply given by the excessive normal stress alone, whereas if it is parallel, that is, when the primary normal stress difference is equal to $-\frac{3}{2\sqrt{6}} p$ and parallel to Π_{xy}, then the dissipation function is determined by the shear stress alone. The dissipation function thus varies between these two extreme conditions of stress tensor components (Fig. 21.2).

21.2.4. *Approximate Solutions for Steady-State Constitutive Equations*

The steady-state solution of the constitutive equations for Π_{xy} and N_1 are given by the equations

$$\frac{p}{\eta_0} \Pi_{xy} q(\kappa) = -2\gamma p - \frac{2}{3} \gamma N_1, \tag{21.78}$$

$$\frac{p}{\eta_0} N_1 q(\kappa) = 4\gamma \Pi_{xy}. \tag{21.79}$$

Since N_1 is second order, at least, with respect to γ, if the second term on the right of Eq. (21.78) is neglected, the approximate solution Π_{xy}^0 for the shear stress obeys

$$\Pi_{xy}^0 q(\kappa_0) = -2\eta_0 \gamma, \tag{21.80}$$

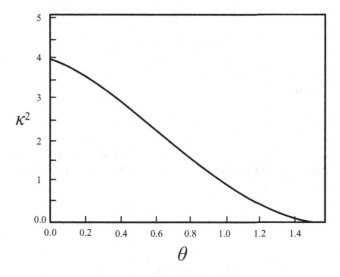

Fig. 21.2. Dissipation function κ^2 is plotted as a function of angle. The ordinate is in the units reduced by $(p^2/4\eta_0)$ and the abscisas is angle in the interval $0 \leq \vartheta \leq \pi/2$.

where

$$q(\kappa_0) = \frac{\sinh \kappa_0}{\kappa_0}, \tag{21.81}$$

and the Rayleigh dissipation function κ_0 under the assumptions made for the evolution equations under consideration is given by the form

$$\kappa_0 = \sqrt{\frac{2}{\eta_0}} \left| \Pi_{xy}^0 \right|. \tag{21.82}$$

Therefore Eq. (21.80) now can be written

$$\sinh \kappa_0 = -(\pm)2\sqrt{2\eta_0}\gamma.$$

Inverting the hyperbolic sine function gives the Rayleigh dissipation function in terms of γ (i.e., velocity gradient):

$$\kappa_0 = \sinh^{-1}[-(\pm)2\sqrt{2\eta_0}\gamma]. \tag{21.83}$$

Finally, with Eq. (21.82) we obtain Π_{xy}^0 in the approximation neglecting the normal stress effect:

$$\Pi_{xy}^0 = -\sqrt{\frac{\eta_0}{2}} \sinh^{-1}(2\sqrt{2\eta_0}\gamma)$$

$$:= -2\eta_0\gamma q_e, \tag{21.84}$$

where

$$q_e = \frac{\sinh^{-1}\left(2\sqrt{2\eta_0}\gamma\right)}{2\sqrt{2\eta_0}\gamma}. \tag{21.85}$$

Thus we see that

$$q_e = q\left(\kappa\right)\big|_{\kappa=2\sqrt{2\eta_0}\gamma}. \tag{21.86}$$

Therefore, if $q(\kappa)$ is approximated to a form in the lowest order of γ, the approximate solutions for the steady-state constitutive equations are obtained in the forms:

$$\Pi_{xy} = -\frac{2\eta_0 q_e}{\left(1 + \frac{8\eta_0^2\gamma^2}{3p^2 q_e^2}\right)}\gamma, \tag{21.87}$$

$$N_1 = -\frac{8(\eta_0 q_e)^2}{p\left(1 + \frac{8\eta_0^2\gamma^2}{3p^2 q_e^2}\right)}\gamma^2. \tag{21.88}$$

Now define the non-Newtonian shear viscosity η and primary normal stress coefficient Ψ_1 by the constitutive relations

$$\Pi_{xy} = -2\eta\gamma, \tag{21.89}$$

$$N_1 = -4\Psi_1\gamma^2, \tag{21.90}$$

where the non-Newtonian shear viscosity and primary normal stress different coefficient are given by the formulas

$$\eta = \frac{\eta_0 q_e}{(1 + 6s^2)}, \tag{21.91}$$

$$\Psi_1 = \frac{2(\eta_0 q_e)^2}{p(1 + 6s^2)}, \tag{21.92}$$

where

$$s = \frac{2\eta_0\gamma}{3p q_e}. \tag{21.93}$$

Since $N_1 = -N_2$ according to Eq. (21.65), if we define the secondary normal stress different coefficient Ψ_2 by

$$N_2 = 4\Psi_2\gamma^2, \tag{21.94}$$

then Ψ_2 is given by the formula

$$\Psi_2 = -\frac{2(\eta_0 q_e)^2}{p(1 + 6s^2)}. \tag{21.95}$$

As is the primary normal stress difference coefficient, it is also, at least, of second order in the shear rate γ, but negative. However, as they stand in Eqs. (21.92) and (21.95), both primary and secondary difference coefficients are of higher than the second order in γ and hence highly non-Newtonian just as the shear viscosity η owing to the presence of the factor $q_e(\gamma)$. It should be noted that $q_e(\gamma) \to 0$ as $\gamma \to \infty$. On the other hand, $q_e(\gamma) \to 1$ as $\gamma \to 0$. Consequently, as $\gamma \to 0$ we have

$$\eta \to \eta_0,$$

$$\Psi_1 \to 2\eta_0^2, \tag{21.96}$$

$$\Psi_2 \to -2\eta_0^2.$$

These limiting behaviors are found consistent with experiments.[18] For related discussion see also the article[19] by Ohr and Eu.

21.3. The Knudsen Problem

21.3.1. *Introductory Remarks*

The Maxwell kinetic theory induced Kundt and Warburg[20] to carry on experiment on rarefied gas viscosity. Their experimental result in turn prompted Maxwell[21] to examine his kinetic theory of gas transport processes deeper than his original theory and to propose the notion of *velocity slip* at the boundaries when a rarefied gas flows through a tube. Influenced by the Kundt–Warburg experiment and Maxwell's theory of slip flow, Knudsen[22] performed a number of experiments on rarefied gases and

---

[18] For closely related discussions on rheological questions, see B. C. Eu and R. E. Khayat, *Rheologica Acta* **30**, 204 (1991), where the evolution equations are developed and tested against experimental data on shear stress and normal stress difference.

[19] Y. G. Ohr and B. C. Eu, *Phys. Lett. A* **101**, 338 (1984) where the normal stress effects are studied for various complex fluids.

[20] A. Kundt and E. Warburg, *Poggendorff's Ann. Phys. (Leipzig)* **155**, 337, (1875); **155**, 525 (1875).

[21] J. C. Maxwell's paper on slip flow in Collected Works of J. C. Maxwell (Cambridge U. P. London, 1927), Vol. 2, p. 682.

[22] M. Knudsen, *Ann. Phys.* **28**, 75 (1909); M. Knudsen, *The Kinetic Theory of Gases* (Methuen, London, 1934).

reported in 1909 on his experimental investigations into the validity of Hagen–Poiseuille (HP) volume flow rate for rarefied gas flow through a long circular tube. According to the experiments, he found that, although the volume flow rate follows the Hagen–Poiseuille flow rate if the mean pressure is in the range of normal gas pressure, it does not obey the HP flow rate, which predicts the flow rate should vanish. But he instead found that it increases after reaching a minimum value as the pressure decreases toward zero. In fact, he was able to confirm his data for the volume flow rate per unit pressure difference Q_k to follow the empirical formula

$$Q_k = ap + b\frac{1 + c_1 p}{1 + c_2 p}, \tag{21.97}$$

where

$$a = \frac{\pi R^4}{8L\eta_0}, \tag{21.98}$$

$$b = \frac{4\sqrt{2\pi}}{3L\sqrt{\rho_1}}, \tag{21.99}$$

c_1 and c_2 are numerical constants depending on the nature of gas molecule, and R and L denote the radius and length of the tube, η_0 the shear viscosity of the gas, ρ_1 the specific density of the gas at temperature T when the pressure is equal to 1 dyn/cm^2, and p the mean gas pressure. The first term on the right of the empirical formula (21.97) represents the HP flow rate and the second term is responsible for giving the minimum of Q_k vs. p. The ultimate validity of the second term as p tends to zero requires further investigation. His experiment thus induced other research workers to investigate the reported phenomenon. For example, Gaede[23] confirmed his prediction of the presence of minimum in flow rate, but Gaede also found that it does not tend to a finite value as p vanishes, but exhibits a logarithmic increase as $p \to 0$. Since Knudsen's experimental discovery contradicts the (Navier–Stokes) hydrodynamic theory prediction, the appearance of the minimum was referred to as the Knudsen paradox. However, it seems more appropriate to call it the Knudsen problem because it may not be a paradox at all from a broader or more precise result in hydrodynamics extending the Navier–Stokes theory of hydrodynamics to flows in the rarefied gas regime where the mean free path is long or, put in another way, the Knudsen number is large. Thus one might say that the gas is displaced far from equilibrium.

---

[23]W. Gaede, *Ann. Phys. Ser.* (4) **41**, 289 (1913).

In this section, we investigate the Knudsen problem[24] from the viewpoint of generalized hydrodynamics described earlier in this work.

21.3.2. *Generalized Hydrodynamics*

We assume that a gas flow is laminar in a circular tube of radius R and length L under a longitudinal pressure gradient p. The pressure difference between the entrance and exit of the tube is denoted by $\Delta p := p_i - p_f$ where p_i is the pressure at the entrance and p_f the pressure at the exit of the tube. The gas is maintained at a constant uniform temperature. Therefore, there is no heat flux involved. We also assume[25] that there is no normal stress differences present.

Since there is axial symmetry present in the flow, the appropriate coordinates to take are cylindrical coordinates. We denote them by (r, θ, z), where the direction of flow is taken z; that is, the axis of tube is parallel to the z-axis of the coordinate system and the flow is cylindrically symmetric around the z-axis. Since the length of the tube is assumed sufficiently long, so that the end effect of the tube is negligible, the flow properties and, in particular, the fluid velocity components are independent of z. They are also independent of polar angle θ because of the axial symmetry of the flow. The flow is assumed to be in the positive direction of the axial coordinate z. The angular component of velocity **u** is equal to zero owing to the absence of a rotary motion of the gas around the z-axis. The radial velocity component u_r is equal to zero for the following reason.

Since we are interested in a steady flow, the steady-state equation of continuity is

$$\nabla \cdot (\rho \mathbf{u}) = 0. \tag{21.100}$$

In the cylindrical coordinate adopted, this equation takes form

$$\frac{1}{r}\frac{\partial}{\partial r}(\rho r u_r) + \frac{1}{r}\frac{\partial}{\partial \theta}(\rho u_\theta) + \frac{\partial}{\partial z}(\rho u_z) = 0. \tag{21.101}$$

Since the density ρ and the velocity components u_θ and u_z do not depend on z and θ for the reasons mentioned earlier, we obtain from Eq. (21.101)

$$\frac{\partial}{\partial r}(\rho r u_r) = 0,$$

---

[24]B. C. Eu, *Phys. Rev.* **40**, 6395 (1989).
[25]This assumption simplifies the analysis of the Knudsen problem. Removal of this assumption will be discussed elsewhere.

which on integration yields

$$\rho r u_r = \text{constant.}$$

Since at the boundary $r = R$ (i.e., tube wall) $u_r = 0$, we conclude that $u_r = 0$ everywhere specially since density $\rho \neq 0$. Since $u_\theta = 0$ as noted earlier, we conclude that the flow velocity $\mathbf{u}$ has a z-component only:

$$\mathbf{u} = (0, 0, u_z). \tag{21.102}$$

We define the normal stress difference N_1 and N_2 in terms of traceless symmetric shear stress tensor $\mathbf{\Pi}$:

$$N_1 = \Pi_{zz} - \Pi_{rr}, \quad N_2 = \Pi_{rr} - \Pi_{\theta\theta}, \tag{21.103}$$

where Π_{rr}, $\Pi_{\theta\theta}$, and Π_{zz} are the normal components of $\mathbf{\Pi}$. We will denote the symmetric off-diagonal elements Π_{rz} and Π_{zr} by a scalar symbol Π:

$$\Pi := \Pi_{rz} = \Pi_{zr}. \tag{21.104}$$

The component Π is the only nonvanishing off-diagonal component of the stress tensor in the flow configuration considered. Then the components of the momentum balance equation are as follows:

$$-\frac{\partial p}{\partial r} - \frac{1}{3r}\frac{\partial}{\partial r}r(N_2 - N_1) - \frac{1}{3r}(2N_2 + N_1) = 0, \tag{21.105}$$

$$-\frac{\partial p}{\partial z} - \frac{1}{r}\frac{\partial}{\partial r}r\Pi = 0. \tag{21.106}$$

Note that these are exact balance equations.

Since the normal stress differences are assumed to be absent for the present consideration, by setting $N_1 = N_2 = 0$ we obtain from Eq. (21.105) the equation

$$\frac{\partial p}{\partial r} = 0. \tag{21.107}$$

Therefore, the pressure is seen independent of r; that is, there is no pressure variation across the tube. Furthermore, since the pressure gradient in the z direction is assumed constant, we find

$$p = -\frac{\Delta p}{L}z + p_i. \tag{21.108}$$

In the adiabatic approximation, the quasilinear evolution equation for the shear stress Π for the flow is given by

$$\Pi q(\kappa) = -2\eta_0\gamma, \tag{21.109}$$

where

$$\gamma = \frac{1}{2}\frac{\partial u_z}{\partial r}, \tag{21.110}$$

$$q(\kappa) = \frac{\sinh\kappa}{\kappa}, \tag{21.111}$$

$$\kappa = \frac{\tau\Pi}{\eta_0}, \tag{21.112}$$

$$\tau = \frac{\sqrt{2\eta_0(m_r k_B T/2)^{1/2}}}{\sqrt{2}n k_B T\sigma}. \tag{21.113}$$

In the equations presented above, η_0 is the Newtonian shear viscosity of the gas. It may be calculated in terms of molecular theory parameters from the first-order Chapman–Enskog formula[26] of the Boltzmann kinetic theory. The symbols n, σ, and m_r stand for number density, size parameter, and reduced mass of the gas molecule.

It is useful to note that the constitutive equation (21.109) for Π was in fact shown in a previous section on non-Newtonian flow to give rise a non-Newtonian viscosity agreeing with experiment in which the normal stress effects are absent. Its experimental relevance has been already established. The present investigation based on the constitutive equation is therefore deemed well founded. We have already shown that the constitutive equation (21.109) together with the momentum balance equation, subject to the boundary conditions

$$u_z(R) = 0, \quad \left(\frac{\partial u_z}{\partial r}\right)_{r=R} = 0, \tag{21.114}$$

gives rise to the velocity profile

$$u_z(r) = \frac{r}{\tau\delta}\left[\cosh\delta - \cosh\left(\frac{r\delta}{R}\right)\right], \tag{21.115}$$

---

[26] See Chapman and Enskog in the footnote cited in earlier chapter.

where

$$\delta = \frac{\tau R \Delta p}{2L\eta_0}. \tag{21.116}$$

We also find that by the boundary condition — which, in fact, is a stick boundary condition — on the velocity derivative at $r = R$

$$\Pi = \frac{r\Delta p}{2L}. \tag{21.117}$$

The velocity profile (21.115), therefore, gives rise to the parabolic HP velocity profile as the parameter δ gets small.

21.3.3. *Flow Rate Through a Tube*

The number of molecules flowing through the tube per unit time Q is given by the formula

$$Q = 2\pi n \int_0^R dr r u_z(r). \tag{21.118}$$

On substitution of the velocity profile (21.115) the integral (21.118) is easily integrated and we obtain the flow rate Q:

$$Q = Q_{HP}\frac{n\pi R^4 \Delta p}{8L\eta_0}(1 + \Delta Q)p,$$

where Q_{HP} is the Hagen–Poiseuille flow rate

$$Q_{HP} = \frac{\pi R^4 \Delta p}{8L\eta_0 \mathcal{R}T} \tag{21.119}$$

and ΔQ is a correction factor to Q_{HP} defined by the formula

$$\Delta Q = 4\delta^2 \left[\sinh^2\left(\frac{\delta}{2}\right) + \left[\cosh\left(\frac{\delta}{2}\right) - \frac{2}{\delta}\sinh\left(\frac{\delta}{2}\right)\right]^2 \right] - 1. \tag{21.120}$$

It is a correction arising from the non-Newtonian flow behavior of the gas. It is a consequence of the constitutive equation assumed for the shear stress Π under the assumption of vanishing normal stress differences. Within the framework of the assumptions taken for the analysis, it is an exact result for the correction factor to the flow rate. It should be recalled that there is no interaction between the boundary wall and the gas molecules — namely, there is no slip taken into account in the calculation of flow profiles. It is important to remark that, although phenomenologically treated here, the constitutive equations including the nonconserved variable evolution equations employed can be derived from the appropriate kinetic equation, if so desired. It is also important to recognize the effect of interaction between the molecules of the boundary wall on the gas molecules, which has been ignored. This is an approximation the effect of which may not be negligible in the low gas density regime. It may affect the limiting behavior of Q as $p \to 0$ shown below.

Now, let us examine the limiting behavior of the result obtained for the flow rate. Since δ is proportional to the parameter τ, which is in turn inversely proportional to pressure, it is easy to infer that, as $p \to 0$,

$$Q \to p \exp\left(\delta_0/p\right), \qquad (21.121)$$

where

$$\delta_0 = \frac{\tau_0 R \Delta p}{2L\eta_0}, \qquad (21.122)$$

$$\tau_0 = \frac{\sqrt{\eta_0 (2m_r k_B T)^{1/2}}}{\sqrt{2}\sigma}. \qquad (21.123)$$

Since Q is linear with respect to p as p increases toward the normal pressure, the limiting behavior implies that there exists a minimum in Q at some value of p in the neighborhood of $p = 0$. The location of the minimum is given by

$$\frac{d}{d\delta}\left\{\delta^{-3}\left[\frac{1}{2}\cosh\delta + \delta^{-2}(\cosh\delta - 1) - \delta^{-1}\sinh\delta\right]\right\} = 0, \qquad (21.124)$$

which yields a transcendental equation for δ_m at the minimum:

$$\delta_m(10 + \delta_m^2)\sinh\delta_m - 5(2 + \delta_m^2)\cosh\delta_m + 10 = 0. \qquad (21.125)$$

Thus the pressure at the minimum is given by

$$p_m = \frac{\delta_0}{\delta_m}. \qquad (21.126)$$

In connection with the limiting behavior of Q given by (21.121), it should be noted that it is not so inconsistent with Gaede's experiment mentioned before, which showed a logarithmic divergence as p tends to zero instead of Knudsen's result.

Despite the feature appearing qualitatively correct in comparison with Knudsen's experimental result, we are not able to make a quantitative comparison with Knudsen's experimental result for the following reason: It is found that the Δp-dependence is quite significant since the flow rate depends on it sensitively. Knudsen, however, reduced the flow rate with Δp in presenting his data without stating the Δp values taken in his experiment. This would be acceptable if the flow depends linearly on Δp and the scaled result is independent of Δp, but in the present theory the flow rate is not linear with respect to Δp, but a rather complicated function of Δp. Therefore, the scaled function cannot be calculated if Δp is not known more precisely. For this reason, we can only make a qualitative numerical comparison of the present theoretical result with his experimental result only if δ_0 is suitably chosen. We find the present theoretical result predicts only the presence of a minimum in the flow rate vs. pressure curves, as Gaede's experiment did. Such a comparison is made in Fig. 21.3 for carbon dioxide gas. In this figure $\overline{Q} := Q\mathcal{R}\mathcal{T}/\Delta\,p$. For this particular comparison, the parameters a, b, c_1, and c_2 of the Knudsen's empirical formula are as follows:

$$a = 0.04880/\text{cm Hg}, \quad b = 0.03489,$$
$$c_1 = 43.13/\text{cm Hg}, \quad c_2 = 53.10/\text{cm Hg}.$$

The figure shows that the present theory clearly predicts the presence of a minimum in flow rate, and we see that it arises from the non-Newtonian

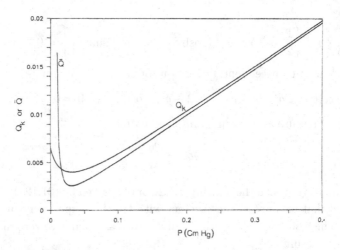

Fig. 21.3.   Reduced flow rate Q_k measured by Knudsen and theoretical reduced flow rate calculated for $\delta_0 = 0.07$. The theoretical flow rate agrees only qualitatively with the experiment. Reproduced with permission from B. C. Eu, *Phys. Rev. A* **40**, 6395 (1989) © American Physical Society.

behavior of the gas when the flow condition is such that the Knudsen number is large and consequently the gas is removed far from equilibrium.

21.3.4. *Application to Light-Induced Viscous Flow in a Gas*

A number of experimental studies have been made in which atoms or molecules in a gas are successfully manipulated by light. In such experiments, a narrow-band laser is tuned within a Doppler-broadened absorption line of an atomic or molecular gas, thereby inducing velocity-selective excitation of atoms or molecules. Thus drifts of particles are produced to give rise to a number of interesting effects, such as an optical piston,[27] light-induced viscous flow arising from collisions in the bulk flow of a gas[28] and with surface.[29] These effects must be distinguished from those originating from the photon pressure, since they are much smaller than the light-induced effects mentioned earlier since the collisions between the matter particles are intimately related to the light-induced effects and much larger than the effects due to the collisions with photons.

In the following discussion, we would like to discuss specifically the experimental result by Hoogeveen *et al.*,[30] who reported on an experiment on a gas flow induced in a capillary by a Doppler-broadened laser beam of radially distributed intensity. The flow generates a pressure difference and a stress in the capillary. By measuring the pressure difference that the flow produces in the capillary by means of a differential manometer and by plotting it against the gas pressure, they were able to show that the normal-pressure-regime behavior follows the classical Hagen–Poiseuille flow rate as expected of gas flow in a capillary under the normal pressure regime. Therefore, light-induced flow can be described by the classical Navier–Stokes hydrodynamics in the normal pressure regime. However, as the gas pressure is decreased down to the rarefied gas density regime, the curves for the

[27]H. G. C. Werij, J. P. Woerdman, J. J. Beenakker and I. Kuscer, *Phys. Rev. Lett.* **52**, 2237 (1984).

[28]V. N. Panfilov, V. P. Strunin and P. L. Chapovskii, *Zh. Eksp. Teor. Fiz.* **85**, 881 (1983) [*Sov. Phys. JETP* **58**, 510 (1983)]; H. G. C. Werij, J. E. M. Haverkort, P. C. M. Planken, E. R. Eliel, J. P. Woerdman, S. N. Atutov, P. L. Chapovskii and F. Kh. Gel'mukhanov, *Phys. Rev. Lett.* **58**, 2660 (1987).

[29]R. W. M. Hoogeveen, R. J. C. Spreeuw and L. J. F. Hermans, *Phys. Rev. Lett.* **59**, 447 (1987).

[30]R. W. M. Hoogeveen, G. J. der Meer, L. J. F. Hermans, A. V. Ghiner and I. Kuscer, *Phys. Rev. A* **39**, 5539 (1989).

pressure difference vs. pressure started to decrease in contrast to the Hagen–Poiseuille theory prediction, which, in fact, exhibits a maximum appearing in the curve at some low pressure value. This unexpected behavior of the pressure difference vs. pressure curve can be adequately accounted for if we apply the generalized hydrodynamics for a non-Newtonian behavior of the gas. If we put the conclusion of this investigation first, *the observed behavior of the pressure difference vs. pressure in question is a manifestation of the Knudsen effect, which can be adequately described by generalized hydrodynamics formalism in the previous section.*

Since the experiment was done in a pure gas under a uniform constant temperature, it is sufficient to consider only the stress evolution equation together with the momentum balance equation and mass balance equation. We assume that the molecule has only two internal states and the molecules of the two different internal states are treated as two different species as far as their statistical mechanical (kinetic theory) treatment of the system goes. Therefore, three species of particles — two matter species in two different internal states and photons — are involved in the kinetic theory of the processes we are going to consider. With this model, we can apply the kinetic theory of radiation and matter[31] we have developed elsewhere along the same line as we have treated in the previous chapters in this book. Then in the first-order cumulant approximation for the dissipation terms in the quasilinear nonconserved variable evolution equations the quasilinear stress tensor evolution equation for species a (a = state 1, state 2, and photon r) is given by

$$\rho \frac{d}{dt}\widehat{\Pi}_a = -2p_a\gamma - \sum_{a=1}^{r} p_a \mathcal{B}_{ab}\Pi_b q(\kappa), \qquad (21.127)$$

where, as in the previous sections, $\widehat{\Pi}_a = \Pi_a/\rho$ and $\mathcal{B}_{ab}$ are collision bracket integrals, which may be treated as empirical parameters characterizing linear transport coefficients of the system consisting of matter and radiation. The symbol γ stands for the traceless symmetric part of the velocity gradient defined by

$$\gamma = \tfrac{1}{2}[\nabla\mathbf{u} + (\nabla\mathbf{u})^t] - \tfrac{1}{3}\delta\nabla\cdot\mathbf{u}, \qquad (21.128)$$

with $\mathbf{u}$ denoting hydrodynamic velocity and δ the unit second rank tensor. As a matter of fact, γ will turn out to be the shear rate. The nonlinear

---

[31]See B. C. Eu and K. Mao, *Physica A* **180**, 65 (1992); *ibid.* **184**, 187 (1992).

factor $q(\kappa)$ is defined by

$$q(\kappa) = \frac{\sinh \kappa}{\kappa} \tag{21.129}$$

with κ denoting a dissipation function defined by

$$\kappa = \frac{\sqrt{\eta_0 \left(2 m_r k_B T\right)^{1/2}}}{n k_B T \sigma} \sqrt{|\mathbf{\Pi} : \mathbf{\Pi}|}, \tag{21.130}$$

since there are no other nonconserved fluxes present. Here m_r is the mean reduced mass and σ the mean size parameter of the molecules. The nonlinear factor is a result of the first-order cumulant approximation for the calortropy production defined in terms of the collision integral of the kinetic equation. In any case, the nonlinear factor and the dissipation function κ may be taken as empirical inputs, but we stress that they have firm kinetic theoretic foundations; see the papers by Eu and Mao cited in footnote 31 cited earlier and the recent monograph by Eu.[32]

In the experiment under consideration here, since the light is in equilibrium and acts as a driving force of the transitions in the internal state of molecules in interaction with it, the evolution equation for the shear stress of radiation is absent in the set of evolution equations (21.127). Therefore, there are only two independent nonconserved variable evolution equations for the shear stress to consider, and the index a in Eq. (21.127) is limited to $a = 1$ and 2. Consequently, the $\mathbf{\Pi}_r$ term in Eq. (21.127) plays the role of the driving force for the internal state transition, being the stress generated by the light (laser), which is proportional to the light intensity distributed radially across the cross section of the capillary. The desired relation of $\mathbf{\Pi}_r$ to the light intensity is provided by a kinetic theory consideration footnote 31 cited earlier. As a matter of fact, $\mathfrak{B}_{ab}$ are related to the gas viscosity η_0 in the following sense:

$$\eta_0 = \frac{\mathfrak{B}_{11} + \mathfrak{B}_{22} - 2\mathfrak{B}_{12}}{\mathfrak{B}_{11}\mathfrak{B}_{22} - \mathfrak{B}_{12}^2} \tag{21.131}$$

and $\mathfrak{B}_{rr}$, $\mathfrak{B}_{1r}$, $\mathfrak{B}_{2r}$ to the photon (radiation) viscosity. Since experiment was performed under steady-state conditions, it is sufficient to consider the steady-state equations (i.e., the adiabatic approximation to the evolution

---

[32]B. C. Eu, *Kinetic Theory of Nonequilibrium Ensembles, Irreversible Thermodynamics, and Generalized Hydrodynamics*, Vols. 1 and 2 (Springer, Switzerland, 2016).

equations)

$$2p_1\gamma + p_1(\mathcal{B}_{11}\Pi_1 + \mathcal{B}_{12}\Pi_2 + \mathcal{B}_{1r}\Pi_r)q(\kappa) = 0, \tag{21.132}$$

$$2p_2\gamma + p_2(\mathcal{B}_{21}\Pi_1 + \mathcal{B}_{22}\Pi_2 + \mathcal{B}_{rr}\Pi_r)q(\kappa) = 0. \tag{21.133}$$

Solving these equations algebraically for Π_1 and Π_2 and adding them, we obtain the total stress Π applied on matter

$$\Pi = \Pi_1 + \Pi_2$$
$$= -2\eta_0\gamma + \Pi_L, \tag{21.134}$$

where

$$\Pi_L = -\frac{[\mathcal{B}_{1r}(\mathcal{B}_{22} - \mathcal{B}_{12}) + \mathcal{B}_{2r}(\mathcal{B}_{11} - \mathcal{B}_{12})]}{\mathcal{B}_{11}\mathcal{B}_{22} - \mathcal{B}_{12}^2}\Pi_r. \tag{21.135}$$

In the case of the flow geometry for the experiment, a cylindrical coordinate is appropriate to take for a circular tube with the gas flow direction taken parallel to the positive axial direction. If we further assume that the normal stress differences are absent to an approximation, then owing to the axial symmetry of flow the steady-state equation of continuity is trivially integrated and we find that the flow velocity has only the z-component:

$$\mathbf{u} = (0, 0, u_z), \tag{21.136}$$

where u_z is a function of r, the radial distance from the axis of flow.

Since the experiment is performed such that the flow generated by the pressure difference induced by the laser is exactly balanced by the opposing pressure difference in the differential manometer, the matter part Π of the stress is equal to zero. Therefore, we obtain from Eq. (21.134) the equation

$$\Pi_L = 2\eta_0\gamma q^{-1}(\kappa), \tag{21.137}$$

where, in view of Eq. (21.136) for the flow velocity, the shear rate γ is now given by the derivative

$$\gamma = \frac{1}{2}\frac{\partial u_z}{\partial r} = \frac{1}{2}\frac{\partial u(r)}{\partial r}. \tag{21.138}$$

Equation (21.137) can be manipulated and solved for Π_L in terms of γ in the same manner as for one-dimensional non-Newtonian flow discussed earlier in the section for the non-Newtonian flow, and the velocity profile of the non-Newtonian flow can be obtained subject to the boundary conditions on the velocity:

$$u(r) = \eta_0^{-1} \int_R^r dr \Pi_L(r). \tag{21.139}$$

Upon substituting Π_L and the resulting $u(r)$ into the flow rate $Q(I)$ defined by the integral

$$Q(I) = 2\pi \int_0^R dr\, r\, u(r)$$

$$= \pi \eta_0^{-1} \int_R^0 dr\, r^2 \Pi_L(r), \tag{21.140}$$

we are able to calculate the flow rate. It is convenient to write the integral (21.140) in the form

$$\frac{\pi R^3 \nu \bar{n}_e}{n} = \pi \eta_0^{-1} \int_R^0 dr\, r^2 \Pi_L(r), \tag{21.141}$$

where $\bar{n}_e$ is the density of the excited state of the molecule, n is the total density, and ν is a parameter of the dimension of inverse time related to the collision frequency, which, in the kinetic theory underlying the present theory, is identifiable by collision bracket integrals. As a matter of fact, this aim can be easily achieved if Π_L in Eq. (21.141) is expressed with the kinetic theory expression given in terms of collision bracket integrals. (See the kinetic theory references cited earlier in footnote 31.) Since we are interested in the pressure dependence of the flow rate, it will be sufficient to treat the integral as a semiempirical parameter.

Balancing the light-induced flow with the pressure difference in the differential manometer is equivalent to opposing the flow with another flow generated by the same pressure difference in the absence of light. Therefore, $Q(I)$ may be equated to the flow rate Q_0 arising in the Knudsen flow in the absence of light. Since Q_0 has been already calculated in the previous section on the Knudsen problem, we may simply use it here. Thereby we obtain the expression

$$Q_0(\Delta p) = \left(\frac{2\pi}{\tau\delta}\right)\left[\frac{1}{2}\cosh\delta + \delta^{-2}(\cosh\delta - 1) - \delta^{-1}\sinh\delta\right]$$

$$:= Q_{\text{HP}}(1 + \Delta Q), \tag{21.142}$$

where various parameters are defined below:

$$\delta = \frac{\tau R \Delta p}{2L\eta_0},$$ (21.143)

$$\tau = \frac{\sqrt{\eta_0}(mk_BT/2)^{1/2}}{nk_BT\sigma},$$ (21.144)

$$Q_{HP} = \frac{\pi R^4}{8L\eta_0}\Delta p,$$ (21.145)

$$\Delta Q = 8\delta^{-2}[\tfrac{1}{2}\cosh\delta + \delta^{-2}(\cosh\delta - 1) - \delta^{-1}\sinh\delta] - 1.$$ (21.146)

Note that Q_{HP} is the Hagen–Poiseuille flow rate, Δp is the pressure difference, m is the reduced mass of the molecules, L is the length of the capillary, and σ is the size parameter of the molecule. It is clear that $Q_0(\Delta p)$ tends to Q_{HP} as δ vanishes; in other words, the Navier–Stokes theory limit. Now equating $Q_0(\Delta p)$ with $Q(I)$, we obtain

$$\frac{\pi R^3 \nu \bar{n}_e}{n} = \left(\frac{2\pi}{\tau\delta}\right)\left[\frac{1}{2}\cosh\delta + \delta^{-2}(\cosh\delta - 1) - \delta^{-1}\sinh\delta\right]$$ (21.147)

$$= Q_{HP}(1 + \Delta Q).$$

To compare the theoretical result with experiment, this equation must be solved for Δp. It may be solved by an iterative method. The lowest-order approximation for Δp is obtained by setting $\Delta Q = 0$. We thereby obtain the lowest-order result

$$\left(\frac{\Delta p}{p}\right)_0 = \frac{8L\eta_0\nu}{R}\left(\frac{\bar{n}_e}{np}\right) := C\left(\frac{\bar{n}_e}{np}\right).$$ (21.148)

This is the classical hydrodynamic result and gives a linear relationship between $(\Delta p/p)_0(\bar{n}_e/n)$ and p^{-1} as observed in the experiment in the normal pressure regime. By using this result (21.148) as the zeroth iterate and substituting it into Eq. (21.147) we obtain the first-order iterate:

$$\left(\frac{\Delta p}{p}\right)_1 = \frac{1}{4}Cb^4\left\{b^2p^3\sinh^2\left(\frac{b}{2p}\right) + p^3\left[b\cosh\left(\frac{b}{2p}\right)\right. \right.$$
$$\left.\left. - 2p\sinh\left(\frac{b}{2p}\right)\right]^2\right\}^{-1},$$ (21.149)

where

$$b = \frac{\sqrt{\eta_0 \left(mk_BT/2\right)^{1/2}}}{nk_BT\sigma} \left(\frac{RC}{2L\eta_0\sigma}\right). \tag{21.150}$$

The process of iteration can be continued, and the sequence converges rapidly. Therefore the first iterate is found sufficient in practice. Formula (21.150) is the main theoretical result of this section. The subscript 1 will be dropped with the understanding that the first iterative solution is meant for $\left(\frac{\Delta p}{p}\right)$.

The theoretical result obtained above is tested by using the experimental data of Hoogeveen *et al.* The parameter C is fixed with the data point in the high pressure regime and Δp is calculated with Eq. (21.149). The numerical result obtained is plotted in Fig. 21.4. In the normal pressure regime it gives the HP flow rate (the broken line), but as the Δp value decreases it starts to deviate from the linear relation and exhibits a maximum around $p \approx 20$ Pa. The appearance of a maximum is intimately related to the non-Newtonian behavior of the gas as the Knudsen number increases and

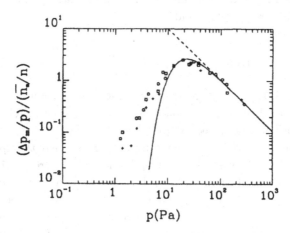

Fig. 21.4. Scaled pressure difference vs. pressure for light-induced flow in a capillary. The broken line is the Navier–Stokes theory prediction and the solid curve is the prediction made using the present theory. The squares, circles, and crosses are the experimental data for the $Q(12,2)$, $Q(12,3)$, $Q(12,2)$ transitions, respectively. The $Q(12,2)$ data were obtained with a stainless-steel capillary, whereas a quartz capillary was used for $Q(12,3)$ and $Q(12,2)$ data. [See R. W. M. Hoogeveen, G. J. der Meer, L. J. F. Hermans, A. V. Ghiner, and I. Kuscer, *Phys. Rev. A* **39**, 5539 (1989).] Reproduced with permission from K. Mao and B. C. Eu, *Phys. Rev. A* **48**, 2471 (1993) © American Physical Society.

thereby the gas is increasingly removed from equilibrium; in other words, we are seeing another example of the Knudsen effect. However, the theory does not exactly agree with the experiment in the post-maximum region except for predicting the existence of a maximum at correct position and magnitude.

There are a number of factors that may be attributed to the discrepancy in the post-maximum region. An important factor may be attributed to the neglect of normal stress differences, but inclusion of the normal stress differences leads to a mathematical complication which could be treated only by a numerical solution method that we have tried to avoid in this investigation. A rough estimate of the normal stress differences appears to give the value of pressure p appearing in the final expression modified. Therefore such a correction might improve the numerical results. Another important source of the deviation experiment may be attributed to the neglect of interaction of the gas molecules with the boundary wall. Nevertheless, we are content with the present result because it enables us to understand in physical terms the light-induced flow studied by means of generalized hydrodynamics. Further improvement may be left for future investigation in which we may eliminate some of the approximations and assumptions taken for the present study made.

21.4. Shock Waves

Shock wave propagation phenomena are another important area of practical applications for the nonlinear (non-Newtonian) constitutive relations within the framework of generalized hydrodynamics. A shock wave is described by a set of conserved variables, denoted by $\{\phi_i\}$ that satisfy the upstream and downstream boundary conditions $\{\phi_i^{us}\}$ and $\{phi_i^{ds}\}$, respectively. Shock wave propagation phenomena have been investigated by employing the Navier–Stokes theory of hydrodynamics, but the theory does not quantitatively predict the observed shock wave structure, but it is important to have a quantitative understanding of the phenomena for various practical engineering purposes. For the reason of theoretical and practical importances of the phenomena, the problem has attracted enduring attention in the literature, some of them with some results worth further consideration, but it is safe to state that most of the theories employed may be said thermodynamically not consistent. For example, the Burnett-order solution by the Chapman–Enskog solution of the Boltzmann equation does yield macroscopic evolution equations that do not obey the second law of thermodynamics, although they appear to give somewhat better numerical results for shock structures. Similar reasons one way or another may be given to other theories.

In a series of papers[33] based on the generalized hydrodynamic theory a thermodynamically consistent theory of shock wave propagation was reported by the present authors. In this subsection, we quickly review the gist of studies made on the subject by using the thermodynamically consistent constitutive equations presented earlier for non-Newtonian fluids. More details of the results presented here are available in the articles reported in the literature. The theory has also been reviewed to a considerable detail in a monograph on nonequilibrium statistical mechanics[34] by Eu to which the reader is referred.

We assume that the flow is in the direction of the x coordinate of a space-fixed coordinate system. Since we are interested in a steady propagation of shock wave, the governing balance equations of the conserved variables $\{\phi_i\} := (\rho, u, \mathcal{E})$ are time-independent and given in the forms

$$\frac{d}{dx}\rho u = 0, \tag{21.151}$$

$$\frac{d}{dx}(\rho u^2 + p + \Pi_{xx}) = 0, \tag{21.152}$$

$$\frac{d}{dx}\left[\rho u\left(\mathcal{E} + \frac{1}{2}u^2\right) + u(p + \Pi_{xx}) + Q_x\right] = 0, \tag{21.153}$$

where the notation is standard and obvious by now. We note that in the case of one-dimensional flow (shock wave)

$$[\nabla \mathbf{u}]_{xx}^{(2)} = \tfrac{2}{3}\partial_x u. \tag{21.154}$$

The balance equations (21.151)–(21.153) are supplemented by the constitutive relations for the nonconserved variables Π_{xx} and Q_x. For evolution equations of nonconserved variables, we take the closure relation

$$\Phi^{(q)} = 0 \quad (4 > q \geq 1), \tag{21.155}$$

[33]M. Al-Ghoul and B. C. Eu, *Phys. Rev. E* **56**, 2981 (1997); *Phys. Rev. Lett.* **86**, 4294 (2001); *Phys. Rev. E* **64**, 046303 (2001); M. Al-Ghoul and B. C. Eu, Generalized Hydrodynamic theory of shock structures in monatomic and diatomic gases, *Proc. First Int. Symp. Advanced Fluid Information (AFI-2001)*, Oct. 4–5, 2001, Tohoku University, Sendai, Japan, pp. 589–594.
[34]B. C. Eu, *Nonequilibrium Statistical Mechanics* (Kluwer, Dordrecht, 1998).

which means that the higher-order moments $\psi^{(1)}$ and $\psi^{(3)}$ appearing in the evolution equations for the shear stress and the heat flux (note that $\psi^{(1)}$ and $\psi^{(3)}$ are either the moments of one-order or more higher than third-order moments) are set equal to zero:

$$\psi^{(1)} = 0; \quad \psi^{(3)} = 0. \tag{21.156}$$

It should be noted that this set of closure is different from those taken in the Grad's theory.[35] The difference between the present and the Grad closures is simply in the manner of taking a linear combinations for higher order moments in terms of lower-order moments. The reason is that if the set of nonconserved moments is truncated, there is no physical reason to include the higher order moments beyond the truncated set. As will be seen, this difference is crucial for the correct behavior of the shock structure and ultimately may be traceable to the question of *thermodynamic consistency of the nonconserved moment set* resulting thereby. After all, we have shown in a number of investigations based on the closures (21.155) and the constitutive equations for steady nonconserved variables in the adiabatic approximation that they give rise to sufficiently accurate nonlinear transport coefficients, such as non-Newtonian viscosities, in comparison with experiments. On the strength of this experimental evidence from the rheological study of non-Newtonian–non-Fourier fluid, we take the closure (21.155) for the shock wave structure study in hand.

With the closures (21.155) the constitutive equations for $\mathbf{\Pi}$ and $\mathbf{Q}$ are given by reduced evolution equations

$$N_{\mathrm{De}}^{-1}\rho\frac{d\widehat{\mathbf{\Pi}}}{dt} = -2p\gamma - 2[\mathbf{\Pi}\cdot\gamma]^{(2)} - [\omega,\mathbf{\Pi}] - \frac{2}{3}\mathbf{\Pi}\nabla\cdot\mathbf{u}$$
$$- \frac{p}{\eta_0}\mathbf{\Pi}q(\kappa), \tag{21.157}$$

$$N_{\mathrm{Q}}^{-1}\rho\frac{d\widehat{\mathbf{Q}}}{dt} = -2p\widehat{C}_p\nabla\ln T - \mathbf{\Pi}\cdot\nabla\widehat{h} + \nabla\cdot(p\mathbf{U}+\mathbf{\Pi})\cdot\widehat{\mathbf{\Pi}} - \frac{2}{3}\mathbf{\Pi}\nabla\cdot\mathbf{u}$$
$$- \mathbf{Q}\cdot\left(\gamma - \omega + \frac{1}{3}\mathbf{U}\nabla\cdot\mathbf{u}\right) - \frac{p\widehat{C}_p T}{\lambda_0}\mathbf{Q}q(\kappa), \tag{21.158}$$

---

[35] H. Grad, *Comm. Pure Appl. Math.* **5**, 257 (1952).

where reduced variables are used in the by-now-familiar notation in which the asterisks are removed from them for brevity. These evolution equations are coupled to the "reduced" balance equations

$$\rho \frac{dv}{dt} = \nabla \cdot \mathbf{u}, \tag{21.159}$$

$$\rho \frac{d\mathbf{u}}{dt} = -\nabla \cdot \mathbf{P} + \rho \widehat{\mathbf{F}}, \tag{21.160}$$

$$\rho \frac{d\mathcal{E}}{dt} = -\nabla \cdot \mathbf{Q} - \mathbf{P} : \nabla \mathbf{u}. \tag{21.161}$$

If the Deborah number N_{De} and the critical number N_δ are large — namely, the fluid is removed far from equilibrium — then the left-hand sides of Eqs. (21.157) and (21.158) are small being $O(N_{\text{De}}^{-1})$ or $O(N_\delta^{-1})$, the time derivatives on their left-hand side can be neglected, and the steady-state equations are obtained for them:

$$-2p\gamma - 2\left[\mathbf{\Pi} \cdot \gamma\right]^{(2)} - [\omega, \mathbf{\Pi}] - \frac{2}{3}\mathbf{\Pi}\nabla \cdot \mathbf{u} - \frac{p}{\eta_0}\mathbf{\Pi}q(\kappa) = 0, \tag{21.162}$$

$$-2p\widehat{C}_p\nabla \ln T - \mathbf{\Pi} \cdot \nabla\widehat{h} + \nabla \cdot (p\mathbf{U} + \mathbf{\Pi}) \cdot \widehat{\mathbf{\Pi}} - \frac{2}{3}\mathbf{\Pi}\nabla \cdot \mathbf{u}$$

$$-\mathbf{Q} \cdot \left(\gamma - \omega + \frac{1}{3}\mathbf{U}\nabla \cdot \mathbf{u}\right) - \frac{p\widehat{C}_pT}{\lambda_0}\mathbf{Q}q(\kappa) = 0. \tag{21.163}$$

Here $\mathbf{\Pi}$ and $\mathbf{Q}$ are, therefore, steady-state solutions. For the shock wave propagation problem in hand here the constitutive equations are given by the equations

$$\frac{p}{\eta_0}\Pi_{xx}q(\kappa) + \frac{4}{3}\Pi_{xx}\frac{\partial u}{\partial x} + \frac{4}{3}p\frac{\partial u}{\partial x} = 0, \tag{21.164}$$

$$\frac{\widehat{h}p}{\lambda_0}Q_xq(\kappa) + Q_x\frac{\partial u}{\partial x} + \Pi_{xx}u\frac{\partial u}{\partial x} + \widehat{h}\left(p + \Pi_{xx}\right)\frac{\partial \ln T}{\partial x} = 0. \tag{21.165}$$

For these equations it should be noted that the flow is one dimensional and the normal stress differences are set equal to zero for the present flow configuration in hand. The nonlinear factor $q(\kappa)$ is defined by the hyperbolic sine function as defined earlier. Note that $\widehat{h} = \widehat{C}_pT$.

The balance equations (21.159)–(21.161) are integrated to obtain the Rankine–Hugoniot relations

$$\rho u = M, \qquad (21.166)$$

$$\rho u^2 + p + \Pi_{xx} = P, \qquad (21.167)$$

$$\rho u \left(\mathcal{E} + \frac{1}{2} u^2 \right) + u(p + \Pi_{xx}) + Q_x = Q, \qquad (21.168)$$

where M, P, and Q are integration constants with the dimensions of momentum per volume, momentum flux per volume, and energy flux per volume, respectively. These equations are supplemented by the ideal gas equation of state and the caloric equation of state[36]:

$$p = \rho T, \qquad (21.169)$$

$$\mathcal{E} = \frac{3}{2} T. \qquad (21.170)$$

Define new dimensionless variables:

$$v = uMP^{-1}, \quad \theta = TM^2P^{-2},$$

$$\sigma = \Pi_{xx}P^{-1}, \quad \phi = pP^{-1},$$

$$r = \rho PM^{-2}, \quad \varphi = Q_xQ^{-1}, \qquad (21.171)$$

$$\xi = x, \quad \alpha = QMP^{-1}.$$

Since the reduced length x has been already made dimensionless by the mean free path, there is no need to reduce the coordinate x further. The integration constant M may be set equal to the upstream momentum $\rho_1 u_1$: $M = \rho_1 u_1$. Henceforth, we will distinguish the upstream and downstream values of variables with the numeral subscripts 1 and 2. The upstream mean free path is defined by

$$l = \frac{\eta_{01}}{M}, \qquad (21.172)$$

where η_{01} is the upstream Newtonian viscosity at the upstream temperature T_1. Note that this means the viscosity of the gas is dependent on temperature, which is generally the case even for a gas. The transport coefficients

---

[36] Recall that the variables and the equations employed here are already made dimensionless.

are reduced with respect to the upstream transport coefficients:

$$\eta^* = \frac{\eta_0}{\eta_{01}}, \quad \lambda^* = \frac{\lambda_0}{\lambda_{01}}. \tag{21.173}$$

With the reduced variables defined earlier, the conservation laws (balance equation), Eqs. (21.166)–(21.170), read as follows:

$$\phi = r\theta,$$

$$rv = 1,$$

$$rv^2 + \phi + \sigma = 1, \tag{21.174}$$

$$rv^3 + 5\phi v + 2\sigma v + 2\alpha\varphi = \alpha.$$

These equations and the steady-state constitutive equations (21.164) and (21.165) for the shock wave under consideration give rise to the following governing equations for the variables involved:

$$\phi v = \theta, \tag{21.175}$$

$$v + \phi + \sigma = 1, \tag{21.176}$$

$$v^2 + 5\theta + 2\sigma v + 2\alpha\varphi = \alpha, \tag{21.177}$$

$$\frac{1}{\eta^*}\phi\sigma q(\kappa) + \frac{4}{3}(\sigma + \phi)\frac{\partial v}{\partial \xi} = 0, \tag{21.178}$$

$$\frac{\alpha\beta}{\lambda^*}\theta\varphi q(\kappa) + (\alpha\varphi + \sigma v)\frac{\partial v}{\partial \xi} + \frac{5}{2}(\phi + \sigma)\theta\frac{\partial \ln\theta}{\partial \xi} = 0. \tag{21.179}$$

In Eq. (21.179) the new dimensionless parameter β is defined by

$$\beta = \frac{5}{3\theta_1}N_{\text{Pr}}, \tag{21.180}$$

where the Prandtl number N_{Pr} is the Prandtl number with the upstream values of the parameters: $N_{\text{Pr}} = \widehat{C}_p T_1 \eta_{01}/\lambda_{01}$. Note that β defined here should not be confused with the inverse temperature in the kinetic theory.

Since the variable reduction scheme somewhat differs from what is customarily employed in the literature (e.g., Grad's paper already cited.), it would be useful to explain it, especially, with regard to the appearance of the dimensionless number in Eq. (21.179). On multiplication of the mean free path l, the first term in Eq. (21.165) can be reduced, apart from the nonlinear factor $q(\kappa)$, to the form as follows:

$$l\frac{\widehat{hp}}{\lambda_0}Q_x = \frac{5\tau\phi\varphi}{2\lambda^*} \cdot \frac{lP^3Q}{\lambda_{01}M^2},$$

where, if the definitions of the Prandtl number and the reduced temperature are made use of, the second factor on the right can be factored as

$$\frac{lP^3Q}{\lambda_{01}M^2} = \frac{\eta_{01}}{\lambda_{01}} \cdot \frac{P^3Q}{M^2} = \frac{2N_{Pr}}{5\theta_1} \cdot \alpha \cdot \frac{P^3}{M^2}.$$

Thus dividing the resulting equation with P^3/M^2 and using the definition of β given in Eq. (21.180), we obtain Eq. (21.179).

Finally, the dissipation function κ can be expressed in terms of reduced variables and parameters defined above in the following form:

$$\kappa = 4(2\gamma_0)^{-1/4}(5cN_M)^{-1/2}\left(\frac{\theta}{\theta_1}\right)^{1/4}\frac{1}{\phi\sqrt{\eta^*}}\sqrt{\sigma^2 + \frac{8}{15f\theta}\alpha^2\varphi^2},$$

$$(21.181)$$

where

$$c = \frac{l}{l_h},$$

$$l_h = \frac{1}{\sqrt{2}\pi d^2 n}; \text{ hard sphere mean free path,} \qquad (21.182)$$

$$f = \frac{\lambda_{01}}{\frac{3}{2}\widehat{C}_pT_1\eta_{01}}.$$

Therefore, it follows that for a hard sphere interaction model

$$c = 1; \quad f = 1; \quad \eta^* = \sqrt{\frac{\theta}{\theta_1}} \qquad (21.183)$$

and for a Maxwell gas interacting by the r^{-4} force law

$$c = \frac{16}{15\sqrt{2}\pi A_2(5)}\sqrt{\frac{\theta_1}{E_d}}, \qquad (21.184)$$

where $A_2(5) = 0.432\ldots$ and $E_d = V_m M^2 / 4md^4 P^2$ with V_m denoting the potential parameter of the Maxwell model of interaction; see, for example, the monograph[37] by Chapman and Cowling for the Maxwell model. The E_d is a dimensionless Maxwell potential energy at the hypothetical contact point of two hard spheres of radius $d/2$. The dimensionless potential energy E_d is set equal to unity by an appropriate choice of parameters M and P. Note that the choice is arbitrary, but *it yields the dissipation function in a simplest possible form*. With such a choice of reduction parameters, we then obtain the dissipation function in terms of reduced parameters only:

$$\kappa = \left(\frac{3\pi}{5}\right)^{1/4} \sqrt{\frac{3A_2(5)}{N_M}} \cdot \frac{\sqrt{\sigma^2 + \frac{8}{15f\theta}\alpha^2\varphi^2}}{\phi\theta^{1/4}}. \tag{21.185}$$

With a little algebraic manipulation the upstream Mach number can be written as follows:

$$N_M = \sqrt{\frac{1 + \frac{1}{5}\mu}{1 - \frac{1}{3}\mu}}, \tag{21.186}$$

where μ is defined by

$$\mu = \sqrt{25 - 16\alpha}. \tag{21.187}$$

Therefore, since the upstream Mach number is expressed by

$$N_M = \frac{v_1}{\sqrt{\gamma_0 \theta_1}},$$

the parameter μ can be expressed in terms of the upstream Mach number alone:

$$\mu = \frac{3(N_M^2 - 1)}{N_M^2 + \frac{3}{5}}. \tag{21.188}$$

Its range is from 0 to 3, which corresponds to $N_M = \infty$. We will see that the parameter μ determines the boundary values of the variables v, ϕ, and θ, because the boundary conditions are expressible in terms of μ.

To determine the boundary conditions on v, ϕ, and θ, we first observe that as $\xi \to \pm\infty$,

$$\sigma \to 0; \quad \varphi \to 0. \tag{21.189}$$

Since Eqs. (21.178) and (21.179) are identically satisfied in these limits, if v and θ become independent of position ξ at the boundaries, we find the

---

[37]S. Chapman and T. G. Cowling already cited before.

following limits, as $\xi \to \pm\infty$:

$$\sigma, \phi \to 0, \tag{21.190}$$

$$\theta = \phi v, \tag{21.191}$$

$$v + \phi = 1, \tag{21.192}$$

$$v^2 + 5\theta = \alpha. \tag{21.193}$$

The boundary values of the variables v, ϕ, and θ are therefore given by

$$v = \tfrac{1}{8}(5 \pm \mu), \tag{21.194}$$

$$\theta = \tfrac{1}{64}(15 \mp 2\mu - \mu^2), \tag{21.195}$$

$$\phi = \tfrac{1}{8}(3 \mp \mu). \tag{21.196}$$

The upper sign is for the upstream and the lower sign for the downstream. These imply that at the boundaries

$$r = \frac{8}{5 \pm \mu}. \tag{21.197}$$

With the help of Eqs. (21.175)–(21.177), the evolution equations for reduced velocity v and reduced temperature θ are now obtained in the forms

$$\frac{dv}{d\xi} = \frac{3\theta(v^2 - v + \theta)}{4\eta^* v^2 (1 - v)} q(\kappa), \tag{21.198}$$

$$\frac{d\theta}{d\xi} = -\frac{\theta}{5v^2(1-v)^2} \left[\frac{\beta\theta v(1-v)(v^2 - 2v - 3\theta + \alpha)}{\lambda^*} \right.$$
$$\left. + \frac{3(v^2 - v + \theta)(\alpha - v^2 - 5\theta)}{4\eta^*} \right] q(\kappa), \tag{21.199}$$

subject to the boundary conditions (21.194) and (21.195). It should be noted that the argument of the dissipation function κ can be expressed in terms of v and θ. Equations (21.198) and (21.199) describe shock waves for the flow configuration under consideration.

They will be compared with the corresponding equations in the Navier–Stokes–Fourier (NSF) theory of shock waves — the classical hydrodynamic theory of shock waves — after we have solved Eqs. (21.198) and (21.199). In this connection, it should be recalled that in the NSF theory of shock waves, namely, the classical hydrodynamic theory of shock waves, the dissipation terms of the steady-state constitutive equations for Π_{xx} and Q_x are linear with respect to Π_{xx} and Q_x in contrast to the present generalized hydrodynamic theory with nonlinear dissipation terms.

We begin our examination of the governing equations (21.198) and (21.199). For this purpose we construct the direction field equation

$$\frac{dv}{d\theta} = -\frac{\omega(1-v)(v^2 - v + \theta)}{[v(1-v)(v^2 - 2v - 3\theta + \alpha) + \frac{3}{4\beta}(v^2 - v + \theta)(\alpha - v^2 - 5\theta)]},$$

(21.200)

where

$$\omega = \frac{15\lambda^*}{4\beta\eta^*\theta}.$$

(21.201)

We note that for a hard sphere gas

$$\eta^* = \theta^{1/2}, \quad \lambda^* = \theta^{3/2}.$$

(21.202)

The singularities of the direction field are given by the equations

$$1 - v = 0,$$

(21.203)

$$v^2 - v + \theta = 0,$$

(21.204)

$$v(1-v)(v^2 - 2v - 3\theta + \alpha) + \frac{3}{4\beta}(v^2 - v + \theta)(\alpha - v^2 - 5\theta) = 0. \quad (21.205)$$

Equation (21.205) factorizes to two factors

$$\frac{15}{4\beta}(\theta - B + \sqrt{B^2 + A})(\theta - B - \sqrt{B^2 + A}) = 0,$$

(21.206)

where

$$A = \frac{4\beta}{15}v(v-1)\left[\left(1 - \frac{3}{4\beta}\right)\alpha + \left(1 + \frac{3}{4\beta}\right)v^2 - 2v\right],$$

(21.207)

$$B = \frac{2\beta}{5}\left[\left(1 - \frac{1}{\beta}\right)v^2 - \left(1 - \frac{5}{4\beta}\right)v - \frac{3}{4\beta}\alpha\right].$$

(21.208)

Equations (21.203) and (21.204) are for loci of zero slopes whereas Eq. (21.205) is for the loci of infinite slopes. There is an interval of v where the discriminant $B^2 + A$ becomes negative and thus Eq. (21.206) represents an ellipse or rather a loop whereas it represents a pair of parabolas in the intervals where the discriminant is positive. The intersections of the five curves arising from Eqs. (21.203)–(21.205) are the following five points:

$$P_0 : v = \tfrac{1}{8}(5 + \mu), \quad \theta = \tfrac{1}{64}(15 - 2\mu - \mu^2),$$

$$P_1 : v = \tfrac{1}{8}(5 - \mu), \quad \theta = \tfrac{1}{64}(15 + 2\mu - \mu^2),$$

$$P_2 : v = 0, \qquad \theta = 0, \qquad\qquad\qquad (21.209)$$

$$P_3 : v = 1, \qquad \theta = 0,$$

$$P_4 : v = 1, \qquad \theta = \tfrac{1}{5}(\alpha - 1).$$

To compare these direction field singularities with those predicted by the Navier–Stokes–Fourier theory, we now examine the governing equations of the Navier–Stokes–Fourier theory, which yield the direction field equation

$$\frac{dv}{d\theta} = \frac{\omega \left(v^2 - v + \theta \right)}{v \left(3\theta + 2v - v^2 - \alpha \right)}. \qquad (21.210)$$

The singularities of the direction field given by this equation is given by the equations

$$v^2 - v + \theta = 0, \qquad (21.211)$$

$$v^2 - 2v - 3\theta + \alpha = 0, \qquad (21.212)$$

$$v = 0, \qquad (21.213)$$

which yield three singular points:

$$P_0 : v = \tfrac{1}{8}(5 + \mu), \quad \theta = \tfrac{1}{64}(15 - 2\mu - \mu^2),$$

$$P_1 : v = \tfrac{1}{8}(5 - \mu), \quad \theta = \tfrac{1}{64}(15 + 2\mu - \mu^2), \qquad (21.214)$$

$$P_2 : v = 0, \quad \theta = 0.$$

The first two direction field singularities P_0 and P_1 in both the generalized hydrodynamic theory (2.209) and the Navier–Stokes–Fourier theory (21.214) coincide. However, the coincidence of P_2 in both theories are only superficial, because the stability analysis of P_2 provides a totally different picture. An example of the loci of zero and infinite slope for both present and NSF theories are plotted Fig. 21.5. The broken curve in the figure is the locus in the NFS theory whereas the solid curves are the loci in the present generalized hydrodynamic theory. Both theories share the same parabola in bold line which intersects the parabola in broken line and the ellipse at

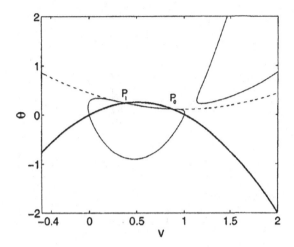

Fig. 21.5. Loci of zeros and infinite slopes in the direction field of the NSF and the generalized hydrodynamic theories. The broken curve is for the NSF theory, and the light curves are for the generalized hydrodynamic theory. The bold curve is shared by both theories, which have common points of intersection P_0 and P_1. There is a parabola in the left corner that is not showing up in the scale used for the figure. Reproduced with permission from M. Al-Ghoul and B. C. Eu, *Phys. Rev. E* **56**, 2981 (1997) © American Physical Society.

the same points P_0 and P_1. This means that both theories share the same boundary conditions at the upstream and downstream. In the case of the direction field equation by the present theory there appear additional singularities P_3 and P_4. It must be noted that the $v = 0$ line is neither the locus of zero slopes nor the locus of infinite slopes. Linear analysis shows that P_3 is neutral in one direction and unstable in the other, whereas P_4 is an unstable focus. As the Mach number increases, the intersections P_0, P_3, and P_4 coalesce at $v = 1$, the point that corresponds to the boundary value for velocity at infinite Mach number. This condition is already almost attained at $N_M = 10$ as shown in Fig. 21.6.

The shock solution must connect points P_0 and P_1. *The fact that the intersections P_0 and P_1 are shared by both theories and there is an intersection of domains where the slopes are negative strongly indicates that a shock solution must exist for the governing equations (21.198) and (21.199) of the generalized hydrodynamic theory as it does for the classical hydrodynamic theory of Navier, Stokes, and Fourier.*

Such solutions can be demonstrated for all the cases of Mach number studied. The examples of velocity profiles are shown in Fig. 21.7, where

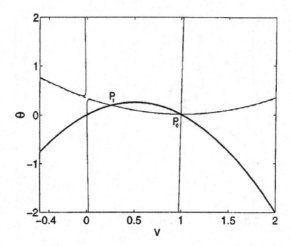

Fig. 21.6.   Same as Fig. 21.5. except for $N_M = 10$. The curve $(0P_1P_01)$ is a part of a closed loop that is not fully shown in the scale used for the figure. Reproduced with permission from M. Al-Ghoul and B. C. Eu, *Phys. Rev. E* **56**, 2981 (1997) © American Physical Society.

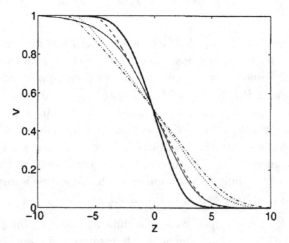

Fig. 21.7.   Velocity profiles in one-dimensional shock waves for various Mach numbers calculated for a variable hard sphere model for the potential energy. Solid line $N_M = 1.5$; bold solid line $N_M = 2$; dashed line $N_M = 5$; dotted line $N_M = 8$; dash-dotted line $N_M = 10$. Reproduced with permission from M. Al-Ghoul and B. C. Eu, *Phys. Rev. E* **56**, 2981 (1997) © American Physical Society.

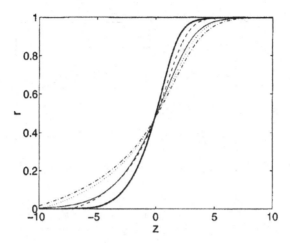

Fig. 21.8. Density profiles in one-dimensional shock for various cases of Mach number in the case of a variable hard sphere (s=0.75) Solid line: $N_M = 1.5$; bold solid line: $N_M = 2$; dotted line: $N_M = 8$; dash-dotted line: $N_M = 10$. Reproduced with permission from M. Al-Ghoul and B. C. Eu, *Phys. Rev. E* **56**, 2981 (1997) © American Physical Society.

the reduced distance z is related to the reduced distance ξ used for the governing equations by the relation

$$\xi = zB\sqrt{5\pi/6N_M}, \qquad (21.215)$$

where $B = 1$ for the Maxwell model of interaction and $B = (7 - 2s)(5 - 2s)/24$ for a variable hard sphere model[38] for which the viscosity may be taken in the form

$$\eta_0 = \mu_0 \left(\frac{T}{T_0}\right)^s.$$

In this formula for η_0, the parameters μ_0, T_0, and s depend on the substance and their conditions. In Fig. 21.8 the density profile is plotted in similar manner to Fig. 21.7.

In shock wave experiments, the shock structures are customarily expressed by means of shock wave width, which is defined by the formula

$$\delta = \frac{n_2 - n_1}{(dn/dz)_{\max}}, \qquad (21.216)$$

where $(dn/dz)_{\max}$ is the maximum value of the density derivative occurring at the transition point. This definition of shock width is not the most effective of definitions characterizing the shock width, but widely used to present

---

[38] G. A. Bird, *Phys. Fluids* **26**, 3222 (1983).

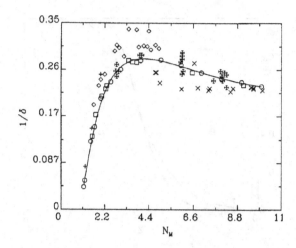

Fig. 21.9.　Theoretical inverse shock width vs. Mach number in the case of argon. A variable hard sphere model is used for the interaction potential with $s = 0.75$. In the figure, the theoretical values are represented by hexagons, which are roughly connected by the solid curve over the entire Mach number range, and other symbols represent experimental data by various authors described in the text. Reproduced with permission from M. Al-Ghoul and B. C. Eu, *Phys. Rev. E* **56**, 2981 (1997) © American Physical Society.

the experimental data in the literature. Examples for shock structure is given by means of the velocity profiles around the shock transition point in Fig. 21.7 for argon. In Fig. 21.9 the theoretical inverse shock width, calculated with a variable hard sphere model for argon in the case of $s = 0.75$, is plotted against Mach number N_M and compared with experimental data available up to $N_M \approx 10$. The value of $s = 0.75$ lies between for the shock tube value and $s = 0.81$ for the wind tunnel value suggested by Pham-van Diep *et al.*[39] This model was tested in connection with shock widths for argon and helium by the aforementioned authors. The experimental data in question are based on the definition of mean free path defined by

$$l_{\text{PEM}} = (\eta_{01}/\rho_1)[(7 - 2s)(5 - 2s)/24]\sqrt{\pi/2\mathcal{R}T_1}. \qquad (21.217)$$

For this definition of mean free path, the reduced distance ξ in the present theory is related to the reduced distance used for the experimental data

---

[39] G. C. Pham-van Diep, D. A. Erwin and E. P. Munz, *J. Fluid Mech.* **232**, 403 (1991); D. A. Erwin, D. C. Pham-van Diep and E. P. Munz, *Phys. Fluids A* **3**, 697 (1991).

considered here as follows:

$$\zeta = [(7 - 2s)(5 - 2s)/24]\,\xi = z\,[(7 - 2s)(5 - 2s)/24]\,\sqrt{\frac{5\pi}{6}}N_M \quad (21.218)$$

and hence the shock widths δ are calculated with the formula

$$\delta = \frac{(7 - 2s)(5 - 2s)}{24}\sqrt{\frac{5\pi}{6}}N_M\frac{(n_2 - n_1)}{(dn/d\xi)_{\max}}. \quad (21.219)$$

The agreement with experiments is judged to be good, especially, in the range of $N_M \lesssim 3$, but there is considerable scattering of experimental data around the theoretical prediction (solid curve), and the scattered experimental data may be attributable to experimental uncertainty, especially, in the high Mach number regime. The meanings of symbols are described in the footnote.[40]

Based on the comparison made between the theoretical values and the experimental data, the present theory seems to provide adequate results consistent with a collection of experimental data gathered by a number of authors. We, especially, note that there is a convergence of theoretical and experimental values in the low and high Mach number regimes. This convergence pattern seems to be a quite encouraging evidence for the theory, and one may cautiously conclude that the thermodynamically consistent generalized hydrodynamic theory raises the expectation of achieving a satisfactory solution to the problem of shock structures even to the entire range of hypersonic flows studied by experiment $N_M \lesssim 10$.

Let us now examine what happens in the shock wave from the viewpoint of thermodynamics and, especially, with regard to the energy dissipation in the flow. The irreversible thermodynamics analysis given below shows what happens to the flow on transit between direction field singularities P_0 and P_1. We can gain the desired information on this particular aspect, if we examine the calortropy production during the flow, which is given by

---

[40]Hexagons = present theory; ($\square$) = H. Alsmeyer, *J. Fluid Mech.* **74**, 497 (1976); (✹) = B. Schmidt, *J. Fluid Mech.* **39**, 361 (1969); (+) = W. Garen *et al.*, *AIAA J.* **12**, 1132 (1974); ($\diamond$) = M. Linzer and D. F. Honig, *Phys. Fluids* **6**, 1661 (1963); × = M. Camac, *Adv. Appl. Mech.* **1**, 240 (1965).

the formula

$$\sigma_c = k_B g \kappa(\Pi, Q) \sinh \kappa(\Pi, Q), \qquad (21.220)$$

where

$$g = \sqrt{\frac{m}{k_B T}} \frac{1}{2n^2 d^2}.$$

In terms of reduced variables employed for the present problem in the generalized hydrodynamic theory, the reduced calortropy production is given by

$$\widehat{\sigma}_c = \sqrt{\frac{\theta}{\theta_1}} \left(\frac{r_1}{r}\right)^2 \kappa(\sigma, \varphi) \sinh \kappa(\sigma, \varphi). \qquad (21.221)$$

Subscript 1 here stands for the upstream value of flow. If $\widehat{\sigma}_c$ is plotted as a function of z, it is peaked at the transition point $z = 0$, as shown in Fig. 21.7. The calortropy production furnishes a measure of energy dissipation accompanying an irreversible process in the system, which in the present case is an evolution of a shock wave. The shock wave in the present case passes through the region between the singular points P_0 and P_1 of the direction field of the shock wave governing equations. This figure provides us of an interesting and valuable physical picture of creation of shock waves as follows.

During the passage of wave between direction field singular points P_0 and P_1 the wave dissipates most of its energy, the energy dissipation reaching a maximum at the transition point and, as a consequence, the wave velocity collapses to a rather negligible magnitude, and this is the physical reason why the wave velocity collapses to rather small value as indicated by the velocity profiles shown in Fig. 21.7 and consequently a shock wave appears: *A shock wave appears because of exhausted energy by the flow after passing through the region between direction field singular points P_0 and P_1.*

We believe that this is the thermodynamic reason for the appearance[41] of shock waves from the view point of energy.

---

[41] In *Phys. Fluids* **26**, 056102 (2014), R. S. Myong discusses on a mathematical reason the Grad theory of shock waves by means of the Maxwell model using the moment method fails to predict a correct Mach number dependence of shock width, but even if his so-called *shock singularity* is removed, the shock width is not guaranteed to come out correctly, unless the energy dissipation is correctly described by the model as the present theory indicates. Therefore, his suggestion to remove the shock singularity only partly and inadequately solves the shock structure problem. One needs to describe a correctly behaved mode of energy dissipation by the wave.

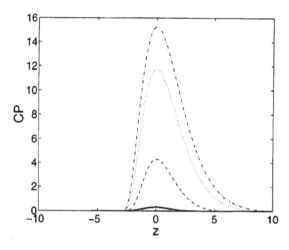

Fig. 21.10. Calortropy production around the shock transition point in the case of argon. It is calculated for the variable hard sphere model as for Fig. 21.9. Reproduced with permission from M. Al-Ghoul and B. C. Eu, *Phys. Rev. E* **56**, 2981 (1997) © American Physical Society.

If the calortropy production in Eq. (21.221) is integrated for z then a reduced integral (global) calortropy production is obtained:

$$\langle \widehat{\sigma}_c \rangle = \int_{-\infty}^{\infty} d\xi \sqrt{\frac{\theta}{\theta_1}} \left(\frac{r_1}{r} \right)^2 \kappa\left(\sigma, \varphi\right) \sinh \kappa(\sigma, \varphi). \qquad (21.222)$$

This was calculated for the Maxwell model of interaction potential and variable hard sphere model as a function of Mach number. Its N_M-dependence is found as follows:

$$\langle \widehat{\sigma}_c \rangle = \langle \widehat{\sigma}_c \rangle_0 (N_M - a)^\epsilon, \qquad (21.223)$$

where $\langle \widehat{\sigma}_c \rangle_0$, a, and ϵ are constants. In the case of the Maxwell model $a = 0.85$ and $\epsilon = 3.14$ and for the variable hard sphere model $a = 0.87$ and $\epsilon = 2.98$. In both cases, the global calortropy production grows linearly with respect to $(N_M - a)^\epsilon$ as the Mach number increases (Fig. 21.10).

Chapter 22

Oscillatory Structures in Nonlinear Systems

We now have seen that the generalized hydrodynamic equations can correctly describe nonlinear macroscopic wave phenomena outside the range of validity of the classical Navier–Stokes–Fourier theory. It is interesting to venture into the range of nonlinear phenomena involving oscillatory behaviors of macroscopic material flows. Such oscillatory or periodic wave motions are experimentally observed in macroscopic processes associated with chemical reactions, especially, with a feedback loop and other natural nonlinear phenomena occurring far removed from equilibrium. In this chapter, we would like to examine some examples of oscillatory phenomena and appearance of periodic structures such as Belousof and Zhaboutinsky and show the potential of the generalized hydrodynamics theory presented in this work as a theoretical tool to investigate such phenomena. Inclusion of such phenomena in the discussion would make the scope of the applicability of the generalized hydrodynamics and the attendant irreversible thermodynamics wider and more comprehensive.

22.1. Chemical Oscillations and Formation of Periodic Structures or Patterns

We have studied the thermodynamics of chemical equilibria and learned how we may calculate the equilibrium constants in terms of thermodynamic quantities. Chemical reactions at the equilibrium level, however, do not reveal dynamical aspects of chemical reactions. We can learn about the dynamical aspects, only if the kinetic evolution of reactions is studied. Potentially, they can provide rather rich features of many aspects of natural and biological phenomena involving chemical reactions.

Such kinetic evolutions can be studied from the standpoint of irreversible thermodynamics, and a general outline of the theory is presented in this chapter, with a particular example that emphasizes the type of wave equations arising in dynamical study of chemical reactions. Numerous concurrently occurring hydrodynamic processes in the system may interfere with the chemical evolution, and many experiments in recent years show that some fascinating phenomena can be observed.

Chemical reactions can happen when reactive molecules come sufficiently close to each other in space within the range of chemical forces on the order of a few bond lengths at most. They therefore occur in shorter spatial scales than the usual hydrodynamic scales, which are usually much longer than a mean free path. And, moreover, they are usually faster than most of hydrodynamic processes such as flow of matter arising from the stress and heat flow present in the fluid, but can occur roughly on the same time scale as molecular collisions. The chemical evolutions can get coupled with diffusion phenomena first of all and fluid flow through the stress generated within the fluid. In recent years, such coupling phenomena have been experimentally observed in connection with chemical oscillations, initially through the famous Belusov–Zhabotinsky (BZ) reaction,[1] which harbingered numerous subsequent studies of oscillatory chemical reactions and pattern formations in connection with chemistry, biochemistry, and biology. However, it should be noted here that the chemical oscillation phenomena associated with biological processes were anteceded by periodic structures first observed by Liesegang[2] and many other researchers[3] following him on inanimate chemical systems.

---

[1]B. P. Belusov, *Sborn Referat. Radiat. Meditsin Za* 1958: *Collection of Abstracts on Radiation Medicine* (Medgiz, Moscow, 1959), p. 145; A. M. Zhabotinsky, *Biofizika* **2**, 306 (1964); *Russ. J. Phys. Chem.* **42**, 1649 (1968). For the treatment of the chemical dynamics of the Brusselator by using the generalized hydrodynamic method described below, see M. Al-Ghoul and B. C. Eu, *J. Phys. Chem.* **100**, 18908 (1996).

[2]R. E. Liesegang, *Chemische Reacktionen in Gallerten* (Düsseldorf, 1889).

[3]W. Ostwald, *Z. Phys. Chem.* **27**, 365 (1897); E. S. Hedges, *Colloids* (Edward Arnold Co., London, 1931); H. K. Henisch, *Crystal Growth in Gels* (The Pennsylvania State University Press, University Park, 1970).

The BZ reaction was initially examined by I. Prigogine and his collaborators[4] by using a linear irreversible thermodynamics formalism under the assumption of local equilibrium hypothesis. This formalism utilizes steady-state constitutive equations for diffusion fluxes, which are linear with respect to the diffusion fluxes. When used in the concentration evolution equations for components undergoing chemical reactions, it necessarily gives rise to a system of reaction–diffusion equations. More generally than the BZ reaction mentioned, reaction–diffusion reactions occur in a model for morphogenesis according to A. M. Turing's theory.[5] In his theory, reaction–diffusion equations appear giving rise to forms or patterns. And they occur beyond the regime of stability of a set of coupled chemical reactions. For this reason alone there is a considerable interest studying reaction–diffusion equations in connection with biological phenomena. As a matter of fact, such forms, patterns, and, more generally, structures appear in diverse natural phenomena, besides biology, for example, in geology and other inanimate natural phenomena such as the Liesegang phenomena.

The reaction–diffusion equations appearing in the theories of Turing in morphogenesis, of Prigogine and his school, and of Liesegang phenomena are all nonlinear parabolic[6] partial differential equations, which are expected to describe wave phenomena associated with the chemical waves accompanying the chemical reactions. One characteristic of parabolic partial differential equations is that the speed of wave propagation is infinite, suggesting that a wave initiating at a point in space instantly reaches points away from the source and hence the boundaries. This generally contradicts the experience with the chemical waves studied in the laboratory. Such parabolic equations inevitably arise because of using the linear theory of irreversible thermodynamics. In this section, we will show that according the generalized hydrodynamic theory presented earlier in Part II of this book the governing equations for the chemical wave propagation is generally *hyperbolic partial differential equations*, and it means that the speed of

---

[4] P. Glansdorff and I. Prigogine, *Thermodynamic Theory of Structure, Stability and Fluctuations* (Wiley-Interscience, New York, 1971).

[5] A. M. Turing, *Phil. Trans. Roy. Soc. London B* **237**, 37 (1952).

[6] For example, the time-dependent Schrödinger equation, a wave equation that has a first-order time derivative term, is a parabolic partial differential equation, whereas a classical wave equation, which has a second-order time derivative term in addition to a second-order spatial derivative term, is a hyperbolic partial differential equation. For more precise mathematical definitions, see P. M. Morse and H. Feshbach, *Methods of Theoretical Physics* Vol. I, Chapter 6, (McGraw-Hill, New York, 1953).

wave propagation is finite and thus should give rise to qualitatively different results than the case of *parabolic partial differential equations.*

The subject matter discussed in this section ranges such a wide area that a section or a chapter in a book would not do justice to the whole subject area that can easily take up a monograph. It is therefore all the more important to provide the basic key idea of how the dynamical theory incorporating the principles of irreversible thermodynamics can be formulated. Our aim here is just to introduce the reader to the notion that the subject matters can come under the purview of irreversible thermodynamics principles and indeed can be studied with the thermodynamics formalism based on generalized hydrodynamics described earlier in this work. The kind of wave phenomena mentioned earlier can be studied more appropriately by using hyperbolic reaction–diffusion equations than the conventionally employed parabolic reaction–diffusion equations.

If there are chemical reactions undergoing in the fluid, the mass fraction evolution equations (19.6) should be modified. It would be necessary to add the reaction rate term to them. Assume that there are m chemical reactions undergoing in the fluid. If we denote the kth chemical reaction rate by R_k and ν_{ak} is the stoichiometric coefficient times the mass of species a of the reaction k, then the overall reaction rate for species a is given by

$$\Lambda_a^{(r)} = \sum_{k=1}^{m} \nu_{ak} R_k. \tag{22.1}$$

This term $\Lambda_a^{(r)}$ serves as the dissipation term—a source term—of the mass fraction evolution equation of mass fraction c_a of species a, which should now read

$$\rho \frac{dc_a}{dt} = -\nabla \cdot \mathbf{J}_a + \Lambda_a^{(r)}. \tag{22.2}$$

In this equation, $\mathbf{J}_a$ is the mass diffusion flux. In the sense that the overall reaction rate term $\Lambda_a^{(r)}$ appears as a source term in the mass fraction balance equation, mass fractions c_a are no longer conserved quantities if a chemical reaction is present in the system.

Now, if linear constitutive equations are assumed for steady diffusion fluxes, as is conventionally done,[7] diffusion fluxes are assumed to be

---

[7]This assumption is tantamount to the assumption that the time evolution of diffusion fluxes is much faster than the concentration variable evolution comparable to the evolutions of other hydrodynamic variables such as $\mathbf{u}$, c_a, and T.

governed by the linear thermodynamic force–flux relation — a constitutive relation:

$$\sum_{b=1}^{m} L_{ab}\mathbf{J}_b + \frac{k_B T}{m_a}\nabla\rho_b = 0, \tag{22.3}$$

where ρ_b is the mass density of b and L_{ab} are phenomenological coefficients obeying the Onsager reciprocal relations $L_{ab} = L_{ba}$. And if the algebraic solutions[8] $\mathbf{J}_a$ of Eq. (22.3) are substituted into Eq. (22.2), then a coupled set of parabolic partial differential equations arise for the evolution equations for mass fractions c_a. They are, as a matter of fact, diffusion equations. Such parabolic wave equations predict an infinite speed of wave propagation as mentioned earlier. In other words, in the context of the chemical reactions considered, the effects of chemical reactions occurring at a point in a local volume will instantly propagate throughout the fluid. This idea of instant propagation of disturbance (i.e., chemical reactions) through the medium, intuitively, does not appears to be realistic, and indeed experimental evidence does not support it generally. Nevertheless, such parabolic reaction–diffusion equations have been conventionally employed to study reaction–diffusion phenomena.

In this section, we show that the thermodynamically consistent generalized hydrodynamic theory introduced earlier produces hyperbolic reaction–diffusion equations in contrast to the conventional approach to the reaction–diffusion phenomena. It opens up a new vista to the study of reaction–diffusion phenomena and, perhaps, morphogenesis, which chemical reactions and pattern formation underlie according to the concept of Turing, and, generally, diverse natural phenomena.

To be specific, we will take an example of typical coupled chemical reactions involving a feedback loop (or an autocatalytic step in more conventional chemical kinetics terminology): a modified Selkov model, which consists of three steps of chemical reactions[9]

$$A \underset{k_{-1}}{\overset{k_1}{\rightleftharpoons}} S,$$

---

[8]The algebraic solutions of Eq. (22.3) are easily obtained by inverting the matrix (L_{ab}) of phenomenological coefficients L_{ab}.

[9]Y. Termonia and J. Ross, *Proc. Natl. Acad. Sci. USA* **78**, 2952, 3563 (1982); P. Richter, P. Rehmus and J. Ross, *Prog. Theor. Phys.* **66**, 385 (1981).

$$S + 2P \underset{k_{-2}}{\overset{k_2}{\rightleftharpoons}} 3P,$$

$$P \underset{k_{-3}}{\overset{k_3}{\rightleftharpoons}} B.$$

These reactions have been studied[10] for some aspects of glycolysis. In these reactions, A and B are kept at fixed concentrations and the intermediates S and P change in time. Note that the second reaction of the set above is the autocatalytic reaction. According to the mass action law, the reaction rates for the two intermediate species S and P are given by

$$R_S = k_1\rho_A - k_{-1}\rho_S - k_2\rho_S\rho_P^2 + k_{-2}\rho_P^3, \tag{22.4}$$

$$R_P = k_2\rho_S\rho_P^2 - k_{-2}\rho_P^3 - k_3\rho_P + k_{-3}\rho_B, \tag{22.5}$$

where k_i and k_{-i} are the reaction rate constants for the forward and reverse reactions in the ith step in the coupled chemical reactions under consideration. To make the equations simpler we assume that the off-diagonal phenomenological coefficients $L_{SP} = L_{PS}$ are equal to zero. Therefore, there are only self-diffusion processes taken into account for species P and S. This assumption is generally not valid for real systems, but it is taken here to make the equations as simple as possible. The assumption that there are only self-diffusion processes involved is, in fact, taken in virtually all works on chemical waves in the literature. We follow the practice in the literature, but the discussion given represents only an aspect of chemical oscillatory phenomena reported in the literature.

Since it is convenient to work with dimensionless equations, the following reduced variables are introduced:

$$\tau = k_3t, \qquad\qquad \xi = \mathbf{r}/L,$$
$$\mathbf{u} = \mathbf{J}_P/Lk_3\sqrt{k_3/k_2}, \qquad \mathbf{v} = \mathbf{J}_S/Lk_3\sqrt{k_3/k_2},$$
$$X = \rho_P/\sqrt{k_3/k_2}, \qquad Y = \rho_S/\sqrt{k_3/k_2},$$
$$A = (k_1/k_3\sqrt{k_3/k_2})\rho_A, \quad B = (k_{-3}/k_3\sqrt{k_3/k_2})\rho_B,$$
$$K = k_{-2}/k_2, \qquad\qquad R = k_{-1}/k_3.$$

Since the chemical reactions evolve in competition with diffusions of chemical species and fluid motion, it is reasonable to consider a dimensionless number made up of reaction time and diffusion times of species and

[10]Y. Termonia and J. Ross, *Proc. Natl. Acad. Sci. USA* **78**, 2952, 3563 (1982); P. Richter, P. Rehmus and J. Ross, *Proc. Theor. Phys.* **66**, 385 (1981).

hydrodynamic speed. Therefore, more specifically, we find it convenient to make use of the dimensionless reaction–diffusion number defined by[11]

$$N_{\rm rd} = \frac{k_B T}{\sqrt{m_P m_S}} k_3^{-1} \left(D_S D_P\right)^{-1/2}, \tag{22.6}$$

where D_S and D_P are essentially the self-diffusion coefficients defined by the diagonal parts of the diffusion constant matrix in the S and P species space. *The reaction–diffusion number is, in essence, the ratio of reaction time × diffusion time to the time scale of hydrodynamic speed squared.* We also define the additional dimensionless parameters

$$f = \sqrt{\frac{m_S D_S}{m_P D_P}}, \quad \widehat{D}_X = \frac{D_P}{k_3 L^2}, \quad \widehat{D}_Y = \frac{D_S}{k_3 L^2}. \tag{22.7}$$

With these reduced variables, the evolution equations for variables $(X, Y, \mathbf{u}, \mathbf{v})$ take the forms

$$\frac{\partial X}{\partial \tau} = -\frac{\partial}{\partial \xi} \cdot \mathbf{u} + B - X + X^2 Y - K X^3, \tag{22.8}$$

$$\frac{\partial Y}{\partial \tau} = -\frac{\partial}{\partial \xi} \cdot \mathbf{v} + A - R Y - X^2 Y + K X^3, \tag{22.9}$$

$$\frac{\partial \mathbf{u}}{\partial \tau} = -N_{\rm rd} f \left(\widehat{D}_X \frac{\partial X}{\partial \xi} + \mathbf{u} \right), \tag{22.10}$$

$$\frac{\partial \mathbf{v}}{\partial \tau} = -N_{\rm rd} f^{-1} \left(\widehat{D}_Y \frac{\partial Y}{\partial \xi} + \mathbf{v} \right). \tag{22.11}$$

The first two of the equations presented above are concentration balance equations whose source terms are reaction rates and last two are the evolution equations for reduced diffusion fluxes $\mathbf{u}$ and $\mathbf{v}$ of species X (i.e., reduced density of P) and Y (i.e., reduced density of S). Recall $R = k_{-1}/k_3$. (See table for the definition of reduced variables.) Therefore, put in the context of the present reactions, *the reaction–diffusion number $N_{\rm rd}$ gives a measure of relative time scales of the chemical evolution of material species X and Y to the ratio of the hydrodynamic time squared to the diffusion time,* when the reduced time is reckoned in the relative scale of the rate constant k_3 and the mean diffusion constant to the sound speed.

---

[11]The reaction–diffusion number should rank among various nondimensional fluid dynamic numbers such as Mach, Reynolds, Prandtl numbers, etc. It characterizes the relative time and spatial scales of a chemical reaction and hydrodynamic processes.

To determine the character of this set of differential equations the eigenvalues of the characteristic matrix of this system are sought. They are determined by the characteristic determinant

$$
\begin{vmatrix}
-\lambda & 0 & 1 & 0 \\
0 & -\lambda & 0 & 1 \\
N_{rd}\widehat{D}_X f & 0 & -\lambda & 0 \\
0 & N_{rd}\widehat{D}_Y/f & 0 & -\lambda
\end{vmatrix} = 0.
\qquad (22.12)
$$

There are four eigenvalues $\lambda_1, \lambda_2, \lambda_3, \lambda_4$ for this characteristic determinant, and they are arranged as follows:

$$
\lambda_1, \lambda_2, \lambda_3, \lambda_4
$$

$$
= +\sqrt{N_{rd}\widehat{D}_X f}, \ -\sqrt{N_{rd}\widehat{D}_X f}, \ +\sqrt{N_{rd}\widehat{D}_Y/f}, \ -\sqrt{N_{rd}\widehat{D}_Y/f}.
$$
$$(22.13)$$

Since the eigenvalues are all real, the system of Eqs. (22.8)–(22.11) is seen to be hyperbolic.[12] It would be interesting to see in which regime of parameter values the hyperbolic behavior is manifest. It would be interesting to see in which regime of parameter values the hyperbolic behavior is manifest.

For this purpose, we estimate N_{rd} for typical values for the parameters involved. If the mean diffusion coefficient is of the order of $10^{-9}\,\mathrm{m^2 s^{-1}}$ and the rate constant is of the order of $10^{12}\,\mathrm{s^{-1}}$, then N_{rd} is of the order of 10^2 at the room temperature. In this case, in the regime of $N_{rd} \lesssim 10^2$ the evolution equations must be hyperbolic, while in the longtime regime the system behaves as if it follows parabolic evolution equations: *Thus the transient regime of evolution and beyond must be described by a hyperbolic system of evolution equations*, and it is the regime of interest in this work.

It is helpful to put the equations into a form of wave equation that will enable us to compare with the classical wave equations and thereby give us better insight. Eliminating the variables (reduced diffusion fluxes) **u** and **v** from the set (22.8)–(22.11) we obtain the coupled second-order partial differential equations for reduced concentrations X and Y:

$$
\frac{\partial^2 \mathbf{Z}}{\partial \tau^2} + \overline{N}_{rd}\mathbf{H}\frac{\partial \mathbf{Z}}{\partial \tau} - \overline{N}_{rd}\mathbf{D}\frac{\partial^2 \mathbf{Z}}{\partial \xi^2} = \overline{N}_{rd}\mathbf{R}(X, Y),
\qquad (22.14)
$$

---

[12]See, for example, A. Jeffrey, *Quasilinear Hyperbolic Systems and Waves* (Pitman, London, 1976).

where $\mathbf{Z}$ is a column vector made up of X and Y

$$\mathbf{Z} = \begin{pmatrix} X \\ Y \end{pmatrix}; \tag{22.15}$$

$\mathbf{R}$ is a column vector consisting of reaction rates

$$\mathbf{R} = \begin{pmatrix} B - X + X^2Y - KX^2 \\ (A - RY - X^2Y + KX^3)\, f^{-2} \end{pmatrix}; \tag{22.16}$$

and $\mathbf{D}$ and $\mathbf{H}$ are square matrices

$$\mathbf{D} = \begin{pmatrix} \widehat{D}_X & 0 \\ 0 & \widehat{D}_Y f^{-2} \end{pmatrix} = \begin{pmatrix} 1 & 0 \\ 0 & f^{-2} \end{pmatrix} \begin{pmatrix} \widehat{D}_X & 0 \\ 0 & \widehat{D}_Y \end{pmatrix}, \tag{22.17}$$

$$\mathbf{H} = \begin{pmatrix} H_{xx} & H_{xy} \\ H_{yx} & H_{yy} \end{pmatrix} \tag{22.18}$$

with the definitions

$$H_{xx} = 1 + \left(1 - 2XY + 3KX^2\right)/\overline{N}_{\mathrm{rd}},$$
$$H_{xy} = -X^2/\overline{N}_{\mathrm{rd}},$$
$$H_{yx} = \left(2XY - 3KX^2\right)/\overline{N}_{\mathrm{rd}},$$
$$H_{yy} = f^{-2} + \left(R + X^2\right)/\overline{N}_{\mathrm{rd}}. \tag{22.19}$$

It is convenient to isolate out the f^{-2} factor from matrix $\mathbf{D}$ and write it in terms of reduced diffusion coefficient matrix $\widehat{\mathbf{D}}$

$$\mathbf{D} = \mathbf{f}_2\widehat{\mathbf{D}}, \tag{22.20}$$

where $\mathbf{f}_2$ and $\widehat{\mathbf{D}}$ are diagonal matrices defined by

$$\mathbf{f}_2 = \begin{pmatrix} 1 & 0 \\ 0 & f^{-2} \end{pmatrix}, \tag{22.21}$$

$$\widehat{\mathbf{D}} = \begin{pmatrix} \widehat{D}_X & 0 \\ 0 & \widehat{D}_Y \end{pmatrix}. \tag{22.22}$$

Then the matrix $\mathbf{R}$ is also factored as follows:

$$\mathbf{R} = \mathbf{f}_2\mathbf{R}', \tag{22.23}$$

where

$$\mathbf{R}' = \begin{pmatrix} B - X + X^2Y - KX^2 \\ A - RY - X^2Y + KX^3 \end{pmatrix}. \tag{22.24}$$

The matrix nonlinear wave equation (22.14) is a pair of coupled nonlinear telegraphist equations, which are coupled hyperbolic partial differential equations. Evidently, such telegraphist equations are generic to the reaction–diffusion equations. Linear telegraphist equations are well known in the field of wave equations. Roughly speaking, *the inverses of the eigenvalues of the coefficient matrix to the second derivative term in the wave equation Eq. (22.14) give the wave velocities (more precisely, group velocities)*,

$$u_X = \frac{C_X}{\sqrt{N_{rd}D_X}}; \quad u_Y = \frac{C_Y}{\sqrt{N_{rd}D_Y}}. \tag{22.25}$$

Therefore, for the present case of reaction–diffusion system the wave equations become parabolic in the limit of $N_{rd}D_X$, $N_{rd}D_Y \to \infty$. Precisely in such limits, there follow the reaction–diffusion equations taken in the conventional approach (i.e., linear irreversible thermodynamic approach) to description of chemical oscillation phenomena:

$$\frac{\partial \mathbf{Z}}{\partial \tau} - \hat{\mathbf{D}}\frac{\partial^2 \mathbf{Z}}{\partial \xi^2} = \mathbf{R}'(X, Y), \tag{22.26}$$

for which we have used

$$\mathbf{H}\left(\overline{N}_{rd} = \infty\right) = \mathbf{f}_2.$$

Therefore, Eq. (22.26) is the infinite wave speed limit of hyperbolic (i.e., telegraphist) equation (22.14). Consequently, if $N_{rd}D_X$ is finite, the equations (22.14) are hyperbolic, and the modes of evolution of the waves predicted by the hyperbolic wave equations are expected to be different from those in the limit of $N_{rd}D_X \to \infty$, at which parabolic wave equations arise. Therefore, it would be interesting to compare their differences. It is an important objective of investigations in the present section.

It is not possible to obtain analytic solutions in closed forms for the nonlinear wave equations (22.14). The numerical solutions have been studied by Al-Ghoul and Eu.[13] Numerical solutions reveal rather rich structures, which can range a wide spectrum of patterns and modes of oscillation, which are different from those observed with parabolic wave equations. We will show some examples for such wave structures later.

Before implementing numerical solutions and a general procedure of a solution method, it is preferable to examine the steady state of the equations and the stability of the steady state.

---

[13]See M. Al-Ghoul and B. C. Eu, *Physica D* **90**, 119, (1996); *ibid.* **97**, 531 (1996).

22.2. Steady States

22.2.1. *Homogeneous Steady States*

If the system is spatially homogeneous, then the spatial derivative of $\mathbf{Z}_h$ vanishes, namely,

$$\nabla_\xi^2 \mathbf{Z}_h = 0, \tag{22.27}$$

and the telegraphist equation becomes a differential equation of time only:

$$\frac{\partial^2 \mathbf{Z}_h}{\partial \tau^2} + \overline{N}_{\mathrm{rd}} \mathbf{H}_h \frac{\partial \mathbf{Z}_h}{\partial \tau} = \overline{N}_{\mathrm{rd}} \mathbf{R} \left(X_h, Y_h \right), \tag{22.28}$$

where $\mathbf{Z}_h$, X_h, Y_h denote spatially homogeneous solutions and $\mathbf{H}_h$ stands for $\mathbf{H}$ evaluated with X_h and Y_h. The homogeneous steady-state solution is then defined by

$$\frac{\partial \mathbf{Z}_h}{\partial \tau} = 0 \tag{22.29}$$

or

$$\mathbf{R} \left(X_0, Y_0 \right) |_{X_h = X_0, Y_h = Y_0} = 0. \tag{22.30}$$

Written out explicitly, this equation yields a pair of the algebraic equations

$$B - X_0 + X_0^2 Y_0 - K X_0^2 = 0, \tag{22.31}$$

$$A - R Y_0 - X_0^2 Y_0 + K X_0^3 = 0. \tag{22.32}$$

This algebraic set is rearranged to the equations

$$K X_0^3 + a_2 X_0^2 + a_1 X_0 + a_0 = 0, \tag{22.33}$$

$$\frac{1}{R} \left(A + B - X_0 \right) = Y_0, \tag{22.34}$$

where

$$a_2 = -\frac{A + B}{1 - KR} := \alpha, \tag{22.35}$$

$$a_1 = \frac{R}{1 - KR}, \tag{22.36}$$

$$a_0 = -\frac{RB}{1 - KR} = -Ba_1. \tag{22.37}$$

The variables A and B are experimental. The cubic equation (22.33) admits three different sets of solution depending on the parameter values of A, B, R, and K. To determine the characters of the roots, it is convenient to define new parameter set (B, C) where $C = A + B$. Then the discriminant of the cubic equation (22.33) takes the form

$$f(B, C) = 4\alpha^{-2} \left[B^2 + 4\lambda B \alpha^{-1} + 4 \left(\lambda^2 - q^2 \right) \alpha^{-2} \right], \qquad (22.38)$$

where

$$\lambda(C) = -\frac{\alpha^2 C}{6R} + \frac{\alpha^2 C^3}{27R^3}, \qquad (22.39)$$

$$q(c) = \frac{\alpha}{3} \left(\frac{\alpha C^2}{3R^2} - 1 \right). \qquad (22.40)$$

Therefore, there is one real and two complex conjugate roots, if $f(B, C) > 0$, which yields the conditions

$$B + 2\alpha^{-1}[\lambda(C) \pm q^{3/2}(C)] > 0 \qquad (22.41)$$

or

$$B + 2\alpha^{-1}[\lambda(C) \pm q^{3/2}(C)] < 0. \qquad (22.42)$$

If $f(B, C) < 0$, there are three real roots for which the following condition for B holds:

$$-2\alpha^{-1}[\lambda(C) + q^{3/2}(C)] < B < -2\alpha^{-1}[\lambda(C) - q^{3/2}(C)]. \qquad (22.43)$$

If $f(B, C) = 0$, then

$$B = -2\alpha^{-1}[\lambda(C) \pm q^{3/2}(C)] \qquad (22.44)$$

and two of the three real roots are equal. In Fig. 22.1 the curve $f(B, C) = 0$ is plotted in the (B, C) plane where domains of positive and negative $f(B, C)$ are indicated: if $f(B, C) > 0$, there is one real root; if $f(B, C) = 0$, i.e., on the curve, then there are three real roots, of which two are double roots; and if $f(B, C) < 0$, there are three distinct real roots.

To analyze the stability of homogeneous steady state (X_0, Y_0) in the time domain, we define fluctuations of the solution for the linearized equation (22.28) from the homogeneous steady state. Thus we define new variables

$$x = X - X_0, \quad y = Y - Y_0, \qquad (22.45)$$

which now obey the equation

$$\frac{\partial^2 \mathbf{Z}_f}{\partial \tau^2} + \overline{N}_{\text{rd}} \mathbf{H}_0 \frac{\partial \mathbf{Z}_f}{\partial \tau} = \overline{N}_{\text{rd}} \mathbf{R}_0 \mathbf{Z}_f, \qquad (22.46)$$

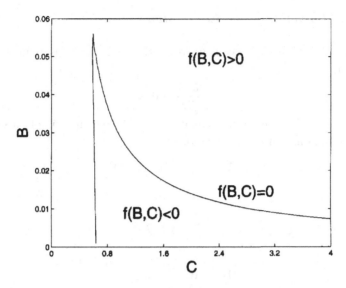

Fig. 22.1. Discriminant of a cubic equation $f(B, C)$ in the (B, C) plane. Reproduced with permission from M. Al-Ghoul and B. C. Eu, *Physica D* **97**, 531 (1996) © Elsevier, B.V.

where

$$\mathbf{Z}_f = \begin{pmatrix} x \\ y \end{pmatrix}, \tag{22.47}$$

$$\mathbf{H}_0 = \begin{pmatrix} H_{xx}^0 & H_{xy}^0 \\ H_{yx}^0 & H_{yy}^0 \end{pmatrix}, \tag{22.48}$$

$$\mathbf{R}_0 = \begin{pmatrix} -1 + 2X_0Y_0 - 3KX_0^2 & X_0^2 \\ \left(-2X_0Y_0 + 3KX_0^2\right)f^{-2} & -\left(R - X_0^2\right)f^{-2} \end{pmatrix}, \tag{22.49}$$

$$H_{xx}^0 = 1 + (1 - 2X_0Y_0 + 3KX_0^2)/\overline{N}_{\mathrm{rd}}, \tag{22.50}$$

$$H_{xy}^0 = -X_0^2/\overline{N}_{\mathrm{rd}}, \tag{22.51}$$

$$H_{yx}^0 = (2X_0Y_0 - 3KX_0^2)/\overline{N}_{\mathrm{rd}}, \tag{22.52}$$

$$H_{yy}^0 = f^{-2} + (R + X_0^2)/\overline{N}_{\mathrm{rd}}. \tag{22.53}$$

Equation (22.46) can be analyzed for the stability of steady states together with inhomogeneous steady state presented in next section, where analysis of the inhomogeneous steady state is made.

22.2.2.   *Inhomogeneous Steady States*

If the system is inhomogeneous in space, the steady state is defined by the condition

$$\frac{\partial \mathbf{Z}_t}{\partial \tau} = 0 \tag{22.54}$$

in which case the telegraphist equation (22.14) becomes simply a second-order partial differential equation of ξ

$$-\mathbf{D}\nabla_\xi^2 \mathbf{Z}_t = \mathbf{R}_t\left(X_t, Y_t\right), \tag{22.55}$$

where

$$\mathbf{Z}_t = \begin{pmatrix} X_t \\ Y_t \end{pmatrix} \tag{22.56}$$

and

$$\mathbf{R}_t = \mathbf{R}|_{X=X_t, Y=Y_t}. \tag{22.57}$$

At the steady state, the parabolic and hyperbolic differential equations share the same steady state, but the temporal evolutions of the two different differential equations differ significantly, since the manner in which the inhomogeneous steady state is asymptotically reached can be qualitatively different. The difference will be evident from their numerical solutions as will be shown.

Evolution equation (22.55) describes the steady-state Turing structures of the system. Since it is useful to examine the linear stability of the structures, Eq. (22.55) is linearized with respect to the state of vanishing reaction rates, namely, (X_0, Y_0) defined earlier. Then with the definition of new variables representing fluctuations from (X_0, Y_0)

$$x = X_t - X_0, \quad y = Y_t - Y_0, \tag{22.58}$$

there follows the linearized inhomogeneous set of telegraphist equations

$$\frac{\partial^2 \mathbf{Z}_l}{\partial \tau^2} + \overline{N}_{\mathrm{rd}}\mathbf{H}_0 \frac{\partial \mathbf{Z}_l}{\partial \tau} - \overline{N}_{\mathrm{rd}}\mathbf{D}\nabla_\xi^2 \mathbf{Z}_l = -\overline{N}_{\mathrm{rd}}\mathbf{D}\mathbf{M}\mathbf{Z}_l, \tag{22.59}$$

where

$$\mathbf{M} = -\mathbf{D}^{-1}\mathbf{R}_0. \tag{22.60}$$

Equation (22.59) can be solved by Fourier transform

$$\mathbf{Z}_l = \sum_\omega \sum_{\mathbf{k}} \Phi(\omega, \mathbf{k}) \exp\left[i(\mathbf{k}\cdot\xi - \omega\tau)\right]. \tag{22.61}$$

Inserting this Fourier expansion, we obtain the linear algebraic set from Eq. (22.59)

$$\left(-\omega^2 \mathbf{I} - i\omega \overline{N}_{\text{rd}} \mathbf{H}_0 + k^2 \overline{N}_{\text{rd}} \mathbf{D} + \overline{N}_{\text{rd}} \mathbf{DM}\right) \Phi(\omega, \mathbf{k}) = 0. \tag{22.62}$$

The dispersion relation is provided by the secular determinant constructed from this linear set

$$\det \left| -\omega^2 \mathbf{I} - i\omega \overline{N}_{\text{rd}} \mathbf{H}_0 + k^2 \overline{N}_{\text{rd}} \mathbf{D} + \overline{N}_{\text{rd}} \mathbf{DM} \right| = 0. \tag{22.63}$$

Note that this is the solvability condition for the linear set (22.62). It gives rise to a fourth-order polynomial of $z = -i\omega$ and k:

$$P_4(z, k) = z^4 + P z^3 + Q z^2 + T z + S = 0, \tag{22.64}$$

where the coefficients are defined by the expressions

$$K_{xx} = k^2 D_{xx} + (\mathbf{DM})_{xx}, \tag{22.65}$$

$$K_{yy} = k^2 D_{yy} + (\mathbf{DM})_{yy}, \tag{22.66}$$

$$P = \overline{N}_{\text{rd}}(H_{xx}^0 + H_{yy}^0), \tag{22.67}$$

$$Q = \overline{N}_{\text{rd}}(K_{xx} + K_{yy}) + \overline{N}_{\text{rd}}^2(H_{xx}^0 H_{yy}^0 - H_{xy}^0 H_{yx}^0), \tag{22.68}$$

$$T = \overline{N}_{\text{rd}}^2 [K_{xx} H_{yy}^0 + K_{yy} H_{xx}^0 - H_{xy}^0 (\mathbf{DM})_{yx} - H_{yx}^0 (\mathbf{DM})_{xy}], \tag{22.69}$$

$$S = \overline{N}_{\text{rd}}^2 (K_{xx} K_{yy} - (\mathbf{DM})_{xy} (\mathbf{DM})_{yx}). \tag{22.70}$$

For the polynomial (22.64) to have all the roots with the negative imaginary part the following Hurwitz conditions[14] must be fulfilled:

$$P > 0, \tag{22.71}$$

$$PQ - T > 0, \tag{22.72}$$

$$PQT - T^2 - P^2 C > 0. \tag{22.73}$$

The first condition, Eq. (22.71), is independent of wave number k and a function of B and C defined earlier. The level curve $P = 0$ coincides with $f(B, C) = 0$ for the homogeneous steady state, Eq. (22.38). The second condition $f_2(k, C, B) := PQ - T$, Eq. (22.72), and the third condition $f_3(k, C, B) := PQT - T^2 - P^2 S$, Eq. (22.73), are even functions of k.

---

[14]F. R. Gantmacher, *Theorie des Matrices*, Vol. 2 (Dunod, Paris, 1966).

Since the polynomial $P_4(z,k)$ is real, it has two pairs of complex conjugate roots. The level curves of the Hurwitz conditions $f_2(k,C,B) = 0$ and $f_3(k,C,B) = 0$ are drawn for a given value of B. Here we remark that if the system were spatially homogeneous the stability of homogeneous steady state would have been determined by the analysis of the curve $f(B,C) = 0$, which gives rise to Fig. 22.1 presented earlier.

The level curves of the Hurwitz conditions are presented in Fig. 22.1 for $f(B,C) = 0$, which coincides with the level curve for the homogeneous steady state, in Fig. 22.2 in the (C,k) and Fig. 22.3 in the (D_y,k) plane, respectively, where k is the frequency of spatial oscillation. The diagrams in Figs. 22.2 and 22.3 are constructed by using the Hurwitz conditions in the case of $R = 0.1$, $K = 1$, $f = 1$, and $B = 0.09$ for the hyperbolic system. All modes of frequency k can give rise to unstable steady states. This stability diagram is constructed in another way by using the Hurwitz condition in the $(k, \widehat{D}_y)$ plane for a hyperbolic system in Fig. 22.3. In this figure, the shaded region is for the stable phase. For this figure $\widehat{D}_x$ is fixed at $\widehat{D}_x = 0.06$ together with $f = 1$ and the ordinate is in the units of 10^{-4}.

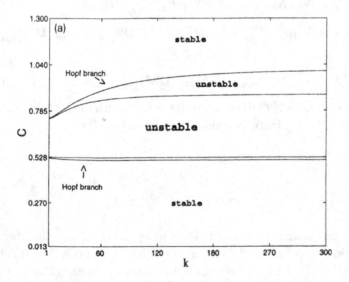

Fig. 22.2. Stability phase diagram for homogeneous steady states. In the domain where $f(B,C) > 0$, there are one real and two complex conjugate roots. In the domain $f(B,C) < 0$, there are three distinct roots. One root is unstable and the other two are stable. Therefore the system is bistable. On the curve $f(B,C) = 0$, there are one single root and a pair of double roots, all of which are real. Reproduced with permission from M. Al-Ghoul and B. C. Eu, *Physica D* **97**, 531 (1996) © Elsevier, B.V.

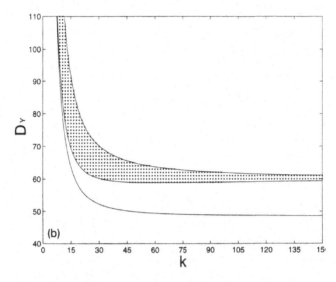

Fig. 22.3.   Stability phase diagram for the hyperbolic system. Reproduced with permission from M. Al-Ghoul and B. C. Eu, *Physica D* **97**, 531 (1996) © Elsevier, B.V.

The feature of the phase diagram in Fig. 22.3 is distinctive from the parabolic case, for which the phase diagram is given in Fig. 22.4. According to the phase diagram for the parabolic case, the unstable steady states are confined to a limited domain in the (C, k) plane bounded by a parabola, the rest of the plane being the domain of stable steady states. This is in a sharp contrast to the case of hyperbolic evolution equations. This distinction is the cause of the difference in the pattern evolutions of the hyperbolic and parabolic systems, although the chemical reactions involved are the same for both cases. This implies that nonlinearity inherent to chemical reactions alone would not determine the patterns predicted of the differential equation system.

22.3.   Numerical Solutions in One Dimension

The telegraphist equations for the cubic reversible reaction model have been numerically solved[15] under the guide of the linear stability theory described earlier. A plethora of oscillatory structures and patterns emerge

---

[15]M. Al-Ghoul, *PhD Thesis*, McGill University, Montreal, 1997.

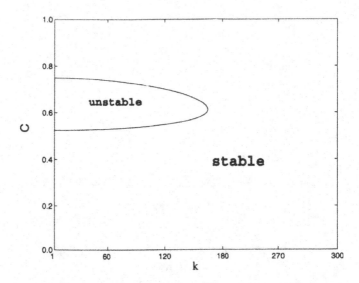

Fig. 22.4. Stability phase diagram for the parabolic system in the case of $R = 0.1$, $K = 1$, $f = 1$, and $B = 0.09$. Reproduced with permission from M. Al-Ghoul and B. C. Eu, *Physica D* **97**, 531 (1996) © Elsevier, B.V.

as solutions, given different initial and boundary conditions on the concentrations and velocities of fluid components: for example,

$$X(0,t) = X(1,t) = C_x, Y(0,t) = Y(1,t) = C_y,$$

$$X(\xi,0) = C_x, Y(\xi,0) = C_y,$$

$$\mathbf{u}(\xi,0) = \mathbf{v}(\xi,0),$$

where C_x and C_y are constants chosen equal to 0.2 in reduced units for most of cases studied, mimicking experimental conditions. The reliability of the numerical results has been checked by using two different numerical solution methods IMSL subroutine MOLCH and the Fourier spectral method[16] combined with Gear's method for stiff ordinary differential equations. The Fourier spectral method took 128 co-location points. The further details of numerical solution methods and results are referred to M. Al-Ghoul's PhD dissertation cited earlier.

Here we would like to show some examples of numerical solutions of the telegraphist equations and the patterns exhibited by them: There are

---

[16]See, for example, C. Canuto, M. Y. Hussaini, A. Quarteroni and T. A. Zang, *Spectral Methods in Fluid Dynamics* (Springer-Verlag, Berlin, 1988).

observed stable limit cycles, traveling wave patterns, and evolutions of stable patterns, all of which are accompanied by either oscillatory or sharp increase in calortropy production, a phenomenon which seems to indicate that organized patterns accompany an increased energy dissipation. It should be recalled that at equilibrium (namely, reversible thermodynamic processes) the calortropy reduces to the Clausius entropy, which is not equal to the calortropy if the processes are of nonequilibrium. It is emphasized here that the entropy is only for reversible processes described by equilibrium thermodynamics.

22.3.1. *Effects of Diffusion*

The effects of diffusion have been examined in the case of no chemical reaction. The telegraphist equations then become linear wave equations in the case. It is found that the steady state is stable and the amplitudes of waves get damped owing to the effect of diffusion. This effect is shown in Fig. 22.5, where in the case of no diffusion a limit cycle appears as shown in panel (a) of Fig. 22.5; two different initial conditions tend to the same limit cycles in long time. When diffusion is turned on, the temporal oscillation gets affected. In the case of $\overline{N}_{rd} = 0.1$, the oscillation becomes quasi-periodic in the XY plane, winding a torus as shown in panel (b) for the hyperbolic system and in panel (c) for the parabolic system. The manner in which the trajectories wind the torus does not look the same for the two systems. This difference is manifestly exhibited by their power spectra.

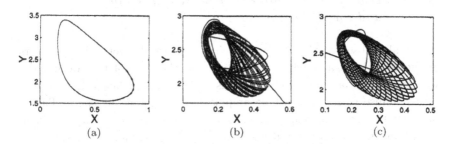

Fig. 22.5. In the case of no diffusion. Two different initial conditions tend to the same limit cycle. (b) A quasiperiodic motion shown by the hyperbolic system at $N_{rd} = 0.1$ when diffusion is turned on. (c) A quasiperiodic motion exhibited by the parabolic system when diffusion is present as for panel (b). Reproduced with permission from M. Al-Ghoul and B. C. Eu, *Physica D* **97**, 531 (1996) © Elsevier, B.V.

The power spectrum analysis of oscillations indicates there are three fundamental frequencies and their subharmonics. A quasi-periodic motion exhibited by the parabolic system shown on panel (c) shows a somewhat different pattern from that of the corresponding hyperbolic system. Its power spectrum is found definitely different from that of the hyperbolic system; see the paper by Al-Ghoul and Eu[17] for the details of power spectra. As the reaction–diffusion number is varied to large values (e.g., $N_{rd} \gtrsim 20$) the system becomes parabolic.

From the investigation on numerical solutions of the telegraphist equations made with varying values of N_{rd} a conclusion is drawn that for the particular case of parameters chosen for the system, if there is no diffusion, there appears a single frequency limit cycle, but *as diffusion is turned on at a very large value of N_{rd} so that the system is essentially parabolic and thus the disturbance propagates at infinite speed, there appears a quasi-periodic motion of two-fundamental frequency — it is a period doubling phenomenon. As the reaction–diffusion number is reduced so that the system thus becomes hyperbolic and the disturbance propagates at a finite speed, the number of frequencies is increased to three or more and eventually to infinity, and the motion becomes chaotic.* Thus this scenario seems to provide a route to appearance of chaotic motions in the case of telegraphist equations. Since the telegraphist equations are generic for reaction–diffusion equations within the generalized hydrodynamic equations, the conclusion drawn may be also generic and diffusion is seen to play a very important role in producing patterns.

22.3.2. *Difference in Patterns between Hyperbolic and Parabolic Systems*

It was also observed that there is a characteristic difference in behaviors of the hyperbolic and parabolic systems. In the case of the former, the wave maintains a sharp front whereas in the case of latter the wave is rather diffuse with nonvanishing amplitude throughout the space. This is probably the most noticeable feature from the experimental viewpoint when a traveling wave is involved. It is also noticed that after waves travel toward the middle and merge, they split and travel back toward the boundaries, and then get reflected and travel back toward the middle, and this process

---

[17]M. Al-Ghoul and B. C. Eu, *Physical D* **90**, 119 (1996).

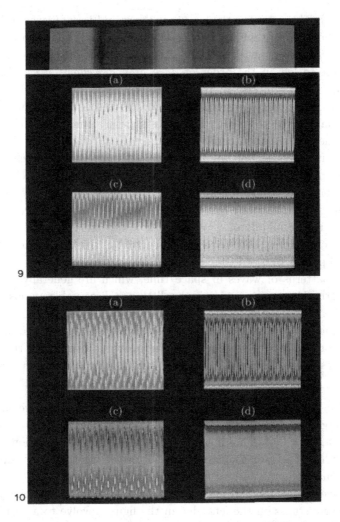

Fig. 22.6. The color code is given in the top of the color strip. This color code applies to other figures shown. Panel 9: space–time plot of concentration waves in the case of the parabolic system for $X = Y = 0.2$ at the boundaries. (a) X; (b) Y; (c) **u**; and (d) **v**. These results coincide with those of the hyperbolic system at $N_{\mathrm{rd}} = 50$. Panel 10: space–time of concentration waves in the case of hyperbolic system for $X = Y = 0.2$ and at the boundaries and $N_{\mathrm{rd}} = 0.1$. (a) X; (b) Y; (c) **u**; and (d) **v**. Reproduced with permission from M. Al-Ghoul and B. C. Eu, *Physica D* **97**, 531 (1996) © Elsevier, B.V.

is repeated. This aspect of the feature is shown in Fig. 22.6, where the color code of the figures presented in this section is given in the color code strip shown at the top panel of the figure: the amplitude of the wave increases from right to left in the order of *dark orange, yellow, green, light blue, dark blue, pink, and red*. In the figures the waves are plotted in space–time over 1000 units of τ (reduced time); the abscissa is for time (τ) and the ordinate is for space (ξ). The wave splitting behaviors are shown in space–time in Figs. 22.6 panels 9(a) and 9(b) for the case of $N_{rd} = 50$. At the value of $N_{rd} = 50$ the hyperbolic system practically behaves as if it is parabolic, and gives the numerical results virtually coinciding with the parabolic reaction diffusion in all qualitative aspects. We note that there are more than one wavelength discernable and the waves merge and split in the middle of $\xi = [0, 1]$. At lower values of N_{rd} — namely, hyperbolic system — the space–time structure of the hyperbolic system in Fig. 22.6 panels 10(a)–(d), for example, at $N_{rd} = 0.1$, exhibits more complicated patterns of waves in space–time, which are generally shorter in wavelength. The wave merging and splitting is discernable at this value of $N_{rd} = 0.1$. It was noted in Ref. 18 that these patterns may be regarded as an example of one-dimensional patterns where spots appear and disappear as time progresses. It is noted that the behavior reminds of the experimental results by Lee *et al.*[19] on birth and demise of spots in a reacting system.

In Fig. 22.7 panels 11(a)–(d) and panels 12(a)–(d), the aforementioned wave merging and splitting phenomenon is still noticeable in the hyperbolic system at $N_{rd} = 0.01$. Thus it seems that the telegraphist equations in question may serve as a mathematical model to investigate such experimental phenomena involving pattern formations. In panel 12 of Fig. 22.7, the patterns have changed significantly. It is observed that nonstationary waves in the early stages on the left edge in the figures evolve to a steady peak in the center that is flanked by steady waves of long fixed wave lengths, which may be regarded as a pattern in space–time. They appear to satisfy the necessary requirements of Turing patterns that *the pattern is stationary, symmetry is broken, and wavelength is intrinsic.*[20] In particular, the central

---

[18]M. Al-Ghoul and B. C. Eu, *Physica D* **97**, 531 (1996).
[19]K. Lee, W. D. McCormick, J. E. Pearson and H. L. Swinney, *Nature* **369**, 631 (1994).
[20]For example, see, P. De Kepper, V. Castets, E. Dulos and J. Boissonade, *Physica D* **49**, 161 (1991).

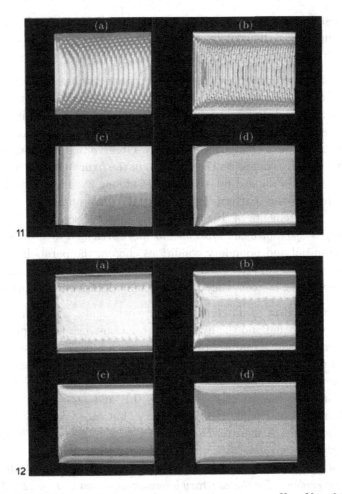

Fig. 22.7.   Panel 11: space–time plot of concentration waves at $X = Y = 0.2$ at the
boundaries in the case of hyperbolic system and $N_{rd} = 0.01$. (a) X; (b) Y; (c) **u**; and
(d) **v**. Panel 12: space–time plot of concentration waves at $X = 1.5$ and $Y = 2.5$ at the
boundaries in the case of hyperbolic system and $N_{rd} = 0.02$. (a) X; (b) Y; (c) **u**; and
(d) **v**. Reproduced with permission from M. Al-Ghoul and B. C. Eu, *Physica D* **97**, 531
(1996) © Elsevier, B.V.

peak may be regarded as what is termed as the Turing hole in the literature.[21] For the same initial and boundary conditions chosen, the parabolic

---

[21] J. J. Perraude, A. De Wit, E. Dulos P. De Kepper, G. Dewel and P. Borckmann, *Phys.
Rev. Lett.* **71**, 1272 (1993); A. De Wit, G. Dewel and P. Borckmann, *Phys. Rev. E* **48**,
R4191 (1993).

reaction–diffusion equations did not produce this kind of patterns where the Turing hole is flanked by waves; they produce only diffusive waves in space.

The effects of difference in the boundary conditions have been studied with the hyperbolic equations, which were solved for different boundary conditions $X = 1.5$ and $Y = 2.5$. The patterns obtained are quite different from those shown in the previous figures. As shown in Fig. 22.8 panels A and B, the patterns have changed significantly. Observe that the nonstationary waves in the early stages (on the left edge in the figures) evolve to a steady peak in the center flanked by steady waves of long fixed wavelengths. They may be regarded as patterns in space–time. They appear to satisfy the necessary requirements of Turing patterns that *the pattern is stationary, symmetry is spontaneously broken, and the wavelength is intrinsic.*[22] For the same initial and boundary conditions chosen, the parabolic reaction–diffusion equations did not produce such patterns where the Turing hole is flanked by waves; instead, they produced only diffusive waves in space.

To see the effect of different diffusion coefficients, the diffusion coefficients were set equal and the hyperbolic equations (i.e., telegraphist, equations) were solved. Examples of the patterns produced by the numerical solutions are shown in Fig. 22.8, where space–time plots of concentration waves are displayed in the case of the hyperbolic system at $X = 1.5$ and $Y = 2.5$ for the case of $N_{rd} = 0.1$, $f = 0.1$, and $\widehat{D}_x = \widehat{D}_y = 0.006$ for panel A figures, and panel B figures $f = 1$, but other parameters are the same as for panel A figures. *The Turing theory* [23] *predicts that the diffusion coefficients must be different in order for patterns to appear, but the hyperbolic system studied indicates otherwise as clearly demonstrated by the numerical evidence for the contrary behavior as shown in Fig.* 22.8: It is not necessary for the diffusion coefficients to be different in order for the Turing structure to appear. For the figures shown in both panels A and B for $\widehat{D}_x = \widehat{D}_y = 0.006$, which, however, yields $f = 0.1$ for panel A, whereas $f = 1$ for panel B, notice that a more structured pattern appears in the early time, but it is transformed to another form of different symmetry for the rest of time. These patterns were found rather sensitive to the parameters chosen. Therefore more extensive numerical studies of different classes of parameter sets and boundary and initial conditions might be employed

---

[22]P. De Kepper, V. Castets, E. Dulos and J. Boissonade, *Physica D* **49**, 161 (1991).
[23]See the reference cited already.

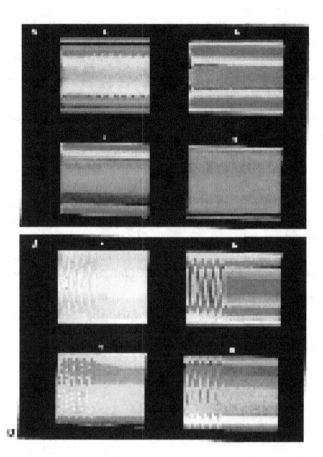

Fig. 22.8. Space–time plot of concentration waves for X in the case of the hyperbolic system at $X = 1.5$ and $Y = 2.5$ at $N_{rd} = 0.1$, $f = 0.1$ with $\widehat{D}_x = \widehat{D}_y = 0.006$. Panel A: (a) X, (b) Y, (c) **u**, and (d) **v**. Panel in B: (a), (b), (c), and (d) are the same as in Panel B except that $f = 1$. Reproduced with permission from M. Al-Ghoul and B. C. Eu, *Physica D* **97**, 531 (1996) © Elsevier, B.V.

to classify patterns according to the classes of parameters and boundary and initial conditions. This might be an object of studies in the future.

22.3.3. *Energy and Matter Dissipation — Calortropy Production*

Macroscopic irreversible processes evolve in space–time, and structures are self-organized, subjected to the laws of thermodynamics and accompanied by the calortropy production at the expense of energy and matter within the system itself and the surroundings. The mode of energy and matter dissipation to create a structure therefore is of considerable interest, and understanding it should provide considerable insight into how natural systems in nature get organized into being.

The hyperbolic reaction–diffusion equations discussed in this section are linear except for chemical reactions. Therefore, the calortropy calculated is a linear irreversible theory approximation to the Boltzmann entropy, and it also provides how the Clausius entropy behaves away from equilibrium. Consequently, it may be regarded as a near-equilibrium behavior of the Clausius entropy, and in that regime we may simply use the term entropy production interchangeably with the calortropy production in the following. Under this understanding, we may examine the entropy production—more precisely, calortropy production—as a function of frequency and wave number of a pattern evolved from a given set of initial and boundary conditions.

Since patterns change as the reaction–diffusion number N_{rd} is varied, the entropy production changes with the pattern. In Fig. 22.9

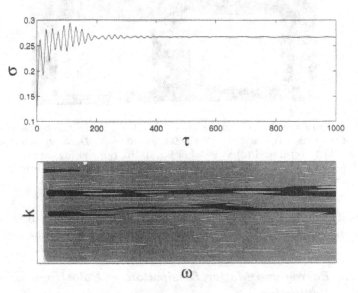

Fig. 22.9. Top panel: space-integrated entropy production corresponding to Fig. 22.8 panel (13). The bottom panel is the Fourier transform of the integrated entropy production plotted in the (ω, k) plane. Reproduced with permission from M. Al-Ghoul and B. C. Eu, *Physica D* **97**, 531 (1996) © Elsevier, B.V.

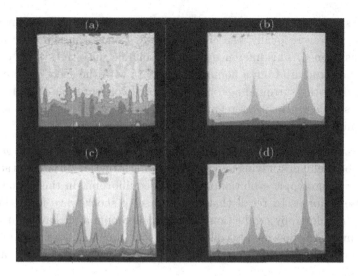

Fig. 22.10. Two-dimensional entropy production spectrum for the hyperbolic system at $N_{rd} = 0.1$, $N_{rd} = 2$, (c) $N_{rd} = 4$, and (d) $N_{rd} = 50$ at the boundary conditions $X = Y = 0.2$ and $\widehat{D}_x \neq \widehat{D}_y$. Reproduced with permission from M. Al-Ghoul and B. C. Eu, *Physica D* **97**, 531 (1996) © Elsevier, B.V.

the space-integrated entropy production corresponding to the pattern in Fig. 22.8 panel 13 is shown.

In Fig. 22.10, the logarithm of the two-dimensional Fourier spectrum of the calortropy spectrum of patterns shown in panel 12 of Fig. 22.7 is plotted in the same color code in the (ω, k) plane for various cases of N_{rd} for the hyperbolic system: (a) $N_{rd} = 0.1$, (b) $N_{rd} = 2$, (c) $N_{rd} = 4$, and (d) $N_{rd} = 50$. Boundary conditions are: $X = 0.2$ and $Y = 0.2$ with $\widehat{D}_x \neq \widehat{D}_y$. The figure is a logarithmic plot in the $\omega - k$ plane in the same color code as for other figures presented. The abscissa is for k and the ordinate is for ω. The figure shows how intense calortropy production (entropy production) is for different modes of ω and k; the low ω modes have the highest intensities for all k and thus favored by the system at the expense of energy and matter. Note that there are certain k values where patterns of high frequency ω modes are formed.

Through the line of investigations presented above, Al-Ghoul and Eu infer that chaotic motions have lower entropy (calortropy) productions than the structure-producing motions, and the global energy–matter dissipation by the system is higher when there is an organized pattern than a chaotic state, but the entropy production is almost independent of the organized patterns formed in the system for a given set of initial and boundary conditions. However, locally, the entropy production is characteristic to the patterns formed. This inference seems reasonable since the system evolves to a local structure from a homogeneous state with the energy and matter provided by the surroundings, but the global consumption of energy and matter should be the same regardless of which structure is formed locally, if the initial and boundary conditions are the same.

This investigation leads us to an inference that the calortropy in general should be useful in studying and understanding structure formations accompanying macroscopic evolution away from equilibrium. In this connection, it is useful for us to recall that in the case of shock wave formation, in which the calortropy production reaches a maximum after having attained a maximum, the flow simply collapses to a low value and, as a consequence, it exhibits a shock wave. See the section on shock wave about this effect.

22.3.4. *Two-Dimensional Model*

The telegraphist equations for the cubic model studied in the previous section can be solved for the case of two spatial dimensions. Before implementing the numerical solution procedure, stability analysis was performed of the steady state in same manner as described for the one-dimensional telegraphist equations discussed earlier. Their solutions, obtained under the guidance of stability analysis made, are found to give rise to a plethora of patterns evolving into a more and more complicated and symmetry breaking patterns, especially, in the low reaction–diffusion number regime. Furthermore, *distinction between the hyperbolic and parabolic systems becomes more noticeable* as the reaction-diffusion number decreases. The limitation of space does not permit us to describe here all the patterns evolved from either random or nonrandom initial and boundary conditions and the distinctions of patterns predicted by either parabolic or hyperbolic systems.

As has been observed with parabolic nonlinear partial differential equations of reaction–diffusion systems studied in the literature, hexagonal structures, spirals, and chaotic structures among other patterns have been observed, especially, as the low reaction–diffusion number regime is approached — namely, the hyperbolic system regime. Unlike the parabolic counterparts, the hyperbolic differential equations are capable of describing transient behaviors of the system. For example, they can describe a phenomenon akin to cell divisions observed in the simulations of the Chlorite–Iodide–Malonic Acid (CIMA) reaction[24] and the Lengyel–Epstein model[25] These results are described in the paper[26] by Al-Ghoul and Eu to which the interested reader is referred.

The point we would like to emphasize here is that the generalized hydrodynamic equations, of which the telegraphist equations are particular examples, appear to be capable of describing numerous nonlinear wave phenomena and pattern formations observed in nature and laboratory, such as spirals, hexagonal structures, chaotic structures as mentioned earlier. Particularly, since the equations are in conformation to the thermodynamic laws, they hold up a promise for us to come up in the future with a thermodynamically consistent comprehensive theories of pattern formations. In this connection, it should be noted that other reaction models can be taken to represent the system of interest and the diffusion flux evolution equations can be made fully nonlinear by including nonlinear chemical reaction terms. Hydrodynamic flows can be fully incorporated, so that fluid flow is fully described by adding the momentum balance equation to the evolution equations. These aspects of formulation, however, must be left to the studies to be taken up in the future. We close this section by showing some of patterns observed with the two-dimensional telegraphist equations for the reversible cubic reaction model.

22.3.4.1.　*Examples of Patterns Formed*

The two space-dimensional telegraphist equations for the reversible cubic reaction equations were solved for fixed boundary conditions and with either random or nonrandom initial conditions. The case of random initial conditions gives rise to rather interesting patterns which have not been seen in the case of nonrandom initial conditions. A combination of spectral method and Gear's method for stiff differential equations was used. As

---

[24]See V. Castets, E. Dulos, J. Boissonade and P. De Kepper, *Phys. Rev. Lett.* **64**, 2953 (1991).

[25]I. Lengyel, G. Rabai and I. Epstein, *J. Am. Chem. Soc.* **112**, 4606 (1990); I. R. Epstein and I. Lengyel, *Physica D* **84**, 212 (1995).

[26]M. Al-Ghoul and B. C. Eu, *Physica D* **97**, 531 (1996).

the reaction–diffusion number is changed, traveling waves, hexagonal structures, stripes, squares, and turbulent patterns emerge in a manner consistent with the study of the one-dimensional hyperbolic reaction–diffusion
equations described in the previous section. A decrease in the reaction diffusion number to small values of $O(10^{-3})$ can destabilize patterns and give
rise to loss of their synchronizations which would eventually produce a turbulent pattern. To obtain the parameters in the equations necessary for
implementing the numerical solution method the phase diagram was constructed similarly to the case of one-dimensional systems. The parameters
have been mostly chosen in the unstable region away from the threshold in
the phase diagram in order to ensure fully developed patterns and reveal the
complexity and possible difference in the dynamics of pattern formations
produced by the corresponding parabolic reaction–diffusion equations. As
indicated by the patterns shown, the two systems of differential equations
produce different patterns. The range of N_{rd} studied was from 0.1 to 10^{-3}.

The boundary conditions were chosen as follows:

$$X(\xi_1 = 0, \tau) = X(\xi_1 = 1, \tau) = C_x, \ X(\xi_2 = 0, \tau) = X(\xi_2 = 1, \tau) = C_x',$$

$$Y(\xi_1 = 0, \tau) = Y(\xi_1 = 1, \tau) = C_y, \ Y(\xi_2 = 0, \tau) = Y(\xi_2 = 1, \tau) = C_y',$$

$$X(\xi_1, \tau = 0) = C_x, \ X(\xi_2 = 0, \tau) = C_x', \ Y(\xi_1, \tau = 0) = C_y,$$

$$Y(\xi, \tau = 0) = C_y', \ \mathbf{u}(\xi, 0) = \mathbf{v}(\xi, 0) = 0,$$

where ξ_1 and ξ_2 are the two reduced Cartesian coordinates. In another
set of calculations, random initial conditions were chosen. Diverse types of
patterns formed for the hyperbolic system are presented with comments in
the following.

(a)   *Division of Patterns and Competition between Stripes
and Hexagons*

In the regime of $N_{\mathrm{rd}} = O(10^{-1})$ and the parameter regime where only an
unstable steady state is available (i.e., $A = 0.5$ and $B = 0.09$) a phenomenon
akin to cell division in biology was observed. Regardless of whether the
initial conditions are random or nonrandom, the system initially had no
clearly recognizable organized patterns. As time reaches the intermediate
time regime, the initially random patterns get synchronized to form a circular spot of high concentration in the middle of the square. If the initial
conditions were random, the patterns oscillate and continue to do so for a
long time. This behavior and patterns are similar to the case of the parabolic

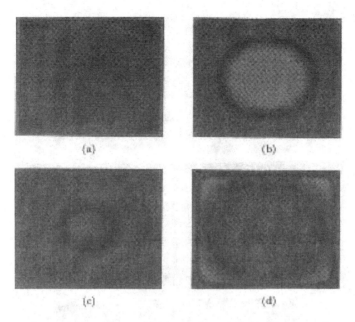

Fig. 22.11. Patterns formed by parabolic reaction–diffusion equations with random initial conditions. The hyperbolic system at high N_{rd} or with nonrandom initial conditions shows very similar patterns. Thus not shown here. Reproduced with permission from M. Al-Ghoul and B. C. Eu, *Physica D* **97**, 531 (1996) © Elsevier, B.V.

system for the same value of A and B. They are shown in Fig. 22.11. In the case of the parabolic system, the patterns were stable with respect to all sorts of small perturbations to the initial conditions with random noise, for it was able to synchronize in space and time to produce coherent oscillatory patterns shown in Fig. 22.11.

For the hyperbolic system, if the initial conditions were random, the circular pattern eventually split into two and then into four circular patterns which moved to the four corners of the square. The amplitudes of the patterns oscillated synchronized, thus producing patterns of a depressed concentration. These patterns then merged and divided into more circular patterns which either elongated or vanished, but they maintained symmetry while becoming more intricate. There can also occur patches of hexagons and stripes oriented differently. This process of organization and synchronized oscillations of local structures continues until the local symmetric patterns lose their stability and develop irregular stripe structures.

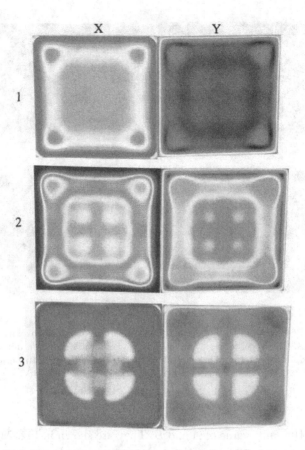

Fig. 22.12.   Evolution of patterns in the hyperbolic reaction–diffusion system in the case of $N_{\mathrm{rd}} = 0.1$ at random initial conditions. Reproduced with permission from M. Al-Ghoul and B. C. Eu, *Physica D* **97**, 531 (1996) © Elsevier, B.V.

This series of evolution of patterns from the stage of patterns similar to those shown in Fig. 22.11 is shown in the sequence of patterns shown in Figs. 22.12. In these figures, the blue means a low concentration of species X and Y whereas the brown means a high concentration of X and Y, the colors in between meaning intermediate concentrations in the order of the colors from blue to red. Therefore, where X is low in concentration, Y is high in concentration, for example. Observe that although the symmetry of the patterns are basically the same for X and Y except that they are at the opposite end of the color code spectrum for the magnitudes of concentration, the fine structures are not the same, and it reflects that their governing equations are not symmetric with respect to X and Y. We also observe that the patterns are created on the basic hexagonal lattice arranged on mutually crossing straight lines as the spots on the lattice points get larger than the lattice spacing and then merge together. The crossing lines may be regarded as the intersections of characteristic planes with the surface of the figure at a given time τ and the basic hexagonal structure created is periodic oscillations of concentrations in the characteristic surfaces of the two-dimensional hyperbolic reaction–diffusion equations. The pattern splitting-oscillation behavior continues in a perfectly symmetrical fashion as long as the system is kept out of equilibrium. The system eventually reaches a saturation point which is a state in which the calortropy production reaches an asymptotic value according to the computation of calortropy production simultaneously performed with the patterns. The pattern therefore is maintained at a high but asymptotic state of energy and matter consumption. At this stage, continuous competition between stripes and hexagonal structures is active; sometimes one symmetry pattern wins over the other and vice versa, cannibalizing each other. An example of this phenomenon can be seen in the figures in the last stage of evolution shown in Fig. 22.14 panel 9. This competition is a result of interactions of different frequency modes and is not observable in the parabolic system.

However, if the state of system is near the critical (Hopf) point, then one pattern completely wins over the other. The same situation arises if the evolution is started from a 16×16 square in the middle of the grid of 128×128 mesh points and if the system is perturbed by a 1% random noise added to the initial conditions. It also must be noted that even if the same parameter set as used for hyperbolic system is used for the parabolic system, it has not produced the patterns exhibited by the hyperbolic system shown in Figs. 22.12–22.13. Neither does there appear a hexagonal structure. Although a circular spot appears and eventually splits up into four spots which move to the four corners in a way similar to Fig. 22.11 — and this behavior repeats over time — none of the transient patterns seen in the hyperbolic system could be observed in the case of the parabolic system. It must be also noted that the transient patterns of the hyperbolic system are not so transient as the term implies, but they are fairly long lived and the lifetime of patterns is fairly long and comparable to those of the patterns of the parabolic system shown in Fig. 22.11.

In summary, in the case of nonrandom initial conditions the evolution of patterns from initially random patterns is quite different from and less

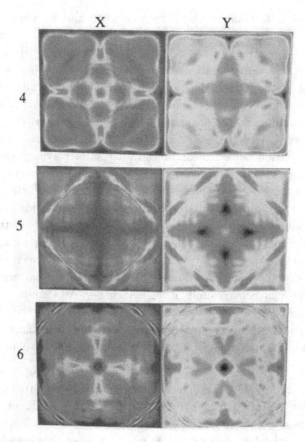

Fig. 22.13.   Figure 22.12 continued. Reproduced with permission from M. Al-Ghoul and B. C. Eu, *Physica D* **97**, 531 (1996) © Elsevier, B.V.

intricate than the sequence presented in Figs. 22.12–22.13 for random initial conditions. In fact, the pattern evolution was found to be rather similar to those parabolic system, and Fig. 22.12 may well substitute for the pattern evolution for the hyperbolic system subjected to the boundary and nonrandom initial conditions comparable with those for the parabolic system.

(b)   *Superposition of Hexagons and Spiral Waves*

As the reaction–diffusion number is further lowered from $N_{rd} = 0.1$, more interesting and complicated phases and patterns appear. For example, if N_{rd} is set equal to 0.01 with nonrandom initial conditions, hexagonal structures begin to form initially, but defects begin to develop and minute spirals

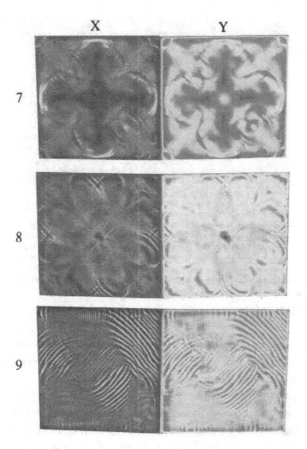

Fig. 22.14. Evolution of spirals in the hyperbolic reaction–diffusion system at $N_{rd} =$ 0.01. Reproduced with permission from M. Al-Ghoul and B. C. Eu, *Physica D* **97**, 531 (1996) © Elsevier, B.V.

around the defects begin to form on the basic hexagonal–rhombic texture as shown in Fig. 22.15.

In the case of nonrandom initial conditions, spirals appear, but do not grow. In contrast to this, if the initial conditions are made random, the chemical inhomogeneity tend to be redistributed to patterns in an ordered manner and then hexagonal–rhombic structures emerge as shown in Fig. 22.16 panel b. The boundary conditions are the same as for Fig. 22.15. The sequence of evolution is shown in Figs. 22.16–22.18. In the regime of $N_{rd} = 0.01$ a mixture of hexagonal, rhombic, and square structures is

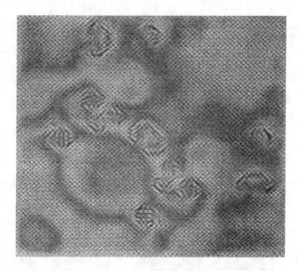

Fig. 22.15. Evolution of spirals in the hyperbolic reaction–diffusion system at $N_{rd} = 0.01$. Reproduced with permission, from M. Al-Ghoul and B. C. Eu, *Physica D* **97**, 531 (1996) © Elsevier, B.V.

organized along the lines which appear to be intersections of characteristic surfaces with the plane of the figure (at $\tau = 500$). There then develop structures at the four corners and along the boundaries. The boundary structures appear to be symmetrical propagating waves, and structures at the four corners to be also symmetrical. Subsequently, a pair of concentration defects of high concentration (black holes) appear inside as shown in the panel a of Fig. 22.16. Panel b of Fig. 22.16 is a magnification of a hexagonal–rhombic–square region in the panel a. The tips then develop into a pair of spirals of opposite chirality as shown in panel a of Fig. 22.17 at $\tau = 550$ and in panel b at $\tau = 600$, while the structures at the four corners and along the boundaries maintain the wavy patterns although with some minor changes in them. The spiral grow and collide with each other with the boundary structures as shown in Fig. 22.18 panel a at $\tau = 800$. Eventually, the pattern develops into an irregular mixture of hexagonal patterns, stripes, and maize-like structures as shown in panel b. This happens at $\tau = 1000$. We note that these structures of various forms grew out of basic texture of a hexagonal structure, when spots along the characteristic lines merge together; see Fig. 22.16.

The boundary conditions for these three set of figures are the same as for Fig. 22.15. In the regime of $N_{rd} = 0.01$, a mixture of hexagonal,

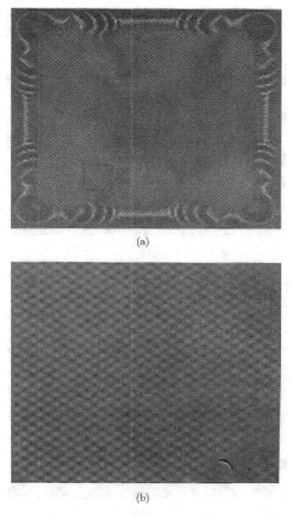

Fig. 22.16.   Evolution of spirals in the hyperbolic system. The boundary conditions are the same as in Fig. 22.13. Reproduced with permission from M. Al-Ghoul and B. C. Eu, *Physica D* **97**, 531 (1996) © Elsevier, B.V.

rhombic, and square structures is organized along the lines which appear to be intersections of characteristic surfaces with plane of figure at time $\tau = 500$. Then there develop structures at the four corners and along the boundaries. These boundary structures appear to be propagating waves; see Fig. 22.16. Subsequently, a pair of concentration defects of high concentration (black holes) appear inside the square in panel a. Panel b is a

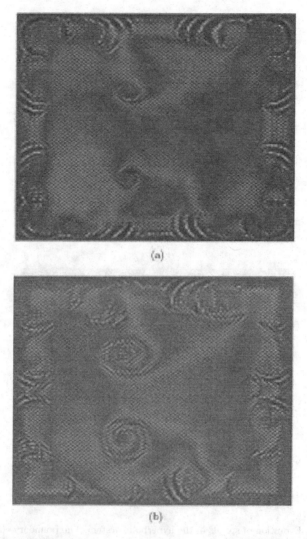

(a)

(b)

Fig. 22.17.   Evolution of spirals in the hyperbolic reaction–diffusion system (continued from Fig. 22.6). Reproduced with permission from M. Al-Ghoul and B. C. Eu, *Physica D* **97**, 531 (1996) © Elsevier, B.V.

magnification of a hexagonal–rhombic–square region in panel a. The tips then develop into a pair of spirals of opposite chirality shown in Fig. 22.17 panel a ($\tau = 550$) and panel b ($\tau = 600$) while the structures at the four corners and along the boundaries maintain the wavy patterns although there occurs some minor changes in them. The spirals grow and collide with each

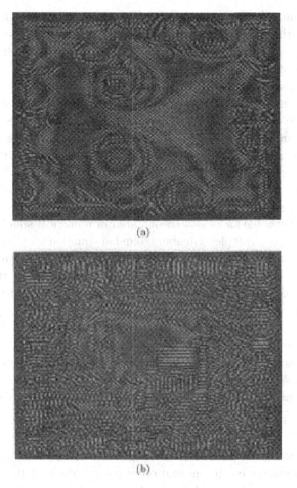

Fig. 22.18. Evolution of spirals in the hyperbolic reaction–diffusion system (continued from Fig. 22.7). Reproduced with permission from M. Al-Ghoul and B. C. Eu, *Physica D* **97**, 531 (1996) © Elsevier, B.V.

other and with the boundary structures as shown in Fig. 22.18 panel a ($\tau = 800$). The pattern, eventually, develops into an irregular mixture of hexagonal patterns, stripes, and maze-like structures shown in Fig. 22.18 panel b ($\tau = 1000$). We observe that the various patterns are grown out of the basic texture of a hexagonal structure, when the spots along the characteristic lines merge together. This sequence seems to suggest a scenario for pattern formation and a genesis of patterns.

To summarize the aforementioned sequence, the hexagonal–rhombic phase was time-dependent and their amplitudes are oscillating. Before spirals appear in the middle of the square, two spots of high concentration are visible in the middle of the hexagonal region, and then an almost symmetrical pair of spirals of opposite chirality begins to form and therefrom grow on the hexagonal texture as shown in panels in the figures shown above. These spirals are multiarmed, having two or three arms. They have a minimum of wavelength approximately 3–4 times larger than that of the hexagonal structure. They then begin to interfere and destroy each other, leaving a kind of turbulent spots behind until the whole pattern becomes chaotic and remains so, as shown in Fig. 22.18 panel b. It was noticed that spirals and incomplete rings develop near the boundaries. We note that this behavior is also encountered in numerical simulations for a system obeying a complex Ginzburg–Landau equation with complex coefficients in a square geometry by other authors.[27] If N_{rd} is varied slightly around 0.01 while other parameters are kept the same as for the previous figures, spirals still emerged in a pair of opposite chirality, but their spatial orientations were altered, for example, by 90 degree with respect to the spirals of the previous value of N_{rd}. *This seems to suggest that the evolution of spirals is sensitive to the value of the reaction–diffusion number. When the magnitude of the interaction increased, the pattern decayed to a very complicated and irregular one. Experimentally, the CIMA (chlorine dioxide-iodine-malonic acid) reaction exhibited a very similar pattern.*[28]

(c)   *Chaotic Patterns at Low Reaction–Diffusion Numbers*

When N_{rd} is further lowered to 0.001 at the nonrandom initial conditions with the boundary conditions and other parameters kept the same as for the previous cases, the wave behavior disappeared and no disturbances were observed. Only homogeneous oscillations were encountered. As the initial conditions were made random, there emerged random patterns which did not get organized to a steady regular pattern over a sufficiently long time span. Examples of such patterns are shown in Figs. 22.19–22.22. These irregular patterns persisted in the intermediate time regime, but later decayed into minor structures (not shown here) which were randomly dispersed all

---

[27] J. A. Sepulchre and A. Babloyantz, *Phys. Rev. E* **48**, 187 (1993).

[28] B. Rudovics, J.-J. Perraud, P. De Kepper and E. Dulos, in *Far-From-Equilibrium Dynamics of Chemical Systems*, eds. J. Gorecki, A. S. Cukrowski, A. L. Kawcynski, and B. Nowakowski (World Scientific, Singapore, 1994).

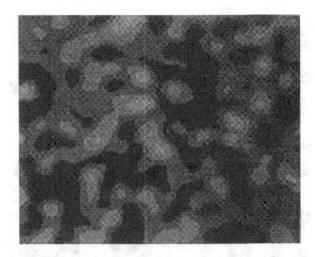

Fig. 22.19. Chaotic irregular patterns in the hyperbolic reaction–diffusion system. Reproduced with permission from M. Al-Ghoul and B. C. Eu, *Physica D* **97**, 531 (1996) © Elsevier, B.V.

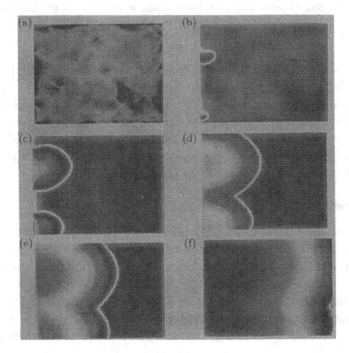

Fig. 22.20. Solitary waves in the bisatble region in the case of $N_{rd} = 0.7$. Reproduced with permission from M. Al-Ghoul and B. C. Eu, *Physica D* **97**, 531 (1996) © Elsevier, B.V.

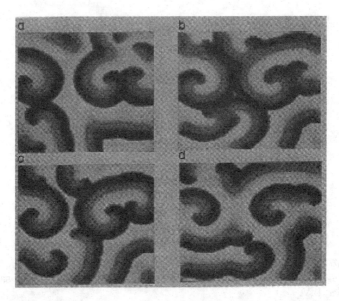

Fig. 22.21. Spirals originating from the unstable steady state in the bistable region perturbed by a 1% random noise. Reproduced with permission from M. Al-Ghoul and B. C. Eu, *Physica D* **97**, 531 (1996) © Elsevier, B.V.

over the space. Their oscillations were not regular in time. And this is consistent with the mode in which the one-dimensional system become temporally chaotic. We have found in the one-dimensional case that the hyperbolic system loses its stability and temporal chaos appears as the reaction–diffusion number is lowered to $N_{rd} \approx 0.001$. The power spectrum, calculated clearly indicated spatial chaos. For the power spectrum, see Fig. 6 in a paper by M. Al-Ghoul and B. C. Eu, *Physica D* **97**, 531 (1996).

(d) *Patterns in the Bistable Region*

When the parameters (parameter values: $A = 0.6202$ and $B = 0.02$) are taken such that the system is in the bistable region, some interesting solutions were obtained such as solitons and spirals when the other parameters such as N_{rd} and f are changed. When random initial conditions were taken for the aforementioned parameter set where the three steady states of concentrations have the values: 0.2920, 0.2667, and 0.02335, and $f = 1.31$ and $N_{rd} = 0.7$, the system was able to remove and distribute the chemicals to

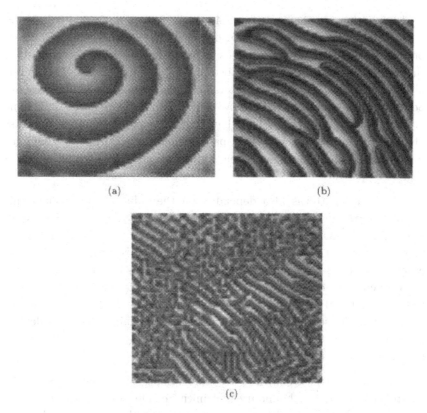

(a) (b)

(c)

Fig. 22.22. Spirals at N_{rd} lowered to 0.5. Reproduced with permission from M. Al-Ghoul and B. C. Eu, *Physica D* **97**, 531 (1996) © Elsevier, B.V.

reach a homogeneous state from which defects appeared on the boundary of the square.

These defects grew until they collided and merged together. Once they coalesced with each other, they formed a packet which then traveled at a constant speed to the other side of the square and vanished. See Fig. 22.20 where the sample of the sequence is shown. These are solitary waves. Such a solitary wave was observed and reported in the literature.[29]

---

[29]S. Kai and H. Miike, *Physica A* **204**, 346 (1994).

For another set of parameters ($A = 0.6202$, $B = 0.02$, $f = 0.8$, $N_{rd} = 0.8$) spirals were observed as shown in Fig. 22.21, when the initial and boundary conditions were so chosen as to make them correspond to the unstable steady state perturbed by 1% random noise. The spirals obtained do not have fixed centers, but the centers meander around. Some of spirals merge when they collide and form a cusp, and some others repulse each other when they come close to each other. In most of cases, spirals were created in pairs of opposite chirality. What was noticeable was that after some time the spirals got elongated as shown in Fig. 22.21 panel d.

The spiral patterns also depended on the value of N_{rd}. When this parameter was lowered, the radius of curvature of the spirals increased. Figure 22.22 panel a shows only one spiral at $N_{rd} = 0.5$, which meanders in the square. At $N_{rd} = 0.35$ the pattern looks like stripes. In fact, at early stages the pattern is nothing but two patches of stripes, which are perpendicular to each other. With time, the stripes tend to get parallel to each other and end up like the one shown in Fig. 22.22 panel b, where roughly parallel stripes (lamellar structures) are maintained, but defects move around continuously. A further decrease of N_{rd} to 0.1 resulted in an irregular pattern shown in Fig. 22.22 panel c.

The spiral patterns shown above, therefore, seem to suggest that with controlled selection of boundary and initial conditions and the reaction-diffusion number N_{rd} the telegraphist equations for the reaction model (i.e., nonlinearity) spiral patterns and patterns in general can be produced in a particular form.

(e) *Calortropy Production and Pattern Formation*

The system maintains patterns — e.g., oscillatory structures — at the expense of energy and matter supplied by the surroundings and within the system. Particularly, oscillatory structures in calortropy are accompanied by temporal oscillatory spectra characteristic to the patterns. Such spectra indicate that patterns formed require energy and matter to be formed and maintained, the amount of which appears to be characteristic to the type and form of the pattern in question. The study of this aspect of energy and matter consumption is quite sketchy at present, and further thorough study in the future seems to be in order, if we would like to understand physically why a particular pattern is formed instead of another from the thermodynamic and energy point of view. The reader is referred to the references cited for the present discussion in connection with the calortropy production, which we have not taken up in the present chapter because of the space limitation.

22.3.4.2. *Remarks on the Cubic Reversible Reaction Model*

The cubic reversible reaction model we have discussed consists of a set of chemical reactions *with an autocatalytic step providing a feedback loop*. It should be noted that chemical reactions studied and known to give rise to oscillatory structures usually involve a feedback loop in the nest of chemical reactions. Mathematically, such a feedback loop is necessary for nonlinear oscillations. In the sense that such a feedback loop is incorporated into the reaction–diffusion equations discussed in this work, the telegraphist equations underlying the evolution equations made use of in the present study could be said to be generic evolution equations representative of diverse nonlinear chemical oscillation phenomena investigated in the literature. Since they are, nevertheless, generalized hydrodynamic equations within the framework of the irreversible thermodynamics described in the previous chapters, it can be safely stated that we have *a generalized hydrodynamics road map* to describe, within the framework of irreversible thermodynamics theory that is thermodynamically consistent with the laws of thermodynamics, diverse chemical oscillatory phenomena observed in nature; they may include chemical, biochemical, geochemical–geological oscillatory (or periodic structure producing) phenomena that surely include the experimentally much studied Liesegang phenomena, for example. The patterns shown in the previous sections of this chapter arise when parameters of reactive systems (e.g., the reaction–diffusion number) and initial and boundary conditions are changed. Therefore to make the theory predictive of the patterns formed, the first step of studies must be devoted to classifying various conditions according to the types of patterns (e.g., hexagons, spirals, stripes, etc.) by systematically studying their range of parameters and initial and boundary conditions. The present chapter is prepared in the hope to offer the readers the range of patterns potentially predictable by the telegraphist equations or, more generally, generalized hydrodynamic equations. The numerical studies presented, we believe, provide a suggestive guidepost for numerical studies of thermodynamically consistent generalized hydrodynamic theory of nonlinear oscillatory phenomena, which are observed in macroscopic processes in nature and laboratory.

22.4.   Liesegang and Related Oscillatory Phenomena

Chemical oscillation phenomena described in the previous section of this chapter have been anteceded by the Liesegang phenomena,[30] in which periodic precipitation patterns (periodic bands) are formed in the wake of a moving chemical reaction front. Such intriguing phenomena were first discovered by the German chemist Raphael Eduard Liesegang in 1896, when he accidentally dropped a solution of silver nitrate ($AgNO_3$) on a thin layer of gelatin containing potassium dichromate ($K_2Cr_2O_7$). Since then, similar banded precipitation structures were observed for numerous systems and studied by many research workers over many years before the Belusov–Zhabotinski oscillation reactions and other kindred chemical oscillation have been studied in the recent years. Experimental observation of Liesegang ring or periodic discrete band formation was a harbinger of experimental and theoretical studies of the phenomena in diverse scientific disciplines[31] related to them. Typical examples are given in Fig. 22.23. Although Liesegang realized that the formation of these patterns was related to the movement of molecules or ions toward each other, he was unable to explain the origin of this surprising and intriguing phenomena, and his finding remained, and still is a cause for, an enduring scientific curiosity. There have been numerous investigations on the Liesegang phenomena, either experimental or theoretical. There have been reviews of the phenomena and related theoretical propositions. Recently, Sadek and Sultan[32] have given a fairly extensive review of literature on the appearance of Liesegang pattern formations observed in nature in various scientific disciplines. It is fair to state that the subject is not closed as yet and there should be a lot to be explored theoretically and experimentally.

There are some theoretical models[33] proposed to comprehend the phenomena on the basis of reaction–diffusion equations for systems of nested

---

[30]R. E. Liesegang, *Photo. Archiv.* **221** (1896); *Chemische Reactionen in Gallerten* (Düsseldorf, 1889); *Colloid Chemistry*, ed. J. Alexander (Chemical Catalog Co., New York, 1926), pp. 783. Also see H. K. Henisch, *Crystal Growth in Gels* (Pennsylvania State University Press, University Park, 1970) for historical review.

[31]On a recent readable account of this subject in the field of chemistry and micro- and nanotechnology, see B. A. Grzybowski, *Chemistry in Motion: Reaction–Diffusion Systems for Micro- and Nanotechnology* (Wiley, New York, 2009).

[32]S. Sadek and R. Sultan, in *Precipitation Patterns in reaction–diffusion Systems*, ed. I. Lagzi, Research Signpost(Kerala, India) 2010, pp. 1–43.

[33]M. Flicker and J. Ross, *J. Chem. Phys.* **60**, 3458 (1974).

Fig. 22.23. Examples of Liesegang patterns grown in gels for several sparingly soluble salts: (a) direct spacing in $Ni(OH)_2$ system; (b) revert spacing in CdS system; (c) fractals in the $La_2(C_2O_4)_3$ system; (d) helicoidal pattern in $CuCrO_4$; (e) spirals; and (f) ripples in HgI_2 system. Reproduced from M. Dayeh, MSc Thesis, American University of Beirut, Lebanon, 2014; courtesy of M. Dayeh.

chemical reactions, and they invariably involve reaction–diffusion equations and some of them involve some sort of a feedback loop or, in the chemical terminology, an autocatalytic step, in which the reaction to form a product is catalyzed by its own kind. There have been only a few numerical solutions of the reaction–diffusion equations, but there is an example of numerical solutions for the theoretical model equation based on the Cahn–Hilliard theory originally developed for another line of investigation in condensed matter physics. We will discuss this aspect later in this section.

The spatiotemporal dynamics of precipitation patterns observed in Liesegang phenomena has been shown to obey several scaling laws irrespective of the geometry of the system and the salt pairs used. There are reported three generic empirical laws regarding the observed periodic structures: They are (1) time law; (2) spacing law and, associated with it, the Matalon–Packter law about the spacing coefficient; and (3) width law.

These laws will be discussed below. A correct theory of the phenomena that one may develop in connection with the Liesegang phenomena should be able to confirm them correctly and, if possible, they should be rigorously derived from the theoretical model employed. It should be an important task of the theory since they are not sufficiently rigorously proved from the theoretical standpoint. It is difficult to do so, because the problem involves nonlinear mathematically, defying an exact theoretical treatment. The theoretical models employed so far employ reaction–diffusion equations with nonlinear source terms posing a great challenge mathematically.

22.4.1. *Empirical Laws of the Phenomena*

22.4.1.1. *Time Law*

The time law is concerned with the position x_n of the nth band, measured from the initial interface of the reagents; the position x_n is proportional to $\sqrt{t_n}$ in which t_n is the time elapsed until appearance of the band. This time law,[34] as a matter of fact, appears to suggest that the phenomenon is intimately related to the diffusive dynamics involved in the phenomenon, because then the relation

$$x_n = A_t \sqrt{t} \tag{22.74}$$

becomes easily comprehensible. Therefore, it is expected to be fulfilled by a model that incorporates a diffusive motion of reaction front. This is probably the motivation to employ a reaction–diffusion equation.

22.4.1.2. *Spacing Law and Matalon–Packter Law*

Liesegang phenomena produce a band structure spacing either in direct or reverse manner, namely, the spacings increase with time or decrease with time of their appearance. In the case of direct spacing, the spacing law[35] is expressible as

$$\lim_{n \to \text{large}} x_n = A_s \left(1 + p\right)^n \quad (p > 0), \tag{22.75}$$

where the positive parameter p is called the spacing coefficient and A_s the amplitude of spacing and n is the position of the band in the sequence

---

[34] H. W. Moarse and G. W. Pierce, *Proc. Acad. Sci.* **38**, 625 (1903).
[35] K. Jablcznski, *Bull. Soc. Chim. France* **33**, 3592 (1923).

of appearance from the origin. The spacing coefficient p is nonuniversal parameter depending on the material (chemicals) of the system: It is expressible by the Matalon–Packter law.[36] It is an empirically deduced law expressible by the formula

$$p = F(b_0) + G(b_0)\frac{b_0}{a_0}, \qquad (22.76)$$

where F and G are decreasing function of b_0 and the parameters a_0 and b_0 are the initial concentrations of chemical species A and B which combine to form the precipitate C in the reaction $A + B \rightarrow C$, forming the band. The empirical parameters should be derived theoretically and expressed in terms of system parameters. There are theories[37] proposed for the Matalon–Packter law on the basis of reaction–diffusion equations, but the formulas obtained by them are not rigorous derivations based on exact solutions of the model evolution equations proposed; they are generally of a qualitative nature based on intuitive arguments. Nevertheless, these empirical laws are useful for comprehending qualitative aspects of the Liesegang phenomena observed. They, particularly, the Matalon–Packter law, should be an object of more rigorous molecular theory investigation, it being not only a proof of the theory being making sense with experiment, but also being intriguingly reminiscent of the laws governing atomic spectral sequences, although its physical origin is obviously different entirely. In this sense, it is an intriguing law from the physical theoretic point of view, because there is a deeper underlying law lurking behind it.

22.4.1.3. *Width Law*

The width law states that the width W_n of the nth band is an increasing function of n and is typically expressible[38] by

$$W_n = wx_n^\alpha \quad (\alpha > 0), \qquad (22.77)$$

where the coefficient w together with the exponent α appears to depend also on materials involved. The parameter α is also an empirical parameter.

---

[36] R. Matalon and A. Packter, *J. Colloid Sci.* **10**, 46 (1955); A. Packter, *Kolloid Z.* **142**, 109 (1955).

[37] C. Wagner, *J. Colloid Sci.* **5**, 85 (1950); Ya. B. Zeldovich, G. I. Barrenblatt and R. L. Salganik, *Sov. Phys. Dokl.* **6**, 869 (1962); L. Galfi and Z. Rácz, *Phys. Rev. A* **38**, 3151 (1988); M. Droz, Z. Rácz, and J. Schmidt, *Phys. Rev. A* **39**, 2141 (1989); M. Droz, *J. Stat. Phys.* **101**, 509 (2000).

[38] See S. C. Müller, S. Kai and J. Ross, *J. Chem. Phys.* **86**, 4078 (1982).

Experimentally, these parameters are difficult to measure with certainty and consequently have not received much attention. Therefore, a reliable theoretical determination should be all the more necessary.

22.4.2. *Models Employed for the Laws*

We will examine a theoretical deduction of the laws, especially, the Matalon–Packter law, reported in the literature, especially, by Antal *et al.*, [39] which is the most challenging of the aforementioned laws. For this purpose, the reaction–diffusion equations in the ion-product supersaturation theory are assumed as follows:

$$\frac{\partial a}{\partial t} = D_a \nabla_x^2 a - k\theta(ab - q^*) - \lambda abd, \tag{22.78}$$

$$\frac{\partial b}{\partial t} = D_b \nabla_x^2 b - k\theta(ab - q^*) - \lambda abd, \tag{22.79}$$

$$\frac{\partial d}{\partial t} = k\theta(ab - q^*) + \lambda abd \tag{22.80}$$

for reaction A + B → D, where a and b stand for the concentration of A and B, D_A and D_B are diffusion coefficients of A and B; and k and λ are rate constants; $\theta(ab - q^*)$ is the Heaviside step function, which is equal to unity if the argument is positive or equal to zero, but the function is equal to zero otherwise; q^* is the maximum of the solubility product ab. The last terms λabd in the equations represent the coagulation rate with the precipitate present — this condition implies that these rate terms become meaningful for a system which has already formed a suspension of d before coagulation (i.e., precipitation).[40] The initial conditions are given by $a(x,0) = a_0\theta(-x)$, $b(x,0) = b_0\theta(x)$, and $d(x,0) = 0$. This theory is based on the assumption that precipitation of nondiffusing precipitate occurs at site x at time if the ion-product exceeds the threshold value q^*. The following is the deduction of the Matalon–Packter law on the basis of (22.78)–(22.80) in the way we understand from the literature cited.

---

[39]T. Antal, M. Droz, J. Magnin, Z. Rácz and M. Zrinyi, *J. Chem. Phys.* **109**, 9479 (1998).
[40]Therefore, this theory of the Liesegang phenomena involves, at least, a two-time scale process, and the last terms might have been introduced as another set of equations holding in the longer time scale of the process.

For the purpose of derivation, Antal *et al.* approximate $a(x,t)$ and $b(x,t)$ just before the appearance of the $(n+1)$th band in the forms

$$a(x,t) = a_0(1 - x/\sqrt{2D_a t}), \tag{22.81}$$

$$b(x,t) = b_0 \frac{(x - x_n)}{\sqrt{2D_b\,(t - t_n)}}. \tag{22.82}$$

These can be deduced from the lowest order solutions of (22.78) and (22.79). Then it is invoked that the condition for the $(n+1)$th band to appear is that the ionic product reaches a maximum value q^* at $x = x_{n+1}$ at $t = t_{n+1}$. Thus there holds the equality

$$a(x_{n+1}, t_{n+1})b(x_{n+1}, t_{n+1}) = q^*. \tag{22.83}$$

Therefore, its spatial derivative at x_{n+1} and t_{n+1} vanishes:

$$\left[\frac{\partial}{\partial x}a(x,t)b(x,t)\right]_{x_{n+1},t_{n+1}} = 0, \tag{22.84}$$

because q^* is independent of x and t. Upon use of (22.81) and (22.82), we obtain the relation between t_{n+1} and peak positions of the bands:

$$\sqrt{2D_a t_{n+1}} = 2x_{n+1} - x_n. \tag{22.85}$$

By making use of the remaining condition (22.83), we find the

$$\frac{q^*}{a_0 b_0}\sqrt{\frac{D_b}{D_a}} = \frac{(x_{n+1}/x_n - 1)\left(1 - \frac{x_{n+1}/x_n}{2x_{n+1}/x_n - 1}\right)}{\sqrt{(2x_{n+1}/x_n - 1)^2 - (x_{n-1}/x_n)^2\,(2x_n/x_{n-1} - 1)^2}}.$$

Let

$$\frac{x_{l+1}}{x_l} = 1 + p_{l+1} \quad (l = 1, 2, \ldots, n, \ldots). \tag{22.86}$$

Then it follows

$$\frac{q^*}{a_0 b_0}\sqrt{\frac{D_b}{D_a}} = \left[\frac{p_{n+1}}{2p_{n+1} + 1}\right]^2 \frac{1}{\sqrt{1 - \frac{(2p_n + 1)^2}{(2p_{n+1} + 1)^2(1 + p_n)^2}}}.$$

If there exists a limit such that

$$\lim_{l \to \infty} p_{l+1} = p,$$

then in the limit of large n

$$\Phi\left(a_0, b_0, q^*, D_a, D_b\right) := \frac{q^*}{a_0 b_0} \sqrt{\frac{D_b}{D_a}}$$

$$= \frac{p^2}{(2p+1)^2 \sqrt{1 - \frac{1}{(1+p)^2}}}. \qquad (22.87)$$

If the right-hand side of this equation is expanded in power series of p, we get

$$\Phi = \frac{1}{\sqrt{2}} p^{3/2} \left[1 + O(p)\right]. \qquad (22.88)$$

However, this is not the Matalon–Packter (MP) law since the a_0-dependence is of $-2/3$ in contrast to -1 by the MP law.

To get out of this quandary posed by (22.87), Antal *et al.* argue that in the region of $0 \lesssim \epsilon < p < 0.5$ the p-dependence of Φ is approximately linear as shown in the below figure. Hence in the aforementioned region of p the MP law holds approximately:

$$p \approx \alpha \Phi = \alpha \frac{q^*}{a_0 b_0} \sqrt{\frac{D_b}{D_a}}. \qquad (22.89)$$

Whether the approximate relation (22.89) represents a derivation of the law or not is up to the mathematical taste of the reader, but it is rationalized

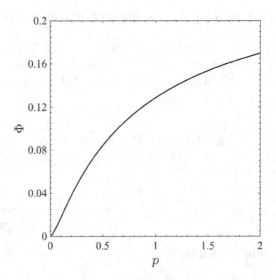

Fig. 22.24.   Plot of Φ vs. p.

as shown on the basis of reaction–diffusion equations in the supersaturation model theory. Other attempts at derivation in the literature are deduced invariably on the basis of qualitative arguments made with a reaction–diffusion equation. Therefore, a mathematically sufficiently rigorous proof of the law still remains an open problem in our opinion.

In the following, given the governing equations (22.78)–(22.80), we formulate a formal theory of determining the properties that govern the precipitation bands including the aforementioned laws. For this purpose, we first show that the governing equations (22.78)–(22.80) can be reduced to a single integral equation for d, solution of which will help a more accurate derivation of the laws. For this purpose, we first define functions $s_a(x,t)$ and $s_b(x,t)$ by transformations

$$s_a = a + d, \tag{22.90}$$

$$s_b = b + d. \tag{22.91}$$

The initial and boundary conditions of s_a and s_b are determined by those of a, b, and d. By using (22.78)–(22.80) we then obtain the decoupled equations for s_a and s_b:

$$\frac{\partial s_a}{\partial t} = D_a \nabla_x^2 s_a, \tag{22.92}$$

$$\frac{\partial s_b}{\partial t} = D_b \nabla_x^2 s_b, \tag{22.93}$$

owing to the assumption that d does not diffuse through space (i.e., immobile) and hence $D_D \nabla_r^2 d = 0$.

The general solutions of (22.92) and (22.93) are

$$s_i(x,t) = \frac{1}{2\sqrt{\pi D_i t}} \int_{-\infty}^{\infty} dx' \, s_i^0(x') e^{-(x-x')^2/4D_a t} \quad (i = a, b), \tag{22.94}$$

where $s_i^0(x)$ is the initial condition at $t = t_n$ of species i:

$$s_a^0(x) = \begin{cases} a_0 & \text{for } 0 \leqslant x < x_n, \ t = t_n, \\ 0 & \text{for } x > x_n, \ t = t_n, \end{cases} \tag{22.95}$$

$$s_b^0(x) = \begin{cases} 0 & \text{for } 0 \leqslant x < x_n, \ t = t_n, \\ b_0 & \text{for } x > x_n, \ t = t_n. \end{cases} \tag{22.96}$$

Note that as the nth band is formed, the ions a and b in the vicinity of $x = x_n$ at time $t = t_n$ are equal to zero. Using the initial condition (22.95), we obtain

$$s_a\left(x,t\right) = \frac{a_0}{2\sqrt{\pi D_a t}} \int_{-\infty}^{0} dx' e^{-\frac{(x-x')^2}{4D_a t}} = \frac{a_0}{2}\left[1 - \phi\left(\frac{x}{2\sqrt{D_a t}}\right)\right], \quad (22.97)$$

where $\phi\left(z\right)$ is the error function:

$$\phi\left(z\right) = \frac{2}{\sqrt{\pi}} \int_{0}^{z} dy\, e^{-y^2} = \frac{2}{\sqrt{\pi}} \sum_{k=0}^{\infty} \frac{(-1)^k z^{2n+1}}{k!(2k+1)}$$

$$= \frac{2}{\sqrt{\pi}} e^{-z^2} \sum_{k=0}^{\infty} \frac{2^k z^{2k+1}}{1 \cdot 3 \cdots (2k+1)}. \qquad (22.98)$$

By using the solution procedure similarly to $s_a(x,t)$ and the initial condition (22.96), we obtain the general solution for (22.93)

$$s_b\left(x,t\right) = \frac{b_0}{2\sqrt{\pi D_b\left(t-t_n\right)}} \int_{0}^{x_n} dx' e^{-\frac{(x-x')^2}{4D_b(t-t_n)}}$$

$$= \frac{b_0}{2}\phi\left(\frac{(x-x_n)}{2\sqrt{D_b\left(t-t_n\right)}}\right). \qquad (22.99)$$

We remark here that solutions $s_a(x,t)$ and $s_b(x,t)$ obtained in (22.97) and (22.99) suggest that the approximations (22.81) and (22.82) are the leading order approximations of the error functions away from the peak position of the band under formation.

Noting that the diffusion dynamics post the nth precipitation band must be started anew past $x = x_n$ at $t = t_n$. Now we consider (22.80). If $ab < q^*$, integration of (22.80)[41] yields

$$d(t) = d_0 \exp\left[\lambda \int_{t_n}^{t} d\tau\, a\left(\tau\right) b\left(\tau\right)\right]. \qquad (22.100)$$

Therefore with (22.97) and (22.99) we find a nonlinear integral equation for $d(t)$:

$$\ln\left[d(t)/d_0\right] = \lambda \int_{t_n}^{t} d\tau \left[\frac{a_0}{2} - \frac{a_0}{2}\phi\left(\frac{x}{2\sqrt{D_a\tau}}\right) - d(\tau)\right]$$

$$\times \left[\frac{b_0}{2}\phi\left(\frac{(x-x_n)}{\sqrt{4D_b\left(\tau-t_n\right)}}\right) - d(\tau)\right]. \qquad (22.101)$$

---

[41]It appears that the first term on the right-hand side of (22.80) is superfluous and causes unwarranted complications. The model seems to be sufficient without the term. We prefer such a minimal model.

The solution of this integral equation provides the description of growth of the $(n + 1)$th precipitate band. Here the rate constant λ for the formation of precipitates may be treated as a small perturbation parameter. If that is possible, then (22.101) may be solved by an iterative procedure. Thus to the lowest order in d on dropping $d(\tau)$ on the right there follows the solution at the center of band position x_{n+1}:

$$
\ln \left[\frac{d(t)}{d_0\left(x_n\right)} \right]_{x=x_{n+1}}^{(0)}
$$
$$
= \frac{\lambda a_0 b_0}{4} \int_{t_n}^{t_{n+1}} d\tau \left\{ \left[1 - \phi \left(\frac{x}{2\sqrt{D_a \tau}} \right) \right] \phi \left(\frac{(x - x_n)}{2\sqrt{D_b\left(\tau - t_n\right)}} \right) \right\}_{x=x_{n+1}} .
$$
(22.102)

On the other hand, if $ab \geqslant q^*$, then integration of (22.80) yields a nonlinear integral equation of $d(t)$:

$$
\ln \left(d/d_0\right)(t) = \lambda \int_{t_n}^{t} d\tau s_a\left(x, \tau\right) s_b\left(x, \tau\right)
$$
$$
- \lambda \int_{t_n}^{t} d\tau \left[s_a\left(x, \tau\right) + s_b\left(x, \tau\right) \right] d\left(\tau\right)
$$
$$
+ \int_{t_n}^{t} d\tau \frac{k}{d\left(\tau\right)} + \lambda \int_{t_n}^{t} d\tau d^2\left(\tau\right) .
$$
(22.103)

On inserting (22.97) and (22.99) we obtain

$$
\ln \left(d/d_0\right)(t) = \lambda \frac{a_0 b_0}{4} \int_{t_n}^{t} d\tau \left\{ \left[1 - \phi \left(\frac{x}{2\sqrt{D_a \tau}} \right) \right] \phi \left(\frac{(x - x_n)}{2\sqrt{D_b\left(\tau - t_n\right)}} \right) \right\}
$$
$$
- \lambda \int_{t_n}^{t} d\tau \left[1 - \phi \left(\frac{x}{2\sqrt{D_a \tau}} \right) + \phi \left(\frac{(x - x_n)}{2\sqrt{D_b\left(\tau - t_n\right)}} \right) \right] d\left(\tau\right)
$$
$$
+ \int_{t_n}^{t} d\tau \frac{k}{d\left(\tau\right)} + \lambda \int_{t_n}^{t} d\tau d^2\left(\tau\right) .
$$
(22.104)

The solution of this integral equation provides the temporal profile of (d/d_0) in the neighborhood of t_{n+1} and its relation to $x_{n+1} - x_n$ together with its width. Therefore, the integral equation derived provides with a clearly stated formalism for determining various properties of the periodic precipitation band, but for the reason of limited time permitted we leave the mathematical investigation of the integral equation (22.103) to the future work.

22.4.3.　Cadmium Sulfide System

After Liesegang observed the well-defined visible bands in silver dichromate system, he investigated on different weakly soluble salts of initially segregated reactants that behaved similarly, and reported various spatial patterns observed by him and others. Particularly, Daus and Tower[42] in their study of the cadmium sulfide system, where a solution containing the sulfide ions (referred to as the outer electrolyte in the following) is allowed to diffuse in a hydrogel matrix containing the cadmium ions (also referred to as the inner electrolyte), observed a direct spacing type of periodic precipitation. Kant *et al.*[43] also studied Liesegang rings of cadmium sulfide in three gel media: gelatin, agar, and starch. A similar direct-spacing spatial pattern of cadmium sulfide in agar was obtained, whereas for the gelatin and starch media, they detected both direct and revert spacing type of rings, depending on whether the cadmium ions are allowed to diffuse into the gel containing sulfide or vice versa. In 1985, Ramaswamy and his coworkers reported a new observation made in cadmium sulfide salt that produces revert spacing followed by direct spacing in the same tube under certain specific pH ranges.[44]

22.4.3.1.　Recent Experimental Observations

More recently, the M. Al-Ghoul group reported[45] on the formation of a new precipitation pattern also observed in the cadmium sulfide system. It displayed a transition from parallel rings to spots with square–hexagonal symmetry when thresholds of some experimental parameters, such as the inner and outer concentrations, are crossed. This kind of transition has not been observed before.

In the experiment, the sulfide ions hydrolyze in water to produce hydroxide ions, which, owing to their higher diffusion coefficient, diffuse

[42]W. Daus and O. F. Tower, *J. Phys. Chem.* **33**, 605 (1929).

[43]B. M. Mehta and K. Kant, *Proc. Natl. Acad. Sci. India, Sec. A* **253** (1970).

[44]N. Palaniandavar, N. Kanniah, F. D. Gnanam and P. Ramaswamy, *Bull. Mater. Sci.* **7**, 105 (1985).

[45]M. Dayeh, M. Ammar and M. Al-Ghoul, *RSC Adv.* **4**, 60034 (2014).

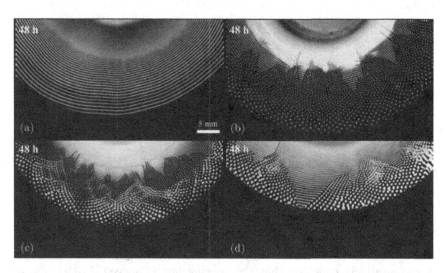

Fig. 22.25. The evolution of precipitation patterns within 48 h at constant outer concentration $[S^{2-}]_0 = 400$ mM and various inner concentrations $[Cd^{2+}]_0$: (A) $= 30$ mM; (B) $= 60$ mM; (C) $= 90$ mM; (D) $= 110$ mM; Gelatin $= 5\%$; Temp $= 22°$ C. Reproduced with permission from M. Dayeh, M. Ammar and M. Al-Ghoul, *RSC Adv.* **4**, 60034 (2014) © Royal Society of Chemistry.

faster than the sulfide ions to react with the inner cadmium ions to produce solid white cadmium hydroxide. Yellow cadmium sulfide is subsequently formed by ion exchange between the hydroxide ions of the diffusing sulfide ions. The transition threshold, the prevalence of spots versus rings, the wavelength of the selected pattern, the size of the resulting spots, and the percentage of their coverage are found to be controllable by adjusting the concentrations of the inner and outer electrolytes, temperature, gel thickness, addition of capping agent, ionic strength in the gel matric, and application of an external electric field. Two-dimensional patterns consisting of rings of dots formed are shown in Fig. 22.25 To shed more light on the control of spots and laws known to be obeyed by Liesegang precipitation patterns, the spacing between consecutive spots on neighboring rings are measured for two sets of plates (i.e., experiments). Set I is prepared at a constant outer concentration of $[S^{2-}]_0 = 250$ mM and the inner concentration of $[Cd^{2+}]_0$ ranging between 40 mM and 80 mM, whereas set II is at a constant inner concentration of $[Cd^{2+}]_0$ at 60 mM with different outer concentration of $[S^{2-}]_0$ ranging between 200 mM and 350 mM.

We denote the spacing between spots on consecutive rings by λ_n defined by

$$\lambda_n = x_{n+1} - x_n. \tag{22.105}$$

Since the spacing law is given by $x_n = A_s \left(1 + p\right)^n$, where p is the spacing coefficient (see (22.75)), it is convenient to cast it in the form

$$\ln \lambda_n = \sigma(p) \left[1 + \chi(p)n\right], \tag{22.106}$$

where

$$\sigma(p) = \ln(A_s p), \quad \chi(p) = \frac{\ln(1 + p)}{\ln(A_s p)}. \tag{22.107}$$

The linear dependence of $\ln \lambda_n$ is tested in Fig. 22.26. As shown in Fig. 22.26, the relationship may be deemed linear between λ_n and n within experimental uncertainty as suggested by the direct spacing law. Thus we may tentatively conclude that the spacing law is obeyed, in the case of direct spacing, by the cadmium sulfide system. According to the Matalon–Packter law, p is inversely dependent on the initial density of diffusing ions a_0, but this aspect has not been fully examined as yet.

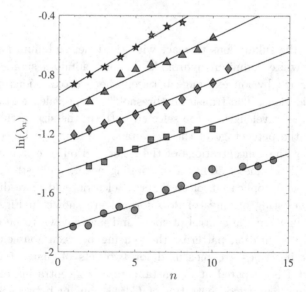

Fig. 22.26. Plot of spacing between two consecutive spots vs. spot number (n) for set I with fixed outer concentration $[S^{2-}]_0 = 250\,\text{mM}$ and different inner concentrations. The spacing coefficient p ranges from $p = 0.09$ (filled circle), 0.11 (■), 0.12 (◆), 0.13 (▲), 0.16 (★).

22.4.3.2. *Spinodal Decomposition Model*

Two theoretical models have been proposed for periodic precipitation phenomena in the literature: (1) the pre-nucleation model based on the general idea of the Ostwald supersaturation–nucleation–depletion cycle.[46] and (2) the spinodal decomposition model adapted to precipitation pattern formation.[47] The latter model is used for the cadmium sulfate system described below, which we find excellently describe the direct spacing law, as mentioned earlier. This aspect and others will be discussed later.

The spinodal decomposition model theory is based on the assumption of the formation of an intermediate compound (C) owing to the reaction of the initially separated outer electrolyte (A) and the inner electrolyte (B) diffusing toward each other. The evolution equations for A and B are thus given by the following reaction–diffusion equations:

$$\frac{\partial a}{\partial t} = D_a \Delta_a - kab, \tag{22.108}$$

$$\frac{\partial b}{\partial t} = D_b \Delta_b - kab, \tag{22.109}$$

where a and b denote the concentrations of the inner cadmium and outer hydroxide ions, respectively; D_a and D_b are their respective diffusion coefficients; Δ is the two-dimensional Laplacian operator; and k is the rate constant for formation of C, the intermediate precipitate. It is assumed for a broad class of reactions that reaction $A + B \to C$ gives rise to a constant density (c_0) of the product C behind a propagating reaction pulse. Then as time progresses, small clusters of the intermediate particles (C) nucleate and aggregate behind the front. This mechanism is known as nucleation and growth. In this model, it is believed that the characteristic time scale for nucleation is much longer than the time needed for the front of the local concentration c_0 of the product C to get into a quasi-equilibrium state, which subsequently undergoes a phase separation by spinodal decomposition. The phases here are a solid phase represented by a high density c_h (precipitate) and a clear phase represented by a low density c_l of the sol formed of c_0. Such a phase separation in the quasi-equilibrium sol is

[46]W. Ostwald, *Lehrbuch der Allgemeinen Chemie* (Englemann, Leipzig, 1897); S. Prager, *J. Chem. Phys.* **25**, 279 (1956).

[47]Z. Rácz, *Physica A* **274**, 50, (1999); J. W. Cahn, J. E. Hilliard, *J. Chem. Phys.* **28**, 258 (1958); J. H. Cahn, *Acta Metallaurgica* **9**, 795 (1961).

assumed and described by the Cahn–Hilliard (CH) equation,[48] which may be regarded as a density evolution equation of the quasi-equilibrium (i.e., a nonequilibrium liquid) suspension of C.

It should be noted that the diffusion flux in the concentration evolution equation relies on the assumption of free energy density functional $F[c]$ of the Ginzburg–Landau theory. Therefore, the quasi-equilibrium state is locally inhomogeneous, rendering the model the simplest hydrodynamic equation that obeys the mass conservation of C. In the quasi-equilibrium state, $F[c]$ should have two minima corresponding to c_l and c_h. Under the assumptions made, the Ginzburg–Landau form for $F[c]$ is symmetric around $\bar{c} := (c_h + c_l)/2$, so that it can be expanded in the form

$$F[c] = -\tfrac{1}{2}\varepsilon(c - \bar{c})^2 + \tfrac{1}{4}\gamma(c - \bar{c})^4 + \tfrac{1}{2}\sigma_0\left(\nabla c\right)^2, \qquad (22.110)$$

where ε, γ, and σ_0 are phenomenological parameters, and the minima of $F[c]$ are fixed at c_h and c_l by setting $\sqrt{\varepsilon/\gamma} = (c_h + c_l)/2 \approx c_h/2$ owing to the fact that $c_h \gg c_l$, namely, the gaps between the precipitation zones have very low steady-state concentration of C. This function can be rewritten in terms of a shifted and rescaled concentration field $\varphi := (2c - c_h - c_l)/(c_h - c_l)$ over the spatial domain Ω:

$$F[\varphi] = \int_{\Omega}\left(-\frac{1}{2}\varepsilon\varphi^2 + \frac{1}{4}\gamma\varphi^4 + \frac{1}{2}\sigma\left(\nabla\varphi\right)^2\right)d\Omega. \qquad (22.111)$$

Here σ is the rescaled surface tension, while ε and γ are the positive constants that define the boundary between the stable and metastable regions ($\varphi = \pm\sqrt{\varepsilon/\gamma}$) and the spinodal line between the metastable and unstable regions ($\varphi = \pm\sqrt{\varepsilon/3\gamma}$) as shown in Fig. 22.27.

When the source term (kab) for the production of C is included, the dynamics of the system is described by the following CH equation with a source term:

$$\frac{\partial c}{\partial t} = -\lambda_0\Delta(\delta F/\delta c) + kab, \qquad (22.112)$$

which, with the change of variable to φ, takes more explicitly the form

$$\frac{\partial \varphi}{\partial t} = -\lambda\Delta\left(\varepsilon\varphi - \gamma\varphi^3 + \sigma\Delta\varphi\right) + ab. \qquad (22.113)$$

---

[48] J. W. Cahn and J. E. Hilliard, *J. Chem. Phys.* **28**, 258 (1958); J. W. Cahn, *Acta Metallurgica* **9**, 795 (1961).

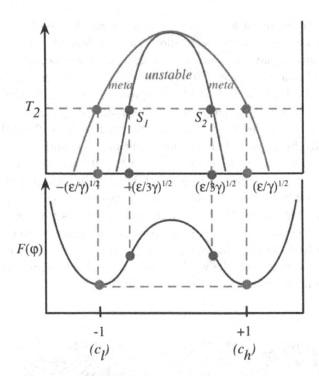

Fig. 22.27. Schematic phase diagram (upper panel) for the spinodal decomposition model and (lower panel) the free energy F as a function of the rescaled concentration field φ. The green curve indicates the binodal curve separating the stable and metastable states with the green binodal points corresponding to the boundaries of the miscibility gap. The red curve defines the spinodal line that separates the metastable and the linearly unstable regions. The red point S_1 and S_2 represent the spinodes.

In this equation, λ and σ are the rescaled kinetic coefficient and surface tension, respectively.[49] The ratio σ/λ gives a characteristic time scale of growth of unstable modes in precipitation.

When the system reaches the unstable region, phase separation or Liesegang band formation takes place in the early stage on a short time scale. This band then acts as a sink for the particles and, in the vicinity of the band, the local concentration of the particles decreases and the front is no longer in the unstable state of the phase space. When the front moves far enough, the depleting effect of the band diminishes. Thus the density

---

[49]T. Antal, M. Droz, J. Magnin and Z. Rácz, *Phys. Rev. Lett.* **83**, 2880 (1999).

of the particles grows again and the process repeats, leading to the formation of Liesegang patterns. In short, the new feature of this scenario is the assumption that the state of the front is quasi-periodically driven into the unstable states domain. This spinodal decomposition scenario described by the (CH) equation is also able to reproduce the transition from bands to spots obtained in the cadmium sulfide/hydroxide precipitation system. The system of coupled equations (22.108), (22.109), and (22.113) is solved numerically by using a finite-element scheme. The initial conditions for a, b and φ are chosen such that $a_0 = a(t = 0) \gg b_0 = b(t = 0)$ and $\varphi_0 = \varphi(t = 0) = -1$, where a_0 is maintained at the inner circular boundary, and no-flux boundary conditions along the outer circular boundary are applied. In Fig. 22.28, the numerical solution reveals the formation of a few Liesegang rings in the early stages of evolution in the wake of a well-organized front moving forward diffusively. This stage is then followed by transition to spots until the whole domain is filled with spots that coarsen with time in agreement with the experiment.

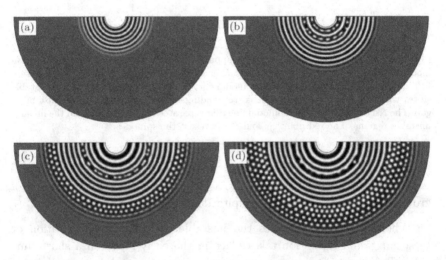

Fig. 22.28. Time evolution of the field φ exhibiting transition from rings to spots: (a) $t = 10$; (b) $t = 20$; (c) $t = 33$; (d) $t = 37$. Parameters are $k = 1$, $D_a = 1$, $D_b = 1$, $\sigma = 1.5$, $\lambda = 0.15$, $\varepsilon = 1$, $\gamma = 0.15$. Initial conditions: $a_0 = 100$, $b_0 = 0.5$, $\varphi_0 = -1$ perturbed with 1% random noise. No-flux boundary conditions are applied at the external boundaries. The radius of the large circle is taken to be eight times greater than that of the small circle. Reproduced with permission from M. Dayeh, M. Ammar and M. Al-Ghoul, *RSC Adv.* **4**, 60034 (2014) © Royal Society of Chemistry.

Appendix A

The Second Virial Coefficient of Lennard-Jones Fluids

In this appendix, we present the temperature dependence of the second virial coefficient for Lennard-Jones (LJ) fluids derived from the statistical mechanical expression.[1]

The statistical mechanical formula for the second virial coefficient can be easily computed by using a numerical method or, if temperature is sufficiently high, by a series expansion method [1]. As a matter of fact, a numerical table is available [1] for the (LJ) fluid. Therefore, it does not pose a practical problem, although a low temperature expansion method is not available. Nevertheless, it would be interesting from the theoretical and pedagogical as well as aesthetic standpoints and also for the practical utility if exact analytic results valid for the entire range of temperature for realistic interaction potential models were available. In the literature [2, 3], an exact analytic form for the second virial coefficient for the LJ fluid was obtained by using coordinate transformation in the cluster integral and reading off the integral table [4] to obtain such a form. However, such transformed integrals have a deeper underlying dynamics that mimics a dynamical system obeying a harmonic oscillator potential. Therefore, if such a feature is made evident, one can gain a considerable insight into the dynamics of the LJ fluids (liquids) [5]. The method employed also provides a valuable lesson on

---

[1]The material of this appendix is based on a paper of B. C. Eu archived in arXiv: Physics/0909.3326v1, 2009.

how to handle such integrals that might appear in the study of statistical mechanics of simple liquids. This method is not available elsewhere in the literature as far as the present authors are aware of.

In this appendix, we present an exact analytic result for the second virial coefficient of the Lennard–Jones (12-6) fluid, which is obtained without using an expansion method and valid for the entire range of temperature. Exact limiting forms are also deduced therefrom as $T \to 0$ and $T \to \infty$. The second virial coefficient obtained is given in terms of parabolic cylinder functions or confluent hypergeometric functions, which are convergent and well defined for all values of temperature. The form presented for the second virial coefficient therefore is valid for all temperatures. With the result presented for the second virial coefficient, it is possible to carry through a thermodynamic theory of LJ fluid exactly up to the order of the second virial coefficient for all thermodynamic functions described in textbook.

A.1. Statistical Mechanical Formula for the Second Virial Coefficient

The statistical mechanical expression for the second virial coefficient [6] of the LJ fluid may be written in the reduced form

$$B_2 = -12v_0 \int_0^\infty dx\, x^2 \left\{ \exp\left[-4\varepsilon\beta \left(x^{-12} - x^{-6} \right) \right] - 1 \right\} := -12v_0 I, \quad (A.1)$$

where $v_0 = \pi\sigma^3/6$, the volume of the contact sphere of diameter σ (σ = size parameter of the LJ potential), ε is the well depth, and $\beta = 1/k_B T$, inverse temperature, k_B being the Boltzmann constant. The object of interest is the integral I in Eq. (A.1). With transformation of variables

$$\alpha = \sqrt{\varepsilon\beta}, \quad (A.2)$$

$$y = \frac{4\alpha^2}{x^{12}}, \quad (A.3)$$

the integral I can be put into the form

$$I(\alpha) = \frac{\sqrt{2\alpha}}{12} J(\alpha), \quad (A.4)$$

where $J(\alpha)$ is defined by the integral

$$J(\alpha) = \int_0^\infty dy\, y^{-5/4} (e^{-y} e^{2\alpha\sqrt{y}} - 1). \quad (A.5)$$

This integral is usually evaluated either by a series expansion method or by a numerical method. Luckily, this integral is in a form readily available from the integral table [4] in a known functional form. However, it can be analytically evaluated without using an integral table or expansion method, as will be shown below.

To achieve this aim, perform integration by parts once to obtain $J(\alpha)$ in the form

$$J(\alpha) = 4\left[\alpha B_{3/4}(\alpha) - B_{1/4}(\alpha)\right],\tag{A.6}$$

where $B_{1/4}$ and $B_{3/4}$ are defined by the integrals

$$B_{1/4}(\alpha) = \int_0^\infty dy\, y^{-1/4} e^{-y} e^{2\alpha\sqrt{y}},\tag{A.7}$$

$$B_{3/4}(\alpha) = \int_0^\infty dy\, y^{-3/4} e^{-y} e^{2\alpha\sqrt{y}}.\tag{A.8}$$

These integrals are functions of parameter α.

A.2. Differential Equation for Integral $J(\alpha)$

Differentiating the integrals $B_{1/4}(\alpha)$ and $B_{3/4}(\alpha)$ with α, we obtain a pair of first-order differential equations

$$\frac{dB_{1/4}}{d\alpha} = \tfrac{1}{2} B_{3/4}(\alpha) + 2\alpha B_{1/4}(\alpha),\tag{A.9}$$

$$\frac{dB_{3/4}}{d\alpha} = 2B_{1/4}(\alpha).\tag{A.10}$$

This pair of differential equations can be combined to a single homogeneous second-order differential equation:

$$\frac{d^2\psi}{d\alpha^2} - 2\alpha\frac{d\psi}{d\alpha} - 3\psi = 0\tag{A.11}$$

with the simplified notation

$$\psi(\alpha) = B_{1/4}(\alpha).\tag{A.12}$$

With the transformations

$$z = \sqrt{2}\alpha,\tag{A.13}$$

and

$$\psi(z) = e^{\frac{1}{4}z^2}\phi(z),\tag{A.14}$$

the differential equation (A.11) can be transformed into a well-recognizable standard form of second-order differential equation

$$\frac{d^2\phi}{dz^2} - \left(1 + \frac{1}{4}z^2\right)\phi(z) = 0. \tag{A.15}$$

This is akin to the Schrödinger equation for a harmonic oscillator, but of a negative energy eigenvalue in the present case. Therefore, it represents a particle of negative energy subjected to a parabolic potential.

A.3. Analytic Solutions for the Second Virial Coefficient

In fact, it is a differential equation for parabolic cylinder functions [7, 8]. Its two independent solutions, one even and the other odd function of z, are given by confluent hypergeometric functions:

$$\phi_1(z) = e^{-\frac{1}{4}z^2} M\left(\frac{3}{4}, \frac{1}{2}, \frac{1}{2}z^2\right)$$

$$= e^{-\frac{1}{4}z^2} \sum_{n=0}^{\infty} \frac{\left(\frac{3}{4}\right)_n}{\left(\frac{1}{2}\right)_n} \frac{\left(z^2/2\right)^n}{n!}, \tag{A.16}$$

$$\phi_2(z) = ze^{-\frac{1}{4}z^2} M\left(\frac{5}{4}, \frac{3}{2}, \frac{1}{2}z^2\right)$$

$$= ze^{-\frac{1}{4}z^2} \sum_{n=0}^{\infty} \frac{\left(\frac{5}{4}\right)_n}{\left(\frac{3}{2}\right)_n} \frac{\left(z^2/2\right)^n}{n!}. \tag{A.17}$$

Here, $M(a, b, t)$ is a confluent hypergeometric function[2] of Kummer [8]:

$$M(a, b, t) = \sum_{n=0}^{\infty} \frac{(a)_n}{(b)_n} \frac{t^n}{n!}, \tag{A.18}$$

where

$$(a)_0 = 1,$$
$$(a)_n = a(a+1)(a+2)\cdots(a+n-1) \ (n \geq 1). \tag{A.19}$$

---

[2]See, for example, E. T. Whittaker and G. N. Watson, *Modern Analysis* (Cambridge, University Press, London, 1952), 4th edition.

It is convergent for all values of t. Its asymptotic form will be of interest to us later: for positive real t, it is given by

$$M(a, b, t) = \frac{\Gamma(b)}{\Gamma(a)} e^t t^{a-b} \left[\sum_{n=0}^{m-1} \frac{(b-a)_n (1-a)_n}{n!} t^{-n} + O\left(t^{-m}\right) \right]. \quad (A.20)$$

This formula may be used to compute the solutions for a large value of t in the case of fixed a and b values. The solutions $\phi_1(z)$ and $\phi_2(z)$, as a matter of fact, are parabolic cylinder functions, which are generic solutions for the Schrödinger equations for quadratic potentials. This implies that the dynamics of the LJ potential fluid closely resembles that of a harmonic (quadratic) potential.

Therefore, the general solution for $\psi(z)$ may be written

$$B_{1/4}(z) = \psi(z) = c_1 M(\tfrac{3}{4}, \tfrac{1}{2}, \tfrac{1}{2}z^2) + c_2 z M(\tfrac{5}{4}, \tfrac{3}{2}, \tfrac{1}{2}z^2), \quad (A.21)$$

where c_1 and c_2 are constants, which may be determined by considering the boundary conditions.

Noting the recurrence relation

$$\frac{d}{dt} M(a, b, t) = \frac{a}{b} M(a+1, b+1, t), \quad (A.22)$$

we find $J(\alpha)$ is given by the expression

$$J(\alpha) = 4c_1 [6\alpha^2 M(\tfrac{7}{4}, \tfrac{3}{2}, \alpha^2) - (1 + 4\alpha^2) M(\tfrac{3}{4}, \tfrac{1}{2}, \alpha^2)]$$
$$+ 4\sqrt{2} c_2 \alpha [\tfrac{10}{3}\alpha^2 M(\tfrac{9}{4}, \tfrac{5}{2}, \alpha^2) + (1 - 4\alpha^2) M(\tfrac{5}{4}, \tfrac{3}{2}, \alpha^2)]. \quad (A.23)$$

Here, the coefficients c_1 and c_2 can be determined by examining the limiting form of $J(\alpha)$ as $\alpha \to 0$ (a boundary condition). From Eq. (A.23),

$$J(\alpha) = 4[-c_1 + \sqrt{2} c_2 \alpha + O\left(\alpha^2\right)], \quad (A.24)$$

whereas direct evaluation of $J(\alpha)$ by series expansion of the factor $\exp\left(2\alpha\sqrt{y}\right)$ in Eq. (A.23) yields

$$J(\alpha) = -4\Gamma\left(\tfrac{3}{4}\right) + 2\Gamma\left(\tfrac{1}{4}\right)\alpha + O\left(\alpha^2\right), \quad (A.25)$$

where $\Gamma\left(\frac{1}{4}\right)$ and $\Gamma\left(\frac{3}{4}\right)$ are gamma functions: $\Gamma\left(\frac{1}{4}\right) = 3.62560\ldots$ and $\Gamma\left(\frac{3}{4}\right) = 1.22541\ldots$. Comparing Eqs. (A.24) and (A.25), we find

$$c_1 = \Gamma\left(\tfrac{3}{4}\right), \tag{A.26}$$

$$c_2 = \frac{1}{2\sqrt{2}}\Gamma\left(\tfrac{1}{4}\right). \tag{A.27}$$

Thus, $J(\alpha)$ is now fully determined.

Putting together the results produced up to this point, we finally obtain the second virial coefficient in the form

$$-B_2/v_0\sqrt{2}\,(\varepsilon\beta)^{1/4} = 4\Gamma(\tfrac{3}{4})\left[6\varepsilon\beta M(\tfrac{7}{4}, \tfrac{3}{2}, \varepsilon\beta) - (1 + 4\varepsilon\beta)M(\tfrac{3}{4}, \tfrac{1}{2}, \varepsilon\beta)\right]$$
$$+ 2\Gamma(\tfrac{1}{4})\sqrt{\varepsilon\beta}[\tfrac{10}{3}\varepsilon\beta M(\tfrac{9}{4}, \tfrac{5}{2}, \varepsilon\beta)$$
$$+ (1 - 4\varepsilon\beta)\,M(\tfrac{5}{4}, \tfrac{3}{2}, \varepsilon\beta)]. \tag{A.28}$$

This is the result we have set out to show for the LJ (12-6) fluid. One may try to put this result into a simpler form by using the recurrence relations of Kummer's functions, but the present form appears to be an optimum form. Rigorous limiting laws can be deduced for B_2 from Eq. (A.28).

The limiting form of B_2 as $T \to \infty$ or $\varepsilon\beta \to 0$ is easily deduced to be

$$B_2 = 4\sqrt{2}\Gamma(\tfrac{3}{4})v_0\,(\varepsilon\beta)^{1/4}\left[1 + O(\varepsilon\beta)\right]. \tag{A.29}$$

Thus, $B_2 \to +0$ as $T \to \infty$. This means that there is a high temperature regime where B_2 is positive, and as $T \to \infty$, it vanishes on the positive side according to the law indicated.

On the other hand, the limiting form of B_2 as $T \to 0$ or $\varepsilon\beta \to \infty$ is deduced from the asymptotic forms of the confluent hypergeometric functions given in Eq. (A.20). We find

$$B_2\,(T) = -16\sqrt{2\pi}v_0 e^{\varepsilon\beta}\,(\varepsilon\beta)^{\frac{3}{2}}\left[1 + \frac{19}{16\varepsilon\beta} + \frac{105}{512\,(\varepsilon\beta)^2} + \cdots\right]. \tag{A.30}$$

This limiting law shows that $B_2\,(T)$ tends to negative infinity according to the formula indicated and is negative below a certain point in T.

These limiting laws for B_2 are not easily deducible from Eq. (A.1) or Eq. (A.5) or the series expansion form [1] thereof, but they are simple to deduce if the exact analytic solution presented is made use of.

From the limiting behaviors (A.29) and (A.30), we can conclude there must exist a point in T at which $B_2(T)$ crosses the T-axis (i.e., becomes zero), that is, the Boyle temperature $T_B = \beta_B^{-1}/k_B$ is defined, as usual, by

$$B_2(T_B) = 0. \qquad (A.31)$$

According to the analytic result obtained, the Boyle point is determined from a real root of the equation

$$0 = 4\Gamma(\tfrac{3}{4})[6\varepsilon\beta_B M(\tfrac{7}{4},\tfrac{3}{2},\varepsilon\beta_B) - (1 + 4\varepsilon\beta_B)M(\tfrac{3}{4},\tfrac{1}{2},\varepsilon\beta_B)]$$
$$+ 2\Gamma(\tfrac{1}{4})\sqrt{\varepsilon\beta_B}[\tfrac{10}{3}\varepsilon\beta_B M(\tfrac{9}{4},\tfrac{5}{2},\varepsilon\beta_B)$$
$$+ (1 - 4\varepsilon\beta_B)M(\tfrac{5}{4},\tfrac{3}{2},\varepsilon\beta_B)]. \qquad (A.32)$$

Its numerical solution yields

$$T_B^* = (\varepsilon\beta_B)^{-1} = 3.41793, \qquad (A.33)$$

which should be compared with the literature value [1] $T_B^* = 3.42$. This value is practically attained with the truncation of $M(a,b,\varepsilon\beta_B)$ at $n = 3$.

As a conclusion, we have presented an exact analytic second virial coefficient of the LJ fluids, which are valid for the entire temperature range, and its asymptotic behaviors (limiting laws) as $T \to 0$ or $T \to \infty$. In view of the agreement of the Boyle temperature with the literature value deduced from the table for the second virial coefficient [1], it seems to be unnecessary to tabulate the numerical values of the second virial coefficients; it is rather trivial to do so. The utility of the result obtained is self-evident for some deductions one can make about thermodynamic properties of the LJ (12-6) fluid. Compared to the method that simply reads off the integral table upon variable transformation in the integral for B_2, the present method provides considerable insights into the dynamics of the LJ fluid. In this regard, recall Eq. (14.17), which indicates that the second virial coefficient may be viewed as if it represents a "quantum mechanical motion" of a particle of negative energy subjected to a parabolic potential.

Finally, it is useful to note that the present exact analytic result for the second virial coefficient owes its existence to the mathematically favorable combination of the exponents 12 and 6 of the potential that produces the closed form for the differential equation for $B_{1/4}(\alpha)$, Eq. (A.7). For other potential models consisting of repulsive and attractive branches with different exponents, we do not obtain a closed differential equation, but an open hierarchy of first-order differential equations, for integrals making up $J(\alpha)$. The case of exponents, (9-6), namely, the LJ (9-6) potential, produces a closed inhomogeneous second-order differential equation, but its solutions do not seem to be as simple and clean as that of the LJ (9-6) potential.

Appendix B

Various Coefficients
Used in Chapter 10

In this appendix, we have collected various coefficients appearing as expansions employed in the formulation of the algorithms to calculate the subcritical thermodynamic properties of fluids in Chapter 10 on the canonical equation of state. With these coefficients used in various expressions, it is possible to calculate exactly but numerically various quantities developed for the subcritical thermodynamic properties with the models for the GvdW parameters in the canonical equation of state.

B.1. Coefficients of $\Pi^{(i)}(x_{sk}, t)$

To begin with, we observe $\Pi(x_{sk}, t) := \Pi^{(0)}(x_{sk}, t)$ and the derivatives $\Pi^{(i)}(x_{sk}, t)$ of Π are decomposable into cubic polynomials in x_{sk}:

$$\Pi^{(i)}(x_{sk}, t) = \Pi_0^{(i)} + \Pi_1^{(i)} x_{sk} + \Pi_2^{(i)} x_{sk}^2 + \Pi_3^{(i)} x_{sk}^3 \quad (0 \leq i \leq 4). \quad (B.1)$$

The coefficients in this polynomial consist of t-independent and t-dependent parts as follows:

$$\Pi_j^{(i)} = P_{ij} + \widehat{\Pi}_j^{(i)}(t), \quad (B.2)$$

where P_{ij} is the t-independent part made up of the parameters determining the critical point and $\widehat{\Pi}_j^{(i)}(t)$ is the t-dependent part determined by the t-dependent part of the GvdW parameters, namely, $\widehat{a}_i(t)$ and $\widehat{b}_i(t)$. These coefficients, although somewhat complicated but straightforward to obtain from the definition of $\Pi^{(i)}(x, t)$, are listed for $i = 0, \ldots, 4$ in the following.

B.2.   Coefficients P_{ij}

$P_{00} = P = 0,$

$P_{10} = \tau + \zeta \left(\nu - 1\right)\left(a_1 + 2\right) + \nu\left(\zeta + 1\right)\left(b_1 + 1\right) = P_1 = 0,$

$P_{11} = 2\zeta\left(3\nu - 1\right) + \zeta\left(3\nu - 2\right)a_1 + \nu\left(1 + 3\zeta\right)b_1,$

$P_{12} = \zeta\left(3\nu - 1\right)a_1 + 3\zeta\nu\left(b_1 + 1\right),$

$P_{13} = \zeta\nu\left(a_1 + b_1\right),$

$P_{20} = 2\zeta\left(\nu - 1\right)\left(2a_1 + a_2 + 1\right) + 2\nu\left(b_1 + b_2\right)\left(\zeta + 1\right)$
$\qquad + 2\zeta\nu\left(a_1 + 2\right)\left(b_1 + 1\right) = 2!P_2 = 0,$

$P_{21} = 6\zeta\nu + 4\zeta\left(3\nu - 1\right)a_1 + 8\zeta\nu b_1 + 4\zeta\nu a_1 b_1$
$\qquad + 2\zeta\left(3\nu - 2\right)a_2 + 2\nu\left(3\zeta + 1\right)b_2,$

$P_{22} = 10\zeta\nu a_1 + 6\zeta\nu b_1 + 6\zeta\nu a_1 b_1 + 2\zeta\left(5\nu - 1\right)a_2 + 6\zeta\nu b_2,$

$P_{23} = 2\zeta\nu\left(b_2 + a_1 b_1 + a_2\right),$

$P_{30} = 6\zeta\left(\nu - 1\right)\left(a_1 + 2a_2 + a_3\right) + 6\nu\left(\zeta + 1\right)\left(b_2 + b_3\right)$
$\qquad + 6\zeta\nu\left(a_1 + 2\right)\left(b_1 + b_2\right) + 6\zeta\nu\left(2a_1 + a_2 + 1\right)\left(b_1 + 1\right) = 3!P_3 = 0,$

$P_{31} = 18\zeta\nu a_1 + 12\zeta\left(3\nu - 1\right)a_2 + 6\zeta\left(3\nu - 2\right)a_3 + 18\zeta\nu b_1 + 36\zeta\nu b_2$
$\qquad + 6\left(3\zeta + 1\right)\nu b_3 + 18\zeta\nu\left(2a_1 b_1 + a_1 b_2 + a_2 b_1\right),$

$P_{32} = 6\zeta\left(3\nu - 1\right)a_3 + 18\zeta\nu b_3 + 18\zeta\nu\left(a_2 + b_2 + a_1 b_1 + a_1 b_2 + a_2 b_1\right),$

$P_{33} = 6\zeta\nu\left(a_1 b_2 + a_2 b_1\right) + 6\zeta\nu\left(a_3 + b_3\right),$

$P_{40} = 24\nu\left(\zeta + 1\right)\left(b_3 + b_4\right) + 24\zeta\left(\nu - 1\right)\left(a_2 + 2a_3 + a_4\right)$
$\qquad + 24\zeta\nu\left(2a_1 + a_2 + 1\right)\left(b_1 + b_2\right) + 24\zeta\nu\left(2 + a_1\right)\left(b_2 + b_3\right)$
$\qquad + 24\zeta\nu\left(a_1 + 2a_2 + a_3\right)\left(b_1 + 1\right) = 4!P_0 = 0,$

$P_{41} = 72\zeta\nu\left(b_1 + 2b_2 + b_3\right)a_1 + 72\zeta\nu\left(2b_1 + b_2\right)a_2 + 72\zeta\nu b_1 a_3$
$\qquad + 72\zeta\nu b_2 + 72\zeta\nu a_2 + 48\zeta\left(3\nu - 1\right)a_3 + 24\zeta\left(2\nu - 1\right)a_4$
$\qquad + 144\zeta\nu b_3 + 24\left(3\zeta + 1\right)\nu b_4,$

$P_{42} = 24\zeta a_4\left(\nu - 1\right) + 48\zeta\nu\left(a_3 + a_4\right) + 48\zeta\nu b_4 + 24\zeta\nu\left(b_3 + b_4\right)$
$\qquad + 48\zeta\nu b_1\left(a_2 + a_3\right) + 24\zeta\nu a_3\left(1 + b_1\right) + 24\zeta\nu a_1 b_3$

$$+ 24\zeta\nu b_3 \left(2 + a_1\right) + 24\zeta\nu a_1 \left(b_2 + b_3\right) + 48\zeta\nu b_2 \left(a_1 + a_2\right)$$

$$+ 24\zeta\nu a_2 \left(b_1 + b_2\right),$$

$$P_{43} = 24\zeta\nu a_1 b_3 + 24\zeta\nu a_2 b_2 + 24\zeta\nu a_3 b_1 + 24\zeta\nu b_4 + 24\zeta\nu a_4.$$

Note that $\widehat{a}_i^{(k)} = \widehat{a}_i^{(k)}(t)$ and $\widehat{b}_i^{(k)} = \widehat{b}_i^{(k)}(t)$, the t-dependent part of $a_i^{(k)}(t)$ and $b_i^{(k)}(t)$. The t-dependence of $\widehat{a}_i^{(k)}(t)$, etc., are henceforth suppressed for brevity of notation. It should be recalled that $P_{i0} :=. P_i = 0$ by virtue of Proposition 1 at the critical point. These five conditions reduce the seven parameters in the quadratic model to two free parameters, which must also satisfy the stability condition, that is, the fifth density derivative being negative if the critical state is to be stable.

B.3. Coefficients $\widehat{\Pi}_j^{(i)}(t)$

$$\widehat{\Pi}_0^{(0)} = \zeta\left(\nu - 1\right)\widehat{a}_0^{(k)} + \nu\left(1 + \zeta\right)\widehat{b}_0^{(k)} + \zeta\nu\widehat{a}_0^{(k)}\widehat{b}_0^{(k)},$$

$$\widehat{\Pi}_1^{(0)} = \zeta\left(3\nu - 2\right)\widehat{a}_0^{(k)} + \left(1 + 3\zeta\right)\nu\widehat{b}_0^{(k)} + 3\zeta\nu\widehat{a}_0^{(k)}\widehat{b}_0^{(k)},$$

$$\widehat{\Pi}_2^{(0)} = \left(3\nu - 1\right)\zeta\widehat{a}_0^{(k)} + 3\zeta\nu\widehat{b}_0^{(k)} + 3\zeta\nu\widehat{a}_0^{(k)}\widehat{b}_0^{(k)},$$

$$\widehat{\Pi}_3^{(0)} = \zeta\nu\widehat{a}_0^{(k)} + \zeta\nu\widehat{b}_0^{(k)} + \zeta\nu\widehat{a}_0^{(k)}\widehat{b}_0^{(k)},$$

$$\widehat{\Pi}_0^{(1)} = \zeta(\nu - 1)\left(2\widehat{a}_0^{(k)} + \widehat{a}_1^{(k)}\right) + \zeta\nu\left(a_1 + 2\right)\widehat{b}_0^{(k)} + \nu(\zeta + 1)(\widehat{b}_0^{(k)} + \widehat{b}_1^{(k)})$$

$$+ \zeta\nu\left(b_1 + 1\right)\widehat{a}_0^{(k)} + \zeta\nu\widehat{a}_0^{(k)}\left(\widehat{b}_0^{(k)} + \widehat{b}_1^{(k)}\right) + \zeta\nu\widehat{b}_0^{(k)}\left(2\widehat{a}_0^{(k)} + \widehat{a}_1^{(k)}\right) + \tau t,$$

$$\widehat{\Pi}_1^{(1)} = 2\zeta(3\nu - 1)\widehat{a}_0^{(k)} + 6\zeta\nu\widehat{b}_0^{(k)} + \zeta(3\nu - 2)\widehat{a}_1^{(k)} + \nu(1 + 3\zeta)\widehat{b}_1^{(k)}$$

$$+ 3\zeta\nu(\widehat{b}_0^{(k)}a_1 + \widehat{a}_0^{(k)}b_1) + 3\zeta\nu(2\widehat{a}_0^{(k)}\widehat{b}_0^{(k)} + \widehat{a}_0^{(k)}\widehat{b}_1^{(k)} + \widehat{a}_1^{(k)}\widehat{b}_0^{(k)}),$$

$$\widehat{\Pi}_2^{(1)} = 3\zeta\nu(\widehat{a}_0^{(k)} + \widehat{b}_0^{(k)}) + \zeta(3\nu - 1)\widehat{a}_1^{(k)} + 3\zeta\nu\widehat{b}_1^{(k)} + 3\zeta\nu(\widehat{a}_0^{(k)}b_1 + \widehat{b}_0^{(k)}a_1)$$

$$+ 3\zeta\nu(\widehat{a}_0^{(k)}\widehat{b}_0^{(k)} + \widehat{a}_0^{(k)}\widehat{b}_1^{(k)} + \widehat{a}_1^{(k)}\widehat{b}_0^{(k)}),$$

$$\widehat{\Pi}_3^{(1)} = \zeta\nu(\widehat{a}_1^{(k)} + \widehat{b}_1^{(k)}) + \zeta\nu(\widehat{b}_0^{(k)}a_1 + \widehat{a}_0^{(k)}b_1) + \zeta\nu(\widehat{a}_0^{(k)}\widehat{b}_1^{(k)} + \widehat{a}_1^{(k)}\widehat{b}_0^{(k)}),$$

$$\widehat{\Pi}_0^{(2)} = 2\zeta\left[\nu - 1 + \nu\left(2 + 3b_1 + b_2\right)\right]\widehat{a}_0^{(k)} + 2\zeta\nu\left(3 + 3a_1 + a_2\right)\widehat{b}_0^{(k)}$$

$$+ 2\zeta\left(3\nu + \nu b_1 - 2\right)\widehat{a}_1^{(k)} + 2\nu\left(1 + 3\zeta + \zeta a_1\right)\widehat{b}_1^{(k)}$$

$$+ 2\zeta\nu\widehat{a}_0^{(k)}\left(\widehat{b}_1^{(k)} + \widehat{b}_2^{(k)}\right) + 2\zeta\left(\nu - 1\right)\widehat{a}_2^{(k)} + 2\nu\left(\zeta + 1\right)\widehat{b}_2^{(k)} + 2\zeta\nu(\widehat{b}_0^{(k)}$$

$$+ \widehat{b}_1^{(k)})(2\widehat{a}_0^{(k)} + \widehat{a}_1^{(k)}) + 2\zeta\nu\widehat{b}_0^{(k)}(\widehat{a}_0^{(k)} + 2\widehat{a}_1^{(k)} + \widehat{a}_2^{(k)}),$$

$$\widehat{\Pi}_1^{(2)} = 2\zeta\nu\left(3 + 4b_1 + 3b_2\right)\widehat{a}_0^{(k)} + 6\zeta\nu\left(1 + 2a_1 + a_2\right)\widehat{b}_0^{(k)}$$
$$+ 4\zeta\left(3\nu + \nu b_1 - 1\right)\widehat{a}_1^{(k)} + 6\zeta\nu\left(2 + a_1\right)\widehat{b}_1^{(k)} + 2\zeta\left(3\nu - 2\right)\widehat{a}_2^{(k)}$$
$$+ 2\nu\left(3\zeta + 1\right)\widehat{b}_2^{(k)} + 6\zeta\nu a\widehat{b}_0^{(k)} + 12\zeta\nu(\widehat{a}_0^{(k)}\widehat{b}_1^{(k)} + \widehat{a}_1^{(k)}\widehat{b}_0^{(k)})$$
$$+ 6\zeta\nu(\widehat{a}_0^{(k)}\widehat{b}_2^{(k)} + \widehat{a}_1^{(k)}\widehat{b}_1^{(k)} + \widehat{b}_0^{(k)}\widehat{a}_2^{(k)}),$$

$$\widehat{\Pi}_2^{(2)} = 6\zeta\nu(b_1 + b_2)\widehat{a}_0^{(k)} + 10\zeta\nu(a_1 + a_2)\widehat{b}_0^{(k)} + 2\zeta\nu(5 + 3b_1)\widehat{a}_1^{(k)}$$
$$+ 6\zeta\nu\left(1 + a_1\right)\widehat{b}_1^{(k)} + 2\zeta\left(5\nu - 1\right)\widehat{a}_2^{(k)} + 6\zeta\nu\widehat{b}_2^{(k)} + 6\zeta\nu\widehat{a}_1^{(k)}\widehat{b}_1^{(k)}$$
$$+ 10\zeta\nu\widehat{b}_0^{(k)}\widehat{a}_2^{(k)} + 6\zeta\nu\widehat{a}_0^{(k)}\widehat{b}_2^{(k)} + 2\zeta\nu(3\widehat{a}_0^{(k)}\widehat{b}_1^{(k)} + 5\widehat{a}_1^{(k)}\widehat{b}_0^{(k)}),$$

$$\widehat{\Pi}_3^{(2)} = 2\zeta\nu b_2\widehat{a}_0^{(k)} + 2\zeta\nu a_2\widehat{b}_0^{(k)} + 2\zeta\nu b_1\widehat{a}_1^{(k)} + 2\zeta\nu a_1\widehat{b}_1^{(k)} + 2\zeta\nu(\widehat{a}_2^{(k)} + \widehat{b}_2^{(k)})$$
$$+ 2\zeta\nu(\widehat{a}_0^{(k)}\widehat{b}_2^{(k)} + \widehat{a}_1^{(k)}\widehat{b}_1^{(k)} + \widehat{b}_0^{(k)}\widehat{a}_2^{(k)}),$$

$$\widehat{\Pi}_0^{(3)} = 6\zeta\nu\left(1 + 3b_1 + 3b_2 + b_3\right)\widehat{a}_0^{(k)} + 6\zeta\nu\left(1 + 3a_1 + 3a_2 + a_3\right)\widehat{b}_0^{(k)}$$
$$+ 6\zeta\left[\nu - 1 + \nu\left(b_1 + b_2\right) + 2\nu\left(b_1 + 1\right)\right]\widehat{a}_1^{(k)} + 6\zeta\nu\left(3 + 3a_1 + a_2\right)\widehat{b}_1^{(k)}$$
$$+ \left[2\left(\nu - 1\right) + \nu\left(b_1 + 1\right)\right]\widehat{a}_2^{(k)} + 6\nu\left[\zeta + 1 + \zeta\left(a_1 + 2\right)\right]\widehat{b}_2^{(k)}$$
$$+ 6\zeta\left(\nu - 1\right)\widehat{a}_3^{(k)} + 6\nu\left(\zeta + 1\right)\widehat{b}_3^{(k)} + 6\zeta\nu\widehat{a}_0^{(k)}(\widehat{b}_0^{(k)} + \widehat{b}_1^{(k)} + \widehat{b}_2^{(k)} + \widehat{b}_3^{(k)})$$
$$+ 6\zeta\nu(\widehat{b}_1^{(k)} + \widehat{b}_2^{(k)})(2\widehat{a}_0^{(k)} + \widehat{a}_1^{(k)}) + 6\zeta\nu(\widehat{b}_0^{(k)} + \widehat{b}_1^{(k)})(2\widehat{a}_1^{(k)} + \widehat{a}_2^{(k)})$$
$$+ 6\zeta\nu\widehat{b}_0^{(k)}(\widehat{a}_1^{(k)} + 2\widehat{a}_2^{(k)} + \widehat{a}_3^{(k)}),$$

$$\widehat{\Pi}_1^{(3)} = 18\zeta\nu\left(b_1 + 2b_2 + b_3\right)\widehat{a}_0^{(k)} + 18\zeta\nu\left(a_1 + 2a_2 + a_3\right)\widehat{b}_0^{(k)}$$
$$+ 18\zeta\nu\left(2b_1 + b_2 + 1\right)\widehat{a}_1^{(k)} + 18\zeta\nu\left(2a_1 + a_2 + 1\right)\widehat{b}_1^{(k)}$$
$$+ 6\zeta\left(6\nu + 3\nu b_1 - 2\right)\widehat{a}_2^{(k)} + 18\zeta\nu\left(a_1 + 2\right)\widehat{b}_2^{(k)} + 6\zeta\left(3\nu - 2\right)\widehat{a}_3^{(k)}$$
$$+ 6\nu\left(3\zeta + 1\right)\widehat{b}_3^{(k)} + 18\zeta\nu\widehat{a}_0^{(k)}\widehat{b}_1^{(k)} + 18\zeta\nu\widehat{a}_1^{(k)}\widehat{b}_0^{(k)} + 36\zeta\nu\widehat{a}_0^{(k)}\widehat{b}_2^{(k)}$$
$$+ 36\zeta\nu\widehat{a}_1^{(k)}\widehat{b}_1^{(k)} + 36\zeta\nu\widehat{b}_0^{(k)}\widehat{a}_2^{(k)} + 18\zeta\nu\widehat{a}_0^{(k)}\widehat{b}_3^{(k)} + 18\zeta\nu\widehat{a}_1^{(k)}\widehat{b}_2^{(k)}$$
$$+ 18\zeta\nu\widehat{b}_0^{(k)}\widehat{a}_3^{(k)} + 18\zeta\nu\widehat{a}_2^{(k)}\widehat{b}_1^{(k)},$$

$$\widehat{\Pi}_2^{(3)} = 18\zeta\nu\left(b_2 + b_3\right)\widehat{a}_0^{(k)} + 18\zeta\nu\left(a_2 + a_3\right)\widehat{b}_0^{(k)} + 18\zeta\nu\left(b_1 + b_2\right)\widehat{a}_1^{(k)}$$
$$+ 18\zeta\nu\left(a_1 + a_2\right)\widehat{b}_1^{(k)} + 18\zeta\nu\left(1 + b_1\right)\widehat{a}_2^{(k)} + 18\zeta\nu\left(a_1 + 1\right)\widehat{b}_2^{(k)}$$
$$+ 18\zeta\nu\widehat{a}_0^{(k)}\widehat{b}_2^{(k)} + 18\zeta\nu\widehat{b}_0^{(k)}\widehat{a}_2^{(k)} - 6\zeta\left(1 - 3\nu\right)a_3^{(k)} + 18\zeta\nu\widehat{b}_3^{(k)}$$
$$+ 18\zeta\nu\widehat{a}_1^{(k)}\widehat{b}_1^{(k)} + 18\zeta\nu\widehat{a}_1^{(k)}\widehat{b}_2^{(k)} + 18\zeta\nu\widehat{a}_2^{(k)}\widehat{b}_1^{(k)} + 18\zeta\nu\widehat{a}_0^{(k)}\widehat{b}_3^{(k)}$$
$$+ 18\zeta\nu\widehat{b}_0^{(k)}\widehat{a}_3^{(k)},$$

$$\widehat{\Pi}_3^{(3)} = 6\zeta\nu(b_3\widehat{a}_0^{(k)} + b_2\widehat{a}_1^{(k)} + b_1\widehat{a}_2^{(k)} + a_3\widehat{b}_0^{(k)} + a_2\widehat{b}_1^{(k)} + a_1\widehat{b}_2^{(k)})$$
$$+ 6\zeta\nu(\widehat{a}_1^{(k)}\widehat{b}_2^{(k)} + \widehat{a}_2^{(k)}\widehat{b}_1^{(k)}) + 6\zeta\nu(\widehat{a}_0^{(k)}\widehat{b}_3^{(k)} + \widehat{b}_0^{(k)}\widehat{a}_3^{(k)})$$
$$+ 6\zeta\nu(\widehat{a}_3^{(k)} + \widehat{b}_3^{(k)}),$$

$$\widehat{\Pi}_0^{(4)} = 24\zeta\nu\,(b_1 + 3b_2 + 3b_3 + b_4)\,\widehat{a}_0^{(k)} + 24\zeta\nu\,(3b_1 + 3b_2 + b_3 + 1)\,\widehat{a}_1^{(k)}$$
$$+ 24\zeta\,(3\nu + 3\nu b_1 + \nu b_2 - 1)\,\widehat{a}_2^{(k)} + 24\zeta\,(3\nu + \nu b_1 - 2)\,\widehat{a}_3^{(k)}$$
$$+ 24\zeta\,(\nu - 1)\,\widehat{a}_4^{(k)} + 24\zeta\nu\,(a_1 + 3a_2 + 3a_3 + a_4)\,\widehat{b}_0^{(k)}$$
$$+ 24\zeta\nu\,(3a_1 + 3a_2 + a_3 + 1)\,\widehat{b}_1^{(k)} + 24\zeta\nu\,(3a_1 + 3a_2 + a_3 + 1)\,\widehat{b}_2^{(k)}$$
$$+ 24\nu\,(3\zeta a_0 + \zeta a_1 + 1)\,\widehat{b}_3^{(k)} + 24\nu(z a_0 + 1)\widehat{b}_4^{(k)}$$
$$+ 24\zeta\nu\widehat{a}_0^{(k)}(\widehat{b}_3^{(k)} + \widehat{b}_4^{(k)}) + 24\zeta\nu(\widehat{b}_2^{(k)} + \widehat{b}_3^{(k)})(2\widehat{a}_0^{(k)} + \widehat{a}_1^{(k)})$$
$$+ 24\zeta\nu(\widehat{b}_1^{(k)} + \widehat{b}_2^{(k)})(\widehat{a}_0^{(k)} + 2\widehat{a}_1^{(k)} + \widehat{a}_2^{(k)}) + 24\zeta\nu(\widehat{b}_0^{(k)} + \widehat{b}_1^{(k)})(\widehat{a}_1^{(k)}$$
$$+ 2\widehat{a}_2^{(k)} + \widehat{a}_3^{(k)}) + 24\zeta\nu\widehat{b}_0^{(k)}(\widehat{a}_2^{(k)} + 2\widehat{a}_3^{(k)} + \widehat{a}_4^{(k)}),$$

$$\widehat{\Pi}_1^{(4)} = 72\zeta\nu\,(b_2 + 2b_3 + b_4)\,\widehat{a}_0^{(k)} + 24\zeta\nu\,(3a_2 + 6a_3 + 2a_4)\,\widehat{b}_0^{(k)}$$
$$+ 72\zeta\nu\,(b_1 + 2b_2 + b_3)\,\widehat{a}_1^{(k)} + 72\zeta\nu\,(a_1 + 2a_2 + a_3)\,\widehat{b}_1^{(k)} + 72\zeta\nu\widehat{a}_1^{(k)}\widehat{b}_1^{(k)}$$
$$+ 72\zeta\nu\,(2b_1 + b_2 + 1)\,\widehat{a}_2^{(k)} + 72\zeta\nu\,(2a_1 + a_2 + 1)\,\widehat{b}_2^{(k)} + 72\zeta\nu\widehat{a}_0^{(k)}\widehat{b}_2^{(k)}$$
$$+ 72\zeta\nu\widehat{b}_0^{(k)}\widehat{a}_2^{(k)} + 24\zeta\,(6\nu + 3\nu b_1 - 2)\,\widehat{a}_3^{(k)} + 24\zeta\,(2\nu - 1)\,\widehat{a}_4^{(k)}$$
$$+ 72\zeta\nu\,(a_1 + 2)\,\widehat{b}_3^{(k)} + 24\nu\,(3\zeta + 1)\,\widehat{b}_4^{(k)} + 144\zeta\nu\widehat{a}_0^{(k)}\widehat{b}_3^{(k)}$$
$$+ 144\zeta\nu\widehat{a}_1^{(k)}\widehat{b}_2^{(k)} + 144\zeta\nu\widehat{b}_0^{(k)}\widehat{a}_3^{(k)} + 144\zeta\nu\widehat{a}_2^{(k)}\widehat{b}_1^{(k)} + 72\zeta\nu\widehat{a}_0^{(k)}\widehat{b}_4^{(k)}$$
$$+ 72\zeta\nu\widehat{a}_1^{(k)}\widehat{b}_3^{(k)} + 48\zeta\nu\widehat{b}_0^{(k)}\widehat{a}_4^{(k)} + 72\zeta\nu\widehat{a}_2^{(k)}\widehat{b}_2^{(k)} + 72\zeta\nu\widehat{b}_1^{(k)}\widehat{a}_3^{(k)},$$

$$\widehat{\Pi}_2^{(4)} = 72\zeta\nu\,(b_3 + b_4)\,\widehat{a}_0^{(k)} + 72\zeta\nu\,(b_2 + b_3)\,\widehat{a}_1^{(k)} + 72\zeta\nu\,(b_1 + b_2)\,\widehat{a}_2^{(k)}$$
$$+ 72\zeta\nu\,(b_1 + 1)\,\widehat{a}_3^{(k)} + 24\zeta\,(3\nu - 1)\,\widehat{a}_4^{(k)} + 72\zeta\nu\,(a_3 + a_4)\,\widehat{b}_0^{(k)}$$
$$+ 72\zeta\nu\,(a_2 + a_3)\,\widehat{b}_1^{(k)} + 72\zeta\nu\,(a_1 + a_2)\,\widehat{b}_2^{(k)} + 72\zeta\nu\,(a_1 + 1)\,\widehat{b}_3^{(k)}$$
$$+ 72\zeta\nu\widehat{b}_4^{(k)} + 72\zeta\nu(\widehat{b}_3^{(k)} + \widehat{b}_4^{(k)})\widehat{a}_0^{(k)} + 72\zeta\nu(\widehat{b}_2^{(k)} + \widehat{b}_3^{(k)})\widehat{a}_1^{(k)}$$
$$+ 72\zeta\nu(\widehat{b}_1^{(k)} + \widehat{b}_2^{(k)})\widehat{a}_2^{(k)} + 72\zeta\nu(\widehat{b}_0^{(k)} + \widehat{b}_1^{(k)})\widehat{a}_3^{(k)} + 72\zeta\nu\widehat{b}_0^{(k)}\widehat{a}_4^{(k)},$$

$$\widehat{\Pi}_3^{(4)} = 24\zeta\nu(b_4\widehat{a}_0^{(k)} + b_3\widehat{a}_1^{(k)} + b_2\widehat{a}_2^{(k)} + b_1\widehat{a}_3^{(k)} + \widehat{a}_4^{(k)})$$
$$+ 24\zeta\nu(a_4\widehat{b}_0^{(k)} + a_3\widehat{b}_1^{(k)} + a_2\widehat{b}_2^{(k)} + a_1\widehat{b}_3^{(k)} + \widehat{b}_4^{(k)})$$
$$+ 24\zeta\nu(\widehat{a}_0^{(k)}\widehat{b}_4^{(k)} + \widehat{a}_1^{(k)}\widehat{b}_3^{(k)} + \widehat{a}_2^{(k)}\widehat{b}_2^{(k)} + \widehat{b}_1^{(k)}\widehat{a}_3^{(k)} + \widehat{b}_0^{(k)}\widehat{a}_4^{(k)}).$$

B.4. Coefficients of the Spinodal Equations

With P_{ij} and $\widehat{\Pi}_j^{(i)}$ presented earlier, it is now possible to show the coefficients of the spinodal equations (10.80) and (10.82). In the model represented by Eqs. (10.19) and (10.21), they are given as follows:

$$\varphi_{10} = \nu(1 + b_1 + \widehat{b}_0^{(k)} + \widehat{b}_1^{(k)})\,\Pi_0^{(0)} + \Pi_0^{(1)}(1 - \nu - \nu\widehat{b}_0^{(k)}),$$

$$\varphi_{11} = \nu\Pi_0^{(0)}(\widehat{b}_1^{(k)} + b_1) + \nu\Pi_1^{(0)}(\widehat{b}_0^{(k)} + \widehat{b}_1^{(k)} + b_1 + 1)$$
$$+ \Pi_1^{(1)}(1 - \nu\widehat{b}_0^{(k)} - \nu) - \nu\Pi_0^{(1)}(\widehat{b}_0^{(k)} + 1),$$

$$\varphi_{12} = \nu\Pi_1^{(0)}(\widehat{b}_1^{(k)} + b_1) + \nu\Pi_2^{(0)}(\widehat{b}_0^{(k)} + \widehat{b}_1^{(k)} + b_1 + 1)$$
$$+ \Pi_2^{(1)}(1 - \nu\widehat{b}_0^{(k)} - \nu) - \nu\Pi_1^{(1)}(\widehat{b}_0^{(k)} + 1),$$

$$\varphi_{13} = \nu\Pi_2^{(0)}(\widehat{b}_1^{(k)} + b_1) + \nu\Pi_3^{(0)}(\widehat{b}_0^{(k)} + \widehat{b}_1^{(k)} + b_1 + 1)$$
$$+ \Pi_3^{(1)}(1 - \nu\widehat{b}_0^{(k)} - \nu) - \nu\Pi_2^{(1)}(\widehat{b}_0^{(k)} + 1),$$

$$\varphi_{14} = \nu\Pi_3^{(0)}(b_1 + \widehat{b}_1^{(k)}) - \nu\Pi_3^{(1)}(1 + \widehat{b}_0^{(k)}),$$

$$\varphi_{20} = 2\nu\Pi_0^{(0)}(b_1 + b_2 + \widehat{b}_1^{(k)} + \widehat{b}_2^{(k)}) + \Pi_0^{(2)}(1 - \nu - \nu\widehat{b}_0^{(k)}),$$

$$\varphi_{21} = 2\nu\Pi_0^{(0)}(b_2 + \widehat{b}_2^{(k)}) + 2\nu\Pi_1^{(0)}(b_1 + b_2 + \widehat{b}_1^{(k)} + \widehat{b}_2^{(k)})$$
$$+ \Pi_1^{(2)}(1 - \nu - \nu\widehat{b}_0^{(k)}) - \nu\Pi_0^{(2)}(\widehat{b}_0^{(k)} + 1),$$

$$\varphi_{22} = 2\nu\Pi_1^{(0)}(b_2 + \widehat{b}_2^{(k)}) + 2\nu\Pi_2^{(0)}(b_1 + b_2 + \widehat{b}_1^{(k)} + \widehat{b}_2^{(k)})$$
$$+ \Pi_2^{(2)}(1 - \nu - \nu\widehat{b}_1^{(k)}) - \nu\Pi_1^{(2)}(\widehat{b}_0^{(k)} + 1),$$

$$\varphi_{23} = 2\nu\Pi_2^{(0)}(b_2 + \widehat{b}_2^{(k)}) + 2\nu\Pi_3^{(0)}(b_1 + b_2 + \widehat{b}_1^{(k)} + \widehat{b}_2^{(k)})$$
$$+ \Pi_3^{(2)}(1 - \nu - \nu\widehat{b}_0^{(k)}) - \nu\Pi_2^{(2)}(\widehat{b}_0^{(k)} + 1),$$

$$\varphi_{24} = 2\nu\Pi_3^{(0)}(b_2 + \widehat{b}_2^{(k)}) - \nu\Pi_3^{(2)}(\widehat{b}_0^{(k)} + 1),$$

$$\varphi_{30} = 6\nu\Pi_0^{(0)}(\widehat{b}_2^{(k)} + \widehat{b}_3^{(k)} + b_2 + b_3) - \Pi_0^{(3)}(\nu\widehat{b}_0^{(k)} + \nu - 1),$$

$$\varphi_{31} = 6\nu\Pi_1^{(0)}(\widehat{b}_2^{(k)} + \widehat{b}_3^{(k)} + b_2 + b_3) + 6\nu\Pi_0^{(0)}(\widehat{b}_3^{(k)} + b_3)$$
$$- \Pi_1^{(3)}[\nu(\widehat{b}_0^{(k)} + 1) - 1] - \nu\Pi_0^{(3)}(\widehat{b}_0^{(k)} + 1),$$

$$\varphi_{32} = 6\nu\Pi_2^{(0)}(\widehat{b}_2^{(k)} + \widehat{b}_3^{(k)} + b_2 + b_3) + 6\nu\Pi_1^{(0)}(\widehat{b}_3^{(k)} + b_3)$$
$$- \Pi_2^{(3)}[\nu(\widehat{b}_0^{(k)} + 1) - 1] - \nu\Pi_1^{(3)}(\widehat{b}_0^{(k)} + 1),$$

$$\varphi_{33} = \nu(\Pi_3^{(0)}(6\widehat{b}_2^{(k)} + 6\widehat{b}_3^{(k)} + 6b_2 + 6b_3) + \Pi_2^{(0)}(6\widehat{b}_3^{(k)} + 6b_3))\Pi_2^{(3)}$$
$$- \Pi_3^{(3)}(\nu(\widehat{b}_0^{(k)} + 1) - 1) - \nu(\widehat{b}_0^{(k)} + 1),$$

$$\varphi_{34} = 6\nu\Pi_3^{(0)}(\widehat{b}_3^{(k)} + b_3) - \nu\Pi_3^{(3)}(\widehat{b}_0^{(k)} + 1),$$

$$\varphi_{40} = 24\nu\Pi_0^{(0)}(\widehat{b}_3^{(k)} + \widehat{b}_4^{(k)} + b_3 + b_4) - \Pi_0^{(4)}(\nu\widehat{b}_0^{(k)} + \nu - 1),$$

$$\varphi_{41} = 24\nu\Pi_1^{(0)}(\widehat{b}_3^{(k)} + \widehat{b}_4^{(k)} + b_3 + b_4) + 24\nu\Pi_0^{(0)}(\widehat{b}_4^{(k)} + b_4)$$
$$- \Pi_1^{(4)}[\nu(\widehat{b}_0^{(k)} + 1) - 1] - \nu\Pi_0^{(4)}(\widehat{b}_0^{(k)} + 1),$$

$$\varphi_{42} = 24\nu\Pi_2^{(0)}(\widehat{b}_3^{(k)} + \widehat{b}_4^{(k)} + b_3 + b_4) + 24\nu\Pi_1^{(0)}(\widehat{b}_4^{(k)} + b_4)$$
$$- \Pi_2^{(4)}[\nu(\widehat{b}_0^{(k)} + 1) - 1] - \nu\Pi_1^{(4)}(\widehat{b}_0^{(k)} + 1),$$

$$\varphi_{43} = 24\nu\Pi_3^{(0)}(\widehat{b}_3^{(k)} + \widehat{b}_4^{(k)} + b_3 + b_4) + 24\nu\Pi_2^{(0)}(\widehat{b}_4^{(k)} + b_4)$$
$$- \Pi_3^{(4)}[\nu(\widehat{b}_0^{(k)} + 1) - 1] - \nu\Pi_2^{(4)}(\widehat{b}_0^{(k)} + 1),$$

$$\varphi_{44} = 24\nu\Pi_3^{(0)}(\widehat{b}_4^{(k)} + b_4) - \nu\Pi_3^{(4)}(\widehat{b}_0^{(k)} + 1).$$

It is then useful to decompose these coefficients into t-independent and t-dependent parts

$$\varphi_{ij} = \omega_{ij} + \widehat{\varphi}_{ij}(t),$$

where

$$\omega_{i0} = (1 - \nu) P_{i0} = 0 \quad (i = 1, \ldots, 4),$$

whereas the rest of ω_{ij} are given by the following:

$$\omega_{11} = \nu b_1 + \nu(b_1 + 1) P_{01} - \nu P_{10} + (1 - \nu) P_{11},$$

$$\omega_{12} = \nu b_1 P_{01} + \nu(b_1 + 1) P_{02} - \nu P_{11} + (1 - \nu) P_{12},$$

$$\omega_{13} = \nu b_1 P_{02} + \nu(1 + b_1) P_{03} - \nu P_{12} + (1 - \nu) P_{13},$$

$$\omega_{14} = \nu(b_1 P_{03} - P_{13}),$$

$$\omega_{21} = 2\nu(b_1 + b_2) P_{01} + (1 - \nu) P_{21} - \nu P_{20},$$

$$\omega_{22} = 2\nu b_2 P_{01} + 2\nu(b_1 + b_2) P_{02} + (1 - \nu) P_{22} - \nu P_{21},$$

$$\omega_{23} = 2\nu b_2 P_{02} + 2\nu (b_1 + b_2) P_{03} + (1 - \nu) P_{23} - \nu P_{22},$$

$$\omega_{24} = 2\nu b_2 P_{03} - \nu P_{23},$$

$$\omega_{31} = 6\nu (b_2 + b_3) P_{01} - (\nu - 1) P_{31} - \nu P_{30},$$

$$\omega_{32} = 6\nu (b_2 + b_3) P_{02} + 6\nu b_3 P_{01} - (\nu - 1) P_{32} - \nu P_{31},$$

$$\omega_{33} = 6\nu (b_2 + b_3) P_{03} + 6\nu b_3 P_{02} - (\nu - 1) P_{33} - \nu P_{32},$$

$$\omega_{34} = 6\nu b_3 P_{03} - \nu P_{33},$$

$$\omega_{41} = 24\nu (b_3 + b_4) P_{01} - \nu P_{40} - (\nu - 1) P_{41},$$

$$\omega_{42} = 24\nu (b_3 + b_4) P_{02} + 24\nu b_4 P_{01} - (\nu - 1) P_{42} - \nu P_{41},$$

$$\omega_{43} = 24\nu (b_3 + b_4) P_{03} + 24\nu b_4 P_{02} - (\nu - 1) P_{43} - \nu P_{42},$$

$$\omega_{44} = 24\nu b_4 P_{03} - \nu P_{43}.$$

Then $\widehat{\varphi}_{ij}(t)$ can be calculated from the formulas presented for $\varphi_{ij}(t)$ by subtracting ω_{ij}. They are quite bulky, but straightforward to calculate. Here, we list the expressions for $\widehat{\varphi}_{i0}(t)$ since they determine the t-dependence of the spinodal curve in the leading order:

$$\widehat{\varphi}_{10} = \nu (1 + b_1) \widehat{\Pi}_0^{(0)} + \nu(\widehat{b}_0^{(k)} + \widehat{b}_1^{(k)}) \widehat{\Pi}_0^{(0)} + (1 - \nu) \widehat{\Pi}_0^{(1)} - \nu \widehat{b}_0^{(k)} \widehat{\Pi}_0^{(1)},$$

$$\widehat{\varphi}_{20} = 2\nu (b_1 + b_2) \widehat{\Pi}_0^{(0)} + 2\nu \widehat{\Pi}_0^{(0)} (\widehat{b}_1^{(k)} + \widehat{b}_2^{(k)}) + (1 - \nu) \widehat{\Pi}_0^{(2)} - \nu \widehat{b}_0^{(k)} \widehat{\Pi}_0^{(2)},$$

$$\widehat{\varphi}_{30} = (1 - \nu) \widehat{\Pi}_0^{(3)} - \nu \widehat{b}_0^{(k)} \widehat{\Pi}_0^{(3)},$$

$$\widehat{\varphi}_{40} = 24\nu \widehat{\Pi}_0^{(0)} (b_3 + b_4 + \widehat{b}_3^{(k)} + \widehat{b}_4^{(k)}) + (1 - \nu) \widehat{\Pi}_0^{(4)} - \nu \widehat{b}_0^{(k)} \widehat{\Pi}_0^{(4)}.$$

Since $\omega_{i0} = 0$, we find

$$\varphi_{ij}(t) = \widehat{\varphi}_{i0}(t).$$

In the quadratic model, we find $\widehat{\varphi}_{i0}(t)$ as follows:

$$\widehat{\varphi}_{10} = \zeta (\nu - 1) (\nu b_1 - \nu + 2) \widehat{a}_0^{(k)} - [\nu (\nu - 2) + \zeta \nu (3\nu - 4 + \nu a_1 - a_1)] \widehat{b}_0^{(k)}$$
$$- \zeta (1 - \nu)^2 \widehat{a}_1^{(k)} + (\zeta + 1) \nu (1 - \nu) \widehat{b}_1^{(k)} + \text{ second-order terms},$$

$$\widehat{\varphi}_{20} = 2\zeta (1 - \nu) (3\nu + 2\nu b_1 - 1) \widehat{a}_0^{(k)}$$
$$+ 2\nu [\zeta (3a_1 + a_2 + \nu b_1 + \nu b_2 + 3) + \nu (b_1 + b_2)] \widehat{b}_0^{(k)}$$
$$+ 2\zeta (1 - \nu) (3\nu + \nu b_1 - 2) \widehat{a}_1^{(k)} + 2\nu (1 - \nu) (1 + 3\zeta + \zeta a_1) \widehat{b}_1^{(k)}$$
$$- 2\zeta (\nu - 1)^2 \widehat{a}_2^{(k)} + 2\nu (\zeta + 1) (1 - \nu) \widehat{b}_2^{(k)} + \text{ second-order terms},$$

$$\widehat{\varphi}_{30} = 6\zeta\nu(1-\nu)(1+3b_1+3b_2)\widehat{a}_0^{(k)} + 6\zeta(1-\nu)[3\nu-1+\nu(3b_1+b_2)]\widehat{a}_1^{(k)}$$
$$+ (1-\nu)(3\nu-2+\nu b_1)\widehat{a}_2^{(k)} + 6\zeta\nu(1-\nu)(1+3a_1+3a_2)\widehat{b}_0^{(k)}$$
$$+ 6\zeta\nu(1-\nu)(3+3a_1+a_2)\widehat{b}_1^{(k)} + 6\nu(1-\nu)(3\zeta+1+\zeta a_1)\widehat{b}_2^{(k)}$$
$$+ \text{second-order terms.}$$

Appendix C

Local Form of Energy Conservation Law

The first law of thermodynamics may be formulated in local form by using the formalism of continuum mechanics. This form is not used in equilibrium thermodynamics, but it gives us a more precise description of heat and work in local form. This form of conservation law will be useful for fluid dynamic considerations and nonequilibrium (irreversible) thermodynamics.

We consider a fluid composed of r chemically inert components. The components, denoted by $i = 1, 2, \ldots, r$, are subject to external conservative potentials ψ_i on components $i = 1, 2, \ldots, r$. We denote the mass density of component i by ρ_i and the total mass density by ρ:

$$\rho = \sum_{i=1}^{r} \rho_i. \tag{C.1}$$

The total energy density (specific energy) will be denoted E. The total energy $\mathbb{E}$ of the fluid enclosed in volume V is then given by the integral

$$\mathbb{E} = \int_V dV \ \rho E (\mathbf{r}, t). \tag{C.2}$$

The rate of change in energy is

$$\frac{d}{dt} \int_V dV \ \rho E (\mathbf{r}, t) = \int_V dV \frac{\partial}{\partial t} \rho E (\mathbf{r}, t). \tag{C.3}$$

Let us denote the energy flux by $\mathbf{J}_e$. This vector is parallel to the unit vector normal to the surface of the volume whose outward direction is taken as positive. Thus, when the energy flows out of the system, the sign of the flux is positive. Since the energy is conserved, the energy change within the volume must be balanced by the inflow of energy from the outside. Therefore, we find

$$\int_V dV \frac{\partial}{\partial t} \rho E\left(\mathbf{r}, t\right) = -\int_\Omega d\mathbf{\Omega} \cdot \mathbf{J}_e, \tag{C.4}$$

where $d\mathbf{\Omega}$ denotes the surface element and the surface integral is over the entire surface Ω of volume V. By the Gauss theorem, Eq. (C.4) may be written as

$$\int_V dV \left[\frac{\partial}{\partial t} \rho E\left(\mathbf{r}, t\right) + \nabla \cdot \mathbf{J}_e \right] = 0. \tag{C.5}$$

Since the volume is arbitrary, Eq. (C.5) implies the following local equation:

$$\frac{\partial}{\partial t} \rho E\left(\mathbf{r}, t\right) + \nabla \cdot \mathbf{J}_e = 0 \tag{C.6}$$

for the local expression of the energy conservation law. Since the fluid may flow with a velocity $\mathbf{u}$, its kinetic energy density is $\frac{1}{2}u^2$, and if it is subject to an external field with a potential, its potential energy density is

$$\psi = \rho^{-1} \sum_{i=1}^{r} \psi_i \rho_i, \tag{C.7}$$

where ψ_i is the potential energy density of species i. If we denote the internal energy density by $\mathcal{E}$, then the total energy density is

$$E = \tfrac{1}{2}u^2 + \psi + \mathcal{E}. \tag{C.8}$$

The energy flux also consists of various components. They are the convective total energy flow $\rho E \mathbf{u}$ arising from the convective motion of the fluid, an energy flow $\mathbf{P} \cdot \mathbf{u}$ stemming from the mechanical work done on the system where $\mathbf{P}$ is the stress (pressure) tensor, a potential energy flow $\sum_i \psi_i \mathbf{J}_i$ arising from the diffusion of various components relative to the fluid motion in the field of force, where $\mathbf{J}_i$ is the diffusion flow of component i, and the heat flow $\mathbf{J}_h$:

$$\mathbf{J}_e = \rho E \mathbf{u} + \mathbf{P} \cdot \mathbf{u} + \sum_{i=1}^{r} \psi_i \mathbf{J}_i + \mathbf{J}_h. \tag{C.9}$$

The diffusion flow $\mathbf{J}_i$ is defined by

$$\mathbf{J}_i = \rho_i(\mathbf{u}_i - \mathbf{u}), \tag{C.10}$$

where $\mathbf{u}_i$ is the velocity of component i. It is important to recognize that the meaning of heat flow $\mathbf{J}_h$ is given by Eq. (C.9). That is, given the meaning of the first three terms on the right-hand side making up the energy flux in Eq. (C.9), the heat flow is the rest of the energy flow so that the energy conservation law as stated in Eq. (C.5) holds valid. Therefore, the meaning of heat flow will change if there are additional terms that should be taken into account to make up $\mathbf{J}_e$. In this sense, our understanding of $\mathbf{J}_h$ can evolve as our understanding of the energy conservation law broadens over time. *Therefore, thermodynamics may be said to be anthropomorphic.*

Since the mass must conserve, the equation of continuity holds:

$$\frac{\partial}{\partial t}\rho = -\nabla \cdot (\rho\mathbf{u}), \tag{C.11}$$

and the momentum balance equation is

$$\frac{\partial}{\partial t}\rho\mathbf{u} = -\nabla \cdot (\mathbf{P} + \rho\mathbf{u}\mathbf{u}) + \sum_{i=1}^{r} \rho_i\mathbf{F}_i, \tag{C.12}$$

where $\mathbf{F}_i$ is the force density on component i:

$$\mathbf{F}_i = -\nabla\psi_i. \tag{C.13}$$

It can be shown with Eqs. (C.11) and (C.12) that the kinetic and potential energy densities together obey the balance equation

$$\frac{\partial}{\partial t}\rho\left(\frac{1}{2}u^2 + \psi\right) = -\nabla \cdot \left(\frac{1}{2}\rho u^2\mathbf{u} + \rho\psi\mathbf{u} + \mathbf{P}\cdot\mathbf{u} + \sum_{i=1}^{r}\psi_i\mathbf{J}_i\right)$$

$$+ \mathbf{P}{:}\nabla\mathbf{u} - \sum_{i=1}^{r}\mathbf{J}_i \cdot \mathbf{F}_i. \tag{C.14}$$

When Eqs. (C.8) and (C.9) are substituted into Eq. (C.6) and the terms are rearranged with the help of Eq. (C.14), the energy conservation law may be cast in the form

$$\rho\frac{d\mathcal{E}}{dt} = -\nabla \cdot \mathbf{J}_h - p\nabla \cdot \mathbf{u} - \mathbf{\Pi}{:}\nabla\mathbf{u} + \sum_{i=1}^{r}\mathbf{J}_i \cdot \mathbf{F}_i, \tag{C.15}$$

where d/dt means the substantial time derivative

$$\frac{d}{dt} = \frac{\partial}{\partial t} + \mathbf{u} \cdot \nabla,$$

and we have split the pressure tensor $\mathbf{P}$ into the hydrostatic pressure p and a symmetric second-rank tensor[1] $\mathbf{\Pi}$:

$$\mathbf{P} = p\boldsymbol{\delta} + \mathbf{\Pi} \tag{C.16}$$

with $\mathbf{U}$ denoting the unit tensor

$$\boldsymbol{\delta} = \begin{pmatrix} 1 & 0 & 0 \\ 0 & 1 & 0 \\ 0 & 0 & 1 \end{pmatrix}. \tag{C.17}$$

With the definition of the specific volume by $v = \rho^{-1}$, the equation of continuity may be written as

$$\rho\frac{dv}{dt} = \nabla \cdot \mathbf{u}. \tag{C.18}$$

We may also write

$$\rho\frac{dQ}{dt} = -\nabla \cdot \mathbf{J}_h, \tag{C.19}$$

where dQ is the differential heat added to the unit mass of the fluid. With Eqs. (C.18) and (C.19), we may recast Eq. (C.15) in the form

$$\rho\frac{d\mathcal{E}}{dt} = \rho\frac{dQ}{dt} - \rho p\frac{dv}{dt} - \mathbf{\Pi}{:}\nabla\mathbf{u} + \sum_{i=1}^{r} \mathbf{J}_i \cdot \mathbf{F}_i. \tag{C.20}$$

The last three terms on the right-hand side of Eq. (C.20) represent the work done on the system per unit time, the third term on the right is the viscous heating term, and the last term the work per unit time arising from diffusion against the external force. Therefore, we will combine them to write

$$\rho\frac{dW}{dt} = -\rho p\frac{dv}{dt} - \mathbf{\Pi}{:}\nabla\mathbf{u} + \sum_{i=1}^{r} \mathbf{J}_i \cdot \mathbf{F}_i, \tag{C.21}$$

so that

$$\frac{d\mathcal{E}}{dt} = \frac{dQ}{dt} + \frac{dW}{dt}. \tag{C.22}$$

This is the mathematical statement of the first law of thermodynamics we wished to formulate in local form. Note that the last two terms on the right-hand side of Eq. (C.21) represent the work arising from the dissipative effects arising from viscous frictions and mass diffusion caused by the external force, respectively. In the case of electrically charged fluids subject

---

[1] It is not the same tensor as the traceless symmetric tensor $\mathbf{\Pi}$ used in Chapter 19 despite the same symbol.

to an external electric field, the last term is related to the Joule heating. When these dissipative terms vanish, the work per unit time is then given by the pressure–volume work per unit time alone:

$$\frac{dW}{dt} = -p\frac{dv}{dt}. \tag{C.23}$$

More precisely, it is the pressure–volume work performed in infinitesimal time interval dt. Note that $đW = (dW/dt)\,dt$ and similarly for heat. It is useful to remember that when work is expressed without taking into account the internal work — namely, the second and third terms — in Eq. (C.21), the first law of thermodynamics is considered in the case of nondissipative processes.

Finally, we note that the internal energy conservation law considered here excludes the effect of radiative heating. If R denotes the radiative heating, the internal energy conservation law should then read

$$\rho\frac{d\mathcal{E}}{dt} = \rho\frac{dQ}{dt} - \rho p\frac{dv}{dt} - \mathbf{\Pi}\mathbf{:}\nabla\mathbf{u} + \sum_{i=1}^{r}\mathbf{J}_i \cdot \mathbf{F}_i + R. \tag{C.24}$$

This equation, Eq. (C.11), and Eq. (C.12) are coupled to constitutive equations for $\mathbf{\Pi}$, $\mathbf{J}_i$, radiation intensity, and so on, which must be supplied, if a full description of the system is desired.

Bibliography

J. O. Hirschfelder, C. F. Curtiss and R. B. Bird, *Molecular Theory of Gases and Liquids* (Wiley, New York, 1954), p. 163.

A. J. M. Garrett, *J. Math. A: Math. Gen.* **13**, 379 (1980).

M. L. Glasser, *Phys. Lett. A* **300**, 381 (2002).

I. S. Gradshteyn and I. M. Ryzhik, *Tables of Integrals, Series and Products*, 4th edition (Academic, London, 1965).

See B. C. Eu and H. Guerin, *Can. J. Phys.* **49**, 486 (1971) in which the Schrödinger equation for the LJ (10–6) and (12–6) potentials are shown solvable analytically at zero energy in terms of a confluent hypergeometric function. At nonzero energy a perturbation method is applicable to compute the energy eigenvalue in a form reminiscent of the eigenvalues of an anharmonic oscillator.

T. L. Hill, *Statistical Mechanics* (McGraw-Hill, New York, 1956).

A. Erdelyi, ed., *Higher Transcendental Functions*, Vols. 1 and 2 (H. Bateman Manuscript Project) (McGraw-Hill, New York, 1953).

M. Abramowitz and I. Stegun, *Handbook of Mathematical Functions* (NBS, Washington, D.C., 1964).

Index

the third, 128
the first, 36, 38, 86, 94
the first and second
    for cyclic processes, 87
the second, 64
    Clausius principle, 64
    Kelvin principle, 64
    mathematical representation
        of, 85
    Planck principle, 64
the zeroth, 12
Legendre transformation, 102
  for Gibbs free energy, 102, 120
  for work function, 120
  in thermodynamics, 440
Lennard-Jones potential, 184
lever rule, 311
Lewis–Randall rule, 273
Liesegang patterns, 576
linear constitutive equations, 457,
  469
  for nonconserved variable, 457
  for the nonconserved variable,
    469
liquid junction, 397
liquifaction
  condition of, 218
  degree of, 216
lithium ion cell, 407
loss modulus, 462

Mach number
  of a shock wave, 519
magnetic induction, 426
magnetic permeability, 426
magnetocaloric effect, 428
magnetostriction, 427
Margules expansion, 317
mass flux, 432
mass fraction, 432
mass fraction evolution equation, 434,
  534
Matalon-Packter law
  of the spacing coefficient
    for the direct spacing, 579
Maxwell construction, 186

Maxwell equation, 460
  the shear stress, 460
Maxwell relation
  for chemical potentials, 141
Maxwell's relation, 103
mean free path, 455
mean free volume, 227
method
  of Giauque and Debye, 130, 427
model theory
  a formal theory of Liesegang
    pattern formation, 583
modified free volume theory
  of diffusion, 227
molality
  mean
    of an electrolyte, 375
mole fraction
  definition of, 141
  in terms of molality, 303, 305
momentum balance equation, 435,
  613
  for plane Couette flow, 478

Nernst equation
  for a lithium ion cell, 409
  for EMF, 401
    of multielectron transfer,
      402
Newtonian law of viscosity, 478
non-Newtonian flow, 476
nonequilibrium, 5
nonequilibrium effect
  on chemical potential, 444
  on equilibrium constant, 446
  on pressure, 442
  on temperature, 441
nonequilibrium Maxwell relations,
  438
  symmetry relation, 439
nonideality correction
  for equilibrium constant, 299
nonlinear factor, 476
normal stress differences
  primary and secondary, 489

Printed in the United States
By Bookmasters